Synthese Library

Studies in Epistemology, Logic, Methodology, and Philosophy of Science

Volume 448

The aim of *Synthese Library* is to provide a forum for the best current work in the methodology and philosophy of science and in epistemology, all broadly understood. A wide variety of different approaches have traditionally been represented in the Library, and every effort is made to maintain this variety, not for its own sake, but because we believe that there are many fruitful and illuminating approaches to the philosophy of science and related disciplines.

Special attention is paid to methodological studies which illustrate the interplay of empirical and philosophical viewpoints and to contributions to the formal (logical, set-theoretical, mathematical, information-theoretical, decision-theoretical, etc.) methodology of empirical sciences. Likewise, the applications of logical methods to epistemology as well as philosophically and methodologically relevant studies in logic are strongly encouraged. The emphasis on logic will be tempered by interest in the psychological, historical, and sociological aspects of science.

In addition to monographs *Synthese Library* publishes thematically unified anthologies and edited volumes with a well-defined topical focus inside the aim and scope of the book series. The contributions in the volumes are expected to be focused and structurally organized in accordance with the central theme(s), and should be tied together by an extensive editorial introduction or set of introductions if the volume is divided into parts. An extensive bibliography and index are mandatory.

More information about this series at https://link.springer.com/bookseries/6607

Carlo Cellucci

The Making of Mathematics

Heuristic Philosophy of Mathematics

Springer

Carlo Cellucci
Department of Philosophy
Sapienza University of Rome
Rome, Italy

ISSN 0166-6991 ISSN 2542-8292 (electronic)
Synthese Library
ISBN 978-3-030-89730-7 ISBN 978-3-030-89731-4 (eBook)
https://doi.org/10.1007/978-3-030-89731-4

This Springer imprint is published by the registered company Springer Nature Switzerland AG
The registered company address is: Gewerbestrasse 11, 6330 Cham, Switzerland

Preface

Mainstream philosophy of mathematics, namely the philosophy of mathematics that has prevailed for the past century, claims that the philosophy of mathematics cannot concern itself with the making of mathematics, in particular discovery, but only with finished mathematics, namely mathematics presented in finished form. On this basis, mainstream philosophy of mathematics argues that mathematics is theorem proving by the axiomatic method. This, however, is untenable because it is incompatible with Gödel's incompleteness theorems, and cannot account for many features of mathematics.

This book offers an alternative approach, heuristic philosophy of mathematics, according to which the philosophy of mathematics can concern itself with the making of mathematics, in particular discovery. On this basis, the book argues that mathematics is problem solving by the analytic method, and that this can account for all the main features of mathematics: mathematical method, objects, demonstrations, definitions, diagrams, notations, explanations, beauty, applicability, and knowledge.

Rome, Italy | Carlo Cellucci

Acknowledgments

I am grateful to many people for their help in the making of this book. For their help, in whatever form, I am especially indebted to Derek Abbott, Andrew Aberdein, Atocha Aliseda, Jeremy Avigad, Sorin Bangu, Arthur Bierman, Aldo Brancacci, Angela Breitenbach, Terry Bristol, Otávio Bueno, Mirella Capozzi, Jessica Carter, Anjan Chakravartty, Riccardo Chiaradonna, Charles Chihara, Leo Corry, Cesare Cozzo, Brian Davies, Philip J. Davis, Ahmed Djebbar, Catarina Dutilh Novaes, Richard Epstein, Giambattista Formica, Miriam Franchella, Michèle Friend, Maria Carla Galavotti, Paola Giacomoni, Donald Gillies, Norma Goethe, Emily Grosholz, Niccolò Guicciardini, Gila Hanna, Michael Harris, Reuben Hersh, Hansmichael Hohenegger, Grazia Ietto Gillies, Emiliano Ippoliti, Gabriele Lolli, Danielle Macbeth, Giovanni Manetti, Per Martin-Löf, Dan Nesher, Reviel Netz, Thomas Nickles, Alexander Paseau, Volker Peckhaus, Eva Picardi, Florindo Pirone, Dag Prawitz, Marwan Rashed, Andrea Reichenberger, Dirk Schlimm, Wilfrid Sieg, Hourya Sinaceur, Nathalie Sinclair, Emidio Spinelli, Bharath Sriraman, Andreas Stephens, Fabio Sterpetti, William Tait, Robert Thomas, Johan van Benthem, Nicla Vassallo, Francesco Verde, Jan von Plato, Weijia Wang, Alan White, Jan Woleński, and Semir Zeki. This does not mean that they share the views expressed in the book or are in any way responsible for any remaining inaccuracies.

I am also grateful to an anonymous reviewer for comments that have given me the opportunity to correct some errors and develop some aspects of the book further.

Some of the views expressed in the book are a revision and further development of views presented in two of my previous books: *Rethinking logic: Logic in relation to mathematics, evolution, and method*, Springer, Cham 2013; and *Rethinking knowledge: The heuristic view*, Springer, Cham 2017. In particular, the book is a great expansion of Part IV of the latter.

Four chapters of the book are a revised version of previously published articles. Chapter 9 is a revised version of "The nature of mathematical objects," in B. Sriraman (ed.), *Handbook of the history and philosophy of mathematical practice*, Springer, Cham. https://doi.org/10.1007/978-3-030-19071-2_20-1; Chapter 11 of "Definition in mathematics," *European Journal for Philosophy of Science* 8 (2018): 605–629; Chapter 12 of "Diagrams in mathematics," *Foundations of Science* 24 (2019): 583–604; and Chapter 13 of "The role of notations in mathematics," *Philosophia* 48 (2020): 1397–1412.

The final part of the research for the book has been made as part of the project "The criterion of truth: From ancient philosophy to contemporary epistemology" (Sapienza University of Rome / 2019–2021).

Contents

Chapter 1
Introduction

Abstract Throughout history, philosophy and mathematics have been related and relevant to each other. Nevertheless, many contemporary mathematicians believe that philosophy, and specifically the philosophy of mathematics, is irrelevant to mathematics. This opinion is due to the fact that mainstream philosophy of mathematics, namely the philosophy of mathematics that has prevailed for the past century, does not account for the making of mathematics, in particular discovery, so it cannot provide any real understanding of the nature of mathematics, let alone contribute to its advancement. This, however, does not mean that the philosophy of mathematics is irrelevant to mathematics, but only that so is mainstream philosophy of mathematics. What is needed is an alternative approach to the philosophy of mathematics.

Keywords Relevance of mathematics to philosophy · Relevance of philosophy to mathematics · Irrelevance view · Working philosophy of the mathematician · Front and back of mathematics · Mainstream philosophy of mathematics · Heuristic philosophy of mathematics

1.1 The Relevance of Mathematics to Philosophy

Throughout history, philosophy and mathematics have been related and relevant to each other. It is no coincidence that some of the earliest philosophers, notably Thales, Pythagoras, and Democritus, were also among the earliest mathematicians, and some major philosophers of the early modern period, notably Descartes, Pascal, and Leibniz, were also major mathematicians.

Mathematics has been relevant to philosophy from the very beginning. Indeed, it has played an important role in the birth itself of philosophy as discipline.

Philosophy as discipline was not born with the Presocratics, because they did not sharply distinguish philosophy from the magic-religious tradition. Thus, Thales says that "the mind of the world is the god, and the whole is endowed with soul and also full of daemons; and the divine power, penetrating the elementary moisture, moves

C. Cellucci, *The Making of Mathematics*, Synthese Library 448,
https://doi.org/10.1007/978-3-030-89731-4_1

it" (Thales 11 A 23, ed. Diels-Kranz). Pythagoras says that "number" is "the source of the continuing existence of divine natures, gods, and daemons" (Iamblichus, *De Vita Pythagorica*, XXVIII.147, ed. Deubner). Democritus says that "in the universe, there are images endowed with divinity," and "the gods are the elements of mind in that universe," and "they are certain huge images of such a size as to enclose the whole universe externally" (Cicero, *De natura deorum*, I.120).

Philosophy as discipline was not born with the Sophists either, because they did not use philosophy to advance knowledge, but only to earn money by teaching rich young men rhetorical tricks. Thus, Plato says that the sophist is "a paid hunter of rich young men," a "merchant of knowledge about the soul," a "retail-dealer in these very same wares," a "seller of knowledge of his own production," an "athlete in verbal combat, appropriating to himself the art of eristic" (Plato, *Sophista*, 231 d 3–e 2). Xenophon says that "there is a good and a shameful way to dispose of one's beauty and wisdom. If a man sells his beauty for money to anyone who wants it, they call him a prostitute," and, "in the same way, those who sell their wisdom for money to anyone who wants it, they call them sophists, or, as it were, prostitutes of wisdom" (Xenophon, *Memorabilia*, I.6.13). Aristotle says that "the art of the sophist is apparent but not real wisdom, and the sophist is one who makes money from apparent and not real wisdom" (Aristotle, *Sophistici Elenchi,* 165 a 21–23).

Philosophy as discipline was born only with Plato. As Nightingale says, "the discipline of philosophy emerged" in "Athens in the fourth century BCE, when Plato appropriated the term 'philosophy' for a new and specialized discipline – a discipline that was constructed in opposition to the many varieties of 'sophia' or 'wisdom' recognized by Plato's predecessors and contemporaries" (Nightingale 1995, 14). Plato was aware to have given birth to a new discipline. This is apparent from the fact that "Plato makes no mention of philosophic predecessors in the *Republic*," because he "did not consider" the Presocratics and Sophists "to be 'philosophers' in his sense of the term" (ibid., 18 and footnote).

Nightingale, however, fails to mention that Plato gave birth to philosophy as discipline by modelling the method of philosophy on the method of mathematics. Specifically, Plato modelled the method of philosophy on the method used by Hippocrates of Chios to solve problems in mathematics. What is more, Plato gave the first formulation of that method. The same method was used by Hippocrates of Cos to solve problems in medicine, but neither Hippocrates of Chios nor Hippocrates of Cos gave a formulation of the method, they simply used it, Plato gave the first formulation. Today the method is known as the analytic method or method of analysis (see Chap. 5).

Plato modelled the method of philosophy on the method of mathematics, because he believed that mathematics was "a prelude" to "the song that must be learned" (Plato, *Respublica,* VII 531 d 7). Namely, a prelude to philosophy, which "tries, through argument and without using any of the senses, to find the being itself of each thing" (ibid., VII 532 a 5–7). Thus, arithmetic "strongly leads the soul upward, compelling it to consider the numbers themselves" (ibid., VII 525 d 5–6). And "geometry is knowledge of what always is," so it is apt "to draw the soul toward truth" and is a stimulus "to turn the gaze upward" (ibid., VII 527 b 6–9).

Since Plato believed that mathematics was a prelude to philosophy, he demanded that would-be philosophers should first study mathematics. Then, some of them would be selected to study philosophy, by testing "who is able to release himself from the eyes and the rest of sense and, guided by truth, ascend to the being itself" (ibid., VII 537 d 5–7). The demand that would-be philosophers should first study mathematics was also made by the inscription above the entrance to Plato's Academy: "Let no one ignorant of geometry enter [ageometretos medeis eisito]" (Aelius Aristides, *Opera*, III, 464.13, ed. Dindorf). With obvious exaggeration, Aristotle even complained that, for Plato and his Academy, "mathematics has come to be" all of "philosophy" (Aristotle, *Metaphysica*, A 9, 992 a 32–33).

Plato's demand that would-be philosophers should first study mathematics, was considered a valuable one even many centuries later. Thus, Galileo said: "Was not Plato very right when he wished that his pupils should be first of all grounded in mathematics?" (Galilei 1968, VIII, 175). In fact, "if I were to restart my studies, I would follow the advice of Plato and start with mathematics" (ibid., VIII, 134).

1.2 The Continued Relevance of Mathematics to Philosophy

After the birth of philosophy as discipline, mathematics has continued to be relevant to philosophy in many respects. In particular, several philosophers of the early modern period maintained that the method of philosophy is the same as the method of mathematics, identified either with the analytic method or method of analysis (see Chap. 5), or with the analytic-synthetic method or method of analysis and synthesis (see Chap. 6), or with the synthetic method or axiomatic method (see Chap. 6).

Thus, according to Descartes, the method of philosophy is the analytic method. Indeed, he says: "The old geometers only used" the synthetic method "in their writings," because they thought of the analytic method "so highly that they reserved it to themselves as a valuable secret. But I have followed the analytic method alone" in "my *Meditations*," because the synthetic method "cannot be applied so conveniently to these metaphysical matters" (Descartes 1996, VII, 156).

According to Hobbes, the method of philosophy is the analytic-synthetic method. Indeed, he says: "The method of philosophizing" is the investigation "of causes by their known effects" or "of effects by their known causes" (Hobbes 1839–1845, I, 58). Now, the investigation of causes by their known effects is resolution or analysis, and the investigation of effects by their known causes is composition or synthesis. Therefore, "the method of philosophizing" is "partly analytic, partly synthetic" (ibid., I, 66).

According to Wolff, the method of philosophy is the synthetic method. Indeed, he says: "The philosophical method" is "the same as the scientific method and the synthetic method" (Wolff 1728, 634). For, "in philosophy, no terms must be used, but those explained by an accurate definition" (ibid., 53). And "no proposition must be admitted, but that which is legitimately deduced from sufficiently established principles" (ibid., 54).

1.3 The Relevance of Philosophy to Mathematics

The relation between philosophy and mathematics, however, has not been one-way only. As mathematics has been relevant to philosophy, philosophy has been relevant to mathematics. Here are some examples.

(1) Philosophy has provided analyses of mathematical concepts.

Thus, Zeno gave an analysis of the concept of infinite set by saying that "time is composed of instants" (Zeno 29 A 27, ed. Diels–Kranz). In particular, it is composed of infinitely many instants, because "time is infinite" by "division" (Zeno 29 A 25, ed. Diels–Kranz). Also, "half a time is equal to its double" (Zeno 29 A 28, ed. Diels–Kranz). Zeno illustrated these assertions using lengths viewed as infinite sets of points, so he also implied that length is composed of infinitely many points, and half a length is equal to its double.

Now, to say that half a time is equal to its double, or that half a length is equal to its double, amounts to saying that two infinite sets can be equivalent even when one of them is a proper subset of the other. Galileo gave an example of this by pointing out that "the square numbers are as many as all the numbers, because they are as many as their roots, and all numbers are roots" (Galilei 1968, VIII, 78). Then, Dedekind used the property in question as a definition of infinite set: "A system *S* is said to be infinite when it is similar to a proper part of itself" (Dedekind 1996, 806).

(2) Philosophy has exposed the inadequacy of mathematical concepts.

Thus, Berkeley observed that, in the calculus of infinitesimals of Leibniz and Newton, if one arrives at a correct conclusion, it is only because "two errors being equal and contrary destroy each other; the first error of defect being corrected by a second error of excess" (Berkeley 1992, 182). Indeed, "if we remove the veil and look underneath" the basic concepts of the calculus, we "shall discover much emptiness, darkness, and confusion; nay, if I mistake not, direct impossibilities and contradictions" (ibid., 169). The "introducing of things so inconceivable" is "a reproach to mathematics" (ibid., 213).

Berkeley's attack contributed to the development of the calculus, by pointing out some critical questions that had to be addressed to obtain an adequate formulation. Even Robinson, who thought that Leibniz's ideas about infinitesimals could be fully vindicated, says that "the vigorous attack directed by Berkeley against the foundations of the calculus in the forms then proposed is, in the first place, a brilliant exposure of their logical inconsistencies" (Robinson 1966, 280).

(3) Philosophy has formulated new methods of discovery and justification.

Thus, as already mentioned, Plato gave the first formulation of the analytic method (see Chap. 5). Also, in Plato, there is "the only extant example of proof by" complete induction "in the ancient mathematical corpus" (Acerbi 2000, 58).

Aristotle gave the first formulation of the analytic-synthetic method or method of analysis and synthesis, and, as a byproduct, the first formulation of the synthetic method or axiomatic method (see Chap. 6).

In appendix to the *Discours de la Méthode*, Descartes published *La Géométrie* and two other treatises, calling them "essays of this method," namely of the method presented in the *Discours*, because "the things they contain could not be found without it," and "one can know from them what it is worth" (Descartes 1996, I, 349).

Descartes is a glaring example of the fact that, as mathematics has been relevant to philosophy, philosophy has been relevant to mathematics. As Bos says, for Descartes "mathematics was a source of inspiration and an example for his philosophy, and, conversely, his philosophical concerns strongly influenced his style and program in mathematics" (Bos 2001, 228).

1.4 The Irrelevance View

Several contemporary mathematicians, however, believe that philosophy, in particular the philosophy of mathematics, is irrelevant to mathematics.

Thus, Hersh says that, "in books and articles bearing the label 'philosophy of mathematics'," there are only "arguments disconnected from what mathematicians do and think about" (Hersh 2014, 21). Indeed, "the professional philosopher, with hardly any exception, has little to say to the professional mathematician" because "he has only a remote and inadequate notion of what the professional mathematician is doing" (Hersh 1979, 34). In particular, "some philosophers who write about mathematics seem unacquainted with any mathematics more advanced than arithmetic and elementary geometry" (ibid.).

Gowers says: "Suppose a paper were published tomorrow that gave a new and very compelling argument for some position in the philosophy of mathematics," and that the "argument caused many philosophers" to "embrace a whole new -ism" (Gowers 2006, 198). Then, "what would be the effect on mathematics? I contend that there would be almost none" (ibid.). For, "the questions considered fundamental by philosophers are the strange, external ones that seem to make no difference to the real, internal business of doing mathematics" (ibid.).

Cheng says: "The philosophers come up with theories that don't seem to have any impact on what the mathematicians do or think," and "ask questions that have no impact on mathematical practice" (Cheng 2004, 3). So, "daily mathematical practice seems barely affected by the questions the philosophers are considering" (ibid., 1). Indeed, "mathematical practice seems to carry on oblivious of what philosophical theories mathematicians happen to subscribe to" (ibid., 2).

1.5 The Reason of the Irrelevance View

If several contemporary mathematicians believe that philosophy, in particular the philosophy of mathematics, is irrelevant to mathematics, it is not because they think that philosophical questions concerning the nature of mathematics are of no consequence to mathematics. Not only they do not think so, but some of them even say that it is impossible to do mathematics without a philosophy that tells you what mathematics is.

Thus, Byers says that "the nature of mathematics is a fundamental question on a par with the other mathematical questions" (Byers 2007, 348). Indeed, "doing mathematics is not really possible without a philosophy that tells you what mathematics is, even if this philosophy is not consciously embraced but is only implicit in what you do" (Byers 2017, 46).

What these mathematicians really think is not that philosophical questions concerning the nature of mathematics are of no consequence to mathematics, but only that mainstream philosophy of mathematics, namely the philosophy of mathematics that has prevailed for the past century, is unable to give an answer to them. Only mathematicians can give such an answer.

Thus, Lebesgue says that "the philosophy of mathematics can be created only by mathematicians" (Lebesgue 1941, 109). For, the task of the philosophy of mathematics is "to show how mathematics is built" (ibid., 110). And this can be done only by mathematicians.

The attitude of these mathematicians towards mainstream philosophers of mathematics is: Who are these outsiders to tell us what mathematics is? Only mathematicians can say what mathematics is.

Therefore, Smoryński says: "To be accepted by mathematicians, the philosopher of mathematics must not only be conversant in mathematics, but must also be a mathematician," so, "if you are not already a mathematician, you very likely have no future as a philosopher of mathematics" (Smoryński 1983, 14).

1.6 The Working Philosophy of the Mathematician

If only mathematicians can say what mathematics is, then the only genuine philosophy of mathematics is the working philosophy of the mathematician.

Indeed, Hersh declares: "By 'philosophy of mathematics' I mean the working philosophy of the professional mathematician, the philosophical attitude toward his work that is assumed by the researcher, teacher, or user of mathematics" (Hersh 1979, 31).

A problem with the claim that the only genuine philosophy of mathematics is the working philosophy of the mathematician is that the latter is different from period to period, from school to school, and even from mathematician to mathematician.

For example, on the one hand, Connes says that "the scientific life of mathematicians can be pictured as an exploration of the geography of the 'mathematical reality' which they unveil gradually in their own private mental frame" (Connes 2008, 1011). Most mathematicians "see themselves as explorers of this 'mathematical world'," which exists independently of them and whose "structure they uncover by a mixture of intuition and a great deal of rational thought" (ibid., 1012). Each "generation builds a mental picture that reflects their own understanding of this world. They construct mental tools that penetrate more and more deeply into it, so that they can explore aspects of it that were previously hidden" (ibid.).

On the other hand, Gowers says that "modern mathematicians are formalists" (Gowers 2006, 199). When they "discuss unsolved problems, what they are doing is not so much trying to uncover the truth as trying to find proofs" (ibid.). To this purpose, mathematicians "start by writing down some axioms and deduce from them a theorem" (ibid., 183). At the end of this process, what they "know is not that the theorem is 'true'," but "merely that a certain statement can be obtained from certain other statements by means of certain processes of manipulation" (ibid.). So, "what a mathematical proof actually does is" only "show that certain conclusions" follow "from certain premises" (Gowers 2002, 41). As to "the validity of these premises," this is a "matter which can safely be left to philosophers" (ibid.).

Thus, if the only genuine philosophy of mathematics is the working philosophy of the professional mathematician, then the philosophy of mathematics can only give an answer to the question: What is the working philosophy of the mathematicians of a given period, or of a given school, or of individual mathematicians?

When so conceived, however, the philosophy of mathematics reduces to the history, or sociology, or psychology of mathematics. Indeed, Hersh says that he wants "to use the methodologies of history," or "sociology," or "psychology" to "develop a coherent, empirically-based, overall understanding of the nature of mathematical practice and knowledge" (Hersh 2014, 141).

But the methodologies of history, or sociology, or psychology cannot tell you what mathematics is. They can only tell you what the mathematicians of a given period, or of a given school, or individual mathematicians, think that mathematics is.

1.7 Different Skills of Mathematicians and Philosophers

The reason why Hersh declares that by philosophy of mathematics he means the working philosophy of the professional mathematician is that, as we have seen, he says that the professional philosopher, with hardly any exception, has little to say to the professional mathematician, because he has only a remote and inadequate notion of what the professional mathematician is doing. In particular, Hersh says that some philosophers who write about mathematics, seem unacquainted with any mathematics more advanced than arithmetic and elementary geometry.

Hersh is quite right in saying so, and some philosophers frankly admit it.

Thus, Bays says that, "in the late nineteenth and early twentieth centuries," the "issues that concerned philosophers of mathematics were often continuous with developments in mathematics proper," such as "the rigorization of analysis in the nineteenth century," or "the discovery of the set-theoretic paradoxes" (Bays 2012, 424). But today the situation is completely different. The "topics that most exercise philosophers of mathematics" do not connect with the "concerns of practicing mathematicians," their "connections tend to be to other areas of philosophy," in particular "metaphysics" and "epistemology" (ibid.). Further, "the mathematics one needs to address these problems is often quite rudimentary," in many cases "elementary arithmetic and geometry is enough" (ibid.).

Other philosophers, however, disagree. They claim that one can very well interest oneself in the philosophy of mathematics, and understand a good deal of the debates on the subject, even with little knowledge of mathematics.

Thus, Dummett says: "If you have little knowledge of mathematics, you do not need to remedy that defect before interesting yourself in the philosophy of mathematics" (Dummett 1998, 124). For, "you can very well understand a good deal of the debates on the subject and a good deal of the theories advanced concerning it without an extensive knowledge of its subject-matter" (ibid.).

But, if one can very well interest oneself in the philosophy of mathematics, and understand a good deal of the debates on the subject, even with little knowledge of mathematics, it is not because little knowledge of mathematics is enough to say what mathematics is. It is rather because such debates are about artificial issues, which have no connection with, and hence shed no light on, the real mathematical process.

On the other hand, however, that the professional philosopher has serious limitations, does not mean that the professional mathematician understands what he is doing, and hence can say what mathematics is. He may not have the necessary skills.

As Hanna and Larvor observe, "the usual reservations about practitioner-testimony apply to mathematics. Adepts in any practice can fail to understand what they are doing, how they are doing it and what conditions make it possible" (Hanna and Larvor 2020, 1137).

Even several mathematicians admit that the professional mathematician may not have the necessary skills.

Thus, Bourbaki says that "the opinions of mathematicians on topics in philosophy, even when these questions are concerned with their field, are most often opinions received at second or third hand, coming from doubtful sources" (Bourbaki 1994, 11).

Byers says that "most practicing mathematicians have no time for anything that is philosophical. They are too busy living within their paradigm, that is, proving theorems" (Byers 2017, 59).

Hersh says that "the art of philosophical discourse is not well developed today among mathematicians, even among the most brilliant," while "philosophical issues" require "careful argument, fully developed analysis, and due consideration of objections. A bald statement of one's own opinion is not an argument, even in philosophy" (Hersh 1979, 34–35).

The professional mathematician may not have the necessary skills, because his function is to create new mathematics, not to say what mathematics is.

As Hardy says, "the function of a mathematician is to do something, to prove new theorems, to add to mathematics, and not to talk about what he or other mathematicians have done" (Hardy 1992, 61).

Of course, nothing excludes that the professional mathematician may say what mathematics is. But this requires that he be skilled, not only in creating new mathematics, but also in reflecting on what mathematicians are doing, how they are doing it, and what conditions make it possible.

For this reason, Hardy began a lecture on the subject of mathematical proof saying: "I have chosen" mathematical proof as "a subject for this lecture, after much hesitation," because the subject is "not from technical mathematics," so "I shall be setting myself a task for which I have no sufficient qualifications" (Hardy 1929, 1).

1.8 The Front and the Back of Mathematics

In the philosophy of mathematics, an important question is the difference between finished mathematics, namely mathematics presented in finished form in journals, textbooks, or lectures, and the making of mathematics, namely the actual process of mathematical research.

Hersh expresses this difference in terms of Goffman's "concept of 'front' and 'back'" (Hersh 1997, 35). The "front is where the public is admitted," and the "back is where it's excluded" (ibid.). For example, in a restaurant, the front is "the dining area," and the back is "the kitchen," in a theater, the front is the "stage," and the back is the "backstage" (ibid.). But Goffman extends "'front' and 'back' from restaurants and theaters to all institutions" (ibid.). Then, mathematics too has a front and a back, where "the front is mathematics in finished form," as it is presented in "lectures, textbooks, journals," and the "back is mathematics" as it appears "among working mathematicians, told in offices or at cafe tables" (ibid., 36).

This view is opposed by "mainstream philosophy" of mathematics, which "doesn't know that mathematics has a back. Finished, published mathematics – the front – is taken as a self-subsistent entity" (ibid.). But this is absurd. For mainstream philosophy of mathematics, not to know that mathematics has a back, is like "for a restaurant critic not to know there are kitchens, or a theater critic not to know there is backstage" (ibid., 37). It means ignoring that "the performance in front" was "concocted behind the scenes" (ibid.). Therefore, it is "impossible to understand the front while ignoring the back" (ibid.).

Hersh is quite right in saying that it is impossible to understand the front while ignoring the back. But his position has a weakness. As we have seen, he assumes that the back is mathematics as it appears among working mathematicians, told in offices or at café tables.

This allows Greiffenhagen and Sharrock to criticize Hersh, arguing that his "treatment of the 'front' and the 'back' as a contrastive pair downplays the continuity of the two" (Greiffenhagen and Sharrock 2011, 841). The continuity is clear from a comparison between mathematical lectures, as one example of mathematics in the 'front', and "meetings between a supervisor and his doctoral students," as "one example of mathematics in the 'back'" (ibid., 854). The comparison shows that "the difference between the 'front' and the 'back'" is "not between two kinds of proof," but only "between different stages: of working with an incomplete idea of a possible proof as opposed to presenting a (presumably) complete, thoroughly worked-out proof" (ibid., 858). Thus, "the 'finished' product in the 'front'" is only "a later stage and product of the 'currently unfinished' work in the 'back'" (ibid., 841). Therefore, "it should not be expected that increased familiarity with what goes on 'in the mathematical back' will lead to any significant revision of understanding of what is on show 'out front'" (ibid., 861).

This objection, however, entirely depends on Hersh's assumption that the back is mathematics as it appears among working mathematicians, told in offices or at café tables. It is only because of this assumption that Greiffenhagen and Sharrock may claim that the finished product in the front is only a later stage and product of the currently unfinished work in the back.

But Hersh's assumption is invalid. The back is not mathematics as it appears among working mathematicians, told in offices or at café tables. It is instead the making of mathematics, in particular discovery, and the process of discovery is not reflected virtually to any extent in finished mathematics (see Chap. 3).

If the back is the making of mathematics, in particular discovery, then it is invalid to say, as Greiffenhagen and Sharrock do, that the difference between the 'front' and the 'back' is not between two kinds of proof, but only between different stages: of working with an incomplete idea of a possible proof as opposed to presenting a (presumably) complete, thoroughly worked-out proof. Indeed, the difference between the 'front' and the 'back' of mathematics is precisely the difference between two kinds of demonstration: axiomatic demonstration, namely demonstration based on the axiomatic method, the front, and analytic demonstration, namely demonstration based on the analytic method, the back (see Chap. 10). Axiomatic demonstration is only a means to present, justify, and teach already acquired propositions. But, for the working mathematician, demonstration is primarily a means to discover solutions to problems, and only analytic demonstration is such a means.

It is also invalid to say, as Greiffenhagen and Sharrock do, that it should not be expected that increased familiarity with what goes on 'in the mathematical back' will lead to any significant revision of understanding of what is on show 'out front'. Mathematics presented in finished form has little or nothing to do with the way it was discovered (see Chap. 3). So, what is on show 'out front' gives no understanding of the making of mathematics, in particular discovery. Only familiarity with what goes on 'in the mathematical back', hence with analytic demonstration, gives such an understanding, and this leads to a significant revision of understanding of what is on show 'out front', because analytic demonstration is explanatory (see Chap. 14).

1.9 The Need for an Alternative Approach

Since, as Hersh says, mainstream philosophy of mathematics recognizes only the front, it does not account for the making of mathematics. This justifies the belief of several contemporary mathematicians, that philosophy, and specifically the philosophy of mathematics, is irrelevant to mathematics. A philosophy that does not account for the making of mathematics, namely for the actual process of mathematical research, cannot provide any real understanding of mathematics, let alone contribute to its advancement.

Indeed, as Byers says, "it is not possible to do justice to mathematics" by "separating the content of mathematical theory from the process through which that theory is developed and understood," so "it is of the utmost importance" to "develop a way of talking about mathematics that contains the entire mathematical experience, not just some formalized version of the results of that experience" (Byers 2007, 5).

Even some philosophers agree that mainstream philosophy of mathematics cannot provide any real understanding of the nature of mathematics, let alone contribute to its advancement.

Thus, Kreisel says that, even those who have "high hopes for (the subject of) philosophy, especially of mathematics," can have "little trust" in "the work of contemporary professional philosophers," in particular in "the logic chopping and obviously minor distinctions of which contemporary (Anglo-Saxon) philosophy is full" (Kreisel 1967, 212). Such work "is intended to clarify ideas in the Socratic manner; but it only keeps the outer forms including the banter of Plato, not the substance, namely the serious search for general definitions" (ibid.). Moreover, it focuses on insignificant aspects, and "there is no evidence that careful work on insignificant aspects leads one" to "recognize what is essential" (ibid.).

Putnam says that today in the philosophy of mathematics "nothing works," indeed "every philosophy seems to fail when it comes to explaining the phenomenon of mathematical knowledge" (Putnam 1994, 499). The "much touted problems in the philosophy of mathematics" are only "problems internal to the thought of various system builders. The systems are" only "intellectual exercises," and they, "without exception, need not be taken seriously" (Putnam 1975–1983, I, 43).

Kitcher says that, if one asks "what the philosophy of mathematics is" today, "many practicing mathematicians and historians of mathematics will have a brusque reply to" this "question: a subject noted as much for its irrelevance as for its vaunted rigor, carried out with minute attention to a small number of atypical parts of mathematics and with enormous neglect of what most mathematicians spend most of their time doing" (Kitcher 1988, 293).

Corfield says that "by far the larger part of activity in what goes by the name 'philosophy of mathematics' is dead to what mathematicians think and have thought, aside from an unbalanced interest in the 'foundational' ideas of the 1880–1930 period, yielding too often a distorted picture of that time" (Corfield 2003, 5).

But the inability to provide any real understanding of the nature of mathematics, let alone to contribute to its advancement, is a failure of mainstream philosophy of mathematics, not of philosophy as such. Indeed, as argued above, in the past, philosophy has been relevant to mathematics. What is needed is an alternative approach.

1.10 Aim of the Book

The aim of the book is to highlight the limitations of mainstream philosophy of mathematics, and to offer an alternative approach to the philosophy of mathematics. The alternative approach should satisfy two demands.

First, the alternative approach should account for the making of mathematics, in particular discovery. This contrasts with mainstream philosophy of mathematics, according to which the philosophy of mathematics cannot concern itself with the making of mathematics, but only with finished mathematics.

Second, the alternative approach should possibly contribute to the advancement of mathematics, as philosophy has done in the past. This contrasts with mainstream philosophy of mathematics, according to which the philosophy of mathematics cannot contribute to the advancement of mathematics.

Since the alternative approach should account for the making of mathematics, in particular discovery, it can be called ‘heuristic philosophy of mathematics’. For, ‘heuristic’ comes from the Greek ‘heuriskein’, which means ‘to discover’.

1.11 Organization of the Book

The book is divided into five parts after the present Introduction, which occurs as Chap. 1.

Part I, ‘Heuristic vs. Mainstream’, proposes heuristic philosophy of mathematics as an alternative to mainstream philosophy of mathematics. In particular, Chap. 2 describes characters, origin, and aim of mainstream philosophy of mathematics, and argues that it does not provide an adequate account of mathematics. Chapter 3 describes characters, origin, and aim of heuristic philosophy of mathematics, and argues that it provides an adequate account.

Part II, ‘Discourse on Method’, describes the basic methods that have been devised for mathematics. In particular, Chap. 4 describes the ancient origin of method, the oblivion of method, the role of method in the rise of modern science, the decline and end of method and its negative effects. Chapter 5 describes the analytic method, its origin, characters, and fortune, and points out the differences between the analytic method, on the one hand, abduction and reductio ad absurdum, on the other hand. Chapter 6 describes Aristotle’s analytic-synthetic method, its difference from the analytic method, Pappus’s analytic-synthetic method and its relation to reductio ad absurdum, the material axiomatic method, and the formal axiomatic method. Chapter 7 describes several rules of discovery: several kinds of induction, and several kinds of analogy, metaphor, metonymy, generalization, and specialization. Chapter 8 describes two views of theories and theory building: the axiomatic view and the analytic view.

Part III, 'The Mathematical Process', deals with the main elements of the mathematical process. In particular, Chap. 9 discusses several views of mathematical objects: mathematical objects as logical objects, simplifications, mental constructions, independently existing entities, abstractions, structures, fictions, idealizations of physical bodies, idealizations of operations, and hypotheses. Chapter 10 discusses four views of mathematical demonstration: mathematical demonstration as axiomatic demonstration, formal demonstration, deductive demonstration, and analytic demonstration. Chapter 11 discusses two views of mathematical definition: the stipulative view, and the heuristic view. Chapter 12 discusses two views of mathematical diagrams: the axiomatic view, and the heuristic view. Chapter 13 discusses two views of mathematical notations: the precision-conciseness view, and the heuristic view.

Part IV, 'The Functionality of Mathematics', deals with the capacity of mathematics to explain facts, to give aesthetic satisfaction, and to be applicable to the world. In particular, Chap. 14 discusses the question of mathematical explanations, arguing that, while there are mathematical explanations of mathematical facts, there are no genuine mathematical explanations of empirical facts. Chapter 15 discusses the nature and role of mathematical beauty. Chapter 16 discusses the question of the applicability of mathematics to the world.

Part V, 'Conclusion', completes and ends the book. In particular, Chap. 17 discusses the question in what sense mathematics is knowledge. Chapter 18 sums up the aims of the book.

1.12 Some General Remarks

The following general remarks about some features of the book may be useful.

(1) Throughout the book, mainly elementary mathematical problems are used to illustrate questions. This is no limitation because, as Pólya says, "elementary mathematical problems present all the desirable variety, and the study of their solution is particularly accessible and interesting" (Pólya 2004, 134). Moreover, "the expert mathematician who has some interest for this sort of study can easily add examples from his own experience to elucidate the points illustrated by elementary examples here" (ibid.). An attempt is made to consider historically significant mathematical problems, to compensate their elementariness with their historical significance.

(2) Throughout the book, the term 'demonstration' is used instead of 'proof', except when the term 'proof' occurs in quotations or in text referring to those quotations. This is because 'proof' is commonly employed to indicate deductive arguments, while 'demonstration' does not have that strict connotation. So, 'demonstration' seems more suitable as a general term for both axiomatic demonstration, which is purely deductive, and analytic demonstration, in which non-deductive inferences also play an essential role.

(3) Throughout the book, reference is made to Gödel's incompleteness theorems and other limitative results. Their statements and demonstrations can be found in several mathematical logic textbooks, in any case they can be found in Cellucci 2007, Chapter V.

(4) When quoting some ancient Greek phrase, the transliteration from the Greek to the Latin alphabet is used.

(5) Where no English translation is listed in the references, the translations are my own. Even where English translations exist, I have often used my own translations for two reasons. First, every translation is an interpretation, and the interpretations assumed in this book are often different from those of the existing English translations. Second, the existing English translations of different works of the same author have often been made by different translators and are non-uniform, so quoting from them may lead to misunderstandings. My own translations ensure, if not correctness, at least uniformity.

(6) Constant use of 'he or she' is clumsy, while constant use of 'she' may lead to misunderstandings. Therefore, I use the generic 'he', while stipulating here that I mean it to refer to persons of both genders.

References

Acerbi, Fabio. 2000. Plato: *Parmenides* 149 a 7–c 3. A proof by complete induction? *Archive for History of Exact Science* 55: 57–76.

Bays, Timothy. 2012. Review of Paolo Mancosu (ed.), *The philosophy of mathematical practice*. *Notices of the American Mathematical Society* 59 (3): 424–428.

Berkeley, George. 1992. *De motu and the analyst*. Dordrecht: Springer.

Bos, Henk J.M. 2001. *Redefining geometrical exactness: Descartes' transformation of the early modern concept of construction*. New York: Springer.

Bourbaki, Nicolas. 1994. *Elements of the history of mathematics*. Berlin: Springer.

Byers, William. 2007. *How mathematicians think: Using ambiguity, contradiction, and paradox to create mathematics*. Princeton: Princeton University Press.

———. 2017. Can you say what mathematics is? In *Humanizing mathematics and its philosophy: Essays celebrating the 90th birthday of Reuben Hersh*, ed. Bharath Sriraman, 45–60. Cham: Springer.

Cellucci, Carlo. 2007. *La filosofia della matematica del Novecento*. Rome: Laterza.

Cheng, Eugenia. 2004. *Mathematics, morally*. http://cheng.staff.shef.ac.uk/morality/morality.pdf.

Connes, Alain. 2008. Advice to a young mathematician III. In *The Princeton companion to mathematics*, ed. Timothy Gowers, 1011–1013. Princeton: Princeton University Press.

Corfield, David. 2003. *Towards a philosophy of real mathematics*. Cambridge: Cambridge University Press.

Dedekind, Julius Wilhelm Richard. 1996. Was sind und was sollen die Zahlen? In *From Kant to Hilbert: A source book in the foundations of mathematics*, ed. William Ewald, Vol. 2, 787–833. Oxford: Oxford University Press.

Descartes, René. 1996. *Oeuvres*. Paris: Vrin.

Dummett, Michael. 1998. The philosophy of mathematics. In *Philosophy 2: Further through the subject*, ed. Anthony C. Grayling, 122–196. Oxford: Oxford University Press.

Galilei, Galileo. 1968. *Opere*. Barbera: Firenze.

Gowers, Timothy. 2002. *Mathematics: A very short introduction*. Oxford: Oxford University Press.

———. 2006. Does mathematics need a philosophy? In *18 unconventional essays on the nature of mathematics*, ed. Reuben Hersh, 182–200. New York: Springer.
Greiffenhagen, Christian, and Wes Sharrock. 2011. Does mathematics look certain in the front, but fallible in the back? *Social Studies in Science* 41: 839–866.
Hanna, Gila, and Brendan Larvor. 2020. As Thurston says? On using quotations from famous mathematicians to make points about philosophy and education. *ZDM Mathematics Education* 52: 1137–1147.
Hardy, Godfrey Harold. 1929. Mathematical proof. *Mind* 38: 1–25.
———. 1992. *A mathematician's apology*. Cambridge: Cambridge University Press.
Hersh, Reuben. 1979. Some proposals for reviving the philosophy of mathematics. *Advances in Mathematics* 31: 31–50.
———. 1997. *What is mathematics, really?* Oxford: Oxford University Press.
———. 2014. *Experiencing mathematics: What do we do, when we do mathematics?* Providence: American Mathematical Society.
Hobbes, Thomas. 1839–1845. *Opera philosophica*. London: Bohn.
Kitcher, Philip. 1988. Mathematical naturalism. In *History and philosophy of modern mathematics*, ed. William Aspray and Philip Kitcher, 293–325. Minneapolis: University of Minnesota Press.
Kreisel, Georg. 1967. Mathematical logic: What has it done for the philosophy of mathematics? In *Bertrand Russell, philosopher of the century*, ed. Ralph Schoenman, 201–272. London: Allen & Unwin.
Lebesgue, Henri. 1941. Les controverses sur la théorie des ensembles et la question des fondements. In *Les Entretiens de Zurich sur les fondements et la méthode des sciences mathématiques*, ed. Ferdinand Gonseth, 109–122. Zurich: Leemann.
Nightingale, Andrea Wilson. 1995. *Genres in dialogue: Plato and the construct of philosophy*. Cambridge: Cambridge University Press.
Pólya, George. 2004. *How to solve it: A new aspect of mathematical method*. Princeton: Princeton University Press.
Putnam, Hilary. 1975–1983. *Philosophical papers*. Cambridge: Cambridge University Press.
———. 1994. *Words and life*. Cambridge: Harvard University Press.
Robinson, Abraham. 1966. *Non-standard analysis*. Amsterdam: North-Holland.
Smoryński, Craig. 1983. Mathematics as a cultural system. *The Mathematical Intelligencer* 5 (1): 9–15.
Wolff, Christian. 1728. *Philosophia rationalis sive logica*. Renger: Frankfurt & Leipzig.

Part I
Heuristic vs. Mainstream

Chapter 2
Mainstream Philosophy of Mathematics

Abstract The chapter describes characters, origin, and goal of mainstream philosophy of mathematics, namely the philosophy of mathematics that has prevailed for the past century. According to it, the philosophy of mathematics cannot concern itself with the making of mathematics but only with finished mathematics, namely mathematics presented in finished form, and the method of mathematics is the axiomatic method. The chapter argues that, because of Gödel's incompleteness theorems and for several other reasons, mainstream philosophy of mathematics does not provide an adequate account of mathematics.

Keywords Mainstream philosophy to mathematics · Foundationalist view · Top-down approach · Closed systems · Mathematics as theorem proving · Relevance of incompleteness theorems · Shortcomings of reductionism

2.1 The Fabric of Mainstream Philosophy of Mathematics

In the Introduction, reference has been made to mainstream philosophy of mathematics, namely the philosophy of mathematics that has prevailed for the past century.

Mainstream philosophy of mathematics consists of the three big foundationalist schools, logicism, formalism, and intuitionism, and their direct or indirect descendants.

The direct descendants of the three big foundationalist schools are revised versions of them: neo-logicism, neo-formalism, and neo-intuitionism.

The indirect descendants of the three big foundationalist schools are variations on their themes: platonism, abstractionism, structuralism, fictionalism, phenomenology, and empiricism.

However, the indirect descendants of the three big foundationalist schools also comprises the philosophy of mathematical practice. For, contrary to the widespread view that the latter is alternative to mainstream philosophy of mathematics, as argued in Chap. 3, the philosophy of mathematical practice is continuous with mainstream philosophy of mathematics.

C. Cellucci, *The Making of Mathematics*, Synthese Library 448,
https://doi.org/10.1007/978-3-030-89731-4_2

2.2 The Characters of Mainstream Philosophy of Mathematics

Mainstream philosophy of mathematics has the following characters.

(1) The philosophy of mathematics cannot concern itself with the making of mathematics, in particular discovery, because discovery is a subjective process, so it cannot be accounted for.

Thus, Dieudonné says that the philosophy of mathematics cannot concern itself with the making of mathematics, because it is impossible to explain how mathematicians "arrived at their results" (Dieudonné 1998, 27). For, "what goes on in a creative mind never has a rational 'explanation', in mathematics any more than elsewhere. All that we know is that it" involves "sudden 'illuminations'" (ibid., 27).

Feferman says that the philosophy of mathematics cannot concern itself with the making of mathematics, because the "individual processes of mathematical discovery appear haphazard and illogical" (Feferman 1998, 77). Therefore, "the creative and intuitive aspects of mathematical work evade logical encapsulation" (ibid., 178).

(2) The philosophy of mathematics can concern itself only with finished mathematics, namely mathematics presented in finished form, because only finished mathematics is objective, so it can be completely justified.

Thus, Pólya says that the philosophy of mathematics can concern itself only with "finished mathematics" because only finished mathematics is objective, being "purely demonstrative, consisting of proofs only" (Pólya 1954, I, vi). Finished mathematics can be completely justified, because "we secure our mathematical knowledge by demonstrative reasoning," which "is safe, beyond controversy, and final" (ibid., I, v).

Dummett says that "the philosophy of mathematics" can concern itself only "with the product of mathematical thought," because only the latter is objective and hence can be completely justified, conversely, "the study of the process of production is the concern of psychology, not of philosophy" (Dummett 1991, 305).

(3) Since the philosophy of mathematics cannot concern itself with the making of mathematics, it cannot contribute to the advancement of mathematics.

Thus, Körner says that, "as the philosophy of law does not legislate, or the philosophy of science devise or test scientific hypotheses, so – we must realize from the outset – the philosophy of mathematics does not add to the number of mathematical theorems and theories" (Körner 1986, 9).

Dummett says that, while mathematicians advance knowledge, philosophers of mathematics only cast "light on what we already know from other sources" (Dummett 2010, 7). So, the philosophy of mathematics "does not advance knowledge," it only "clarifies what we already know" (ibid., 21).

(4) The task of the philosophy of mathematics is primarily to give an answer to the question: How do mathematical propositions come to be completely justified? And, subordinately to it, to the question: Do objects exist in virtue of which mathematical propositions are true, and if so what is their nature?

Thus, Lehman says that the task of the philosophy of mathematics is primarily to give an answer to the question of "how mathematical beliefs come to be completely justified," and, subordinately to it, to the question of "whether there are entities in virtue of which the propositions are true," and "if so what their nature is" (Lehman 1979, 1).

Shapiro says that the task of the philosophy of mathematics is primarily to give an answer to the question of what are the true "justifications for mathematical propositions," and, subordinately to it, to the question of whether there are mathematical objects in virtue of which the propositions are true, and if so what is the "underlying nature of mathematical objects" (Shapiro 2004, 37).

(5) The method of mathematics is the axiomatic method. The latter is the method according to which, to demonstrate a proposition, one starts from given principles which are true, in some sense of 'true', and deduces the proposition from them (see Chap. 6).

Thus, Kac and Ulam say that "mathematics owes its unique position to its adherence to the axiomatic method," which "consists in starting with a few statements (axioms) whose truth is taken for granted and then deriving other statements from them by the application of rules of logic alone" (Kac and Ulam 1992, 139).

Rota says that, according to "the accepted description of mathematical truth," a "mathematical statement is held to be true if it is correctly derived from the axioms by application of the rules of inference," because "the truth of all theorems can 'in principle' be 'found' in the axioms" (Rota 1997, 109).

(6) The role of axiomatic demonstration, namely demonstration based on the axiomatic method, is to guarantee the truth of a proposition.

Thus, Bass says that axiomatic demonstration "is the defining source of mathematical truth" (Bass 2015, 129). For, "saying that a mathematical claim is true means, for a mathematician, that there exists" an axiomatic demonstration "of it" (ibid., 132).

Jaffe says that axiomatic demonstration "has the highest degree of certainty possible for man" (Jaffe 1997, 135). While "scientific hypotheses come and go," the truth of a mathematical proposition obtained by an axiomatic demonstration "lasts forever" (ibid., 139).

(7) Since the method of mathematics is the axiomatic method, mathematics is a body of truths, and indeed truths that are certain. Therefore, mathematics is about truth and certainty.

Thus, Feferman says that, since the method of mathematics is the axiomatic method, mathematics is "the paradigm of certain and final knowledge: not fixed, to be sure, but a steadily accumulating coherent body of truths obtained by successive deduction from the most evident truths" (Feferman 1998, 77).

Chihara says that, since the method of mathematics is the axiomatic method, "mathematics is a system of truths and mathematicians are attempting to arrive at truths" (Chihara 1990, 171). For hundreds of years mathematicians "have reasoned and constructed their theories with the tacit belief that" the "principles of mathematics," and the propositions deduced from them, are "true. Many of these mathematical beliefs have been checked and rechecked countless times, and in countless ways," so we "have some strong reasons supporting the belief that mathematics is a body of truths" (ibid., 172–173).

(8) Since the method of mathematics is the axiomatic method, mathematical reasoning consists of deductive reasoning.

Thus, Dummett says that "deductive inference patently plays a salient part in mathematics. The correct observation that the discovery of a theorem does not usually proceed in accordance with the strict rules of deduction has no force: a proof has to be set out in sufficient detail to convince readers, and, indeed, its author, of its deductive cogency" (Dummett 1991, 305).

Prawitz says that "the notion of deduction" is "the key concept for understanding validity in mathematics" because, "when it comes to explaining the remarkable phenomenon that work on a mathematical problem may end in a result that everyone finds definitive and conclusive, the notion of deduction is a central one" (Prawitz 1998, 332).

(9) The philosophy of mathematics is a new independent subject introduced by Frege. Although several philosophers before Frege have said something about the nature of mathematics, Frege has completely revolutionized the field.

Thus, Resnik says that Frege "completely revolutionized the field" of the philosophy of mathematics, "by demolishing the naive views of his contemporaries and predecessors and by furnishing a model for the type of precision needed in treatments of the philosophy of mathematics" (Resnik 1980, 15).

Kenny says that Frege's "work completely antiquated everything previously written. No one can now take seriously the work of even the greatest previous writers on the subject," and "no one henceforth could write on the topic without taking his work as a starting point" (Kenny 1995, 211).

(10) The philosophy of mathematics can be developed independently of experience, because mathematics is an armchair discipline that is the product of thought alone, so it does not depend on experience.

Thus, George and Velleman say that "mathematical knowledge" is "unconstrained by experience," it enters "the world touched only by the hand of reflection," and is "justified by pure ratiocination" (George and Velleman 2002, 1). For, "mathematics is the purest product of conceptual thought, which is a feature of human life" that "sets it apart from all else" (ibid.).

Dummett says that mathematics is an "armchair discipline" which "needs no input from experience: it is the product of thought alone" (Dummett 2010, 4). This shows that "thought, without any specialized input from experience, can advance knowledge in unexpected directions" (ibid., 5).

2.3 Original Formulation of Mainstream Philosophy of Mathematics

As we have just seen, according to mainstream philosophy of mathematics, the philosophy of mathematics is a new independent subject introduced by Frege. In fact, Frege first formulated the characters that mainstream philosophy of mathematics assigns to the philosophy of mathematics. He formulated them as follows.

(1) The philosophy of mathematics cannot concern itself with the making of mathematics, in particular "with the way in which" mathematical propositions "are discovered" (Frege 1960, 23). For, discovery is a subjective and psychological process, being an inner act of one person's mind, and "we do not directly observe the processes in the mind of another," so "we are unable to unite the inner states experienced by different people in one consciousness and so compare them" (Frege 1979, 3–4).

(2) The philosophy of mathematics can concern itself only with finished mathematics, because only finished mathematics is objective, therefore for each judgment of finished mathematics we can give "the justification for making the judgment" (Frege 1960, 3).

(3) Since the philosophy of mathematics cannot concern itself with the making of mathematics, it cannot contribute to the advancement of mathematics, so "there are no new truths in my work" (Frege 1967, 6).

(4) The task of the philosophy of mathematics is primarily to give an answer to the question: For any mathematical proposition, what is "the ultimate ground upon which rests the justification for holding it be true?" (Frege 1960, 3). And, subordinately to it, to the question of whether numbers exist, and if so "what number is" (ibid., xiv)

(5) The method of mathematics is the axiomatic method, namely the method according to which, to demonstrate a theorem, one "starts from propositions that are accepted as true," the principles, and arrives "via chains of inferences to the theorem" (Frege 1979, 204).

(6) The role of axiomatic demonstration is to guarantee the truth of a proposition, by showing "the ultimate ground upon which rests the justification for holding" the proposition "to be true" (Frege 1960, 3). Thus one "can finally provide it with the most secure foundation" (Frege 1967, 5).

(7) Since the method of mathematics is the axiomatic method, mathematics is a body of truths, and indeed of truths that are certain, because mathematics is "a system of truths that are connected with one another by" deductive "inference" (Frege 1979, 205). And "the most reliable way of carrying out a proof, obviously, is to follow" deductive "logic" (Frege 1967, 5).

(8) Since the method of mathematics is the axiomatic method, mathematical reasoning consists of deductive reasoning, because "there is no such thing as a peculiarly" mathematical "mode of inference that cannot be reduced to the general" deductive "inference-modes of logic" (Frege 1984, 113).

(9) The philosophy of mathematics is a new independent subject, which has been made possible by "a renaissance of logic" (Frege 2013, I, xxvi).

(10) The philosophy of mathematics can be developed independently of experience, because mathematics involves an "a priori mode of cognition" (Frege 1979, 277). So mathematics does not depend on experience, but only on a priori "intuition" in "its proofs" (ibid., 278).

2.4 The Foundationalist View of Mathematics

Mainstream philosophy of mathematics puts forward the foundationalist view of mathematics, which is based on the following assumptions.

(I) There is immediately justified knowledge, namely knowledge which is justified without inference, and all other knowledge is deduced from it, therefore it is justified knowledge.

The foundationalist view owes its name to the architectural metaphor, according to which knowledge is an edifice whose foundation consists of the immediately justified knowledge, and whose body consists of all other knowledge deduced from it, hence knowledge anchored to the foundation via deductive inference.

The immediately justified knowledge are the axioms, all other knowledge deduced from it are the theorems. Therefore, mathematics is theorem proving by the axiomatic method.

(II) The immediately justified knowledge is true and certain because it is based on intuition, and hence so is all knowledge deduced from it. Therefore, mathematics is true and certain because it is based on intuition.

(III) There is a part of mathematics such that all other parts of mathematics can be reduced to it. Specifically, there is a mathematical theory such that all other mathematical theories can be reduced to it. This mathematical theory is The Foundation.

2.5 Original Formulation of the Foundationalist View

Aristotle gave the first formulation of the foundationalist view of mathematics, with the exception of assumption (III).

Aristotle argues for assumption (I) of the foundationalist view as follows. Suppose there is no immediately justified knowledge. Then there will be no primitive premisses, the series of the premisses will be infinite, and one will be "led back in an infinite regress" (Aristotle, *Analytica Posteriora*, A 3, 72 b 8–9). But "it is impossible to traverse an infinite series" (ibid., A 3, 72 b 10–11). So, if the series of the premisses is infinite, then "there is no knowledge" (ibid., A 3, 72 b 5–6). But, as a matter of fact, there is knowledge. Contradiction. Therefore, there is immediately justified knowledge, and all other knowledge is deduced from it, hence it is justified knowledge. Aristotle also introduces the architectural metaphor by saying that, since all other knowledge is deduced from it, the immediately justified knowledge is "that from which a thing first arises," therefore it is like "the foundation of a house" (Aristotle, *Metaphysica*, Δ 1, 1013 a 4–5).

Aristotle argues for assumption (II) of the foundationalist view as follows. Since the immediately justified knowledge is that from which all other knowledge is deduced, the immediately justified knowledge is indemonstrable. Then, "it is intuition, and not discursive thinking, that apprehends the primitive things," namely "it is intuition that apprehends the unchanging and first terms in the order of demonstrations" (Aristotle, *Ethica Nicomachea*, Z 11, 1143 a 36–1143 b 3). Now, intuition is intuition of the essence of things, and "when intuition is of the essence of things, it is true" (Aristotle, *De Anima*, Γ 6, 430 b 28). And, about the essence of things, "it is not possible to be mistaken" (Aristotle, *Metaphysica*, Θ 10, 1051 b 31). Therefore, the immediately justified knowledge is true and certain. Also, all knowledge deduced from the immediately justified knowledge is true and certain. For, "a conclusion from truths is always true" (Aristotle, *Analytica Posteriora*, A 6, 75 a 5–6). And a conclusion from premisses which are certain "results by necessity because these things are so" (Aristotle, *Analytica Priora*, A 1, 24 b 19–20).

On the other hand, while arguing for assumptions (I) and (II) of the foundationalist view, Aristotle does not argue for assumption (III). This depends on the fact that "the diagonal" of the square "is incommensurable with the side" (Aristotle, *Topica*, Θ 13, 163 a 11–12). Therefore, geometry cannot be reduced to arithmetic.

The assumption (III) acquired credibility only in the second half of the nineteenth century, when the basis was laid down for a reduction of arithmetic and geometry to set theory.

2.6 A Remark on the Original Formulation of the Foundationalist View

That Aristotle gave the first formulation of the foundationalist view, with the exception of assumption (III), does not mean, however, that for Aristotle the axiomatic method is the method of the making of mathematics.

For Aristotle, the axiomatic method is only the method of finished mathematics, because the purpose of the axiomatic method is not to acquire new knowledge, but only to present, justify, and teach already acquired propositions (see Chap. 6).

Conversely, the method of the making of mathematics is the analytic-synthetic method or method of analysis and synthesis.

The latter is the method according to which, to solve a problem, one looks for some hypothesis that is a sufficient condition for solving the problem, namely such that a solution to the problem can be deduced from the hypothesis. The hypothesis is obtained from the problem, and possibly other data already available, by some non-deductive rule, and must be plausible, namely such that the arguments for the hypothesis are stronger than the arguments against it, on the basis of experience. If the hypothesis so obtained is not a principle (or a proposition deduced from principles), one looks for another hypothesis that is a sufficient condition for solving the problem posed by the previous hypothesis, it is obtained from the latter, and possibly other data already available, by some non-deductive rule, and must be plausible. And so on, until one arrives at a principle (or a proposition already deduced from principles). The principles must be true. When one arrives at a principle (or a proposition deduced from principles), the process terminates. This is analysis.

At this point, one tries to see whether, inverting the order of the steps followed in analysis, one obtains a deduction of the solution of the problem from the principle (or proposition deduced from principles) arrived at in analysis. This is synthesis (see Chap. 6).

2.7 Logicism and the Foundationalist View

The three big foundationalist schools, logicism, formalism, and intuitionism, make all the assumptions (I) – (III) of the foundationalist view, although they differ in their way of implementing them.

As to logicism, Frege says that, "if we start from a theorem and trace the chains of inference backwards," we "must eventually come to an end by arriving at truths which cannot themselves be inferred in turn from other truths" (Frege 1979, 204). These truths are immediately justified knowledge.

Immediately justified knowledge is knowledge such that one cannot be "in doubt about its truth" (ibid., 205). For, it is based on intuition. Specifically, the immediately justified knowledge of arithmetic is based on intellectual intuition, which is "the logical source of knowledge" (ibid., 267). Then, all other knowledge deduced from it will be knowledge such that one cannot be in doubt about its truth.

The character of the immediately justified knowledge of arithmetic is clear from the fact that "we can count just about everything that can be an object of thought: the ideal as well as the real, concepts as well as objects, temporal as well as spatial entities, events as well as bodies" (Frege 1984, 112). Since "the basic propositions on which arithmetic is based," namely the immediately justified knowledge of arithmetic, "must extend to everything that can be thought," surely "we are justified in ascribing such extremely general propositions to logic" (ibid.). Thus, "there is no such thing as a peculiarly arithmetical mode of inference that cannot be reduced to the general inference-modes of logic" (ibid., 113).

Then, "a rigorous establishment of arithmetical laws reduces them to purely logical laws" (ibid., 145). Specifically, it reduces them to the basic logical laws of Frege's logical system. Therefore, Frege's logical system is The Foundation.

2.8 Formalism and the Foundationalist View

As to formalism, Hilbert says that, in mathematics, there are "a few distinguished propositions" which "suffice by themselves for the construction, in accordance with logical principles, of the entire framework" (Hilbert 1996a, 1108). These basic propositions are immediately justified knowledge.

The basic propositions are true and certain because they are based on intuition, specifically, on Kant's pure intuition of space and time. However, a distinction must be made between them, because some of them are based on intuition directly, others indirectly.

The basic propositions which are based on intuition directly are the propositions of finitary mathematics, namely the mathematics which can be expressed without using actual infinite sets. These propositions are "real propositions" (Hilbert 1967b, 470). For, they are about certain "concrete objects that are intuitively present as immediate experience prior to all thought" (ibid., 464). Therefore, they can be based on intuition directly.

The basic propositions which are based on intuition indirectly are the propositions of infinitary mathematics, namely mathematics which cannot be expressed without using actual infinite sets. These propositions are "ideal propositions," because they are about certain abstract objects which are only "ideal objects" (ibid., 470). They are not intuitively present, and hence cannot be based on intuition directly.

Nevertheless, the ideal propositions can be based on intuition indirectly. For, "there is a condition, a single but absolutely necessary one, to which the use" of the ideal propositions "is subject, and that is the proof of consistency" (Hilbert 1967a, 383). Namely, it must be proved that, from the ideal propositions, "it is impossible for us to obtain two logically contradictory propositions, A and $\neg A$" (Hilbert 1967b, 471).

To prove this, the ideal propositions must be formalized. This is necessary because "the ideal propositions, insofar as they do not express finitary assertions, do not mean anything in themselves," so "the logical operations cannot be applied to them in a contentual way," hence "it is necessary to formalize the logical operations and also the mathematical proofs themselves" (Hilbert 1967a, 381). Thus, mathematics becomes "manipulation of signs according to rules" (Hilbert 1967b, 467).

Once the ideal propositions have been formalized, their consistency must be proved, and must be proved by a proof based on the "intuitive mode of thought" (Hilbert 1996d, 1150). For, only the intuitive mode of thought can avoid "any dubious or problematic mode of inference" (Hilbert 1996c, 1139). A consistency proof based on the intuitive mode of thought will warrant that the basic propositions which are ideal propositions are "incontestable and ultimate truths" (Hilbert 1996b, 1121). Indeed, if they "do not contradict one another with all their consequences, then they are true" (Hilbert 1980, 39).

Through a consistency proof based on the intuitive mode of thought, the ideal propositions are based on intuition indirectly, because their consistency "rests on a kind of intuitive insight" (Hilbert 1996e, 1161).

Although there are several mathematical theories, Zermelo set theory "encompasses all mathematical theories (like number theory, analysis, geometry), in the sense that the relations which obtain between the objects of one of these mathematical theories are perfectly represented by the relations which obtain within a subdomain of Zermelo set theory" (Hilbert 2013, 356). Thus, "Zermelo axiom system represents the most comprehensive axiomatic system" (ibid.). Therefore, Zermelo set theory is The Foundation.

2.9 Intuitionism and the Foundationalist View

As to intuitionism, Brouwer says that infinitary mathematics is merely a "linguistic science, operating on meaningless words or symbols by means of logical rules" (Brouwer 1975, 522). By so operating, "no more is obtained than a linguistic structure" which "irrevocably remains separated from mathematics" (ibid., 97). Then, Hilbert's attempt to justify infinitary mathematics by a consistency proof based on the intuitive mode of thought "contains a circulus vitiosus since such justification is based on the (contentual) correctness of the assertion that correctness of a proposition follows from its noncontradictority" (Brouwer 1998, 41).

We "can obtain knowledge of mathematics" only if mathematics is "constructed by direct intuition" (Brouwer 1975, 75). And specifically, only if it is constructed by Kant's pure "intuition of time," which is "the basic intuition of mathematics" (ibid., 71). Therefore infinitary mathematics must be replaced with an alternative mathematics, "intuitionistic mathematics," which is built starting from basic propositions based on intuition, and "deducing theorems" from them "exclusively by means of introspective construction" (ibid., 488). Namely, by means of deductive inferences based on intuition. The basic propositions are immediately justified knowledge.

The basic propositions are true and certain because they are based on intuition, and intuition is "the origin of mathematical certainty" (ibid., 508). All propositions deduced from them are also true and certain, because they are deduced exclusively by means of deductive inferences based on intuition. Therefore, mathematics is true and certain being based on intuition.

By deducing theorems from basic propositions, intuitionistic mathematics proceeds by the axiomatic method. It might be thought that, from the intuitionistic point of view, the axiomatic method is unimportant, but it is not so. As Heyting says, of course, "from the intuitionistic point of view," the axiomatic "method cannot be used in its creative function," because "a mathematical object is considered to exist" only "after its construction" by intuition, so "it cannot be brought into existence by a system of axioms" (Heyting 1962, 239). But nothing prevents from using the axiomatic method in its descriptive function, as a description of constructions already made, and "the descriptive function of a system of axioms is as important intuitionistically as it is classically" (ibid.).

All of mathematics can be reduced to two basic theories, the theory of "more or less freely proceeding infinite sequences of mathematical entities previously acquired," and the theory of "mathematical species, i.e., properties supposable for mathematical entities previously acquired" (Brouwer 1975, 482). Therefore, the theory of more or less freely proceeding infinite sequences, and the theory of species together are The Foundation.

2.10 The Top-Down Approach to Mathematics

Assumption (I) of the foundationalist view, that mathematics is theorem proving by the axiomatic method, is an expression of the top-down approach to mathematics, according to which every part of mathematics must be developed from above, namely starting from general principles, by the axiomatic method.

An example of the top-down approach in the early modern period is the calculus of infinitesimals of Leibniz, as presented by L'Hôpital in the first textbook on the subject.

At the beginning of the textbook, L'Hôpital gives the following two definitions: "Definition I. Those quantities are called 'variable' which increase or decrease continually" (L'Hôpital 2015, 1). And: "Definition II. The infinitely small portion by which a variable quantity continually increases or decreases is called the 'differential'" (ibid., 2).

Then, L'Hôpital introduces a notation for variables and differentials: "In what follows, we will make use of the symbol d to denote the differential of a variable quantity that is expressed by a single letter" (ibid.). So, if x denotes a variable quantity, dx will denote its differential. Generally, we will "suppose that the final letters of the alphabet z, y, x, etc., denote variable quantities," so that "as x becomes $x + dx$, y, z, etc., become $y + dy$, $z + dz$, etc." (ibid., 3).

Next, L'Hôpital gives the following two principles or postulates: "Postulate I. We suppose that two quantities that differ by an infinitely small quantity may be used interchangeably, or (what amounts to the same thing) that a quantity which is increased or decreased by another quantity that is infinitely smaller than it is, may be considered as remaining the same" (ibid.). And "Postulate II. We suppose that a curved line may be considered as an assemblage of infinitely many straight lines, each one being infinitely small, or (what amounts to the same thing) as a polygon with an infinite number of sides, each being infinitely small, which determine the curvature of the line by the angles formed amongst themselves" (ibid.).

About these postulates, L'Hôpital says: "The two postulates" seem "to me so clear, that I do not believe that they can leave any doubt in the minds of attentive readers" (ibid., liv).

On this basis, L'Hôpital demonstrates results of the calculus of infinitesimals of Leibniz. For example, he demonstrates that "the differential of xy is $ydx + xdy$" by arguing that, if we suppose that x becomes $x + dx$ and y becomes $y + dy$, then "xy becomes $xy + ydx + xdy + dxdy$, which is the product of $x + dx$ and $y + dy$," so "the differential is $ydx + xdy + dxdy$" (ibid., 4). But, by Postulate I, the latter quantity may be considered to be equal to $ydx + xdy$, because "the quantity $dxdy$ is infinitely small with respect to the other terms $x + dx$ and $y + dy$" (ibid.). This establishes the result.

The top-down approach has been the paradigm of mathematics in the past century, especially because of the influence of the Göttingen school of mathematics headed by Hilbert, on the one hand, and of the Bourbaki group, on the other hand.

2.11 The Foundationalist View and Closed Systems

Assumption (I) of the foundationalist view, that mathematics is theorem proving by the axiomatic method, implies that mathematical theories are closed systems.

A closed system is a system whose development does not take place by interacting with the environment, receiving inputs from, and delivering outputs to the outside. Its development remains completely internal to the system, so the system is a self-sufficient totality.

Assumption (I) implies that mathematical theories are closed systems, because a mathematical theory is based on principles that are given once for all, and its development consists entirely in deducing propositions from the principles.

Since deduction is non-ampliative, namely the conclusion is implicitly contained in the premisses (see Chap. 5), the propositions deduced from the principles are already implicitly contained in them. So, the development of a mathematical theory remains completely internal to the theory.

That assumption (I) implies that mathematical theories are closed systems, is vividly expressed by Frege by saying that in mathematics there are some primitive truths and "the whole of mathematics is contained in these primitive truths as in a seed" (Frege 1979, 204–205). So "our only concern is to generate the whole of mathematics from this seed. The essence of mathematics has to be defined by this seed of truths" (ibid., 205). Namely, the mathematician's only concern is to deduce the whole of mathematics from some given primitive truths, in which the whole of mathematics is implicitly contained.

2.12 Mathematics as Theorem Proving and Mathematicians

Assumption (I) of the foundationalist view, that mathematics is theorem proving by the axiomatic method, is shared by the majority of mathematicians.

Thus, Kline says that "the mathematician starts with axioms," from "these axioms theorems are derived by deductive arguments, which are unassailable, and thus the mathematician arrives at new truths" (Kline 1981, 236).

Ulam says that "mathematicians start with axioms whose validity they don't question. You might say it is just a game" which "we play according to certain rules" of deduction, "starting with statements which we cannot analyze further" (Ulam 1986, 13–14).

Gowers says that the mathematician "starts with axioms" and "proceeds to the desired conclusion by means of only the most elementary logical rules" (Gowers 2002, 39).

Sternheimer says that "in mathematics one starts with axioms and uses logical deduction therefrom to obtain results that are absolute truth in that framework" (Sternheimer 2011, 42).

2.13 Inadequacy of the Infinite Regress Argument

Although assumption (I) of the foundationalist view is shared by the majority of mathematicians, nevertheless it is invalid.

First, Aristotle's argument for assumption (I), that for any part of mathematics there is immediately justified knowledge, is invalid. This can be seen as follows.

Aristotle argues that, if there is no immediately justified knowledge, then there will be no primitive premisses, so the series of the premisses will be infinite. But it is impossible to traverse an infinite series. So, if the series of the premisses is infinite, then there is no knowledge. But, as a matter of fact, there is knowledge. Contradiction. Therefore, there is immediately justified knowledge.

Now, Aristotle is quite right in saying that it is impossible to traverse an infinite series. But this only means that, if the series of the premisses is infinite, then there is no immediately justified knowledge. It does not mean that there is no knowledge.

There would be no knowledge only if the premisses occurring in the infinite series were arbitrary. But they need not be arbitrary, they can be plausible, namely such that the arguments for them are stronger than the arguments against them, on the basis of experience. Now, if the premisses are plausible, then there is knowledge. Admittedly, such knowledge is not absolutely certain, it is provisional, always in need of further consideration. But, as it will be argued below, this is the only kind of knowledge that is possible for us.

At each stage, we may check whether the premisses used until then are plausible. To have knowledge, it is not necessary that we arrive at immediately justified premisses, but only that, at each stage, the premisses used at that stage are plausible. Therefore, Aristotle's argument is invalid.

2.14 The Foundationalist View and Gödel's Incompleteness Theorems

Not only Aristotle's argument for assumption (I) of the foundationalist view is invalid, but no other argument for it could be valid. For, assumption (I) is refuted by Gödel's incompleteness theorems.

Assumption (I) is refuted by Gödel's first incompleteness theorem. For, by the latter, for any consistent, sufficiently strong, formal system, there are propositions of the system that are true but cannot be deduced from the axioms of the system. This implies that mathematics cannot consist in the deduction of propositions from given axioms because, for any choice of axioms for a given part of mathematics, there will always be true propositions of that part which cannot be deduced from the axioms.

Assumption (I) is also refuted by Gödel's second incompleteness theorem. For, by the latter, for any consistent, sufficiently strong, formal system, it is impossible to demonstrate, by absolutely reliable means, that the axioms of the system are consistent. So there is no guarantee that the propositions deduced from the axioms are justified knowledge.

2.15 Gödel's Attempt to Reaffirm Mathematics as Theorem Proving

Although assumption (I) of the foundationalist view is refuted by Gödel's incompleteness theorems, mathematicians have greatly resisted accepting this conclusion. Gödel himself tries to reaffirm that mathematics is theorem proving by the axiomatic method, by arguing that his incompleteness theorems merely require that, instead of being formalizable in a single formal system, mathematics is formalizable in an infinite sequence of formal systems.

Indeed, Gödel says that, although his first incompleteness theorem implies that it is "impossible to formalize all of mathematics in a single formal system, a fact that intuitionism has asserted all along," nevertheless "everything mathematical is formalizable" (Gödel 1986–2002, I, 389). It is formalizable not in a single formal system, but in "a sequence (continuable into the transfinite) of formal systems" (ibid., I, 237).

However, for this argument to be credible, the transition from a formal system to the next one in the sequence of formal systems should itself be formal. For, if the transition is not formal and requires an appeal to intuition, it will be impossible to say that everything mathematical is formalizable, the appeal to intuition will lead beyond what is formalizable.

But, if the transition from a formal system to the next one in the sequence of formal systems is itself formal, then, as McCarthy argues, for the sequence of formal systems it will be possible to demonstrate a theorem that "is an exact analogue" of "Gödel's first" incompleteness "theorem" (McCarthy 1994, 427). Therefore, not everything mathematical will be formalizable in the continuable sequence of formal systems.

Since not everything mathematical will be formalizable in the continuable sequence of formal systems, then, contrary to Gödel's claim, mathematics cannot consist in the activity of "an idealized mathematician who entertains a sequence of successive" formal systems, and whose choices of formal systems "are effectively determined at each stage" (ibid., 444). Mathematics cannot consist in that, even if one identifies mathematics with the activity of "an idealized mathematician whose epistemic alternatives are effectively determined at each stage, but who may have a choice among these alternatives" (ibid., 446). Therefore, Gödel's argument is invalid.

2.16 Recalcitrant Mathematicians

Already Post stigmatized the great resistance of mathematicians to accept that assumption (I) of the foundationalist view is refuted by Gödel's incompleteness theorems. In 1941 he wrote: "It is to the writer's continuing amazement that ten years after Gödel's remarkable achievement current views on the nature of mathematics are thereby affected only to the point of seeing the need of many formal systems, instead of a universal one" (Post 1965, 345). Instead, "has it seemed to us to be inevitable that these developments will result in a reversal of the entire axiomatic trend of the late nineteenth and early twentieth centuries," and that axiomatic "thinking will then remain as but one phase of mathematical thinking" (ibid.).

A fortiori, it is to our continuing amazement that today mathematicians continue to say that "the axiomatic method is 'the' method of mathematics, in fact, it is mathematics" (Naylor and Sell 2000, 6).

This involves denying the relevance of Gödel's first incompleteness theorems to assumption (I) of the foundationalist view. To this purpose, the following arguments have been used.

(1) Gödel's incompleteness theorems do not refute the assumption that mathematics is theorem proving by the axiomatic method. They merely involve that, instead of being formalizable in a single formal system, mathematics is formalizable in a network of formal systems.

Thus, Curry says that "the propositions of mathematics are the propositions" of "some formal system," so "we have not confined mathematics to a single formal system" (Curry 1954, 231). Of course, by Gödel's first incompleteness theorem, "it is hopeless to find a single formal system which will include all of mathematics," so "the essence of mathematics" cannot lie "in any particular kind of formal system" (Curry 1951, 56). But, instead of lying in any particular kind of formal system, "the essence of mathematics lies in the formal method as such" and hence in formal systems, and "in this sense mathematics is the science of formal systems" (Curry 1977, 14). Since, by Gödel's first incompleteness theorem, "the concept of intuitively valid proof cannot be exhausted by any single formalization," it follows that "mathematical proof is precisely that sort of growing thing which the intuitionists have postulated for certain infinite sets" (ibid., 15).

This argument, however, is the same as Gödel's argument that has been discussed above, therefore it is invalid for the reasons stated there. Moreover, the claim that mathematical proof is precisely that sort of growing thing which the intuitionists have postulated for certain infinite sets, is incompatible with the concept of formal proof, because a formal proof cannot be a growing thing. Therefore, mathematical proof transcends the formal method as such.

(2) Gödel's incompleteness theorems do not refute the assumption that mathematics is theorem proving by the axiomatic method, because the propositions that are true but indemonstrable from the axioms are very artificial. They have no connection with real mathematics, and hence are mathematically insignificant.

Thus, Dieudonné says: "Let us suppose that tomorrow, as if by magic, all the works of logic written after 1925," in particular Gödel's incompleteness paper, "disappeared: well, no mathematician, when he proves a theorem, would notice it" (Dieudonné 1981, 22). For, "the undecidable proposition A described by Gödel appears to be very artificial, without any connection with any other part of the current theory of numbers" (Dieudonné 1998, 231). In fact, "among the numerous classical questions which are not resolved within number theory, it has not yet been established, as far as I know, that any of them is undecidable" (ibid.).

This argument, however, is invalid because there are propositions of the theory of numbers of the usual kind, which are true but cannot be demonstrated in first-order Peano arithmetic PA. An example of such propositions is Goodstein's theorem, which can be stated as follows.

The pure base n representation of a natural number m is the result of writing m as a sum of powers of n, then rewriting the various exponents of n themselves as sums of powers of n, and so on until possible (writing n^0 as simply 1).

For example, the pure base 2 representation of 26 is $26 = 2^{2^{2^1}} + 2^{2^1+1} + 2^1$.

For any natural number m and $n \geq 1$, the Goodstein sequence starting from m is the sequence of natural numbers $g(m, n)$ defined as follows:

$$g(m, 1) = m$$
$$g(m, n+1) = \text{the result of taking the pure base } n+1 \text{ representation of } g(m, n), \text{ then replacing each occurrence of the base base } n+1 \text{ with } n+2, \text{ and finally subtracting } 1.$$

For example, for $m = 3$, $g(3, 1) = 3 = 2^1 + 1$, $g(3, 2) = 3^1 + 1 - 1 = 3^1$, $g(3, 3) = 4^1 - 1 = 3$, $g(3, 4) = 3 - 1 = 2$, $g(3, 5) = 2 - 1 = 1$, $g(3, 6) = 1 - 1 = 0$. Thus, for $m = 3$, there is an n, namely $n = 6$, such that $g(m, n) = 0$.

Goodstein's theorem states that this holds generally: For any natural number m there is an n such that $g(m, n) = 0$.

Goodstein's theorem is a proposition of number theory of the usual kind, and can be expressed in first-order Peano arithmetic PA. Now, by a result of Kirby and Paris, we have: Goodstein's theorem $\Rightarrow$ consistency of PA. On the other hand, by Gödel's second incompleteness theorem, the consistency of PA cannot be proved in PA. Therefore, Goodstein's theorem cannot be proved in PA. So, despite Goodstein's theorem being purely number-theoretic in character and "being expressible in first-order" Peano "arithmetic, we cannot give a proof of it in Peano arithmetic" (Kirby and Paris 1982, 286).

In addition to Goodstein's theorem, there are other propositions of the theory of numbers of the usual kind, which are true but cannot be demonstrated in first-order Peano arithmetic PA.

2.17 The Failure of Deductivism

Assumption (I) of the foundationalist view, that mathematics is theorem proving by the axiomatic method, is an expression of deductivism, the view that mathematical reasoning is either deductive or defective. The mathematical reasoning to which deductivism refers includes not only first-order reasoning but also higher-order reasoning. For, as Hilbert acknowledges, mathematics requires reasoning involving "higher types of variables" (Hilbert 1998, 231).

But deductivism conflicts with the strong incompleteness theorem for second-order logic. By the latter, there is no consistent formal system for second-order logic capable of deducing all second-order logical consequences of any given set of propositions.

Indeed, assume that there is a consistent formal system L^2 for second-order logic capable of deducing all second-order logical consequences of any given set of propositions.

Now, by Gödel's first incompleteness theorem for second-order Peano arithmetic PA^2, there is a proposition G which is true in the intended interpretation of PA^2, but cannot be deduced from the axioms of PA^2 by means of the rules of L^2.

By our assumption about L^2, from this it follows that G cannot be a second-order logical consequence of the axioms of PA^2. This means that there must be some full interpretation in which the axioms of PA^2 are true and G is false. (A full interpretation is an interpretation where the domain of second-order variables consists of all subsets of the domain of first-order variables).

But the axioms of PA^2 are categorical, namely any full interpretation in which the axioms of PA^2 are true is isomorphic to the intended interpretation of PA^2. Then, since there must be some full interpretation in which the axioms of PA^2 are true and G is false, from the fact that G is false in such full interpretation, it follows that G must be false in the intended interpretation of PA^2.

But G is true in the intended interpretation of PA^2. Contradiction. Therefore, there is no consistent formal system for second-order logic capable of deducing all second-order logical consequences of any given set of propositions.

The strong incompleteness theorem for second-order logic means that deduction is not strong enough to obtain all second-order logical consequences of any given set of propositions, therefore deductivism does not account for mathematical reasoning, because mathematics requires reasoning involving higher types of variables. This is another reason why assumption (I) of the foundationalist view is invalid.

2.18 Other Shortcomings of Mathematics as Theorem Proving

In addition to the reasons considered above, there are also other important reasons why assumption (I) of the foundationalist view, that mathematics is theorem proving by the axiomatic method, is invalid. Here are the main ones.

(1) Assumption (I) does not account for how mathematicians solve problems. For example, it does not account for how Ribet, Wiles, and Taylor solved the problem posed by Fermat's conjecture: Show that there are no positive integers x, y, z such that $x^n + y^n = z^n$ for $n > 2$ (see Chap. 5). Therefore, as Friend says, the assumption that mathematics is theorem proving by the axiomatic method "does not reflect the modus operandi of the working mathematician" (Friend 2014, 213–214).

(2) Assumption (I) does not account for the fact that solving a problem of a given part of mathematics may require hypotheses from other parts of mathematics. For example, to solve the problem posed by Fermat's conjecture, which is a problem about the integers, Ribet used a hypothesis about modular forms in hyperbolic space, the Taniyama-Shimura conjecture: Every elliptic curve over the rational numbers is modular. Then, to solve the problem posed by the Taniyama-Shimura conjecture, Wiles and Taylor used hypotheses from various parts of mathematics, from differential geometry to complex analysis. Assumption (I) does not account for this fact because, according to it, a solution to a problem of a given part of mathematics should be deduced from the axioms for that part.

(3) Assumption (I) does not account for the fact that a demonstration may yield something new. For, according to it, a solution to a problem is deduced from the principles. Now, a deduction from the principles cannot yield anything essentially new with respect to them, because deductive rules are non-ampliative, namely their conclusion is implicitly contained in the premisses (see Chap. 5). Therefore, as Rota says, assumption (I) "has led to a widespread prejudice among scientists that mathematics is nothing but a pedantic grammar, suitable only for belaboring the obvious" (Rota 1997, 142).

(4) Assumption (I) does not account for the fact that new solutions, even hundreds of them, are often sought for problems for which a solution is already known. For example, for the Pythagorean theorem, "over four hundred" demonstrations are "known, and their number is still growing" (Maor 2007, xi). As another example, in 1950 a Fields Medal was awarded to Selberg for producing a new demonstration of a theorem, the prime-number theorem, for which a demonstration was already known. In fact, multiple demonstrations of mathematical theorems are almost characteristic of contemporary mathematics. Indeed, as Rota says, leafing through contemporary mathematics journals, one soon realizes that "few published research papers in mathematics present solutions of as yet unsolved problems; fewer still are formulations of new theories. The overwhelming majority of research papers in mathematics is concerned not with proving, but with reproving" (Rota 1997, 116).

Assumption (I) does not account for this fact because, according to it, an axiomatic demonstration establishes the truth of a theorem, and this is its function. Then, once the truth of a theorem has been established, there is no point in establishing it once again by another demonstration, let alone by hundreds of them.

(5) Assumption (I) does not account for the fact that different demonstrations of the same proposition may have different degrees of reliability. For, according to it, all demonstrations are deductions from principles which are true, so they are all equally reliable.

2.19 Mathematics and Intuition

Assumption (II) of the foundationalist view, that mathematics is true and certain being based on intuition, is invalid because intuition is unreliable and inadequate as a basis for mathematics. Indeed, there are many examples of the fact that intuition leads to wrong conclusions.

For example, intuition tells us that a plane figure has non-zero area and finite perimeter. But a counterexample to this is provided by the Sierpiński triangle. The latter is so called because it has been studied in Sierpiński 1915, but was already present in the floors of several medieval churches in Rome, made by craftsmen of the Cosmati family (see Conversano and Tedeschini Lalli 2011). For example, the following is a detail of the thirteenth century pavement of the Basilica of San Lorenzo fuori le Mura in Rome.

The Sierpiński triangle is defined as follows.

Step 0: Start with an equilateral triangle.

Step 1: Divide it into four equal equilateral triangles, and remove the one in the centre.

Step 2: Repeat Step 1 with each of the remaining equilateral triangles indefinitely.

The Sierpiński triangle is the limit of this process.

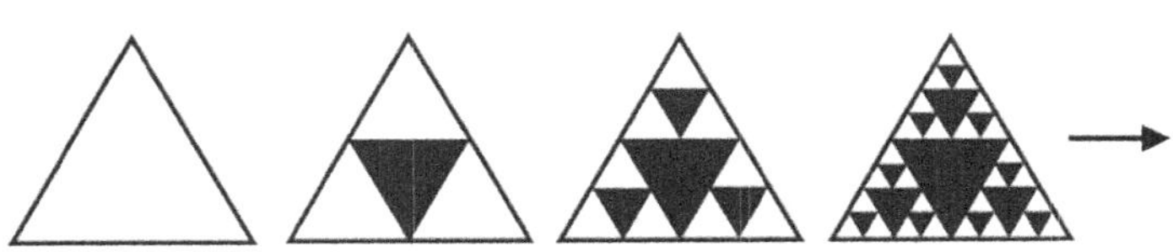

If the equilateral triangle of Step 0 has area a and perimeter p, then the result of the first application of Step 1 has area $(3/4) \times a$ because we must subtract the area of the removed triangle, and perimeter $(3/2) \times p$ because we must add the perimeter of the removed triangle. The result of the second application of Step 1 has area $(3/4)^2 \times a$ and perimeter $(3/2)^2 \times p$. The result of the third application of Step 1 has area $(3/4)^3 \times a$ and perimeter $(3/2)^3 \times p$. Generally, the result of the n-th application of Step 1 has area $(3/4)^n \times a$ and perimeter $(3/2)^n \times p$.

Thus, at each application of Step 1, the area decreases and the perimeter increases. As we continue the process indefinitely, the area converges to zero and the perimeter diverges to infinity. Therefore, the Sierpiński triangle has zero area and infinite perimeter.

The Sierpiński triangle is only one of several counterexamples to the conclusions of intuition. They show that intuition is unreliable and inadequate as a basis for mathematics.

2.20 Foundationalist Programs and Intuition

That assumption (II) of the foundationalist view is invalid because intuition is unreliable and inadequate as a basis for mathematics, is also clear from the fact that the attempts of the three big foundationalist schools to base mathematics on intuition ended in failure.

(1) The main basic logical law of Frege's logical system is a principle concerning extensions of concepts, the Basic Law V: For any concepts *F* and *G*, the extension of *F* is identical to the extension of *G* if and only if, for every object *a*, *F*(*a*) if and only if *G*(*a*). Frege believed that the truth of the Basic Law V was guaranteed by intuition, indeed he claimed that the Basic Law V is what one thinks when "one speaks of extensions of concepts" (Frege 2013, I, vii). Frege had such a confidence in intuition as to claim: "It is from the outset unlikely that," being based on the Basic Law V, my logical system "could be built on an insecure, defective basis," indeed "I could only acknowledge it as a refutation" if "someone proved to me that my basic principles lead to manifestly false conclusions. But no one will succeed in doing so" (ibid., I, xxvi).

But Frege deluded himself. By showing that the Basic Law V leads to a contradiction, Russell just succeeded in doing so. Indeed, let *R* be the concept defined by: for any object *x*, *R*(*x*) if and only if there is a concept *F* such that *x* is the extension of *F* and not-*F*(*x*). Let *r* be the extension of *R*. Now, assume *R*(*r*). Then, by the definition of *R*, there is a concept *F* such that *r* is the extension of *F* and not-*F*(*r*). From this, since *r* is also the extension of *R*, by the Basic Law V, it follows not-*R*(*r*). Thus, if *R*(*r*), then not-*R*(*r*). Conversely, assume not-*R*(*r*). Then there is a concept *F* (namely *R*) such that *r* is the extension of *F* and not-*F*(*r*). Hence, by the definition of *R*, *R*(*r*). Thus, if not-*R*(*r*), then *R*(*r*). Therefore we conclude that *R*(*r*) if and only if not-*R*(*r*). This is a contradiction, known as Russell's paradox.

Russell's paradox confirms that intuition is unreliable as a basis for mathematics. As Gödel says, Russell brought "to light the amazing fact that our logical intuitions (i.e. intuitions concerning such notions as: truth, concept, being, class, etc.) are self-contradictory" (Gödel 1986–2002, II, 124).

(2) Hilbert assumed that the consistency of Zermelo set theory could be proved by a proof based on the intuitive mode of thought. In a paper from 1931, he even went so far as to say: "I believe" that "I have fully attained what I desired and promised: The world has thereby been rid, once and for all, of the question of the foundations of mathematics as such" (Hilbert 1996e, 1157).

But Hilbert deluded himself. By Gödel's second incompleteness theorem, it is impossible to prove the consistency of Zermelo set theory by a proof based on the intuitive mode of thought. Therefore, intuition is inadequate as a basis for mathematics. Moreover, even if it were possible to demonstrate the consistency of Zermelo set theory by a demonstration based on the intuitive mode of thought, this would not guarantee that mathematical theorems are incontestable and ultimate truths. For, by the theorem on the false extensions, any consistent sufficiently strong formal system has a consistent extension in which some false proposition is provable.

The theorem on the false extensions is a corollary of Gödel's first incompleteness theorem. Indeed, by the latter, for any consistent, sufficiently strong, formal system S, there is a proposition G of S which is true but unprovable in S. Then, let T be the formal system obtained from S by adding $\neg G$ as a new axiom. Since G is unprovable in S, the system T is consistent. Trivially $\neg G$, being an axiom of T, is provable in T. On the other hand, since G is true, $\neg G$ is false. Therefore, T is a consistent extension of S in which the false proposition $\neg G$ is provable.

The theorem on the false extensions shows that consistency is no guarantee against falsity. Hilbert need not have waited for Gödel to realize that, he could have learned it from Kant. For, Kant had made it clear that it is, "to be sure, a necessary logical condition" that in "a concept no contradiction must be contained," but this is "far from sufficient for the objective reality of the concept, i.e., for the possibility of such an object as is thought through the concept" (Kant 1998, A220/B268). Indeed, it is not enough to assume, as "condition of all our judgments whatsoever," that "they do not contradict themselves," because "for all that a judgment may be free of any internal contradiction, it can still be either false or groundless" (ibid., A150/B190).

(3) Brouwer assumed that intuition is "sufficient to build up all mathematics" (Brouwer 1975, 61). Therefore, "man builds up pure mathematics out of the basic intuition of the intellect" (ibid., 53).

But this is invalid. On the one hand, the assumption that intuition is sufficient to build up all mathematics conflicts with the fact that it is impossible to construct certain mathematical objects that are important for physics by direct intuition. For example, let f be the function defined by: $f(x) = 0$ if $x = 0$, $f(x) = 1$ if $x \neq 0$, for any real number x. Now, it is impossible to construct this function on the basis of intuition (see Cellucci 2007, Section 3.10).

On the other hand, the assumption that intuition is sufficient to build up all mathematics conflicts with Gödel's first incompleteness theorem. By the latter, for any consistent, sufficiently strong, formal system, there are propositions of the system that are true but cannot be deduced from the axioms of the system. Then, such propositions are mathematical truths that are not based on intuition because, as we have seen above, according to Brouwer, a proposition is a mathematical truth only if it is deduced from basic propositions based on intuition, by means of introspective construction. Thus, even when the axioms are based on intuition, there will be mathematical truths not based on intuition. Therefore, intuition is not sufficient to build up all mathematics.

Kreisel says that Brouwer "devoted the ten years after publication" of Gödel's incompleteness paper "to non-scientific activities" because "he had received an intellectual shock" (Kreisel and Newman 1969, 45). For, he had tried to refute the basic assumptions of Hilbert's formalism but had failed, while Gödel had succeeded in refuting them by his "immensely natural proofs" (ibid.). Then, "seeing the incomparable superiority of" Gödel's incompleteness results in refuting the basic assumptions of Hilbert's formalism, "Brouwer had to face the question" to "what extent he had even begun to master his own logical ideas" (ibid.).

But it seems more likely that the intellectual shock Brouwer had received was due to the fact that Gödel's first incompleteness theorem had shown that intuition is not sufficient to build up all mathematics, thus refuting the basic assumption of Brouwer's intuitionism.

2.21 Foundationalist Programs, the World, the Elephant, and the Tortoise

The failure of the attempts of the three big foundationalist schools to base mathematics on intuition brings to mind the story, mentioned by Locke, of the Indian "who, saying that the world was supported by a great elephant, was asked, what the elephant rested on; to which his answer was, a great tortoise: But being again pressed to know what gave support to the broad-back'd tortoise, replied, something, he knew not what" (Locke 1975, 296).

The world, the elephant and the tortoise are, respectively, in the case of Frege, arithmetic, the purely logical laws, and intellectual intuition; in the case of Hilbert, mathematics, the consistency proof, and Kant's pure intuition; and, in the case of Brouwer, mathematics, constructions, and Kant's pure intuition of time.

One cannot base mathematics on intuition any more than one can base the world on a tortoise, the disproportion between, on the one hand, what is to be supported, and, on the other hand, what is supposed to support it, is too large. For this reason, the attempts of the three big foundationalist schools to base mathematics on intuition were doomed to fail.

Russell, who extended logicism from arithmetic, to which Frege had limited it, to all of pure mathematics, eventually admitted the inevitability of failure. Originally, he had believed that all our knowledge, in particular all our mathematical knowledge, "is either intuitive or inferred" from "intuitive knowledge from which it follows logically" (Russell 1998, 81). And is inferred "by the use of self-evident principles of deduction," so all our knowledge ultimately "depends upon our intuitive knowledge" (ibid., 63). On the other hand, only intuitive knowledge gives "an infallible guarantee of truth" (ibid., 68). Therefore, Russell had attempted to base all of mathematics on intuition.

But, "as the work proceeded, I was continually reminded of the fable about the elephant and the tortoise. Having constructed an elephant upon which the mathematical world could rest, I found the elephant tottering, and proceeded to construct a tortoise to keep the elephant from falling. But the tortoise was no more secure than the elephant" (Russell 1956, 54–55)). So, "after some twenty years of very arduous toil, I came to the conclusion that there was nothing more that I could do in the way of making mathematical knowledge indubitable" (ibid., 55).

2.22 Mathematics, Truth, and Certainty

Assumption (II) of the foundationalist view, that mathematics is true and certain being based on intuition, is shared by several mathematicians.

For example, Byers claims that "mathematics is about truth," it "is a way of using the mind with the goal of knowing the truth, that is, of obtaining certainty" (Byers 2007, 327). For, "truth" is "knowledge that is certain" (ibid., 330). Indeed, "truth in mathematics and the certainty that arises when that truth is made manifest are not two separate phenomena; they are inseparable from one another – different aspects of the same underlying phenomenon" (ibid., 329). The certainty of mathematics is such that one cannot "have the slightest doubt" about it, "mathematical truth has this kind of certainty, this quality of inexorability. This is its essence" (ibid., 328). Mathematics is true and certain because it is based on intuition, for, "without the flash of insight there is no truth just as there is no understanding, which is, after all, another word for this quality of certainty that we are discussing" (ibid., 346).

But these claims are invalid, because they conflict with Gödel's second incompleteness theorem, by which, for any consistent, sufficiently strong, formal system, it is impossible to demonstrate, by absolutely reliable means, that the axioms of the system are consistent, let alone that they are true and certain. Therefore, mathematics cannot be said to be true and certain.

2.23 The Relevance of Gödel's Second Incompleteness Theorem

Some people have denied the relevance of Gödel's second incompleteness theorem to assumption (II) of the foundationalist view, by raising some objections. But the objections are invalid. Here are the main ones.

(1) If mathematics cannot be said to be true and certain, then Gödel's second incompleteness theorem, being a mathematical result, cannot be said to be true and certain. But the claim that, by Gödel's second incompleteness theorem, mathematics cannot be said to be true and certain, depends on the assumption that Gödel's second incompleteness theorem can be said to be true and certain. Therefore, the claim that, by Gödel's second incompleteness theorem, mathematics cannot be said to be true and certain, is invalid. (This objection was raised in correspondence by Hersh, acting as 'advocatus diaboli', not because he shared it).

The objection is invalid because the claim that, by Gödel's second incompleteness theorem, mathematics cannot be said to be true and certain, does not depend on the assumption that Gödel's second incompleteness theorem can be said to be true and certain. It is a reductio ad absurdum, being of the following kind. Let us suppose, for argument's sake, that mathematics can be said to be true and certain. Then Gödel's second incompleteness theorem, being a mathematical result, can be said to be true and certain. But, by Gödel's second incompleteness theorem, mathematics cannot be said to be true and certain. This contradicts our assumption that mathematics can be said to be true and certain. Therefore, by reductio ad absurdum, we conclude that mathematics cannot be said to be true and certain.

(2) Nothing in Gödel's second incompleteness theorem "in any way contradicts the view that there is no doubt whatever about the consistency of any of the formal systems" *T* "that we use in mathematics" (Franzén 2005, 105). For, either we have no doubts about the consistency of *T*, or we do have doubts about the consistency of *T*. Now, "if we have no doubts about the consistency" of *T*, then "there is nothing in the second incompleteness theorem to give rise to any such doubts. And if we do have doubts about the consistency" of *T*, then "we have no reason to believe that a consistency proof" for *T* "formalizable" in *T* "would do anything to remove those doubts" (ibid., 105–106). For, the consistency of *T* "is precisely what is in question" (ibid., 105).

The objection is invalid because, if we have no doubts about the consistency of *T*, we are rationally justified in having no such doubts only if we can demonstrate by absolutely reliable means that *T* is consistent. But, by Gödel's second incompleteness theorem, this is impossible. On the other hand, if we do have doubts about the consistency of *T*, then the question is not whether a consistency proof for *T* formalizable in *T* would do anything to remove those doubts. It is, instead, whether a consistency demonstration for *T* by absolutely reliable means would do anything to remove them, and the answer is yes, definitely.

That, by Gödel's second incompleteness theorem, a demonstration of the consistency of *T* by absolutely reliable means is impossible, implies that the doubts about the consistency of *T* cannot be removed. Then, it is invalid to say that nothing in Gödel's second incompleteness theorem in any way contradicts the view that there is no doubt whatever about the consistency of any of the formal systems *T* that we use in mathematics. By Gödel's second incompleteness theorem we are not entitled to say that.

(3) Gödel's second incompleteness theorem is irrelevant to the question of the consistency of a formal system, because we can adopt Wittgenstein's rule: "Don't draw any conclusions from a contradiction: make that a rule" (Wittgenstein 1976, 209). Then, if we get to a contradiction, we will "simply say: "This is no use – and we won't draw any conclusion from it"" (ibid.). So the contradiction is sealed off, and a "contradiction is harmless if it can be sealed off" (Wittgenstein 1978, III, § 80).

The objection is invalid because contradictions usually arise from the use of flawed concepts, and their discovery is a powerful incentive to reformulate them. Thus, the paradoxes of the calculus of infinitesimals of Leibniz and Newton led to reformulate the concept of infinitesimal, Russell's paradox led to reformulate the concept of set. On the contrary, Wittgenstein's rule leaves everything as it is. It simply prevents the contradiction from being applied to produce arbitrary results, but it does not explain the cause of the contradiction, nor does anything to remove it, so it does not contribute to reformulate flawed concepts.

2.24 The Ironic Status of Gödel's Incompleteness Theorems

The status of Gödel's incompleteness theorems is ironic. On the one hand, they are the greatest successes of mathematical logic. On the other hand, mathematical logic was created by Frege with the purpose to give a secure foundation to mathematics, placing "the truth of a proposition beyond all doubt" (Frege 1960, 2). But, by Gödel's incompleteness theorems, this purpose cannot be achieved. Then, it is ironic that Gödel's incompleteness theorems, which are the greatest successes of mathematical logic, have shown the impossibility of achieving that very purpose for which mathematical logic was created.

It is also ironic that Gödel's incompleteness theorems have been established by a conservative like Gödel, who supported a "'rightward' philosophy of mathematics" (Gödel 1986–2002, III, 379). Namely, a philosophy that upholds the certainty of mathematical knowledge, which "belongs in principle toward the right" (ibid., III, 375). According to Gödel, mathematics "always has, in and of itself, an inclination toward the right" (ibid., III, 375). Conversely, "the antinomies of set theory" were "employed as a pretext for the leftward upheaval," which resulted in that "many or most mathematicians denied that mathematics, as it had developed previously, represents a system of truths" (ibid., III, 377). Then, it is ironic that Gödel, with his incompleteness theorems, provided strong arguments for the view of those mathematicians.

This point is also made by Simpson, who regrets that "Gödel supplied heavy artillery for all would-be assailants of mathematics. Authors such as Kline 1980 cite Gödel with monotonous repetition and devastating effect. The assault rages as never before" (Simpson 1988, 362). Therefore, "the need to defend the integrity of mathematics has not abated. On the contrary, Gödel's theorem made this need more urgent than ever" (ibid.). But what Simpson considers an assault on mathematics is only an attempt to correct a view that does not correspond to real mathematics.

2.25 Mathematics and Error

Mathematics cannot be said to be true and certain not only in principle, as Gödel's second incompleteness theorem implies, but also in practice. For, mathematicians often make errors, and widely acknowledge this themselves.

Thus, Euclid devoted a whole book, entitled *Pseudaria*, to this problem. The book has been lost, but Proclus tells us that in it Euclid "enumerates in order the various modes of fallacious reasoning, and exercises our intelligence on each of them by a variety of theorems, opposing the true to the false, and making the refutation of the error agree with the demonstration" (Proclus, *In primum Euclidis Elementorum librum commentarii*, 70.11–14, ed. Friedlein). Thus, while Euclid's "book of *Elementa* contains an impeccable and complete guide of the science itself of geometrical matters," the book of *Pseudaria* "has the purpose of correcting and exercising" (ibid., 70.15–18).

Lecat lists "about five hundred errors, made by some 330 mathematicians, many of them famous, such as: Abel, Cauchy, Cayley, Chasles, Descartes, Euler, Fermat, Galilei, Gauss, Hermite, Jacobi, Lagrange, Laplace, Legendre, Leibniz, Newton, Poincaré and Sylvester" (Lecat 1935, vii–viii).

Davis says that "a mathematical error of international significance," namely an error that is the "conjunction of a mathematician of great reputation and a problem of great notoriety," may "occur every twenty years or so" (Davis 1972, 262).

It might be objected that, even if mathematicians often make errors, this does not destroy the truth and certainty of mathematics, because mathematicians promptly detect and correct errors.

Thus, Hadamard says: "In our domain," mathematics, "we do not need to ponder on errors. Good mathematicians, when they make them, which is not infrequent, soon perceive and correct them" (Hadamard 1954, 49). As for me, "I make many more of them than my students do; only I always correct them so that no trace of them remains in the final result," because "insight" always "warns me that my calculations do not look as they ought to" (ibid.).

Thom says that "there is no case in the history of mathematics where the mistake of one man has thrown the entire field on the wrong track," since "never has a significant error slipped into a conclusion without almost immediately being discovered" (Thom 1971, 697).

But this is unjustified. There are several historical cases of mathematical demonstrations containing errors that have not been almost immediately discovered. Thus, the errors in Kempe's demonstration of the four colour theorem were detected eleven years after its publication. The errors in Jordan's demonstration of the Jordan curve theorem were detected twenty years after its publication. What is more important, there is no evidence that some of the demonstrations which today are considered to be valid do not contain errors.

2.26 Shortcomings of Reductionism in the Main Foundationalist Programs

Assumption (III) of the foundationalist view, that there is a part of mathematics such that all other parts of mathematics can be reduced to it, specifically, there is a mathematical theory such that all other mathematical theories can be reduced to it, is also invalid.

Assumption (III) is invalid as implemented by the main foundational programs. Indeed:

(1) Assumption (III) is invalid as implemented by logicism because, as we have seen above, Frege's logical system leads to a contradiction, Russell's paradox.

(2) Assumption (III) is invalid as implemented by formalism because Zermelo set theory does not permit to establish several mathematical results, for example, that all Borel sets are determined.

(3) Assumption (III) is invalid as implemented by intuitionism because, as said above, the theory of more or less freely proceeding infinite sequences and the theory of species together do not account for many mathematical results that are important to physics.

2.27 Shortcomings of Set Theoretical Reductionism

It may be objected that, while the fact that Zermelo set theory does not permit to establish several mathematical results, for example, that all Borel sets are determined, poses an obstacle to the reduction of all parts of mathematics to set theory, Zermelo-Fraenkel set theory removes this obstacle. For, it permits to establish all these results. Then, Zermelo-Fraenkel set theory ZF is The Foundation.

In the past century, the credibility of assumption (III) of the foundationalist view, has essentially relied on the conviction that all parts of mathematics can be reduced to ZF.

Thus, Maddy says that ZF provides "a framework in which all classical mathematical objects and structures can be defined and all classical mathematical theorems proved" (Maddy 2000, 344). Indeed, "for all mathematical objects and structures, there are set theoretic surrogates and instantiations, and the set theoretic versions of all classical mathematical theorems can be proved from the standard axioms for the theory of sets" (Maddy 1997, 34).

However, assumption (III) with ZF as The Foundation is invalid for the following reasons.

(1) Assumption (III) conflicts with the fact that, as Mac Lane points out, although set theory provides a standard foundation for mathematics, many interesting "questions cannot be settled on the basis of Zermelo-Fraenkel axioms for set theory" (Mac Lane 1986, 385). As regards these questions, an alternative is provided by "category theory" (ibid., 406). However, while "categories and functors are everywhere in topology and in parts of algebra," they "do not as yet relate well to most of analysis" (ibid., 407). So, many interesting questions cannot be settled on the basis of the axioms for category theory. Therefore, while Zermelo-Fraenkel set theory and category theory are each adequate for certain questions, neither of them "is wholly successful" (ibid., 407). So, today there is no mathematical theory such that all other mathematical theories can be reduced to it.

(2) Assumption (III) is conclusively refuted by Gödel's first incompleteness theorem, by which there are arithmetic truths that cannot be deduced from the axioms of ZF. Therefore, even arithmetic cannot be reduced to ZF.

(3) Assumption (III) is purely ideological. Even Maddy admits that "the average algebraist or geometer loses little time over set theory" (Maddy 1990, 4). In fact, "it cannot be denied that mathematicians from various branches of the subject – algebraists, analysts, number theorists, geometers – have different characteristic modes of thought, and that the subject would be crippled if this variety were somehow curtailed" (Maddy 1997, 33).

(4) Assumption (III) does not provide unique set theoretic surrogates for all mathematical objects, starting with the natural numbers. For example, Zermelo identifies natural numbers 0, 1, 2, 3, ... with the sets $\varnothing$, $\{\varnothing\}$, $\{\{\varnothing\}\}$, $\{\{\{\varnothing\}\}\}$, ..., while von Neumann identifies them with the sets $\varnothing$, $\{\varnothing\}$, $\{\varnothing,\{\varnothing\}\}$, $\{\varnothing,\{\varnothing\},\{\varnothing,\{\varnothing\}\}\}$, ... But, as even Maddy admits, there is no reason "deep enough to motivate a metaphysical argument that one rather than the other uncovers the true identity of the natural numbers. And the other identifications, of integers, rationals, reals, functions, etc., all share this type of arbitrariness" (Maddy 1997, 24).

(5) Assumption (III) does not take into account that the reduction of mathematical objects to sets may attach properties to mathematical objects that say nothing about their nature. For example, on the one hand, Zermelo's identification of 1 with $\{\varnothing\}$, and of 3 with $\{\{\{\varnothing\}\}\}$, attaches the property $1 \notin 3$ to numbers 1 and 3 because $\{\varnothing\}\notin\{\{\{\varnothing\}\}\}$. On the other hand, von Neumann's identification of 1 with $\{\varnothing\}$, and of 3 with $\{\varnothing,\{\varnothing\},\{\varnothing,\{\varnothing\}\}\}$, attaches the property $1 \in 3$ to numbers 1 and 3 because $\{\varnothing\}\in\{\varnothing,\{\varnothing\},\{\varnothing,\{\varnothing\}\}\}$. But this says nothing about the nature of the numbers 1 and 3.

(6) Assumption (III) does not take into account that the reduction of mathematical objects to sets may attach properties to mathematical objects that are inconsistent with each other. For example, as we have just seen, on the one hand, Zermelo's definition attaches the property $1 \notin 3$ to numbers 1 and 3, on the other hand, von Neumann's definition attaches the property $1 \in 3$ to them.

(7) Assumption (III) does not take into account that the most important advances in mathematics do not consist in the reduction of mathematical objects to a single kind of objects, but rather in the introduction of new concepts and new hypotheses to solve problems. Thus, Dedekind says that "every theorem of algebra and higher analysis, no matter how remote, can be expressed as a theorem about natural numbers," but there is "nothing meritorious" in "actually performing this wearisome circumlocution and insisting on the use and recognition of none other than natural numbers" (Dedekind 1996, 792). Indeed, "the greatest and most fruitful advances in mathematics" have "invariably been made by the creation and introduction of new concepts, rendered necessary by the frequent recurrence of complex phenomena which could be mastered by the old notions only with difficulty" (ibid.).

2.28 The Irrelevance of the Existence of Mathematical Objects

That assumptions (I) – (III) of the foundationalist view are all invalid, is not the only shortcoming of mainstream philosophy of mathematics. Another shortcoming arises from the fact that, according to mainstream philosophy of mathematics, a main task of the philosophy of mathematics is to give an answer to the question: Do objects exist in virtue of which mathematical propositions are true, and if so what is their nature?

The question of the existence of mathematical objects has been a central focus of mainstream philosophy of mathematics from the first half of the twentieth century to the present, meaning existence in the metaphysical sense of one of the schools of mainstream philosophy of mathematics. But, in fact, such question is irrelevant to mathematics, because the work of mathematicians on certain mathematical objects does not depend on, and is not affected by, an answer to the question of whether those mathematical objects exist or not, in the metaphysical sense of one of those schools. Mathematicians accept, or reject, mathematical objects not on metaphysical grounds, but because they are, or are not, functional to the advancement of mathematics.

For example, mathematicians eventually accepted imaginary numbers as a legitimate kind of numbers, not because they became convinced that imaginary numbers existed in some metaphysical sense, but because imaginary numbers were functional to the advancement of mathematics. Thus, Gauss said that analysis "would lose immensely in beauty and roundness, and would be forced to add very hampering restrictions to truths which otherwise would hold generally, if these imaginary quantities were to be neglected" (Gauss 1880, 156).

As another example, mathematicians eventually rejected infinitesimals as introduced by Leibniz and Newton, not because they became convinced that infinitesimals did not exist in some metaphysical sense, but because infinitesimals were not functional to the advancement of mathematics. Thus, Abel said that the calculus of infinitesimals of Leibniz and Newton "is so lacking in plan and overall idea, that it is really quite stunning that it can be studied by so many people" (Abel 1902b, 23). In particular, "it is a shame that we dare to base the slightest demonstration on" divergent series, for, "by using them, one may draw any conclusion he pleases, and it is they that have produced so many fallacies and so many paradoxes," therefore, "what is more important in mathematics is without foundation" (Abel 1902a, 16).

That the question of the existence of mathematical objects is irrelevant to mathematics is also clear from the demonstrations by reductio ad absurdum, in which one reasons about what does not exist. As Thomas says, in mathematics "we must be able to reason as dependably about what does not exist – even in a mathematical sense – as about what does, for instance in 'reductio' proofs," so, "whether some things exist or not is not of any practical importance" (Thomas 2014, 248).

A proper approach to the question of the existence of mathematical objects is put forward by Kant, who says: "In mathematical problems the question is not" about "existence as such at all, but about the properties of the objects in themselves, solely insofar as these are" connected "with the concept of them" (Kant 1998, A719/B747).

That the question of the existence of mathematical objects is irrelevant to mathematics, in the sense stated above, has been underlined also by several contemporary mathematicians.

Thus, Nelson says: "Share with me a fantasy: we open our morning newspaper to find a report with a banner headline, 'Numbers Vanish! – Early last night the natural numbers" suddenly "disappeared. Mathematicians have expressed stunned despair. Without numbers, they say, they can no longer prove theorems'" (Nelson 1994, 571). But "this is nonsense," indeed, if natural numbers disappeared, "the newspaper could still put marks 2, 3, etc., on the inside pages," and "we mathematicians could continue to put marks on paper, just as before, and hopefully submit them to editors of mathematical journals" (ibid.).

Rota says: "The existence of mathematical items is a chapter in the philosophy of mathematics that is devoid of consequence" (Rota 1997, 161). If "someone proved beyond any reasonable doubt that mathematical items do not exist," this would not "affect the truth of any mathematical statement" (ibid.). Indeed, "it does not matter whether mathematical items exist," one "can spend a lifetime working on mathematics without ever having any idea whether mathematical items exist, nor does one have to care about such a question" (ibid.).

Gowers says: "There certainly are philosophers who take seriously the question of whether numbers exist," but "this distinguishes them from mathematicians," who "can, and even should, happily ignore this seemingly fundamental question" (Gowers 2002, 17). In fact, "rather than worrying about the existence, or otherwise," of something, mathematicians think "about its properties" (ibid., 70). One might wonder: "How can one consider a set of properties without first establishing that there is something that has those properties? But this is not difficult at all. For example, one can speculate about the character a female president of the United States would be likely to have, even though there is no guarantee that there will ever be one" (ibid., 70–71).

2.29 Other Shortcomings of Mainstream Philosophy of Mathematics

In fact, all the characters of mainstream philosophy of mathematics have shortcomings.

(1) According to mainstream philosophy of mathematics, the philosophy of mathematics cannot concern itself with the making of mathematics, in particular discovery, because discovery is a subjective process, so it cannot be accounted for.

But this is invalid. Already Greek mathematicians invented a method of discovery, the analytic method, they used it as a basis for their mathematical practice, and even reported their processes of discovery by publishing their analyses.

This was possible because discovery is not a subjective process that cannot be accounted for. The hypotheses for solving mathematical problems are obtained by non-deductive rules, and the choice among alternative hypotheses is made by comparing the arguments for and against them on the basis of experience (see Chap. 5). So, discovery can be accounted for.

(2) According to mainstream philosophy of mathematics, the philosophy of mathematics can concern itself only with finished mathematics, because only finished mathematics is objective, so it can be completely justified.

But this is invalid, because justification is not entirely objective, it may involve subjective considerations.

For example, the calculus of infinitesimals of Leibniz and Newton was inconsistent, because infinitesimals were taken to be zero within some demonstrations and non-zero within other demonstrations, and were even taken to be zero at one place and non-zero at another place within the same demonstration. This led to falsities. For example, according to L'Hôpital 2015 presentation of the calculus of infinitesimals of Leibniz described above, by Postulate I two quantities that differ by an infinitely small quantity may be used interchangeably. So $dx + dx = dx$, hence $2dx = dx$, therefore $2 = 1$. Nevertheless, for over 150 years the calculus of infinitesimals of Leibniz and Newton was considered to be justified, because it was very effective in solving problems in science and engineering.

What is more important, by Gödel's second incompleteness theorem, finished mathematics cannot be completely justified.

(3) According to mainstream philosophy of mathematics, since the philosophy of mathematics cannot concern itself with the making of mathematics, it cannot contribute to the advancement of mathematics.

But this is invalid because, as argued in the Introduction, philosophy has contributed to the advancement of mathematics.

(4) According to mainstream philosophy of mathematics, the task of the philosophy of mathematics is primarily to give an answer to the questions: How do mathematical propositions come to be completely justified? And, subordinately to it, to the question: Do objects exist in virtue of which mathematical propositions are true, and if so what is their nature?

But this is invalid. Indeed, the question, 'How do mathematical beliefs come to be completely justified?' is empty because, by Gödel's second incompleteness theorem, mathematical propositions cannot be completely justified. And, as we have argued above, the question, 'Do objects exist in virtue of which mathematical propositions are true, and if so what is their nature? is irrelevant to mathematics.

(5) According to mainstream philosophy of mathematics, the method of mathematics is the axiomatic method.

But this is invalid because, as we have seen above as regards assumption (I) of the foundationalist view of mathematics, such claim is refuted by Gödel's first incompleteness theorem.

(6) According to mainstream philosophy of mathematics, the role of axiomatic demonstration, namely demonstration based on the axiomatic method, is to guarantee the truth of a proposition.

But this is refuted by Gödel's second incompleteness theorem, by which axiomatic demonstration cannot be said to guarantee the truth of a proposition.

(7) According to mainstream philosophy of mathematics, since the method of mathematics is the axiomatic method, mathematics is a body of truths, and indeed of truths that are certain. Therefore, mathematics is about truth and certainty.

But this is refuted by Gödel's second incompleteness theorem, by which mathematics cannot be said to be true and certain. As Kline says, "mathematics is a body of knowledge. But it contains no truths. The contrary belief, namely, that mathematics is an unassailable collection of truths," is "a popular fallacy" (Kline 1964, 9).

(8) According to mainstream philosophy of mathematics, since the method of mathematics is the axiomatic method, mathematical reasoning consists of deductive reasoning.

But this is invalid because, by the strong incompleteness theorem for second-order logic, there is no consistent formal system for second-order logic capable of deducing all second-order logical consequences of any given set of propositions. Then, quantification logic is not an adequate theory of deductive reasoning, and hence is inadequate to account for mathematical reasoning, because much of mathematics requires second-order logic or beyond. Generally, the assumption that mathematical reasoning consists of deductive reasoning conflicts with the fact that "deductive inferences play only a small role in reasoning" (Doyle 1987, 175).

(9) According to mainstream philosophy of mathematics, the philosophy of mathematics is a new independent subject introduced by Frege.

But this is invalid, because the philosophy of mathematics goes back to the beginning of philosophy as discipline, many major philosophers have made substantial contributions to it, and their work remains important even today. Frege only introduced a particular kind of philosophy of mathematics, mainstream philosophy of mathematics.

This involved a basic change with respect to the philosophical tradition. In the latter, the philosophy of mathematics aimed to account for mathematics as a part of human knowledge in general. On the contrary, Frege introduced the philosophy of mathematics as a subject which aims to account for mathematics as a matter unto itself, not as a part of human knowledge in general.

Like Frege, the ensuing mainstream philosophers of mathematics aim to account for mathematics as a matter unto itself. As Hersh says, according to them, "Frege is the first full-time philosopher of mathematics" (Hersh 1997, 141). And mainstream philosophers of mathematics are full-time philosophers of mathematics.

But to be a full-time philosopher of mathematics is not a good thing because it means to have a one-sided and impoverished view of mathematics. For, mathematics is a part of human knowledge in general, and can be understood only within the framework of the latter. Frege says that "a philosopher who has nothing to do with geometry is only half a philosopher" (Frege 1979, 273). Much in the same way we can say that a philosopher who has nothing to do with mathematics as a part of human knowledge in general is only half a philosopher.

(10) According to mainstream philosophy of mathematics, the philosophy of mathematics can be developed independently of experience, because mathematics is an armchair subject that is the product of thought alone, so it does not depend on experience.

But this conflicts with the fact that the solutions of mathematical problems are only plausible, so they depend on experience. The idea that mathematics can be developed independently of experience has had a deleterious effect upon the philosophy of mathematics. As Hersh says, it has made the philosophy of mathematics into "an encapsulated entity, isolated, timeless, ahistorical, inhuman, connected to nothing else in the intellectual or material realms" (Hersh 1997, 25).

Because of these shortcomings, it seems fair to say that mainstream philosophy of mathematics does not provide an adequate account of mathematics.

2.30 Mainstream Philosophy of Mathematics and Mathematical Genius

Three further questions concerning mainstream philosophy of mathematics may be considered. The first question is mathematical genius.

Mainstream philosophy of mathematics adheres to the myth of mathematical genius, put forward by Novalis (see Chap. 4). According to it, the making of mathematics, in particular discovery, is based on leaps of intuition which are a prerogative of the genius.

After Novalis, the myth of mathematical genius has been reaffirmed by several mathematicians.

Thus, Poincaré says that only those who have mathematical genius "may become creators, and try" to discover "with more or less chance of success according as their intuition is more or less developed in them" (Poincaré 2015, 386). Mathematical genius is inborn, since it is the result of the "meeting of two germinal cells, of different sex," containing "the mysterious elements whose mutual reaction must produce the genius," these elements are rare, and "their meeting is still more rare" (ibid., 410).

Halmos says that "to be a scholar of mathematics" you need genius, because "you must be born with talent, insight, concentration, taste, luck, drive and the ability to visualize and guess" (Halmos 1985, 400).

Byers says that there are "people who find a way to transcend their limitations" and "dare to do what appears to be impossible" (Byers 2007, 16). The "impossible is rendered possible through acts of genius," and "mathematics boasts genius in abundance" (ibid.). Genius is a matter of insight, and "insight often reveals itself in a flash" (ibid., 329). Thus "mathematics transcends logic" (ibid., 26).

But the myth of mathematical genius is at odds with facts. As it will be argued in the following chapters, the making of mathematics, in particular discovery, is not an irrational process based on leaps of intuition, but a rational process that can be analyzed in terms of rules. And it is not the result of extraordinary thought processes, which are the hallmark of mathematical genius, but of ordinary thought processes that produce an extraordinary outcome.

Rather than mathematical genius, the discoverer must have enough knowledge to go to the edge of the field, while being flexible enough to go over the border, and must be able to undergo long periods of total absorption in the problem. Indeed, when Newton was asked how he could make his discoveries, "he answered: Nocte dieque incubando [By thinking about it day and night]" (Ortega y Gasset 1957, 47). His discoveries were "due to nothing but industry and patient thought" (Newton 2004, 94). And Einstein said: "It's not that I'm so smart, it's just that I stay with problems longer" (Einstein 2011, 481).

2.31 Mainstream Philosophy of Mathematics and Mathematical Logic

The second question concerning mainstream philosophy of mathematics that may be considered is mathematical logic.

According to mainstream philosophy of mathematics, mathematical logic is a proper and adequate tool for the philosophy of mathematics.

The three big foundationalist schools make this claim.

Thus, Frege says that, by means of mathematical logic, "every gap in the chain of deductions is eliminated with the greatest care," so we can "say with certainty upon what primitive truths the proof depends" (Frege 1960, 4).

Hilbert says that "in the logical calculus we possess a sign language that is capable of representing" all "mathematical propositions in formulas and of expressing" all "logical inference through formal processes" (Hilbert 1967a, 381).

Brouwer says that, admittedly, standard mathematical logic is only "a mathematical study of linguistic symbols" (Brouwer 1975, 96). But "intuitionist mathematics has its general introspective theory of mathematical assertions, a theory which with some right may be called 'intuitionist mathematical logic'" (ibid., 524). By means of it, intuitionism has "built a new structure of mathematics proper with unshakeable certainty" (Brouwer 1998, 42).

The direct and indirect descendants of the three big foundationalist schools also make the claim.

Thus, Kreisel says that the approach to philosophical problems of the philosophical tradition is valid only "at an early stage, when we know too little about the phenomenon involved and about our knowledge of it in order to ask sensible specific questions" (Kreisel 1984, 82). At a mature stage, this approach must be replaced with one based on mathematical logic, which is "a tool in the philosophy of mathematics; just as other mathematics, for example the theory of partial differential equations, is a tool in what used to be called natural philosophy" (Kreisel 1967, 201).

Dummett says that "it is rash to tackle the philosophy of mathematics unless one has" a "reasonable knowledge of mathematical logic," not "so much as part of the object of study as serving as a tool of inquiry" (Dummett 1998, 123–124). Thus, "if you have little knowledge of mathematical logic, you would be strongly advised to acquire some" (ibid., 124).

But the claim that mathematical logic is a proper and adequate tool for the philosophy of mathematics is invalid, because mathematical logic has failed to provide a foundation for mathematics and to give an account of mathematical reasoning.

As Rota says, "mathematical logic has given up all claims of providing a foundation to mathematics," and "very few logicians of our day believe that mathematical logic has anything to do with the way we think" (Rota 1997, 92–93).

For example, Wang admits that, "as we understand the nature of mathematical logic better, we find that the early belief in its philosophical relevance was largely an illusion" (Wang 2016, 28). Gradually, "the inadequacies of mathematical logic as the basic tool for the philosophy of mathematics and for general philosophy have come to be felt" (ibid., 30).

2.32 Mainstream Philosophy of Mathematics and Philosophy

The third question concerning mainstream philosophy of mathematics that may be considered is the attitude towards philosophy. In this respect, there is a difference between the three big foundationalist schools and their direct or indirect descendants.

The attitude of the three big foundationalist schools is anti-philosophical. According to them the justification of mathematics cannot be given by philosophy but only by mathematics itself, so it can only be a self-justification.

For example, Hilbert says that "mathematics is a presuppositionless science" (Hilbert 1967b, 479). The justification of mathematics cannot be given by philosophy, but only by mathematical logic, which is a part of mathematics that enables to "bring mathematical concept-formations and inferences into such a form that they are irrefutable and yet furnish a model of the entire science" (Hilbert 1996d, 1152).

Conversely, the attitude of the direct or indirect descendants of the three big foundationalist schools is not anti-philosophical. However, it consigns philosophy to irrelevance. For, most descendants of the three big foundationalist schools are part of analytic philosophy. And the basic assumption that the philosophy of mathematics cannot concern itself with the making of mathematics and hence cannot contribute to the advancement of mathematics, is part of the assumption of analytic philosophy that, through philosophy, "we do not seek to learn anything new," but only "to understand something that is already in plain view" (Wittgenstein 2009, I, § 89). Now, as Frodeman says, analytic philosophy "has led philosophy, potentially the most relevant of subjects, to become a synonym for irrelevance" (Frodeman 2013, 1918). This is argued for at length in Cellucci (2018, 2019).

Therefore, the attitude of mainstream philosophy of mathematics towards philosophy is either anti-philosophical, or consigns philosophy to irrelevance.

References

Abel, Niels Henrik. 1902a. Letter to Holmboe, 16 January 1826. In *Memorial publié à l'occasion du centenaire de sa naissance: Correpondance*, ed. Niels Henrik Abel, 13–19. Dybwad: Kristiania.

———. 1902b. Letter to Hansteen, 29 march 1826. In *Memorial publié à l'occasion du centenaire de sa naissance: Correpondance*, ed. Niels Henrik Abel, 21–24. Dybwad: Kristiania.

Bass, Hyman. 2015. Mathematics and teaching. In *I, mathematician*, ed. Peter Casazza, Steven G. Krantz, and Randi D. Ruden, 129–139. Washington: The Mathematical Association of America.

Brouwer, Luitzen Egbertus. 1975. *Collected works 1: Philosophy and foundation of mathematics*. Amsterdam: North-Holland.

———. Jan. 1998. Intuitionistic reflections on formalism. In *From Brouwer to Hilbert: The debate on the foundations of mathematics in the 1920s*, ed. Paolo Mancosu, 40–44. Oxford: Oxford University Press.

Byers, William. 2007. *How mathematicians think: Using ambiguity, contradiction, and paradox to create mathematics*. Princeton: Princeton University Press.

Cellucci, Carlo. 2007. *La filosofia della matematica del Novecento*. Roma: Laterza.

———. 2018. Philosophy at a crossroads: Escaping from irrelevance. *Syzetesis* 5 (1): 13–53.

———. 2019. The most urgent task of philosophy today. *Borderless Philosophy* 2: 47–75.

Chihara, Charles S. 1990. *Constructibility and mathematical existence*. Oxford: Oxford University Press.

Conversano, Elisa, and Laura Tedeschini Lalli. 2011. Sierpiński triangles in stone in medieval floors in Rome. *Aplimat Journal of Applied Mathematics* 4: 113–122.

Curry, Haskell Brooks. 1951. *Outlines of a formalist philosophy of mathematics*. Amsterdam: North-Holland.

———. 1954. Remarks on the definition and nature of mathematics. *Dialectica* 8: 228–233.

———. 1977. *Foundations of mathematical logic*. New York: Dover.

Davis, Philip. 1972. Fidelity in mathematical discourse: Is one and one really two? *The American Mathematical Monthly* 79: 252–263.

Dedekind, Julius Wilhelm Richard. 1996. Was sind und was sollen die Zahlen? In *From Kant to Hilbert: A source book in the foundations of mathematics*, ed. William Ewald, vol. 2, 787–833. Oxford: Oxford University Press.

Dieudonné, Jean. 1981. Logica e matematica nel 1980. In *La nuova ragione: Scienza e cultura nella società contemporanea*, ed. Paolo Rossi, 15–25. Bologna: Il Mulino.
———. 1998. *Mathematics: The music of reason*. Berlin: Springer.
Doyle, Jon. 1987. Logic, rationality, and rational psychology. *Computational Intelligence* 3: 175–176.
Dummett, Michael. 1991. *Frege: Philosophy of mathematics*. London: Duckworth.
———. 1998. The philosophy of mathematics. In *Philosophy 2: Further through the subject*, ed. Anthony C. Grayling, 122–196. Oxford: Oxford University Press.
———. 2010. *The nature and future of philosophy*. New York: Columbia University Press.
Einstein, Albert. 2011. *The ultimate quotable Einstein*. Princeton: Princeton University Press.
Feferman, Solomon. 1998. *In the light of logic*. Oxford: Oxford University Press.
Franzén, Torkel. 2005. *Gödel's theorem: An incomplete guide to its use and abuse*. Wellesley: A K Peters.
Frege, Gottlob. 1960. *The foundations of arithmetic: A logico-mathematical enquiry into the concept of number*. New York: Harper.
———. 1967. *Begriffsschrift*, a formula language, modeled upon that of arithmetic, for pure thought. In *From Frege to Gödel: A source book in mathematical logic, 1879–1931*, ed. Jean van Heijenoort, 5–82. Cambridge: Harvard University Press.
———. 1979. *Posthumous writings*. Oxford: Blackwell.
———. 1984. *Collected papers on mathematics, logic, and philosophy*. Oxford: Blackwell.
———. 2013. *Basic laws of arithmetic: Derived using concept-script*. Oxford: Oxford University Press.
Friend, Michèle. 2014. *Pluralism in mathematics: A new position in philosophy of mathematics*. Dordrecht: Springer.
Frodeman, Robert. 2013. Philosophy dedisciplined. *Synthese* 190: 1917–1936.
Gauss, Carl Friedrich. 1880. Letter to Bessel, 18 December 1811. In *Briefwechsel*, ed. Carl Friedrich Gauss and Friedrich Wilhelm Bessel, 155–160. Leipzig: Engelmann.
George, Alexander, and Daniel J. Velleman. 2002. *Philosophies of mathematics*. Oxford: Blackwell.
Gödel, Kurt. 1986–2002. *Collected works*. Oxford: Oxford University Press.
Gowers, Timothy. 2002. *Mathematics: A very short introduction*. Oxford: Oxford University Press.
Hadamard, Jacques. 1954. *An essay on the psychology of invention in the mathematical field*. Mineola: Dover.
Halmos, Paul Richard. 1985. *I want to be a mathematician: An automathography*. Berlin: Springer.
Hersh, Reuben. 1997. *What is mathematics, really?* Oxford: Oxford University Press.
Heyting, Arend. 1962. Axiomatic method and intuitionism. In *Essays on the foundations of mathematics*, ed. Yehoshua Bar-Hillel, E.I.J. Poznanski, Michael O. Rabin, and Abraham Robinson, 237–247. Jerusalem: Magnes Press.
Hilbert, David. 1967a. On the infinite. In *From Frege to Gödel: A source book in mathematical logic, 1879–1931*, ed. Jean van Heijenoort, 367–392. Cambridge: Harvard University Press.
———. 1967b. The foundations of mathematics. In *From Frege to Gödel: A source book in mathematical logic, 1879–1931*, ed. Jean van Heijenoort, 464–479. Cambridge: Harvard University Press.
———. 1980. Letter to Frege, 29 December 1899. In Gottlob Frege, *Philosophical and mathematical correspondence*, 38–41. Oxford: Blackwell.
———. 1996a. Axiomatic thought. In *From Kant to Hilbert: A source book in the foundations of mathematics*, ed. William Ewald, vol. 2, 1107–1115. Oxford: Oxford University Press.
———. 1996b. The new grounding of mathematics: First report. In *From Brouwer to Hilbert*, ed. William Ewald, vol. 2, 1117–1134. Oxford: Oxford University Press.
———. 1996c. The logical foundations of mathematics. In *From Brouwer to Hilbert*, ed. William Ewald, vol. 2, 1134–1148. Oxford: Oxford University Press.

———. 1996d. The grounding of elementary number theory. In *From Kant to Hilbert: A source book in the foundations of mathematics*, ed. William Ewald, vol. 2, 1149–1157. Oxford: Oxford University Press.

———. 1996e. Logic and the knowledge of nature. In *From Kant to Hilbert: A source book in the foundations of mathematics*, ed. William Ewald, vol. 2, 1157–1165. Oxford: Oxford University Press.

———. 1998. Problems of the grounding of mathematics. In *From Brouwer to Hilbert: The debate on the foundations of mathematics in the 1920s*, ed. Paolo Mancosu, 227–233. Oxford: Oxford University Press.

———. 2013. *Lectures on the foundations of arithmetic and logic 1917–1933*. Heidelberg: Springer.

Jaffe, Arthur. 1997. Proof and the evolution of mathematics. *Synthese* 111: 133–146.

Kac, Mark, and Stanislaw M. Ulam. 1992. *Mathematics and logic*. Mineola: Dover.

Kant, Immanuel. 1998. *Critique of pure reason*. Cambridge: Cambridge University Press.

Kenny, Antony. 1995. *Frege: An introduction to the founder of modern analytic philosophy*. London: Penguin Books.

Kirby, Laurie, and Jeff Paris. 1982. Accessible independence results for Peano arithmetic. *Bulletin of the London Mathematical Society* 14: 285–293.

Kline, Morris. 1964. *Mathematics in Western culture*. Oxford: Oxford University Press.

———. 1980. *Mathematics: The loss of certainty*. Oxford: Oxford University Press.

———. 1981. *Mathematics and the physical world*. Mineola: Dover.

Körner, Stephan. 1986. *The philosophy of mathematics: An introductory essay*. Mineola: Dover.

Kreisel, Georg. 1967. Mathematical logic: What has it done for the philosophy of mathematic? In *Bertrand Russell, philosopher of the century*, ed. Ralph Schoenman, 201–272. London: Allen & Unwin.

———. 1984. Frege's foundations and intuitionistic logic. *The Monist* 67: 72–91.

Kreisel, Georg, and Maxwell Newman. 1969. Luitzen Egbertus Jan Brouwer 1881–1966. *Biographical Memoirs of Fellows of the Royal Society* 15: 39–68.

L'Hôpital, Guillaume. 2015. *Analyse des infiniments petits*. Cham: Springer.

Lecat, Maurice. 1935. *Erreurs de mathématiciens des origines à nos jours*. Bruxelles: Castaigne.

Lehman, Hugh. 1979. *Introduction to the philosophy of mathematics*. Oxford: Blackwell.

Locke, John. 1975. *An essay concerning human understanding*. Oxford: Oxford University Press.

Mac Lane, Saunders. 1986. *Mathematics: Form and function*. Berlin: Springer.

Maddy, Penelope. 1990. *Realism in mathematics*. Oxford: Oxford University Press.

———. 1997. *Naturalism in mathematics*. Oxford: Oxford University Press.

———. 2000. Mathematical progress. In *The growth of mathematical knowledge*, ed. Emily Grosholz and Herbert Breger, 341–352. Dordrecht: Springer.

Maor, Eli. 2007. *The Pythagorean theorem: A 4,000-year history*. Princeton: Princeton University Press.

McCarthy, Timothy G. 1994. Self-reference and incompleteness in a non-monotonic setting. *Journal of Philosophical Logic* 23: 423–449.

Naylor, Arch W., and George R. Sell. 2000. *Linear operator theory in engineering and science*. Berlin: Springer.

Nelson, Edward. 1994. Taking formalism seriously. In *Logic, methodology and philosophy of science IX*, ed. Dag Prawitz, Bian Skyrms, and Dag Westerståhl, 571–577. Amsterdam: North-Holland.

Newton, Isaac. 2004. Correspondence with Richard Bentley [1692–3]. In *Philosophical writings*, ed. Isaac Newton, 94–105. Cambridge: Cambridge University Press.

Ortega y Gasset, José. 1957. *On love: Aspects of a single theme*. New York: The World Publishing Company.

Poincaré, Henri. 2015. *The foundations of science: Science and hypothesis, The value of science, Science and method*. Cambridge: Cambridge University Press.

Pólya, George. 1954. *Mathematics and plausible reasoning*. Princeton: Princeton University Press.

Post, Emil Leon. 1965. Absolutely unsolvable problems and relatively undecidable propositions: Account of an anticipation. In *The undecidable*, ed. Martin Davis, 340–433. New York: Raven Press.

Prawitz, Dag. 1998. Comments on the papers. *Theoria* 64: 283–337.

Resnik, Michael D. 1980. *Frege and the philosophy of mathematics*. Ithaca: Cornell University Press.

Rota, Gian-Carlo. 1997. *Indiscrete thoughts*. Cham: Birkhäuser.

Russell, Bertrand. 1956. *Portraits from memory, and other essays*. New York: Simon and Schuster.

———. 1998. *The problems of philosophy*. Oxford: Oxford University Press.

Shapiro, Stewart. 2004. Foundations of mathematics: Metaphysics, epistemology, structure. *The Philosophical Quarterly* 54: 16–37.

Sierpiński, Wacław. 1915. Sur une courbe dont tout point est un point de ramification. *Comptes Rendus de l'Academie des Sciences de Paris* 160: 302–305.

Simpson, Stephen G. 1988. Partial realizations of Hilbert's program. *The Journal of Symbolic Logic* 53: 349–363.

Sternheimer, Daniel. 2011. The deformation philosophy, quantization and noncommutative space-time structures. In *Higher structures in geometry and physics*, ed. Albert S. Cattaneo, Antony Giaquinto, and Ping Xu, 39–56. New York: Springer.

Thom, René. 1971. 'Modern' mathematics: An educational and philosophic error? *American Scientist* 59 (6): 695–699.

Thomas, Robert. 2014. Reflections on the objectivity of mathematics. In *From a heuristic point of view*, ed. Cesare Cozzo and Emiliano Ippoliti, 241–256. Newcastle upon Tyne: Cambridge Scholars Publishing.

Ulam, Stanislaw M. 1986. *Science, computers, and people. From the tree of mathematics*. Boston: Birkhäuser.

Wang, Hao. 2016. *From mathematics to philosophy*. Abingdon: Routledge.

Wittgenstein, Ludwig. 1976. *Lectures on the foundations of mathematics, Cambridge 1939*. Brighton: Harvester Press.

———. 1978. *Remarks on the foundations of mathematics*. Oxford: Blackwell.

———. 2009. *Philosophische Untersuchungen – Philosophical investigations*. Oxford: Wiley-Blackwell.

Chapter 3
Heuristic Philosophy of Mathematics

Abstract An alternative to mainstream philosophy of mathematics is heuristic philosophy of mathematics, according to which the philosophy of mathematics is concerned with the making of mathematics, in particular discovery, and the method of mathematics is the analytic method. Heuristic philosophy of mathematics is not to be confused with other views of mathematics that might seem similar to it, in particular with the philosophy of mathematical practice, which is not concerned with the making of mathematics, in particular discovery, but, like mainstream philosophy of mathematics, only with finished mathematics. The chapter describes characters, origin, goal, and advantages of heuristic philosophy of mathematics.

Keywords Heuristic philosophy of mathematics · Practical heuristics · Philosophy of mathematical practice · Heuristic view of mathematics · Bottom-up approach · Open systems · Mathematics as problem solving · Problems vs. theorems

3.1 The Characters of Heuristic Philosophy of Mathematics

Although mainstream philosophy of mathematics has prevailed for the past century, in the second part of the twentieth century an alternative minority tradition arose, which can be called heuristic philosophy of mathematics.

Heuristic philosophy of mathematics has the following characters, which for comparison are stated parallel to the characters of mainstream philosophy of mathematics in Chap. 2.

(1) The philosophy of mathematics can concern itself with the making of mathematics, in particular discovery, because discovery is an objective process, so it can be accounted for.

(2) The philosophy of mathematics can concern itself also with finished mathematics, namely mathematics presented in finished form. But finished mathematics is never really finished, because every mathematical concept or hypothesis can always be called into question, modified, or reinterpreted.

(3) Since the philosophy of mathematics can concern itself with the making of mathematics, it can possibly contribute to the advancement of mathematics.

C. Cellucci, *The Making of Mathematics*, Synthese Library 448,
https://doi.org/10.1007/978-3-030-89731-4_3

(4) The task of the philosophy of mathematics is primarily to give an answer to the question: How is mathematics made? And, subordinately to it, to the questions: What is the method of mathematics? What is the nature of mathematical objects, demonstrations, definitions, diagrams, notations? What is the nature of mathematical explanations? What is the nature of mathematical beauty? Why is mathematics applicable to the world? In what sense is mathematics knowledge?

(5) The method of mathematics is the analytic method. The latter is the method according to which, to solve a problem, one looks for some hypothesis that is a sufficient condition for solving the problem, namely such that a solution to the problem can be deduced from the hypothesis. The hypothesis is obtained from the problem, and possibly other data, by some non-deductive rule, and must be plausible, namely such that the arguments for the hypothesis are stronger than the arguments against it, on the basis of experience. But the hypothesis is in turn a problem that must be solved, and is solved in the same way. Namely, one looks for another hypothesis that is a sufficient condition for solving the problem posed by the previous hypothesis, it is obtained from the latter, and possibly other data already available including data acquired from mathematical diagrams, by some non-deductive rule, and must be plausible. And so on. Thus, solving a problem is a potentially infinite process (see Chap. 5).

(6) The role of analytic demonstration, namely demonstration based on the analytic method, is to discover a solution to a problem.

(7) Since the method of mathematics is the analytic method, mathematics is a body of problems and solutions to them that are plausible. Therefore, mathematics is about plausibility.

(8) Since the method of mathematics is the analytic method, mathematical reasoning consists of both deductive reasoning and non-deductive reasoning.

(9) The philosophy of mathematics goes back to the beginning of philosophy, many major philosophers have made substantial contributions to it, and their work remains important even today.

(10) The philosophy of mathematics cannot be developed independently of experience. For, several mathematical problems have an extra-mathematical origin, and the solutions to mathematical problems are only plausible, so their evaluation depends on experience.

3.2 Original Formulation of Heuristic Philosophy of Mathematics

The original formulation of heuristic philosophy of mathematics can be credited to Lakatos's Ph.D. dissertation (Lakatos 1961). Unfortunately, the dissertation is still unpublished as a whole. Pieces of it have been published separately in Lakatos 1963–1964, Lakatos 1976, and Lakatos 1978, II, Chap. 5, but from them it is not easy to get an overall picture.

Moreover, in Lakatos 1976, the editors Worrall and Zahar have added notes in which, as Davis and Hersh say, they "correct Lakatos whenever he calls into question the existence of a final solution of the problem of mathematical rigor," because they believe that "a modern formal deductive proof is infallible" (Davis and Hersh 1981, 353). But they "are wrong. What is more surprising, their objection is rooted in the very error which Lakatos attacked so vehemently," namely "the error of identifying mathematics" with its representation in "first-order logic" (ibid.).

Indeed, Lakatos criticizes mainstream philosophy of mathematics because it identifies "mathematics with its formal axiomatic abstraction," in which "mathematical theories are replaced by formal systems, proofs" by derivations in first-order logic, and "definitions by 'abbreviatory devices' which are 'theoretically dispensable' but 'typographically convenient'" (Lakatos 1976, 1). So, mainstream philosophy of mathematics "denies the status of mathematics to most of what has been commonly understood to be mathematics, and can say nothing about its growth," in particular it can say nothing about "the 'creative' periods" and "the 'critical' periods of mathematical theories" (ibid., 2). Therefore, in mainstream philosophy of mathematics "there is no proper place for methodology qua logic of discovery" (ibid., 3).

Contrary to mainstream philosophy of mathematics, heuristic philosophy of mathematics is concerned with methodology qua logic of discovery. According to it, although there is no infallibilist logic of discovery, namely "one which would infallibly lead to results," nevertheless "there is a fallibilist logic of discovery" (ibid., 143–144, footnote 2). The latter consists in "the method of proof and refutations" (ibid., 50). The rules of the method can be found through case studies in the history of mathematics, because there is a strict relation between "the history of mathematics and the logic of mathematical discovery" (ibid., 4). In particular, the rules of the method can be found through the study of the history of Euler's conjecture for polyhedra: The number of vertices V, edges E, and faces F in a convex polyhedron satisfy the equality $V - E + F = 2$.

That Lakatos is the initiator of an alternative to mainstream philosophy of mathematics is widely acknowledged.

Thus, Hersh says that, "starting with Imre Lakatos' 1976 *Proofs and Refutations*, some writers have been turning away from the search for a 'foundation' for mathematics and instead, seeking to understand and clarify the actual practice of mathematics – what 'real mathematicians really do'" (Hersh 2014, 241).

Rota says that Lakatos's views, "published in the book *Proofs and Refutations*, were met with a great deal of anger on the part of the mathematical public who held the axiomatic method to be sacred and inviolable. Lakatos' book became anathema among philosophers of mathematics of the positivistic school. The truth hurts" (Rota 1997, 50).

Nickles says that Lakatos rejected the "reduction of mathematics to formalized mathematics," his *Proofs and Refutations* is "a highly original investigation of creative problem solving and the growth of knowledge in mathematics," which makes Lakatos "the most important philosopher of mathematics" since "the mid-twentieth century" (Nickles 2000, 207).

Admittedly, before Lakatos, Wittgenstein had made some critical remarks against mainstream philosophy of mathematics. Thus, Wittgenstein says that the so-called "foundations are no more the foundations of mathematics than the painted rock is the support of the painted tower" (Wittgenstein 1978, V, § 13). And "logic and mathematics are not based on axioms, any more than a group is based on the elements and operations that define it" (Wittgenstein 2005, 377). But, contrary to Lakatos, Wittgenstein denies the possibility of a methodology qua logic of discovery, because he says that "mathematical discovery is always unmethodical: you have no method for making the discovery" (Wittgenstein 2016, 46).

Therefore, Lakatos has been the first to criticize mainstream philosophy of mathematics for not having a proper place for methodology qua logic of discovery. This justifies the claim that the original formulation of heuristic philosophy of mathematics can be credited to him.

3.3 Shortcomings of the Original Formulation

Despite its merits, however, Lakatos's formulation of heuristic philosophy of mathematics has some serious shortcomings.

Lakatos says that the first step of the method of proof and refutations is the naive conjecture, because discovery "moves from the naive conjecture" (Lakatos 1976, 42). Then, one would expect that the first rule of Lakatos's method of proof and refutations would indicate how to arrive at the naive conjecture. But the rule does nothing of the kind. For, it states: "Rule 1. If you have a conjecture, set out to prove it and to refute it. Inspect the proof carefully to prepare a list of non-trivial lemmas (proof-analysis); find counterexamples both to the conjecture (global counterexamples) and to the suspect lemmas (local counterexamples)" (ibid., 50).

So, the rule assumes that you already have a conjecture. Therefore, Lakatos's method of proof and refutations does not account for how to arrive at the naive conjecture.

Regarding his conjecture $V - E + F = 2$, Euler declares: "From the consideration of many types of solids I have been led to understand that the properties, which I had discerned in them, clearly extended to all solids, even if I was not allowed to show this by rigorous demonstration" (Euler 1758, 141). Thus, Euler states that he arrived at his conjecture by induction from observed cases.

But Lakatos rejects Euler's statement because, following Popper, he claims that "there are no such things as inductive conjectures" (Lakatos 1976, 73). The method of proof and refutations requires "no inductivist starting point at all" (ibid., 72). The "naive conjectures are not inductive conjectures: we arrive at them by trial and error" (ibid., 73).

This, however, conflicts with the fact that, as we will see in Chap. 17, Popper himself says that the success of trials depends very largely on the number and variety of the trials: the more we try, the more likely it is that one of our attempts will be successful. This amounts to admitting that trial and error depends on induction.

Then, it is contradictory to claim, as Lakatos does, that naive conjectures are not inductive conjectures, we arrive at them by trial and error.

In particular, regarding Euler's conjecture, Lakatos says that, "after much trial and error" it was noticed "that for all regular polyhedra $V - E + F = 2$," it was guessed "that this may apply for any polyhedron whatsoever," and this was put forward as a "conjecture" (ibid., 6–7). The "trials and errors" through which the conjecture $V - E + F = 2$ was reached "are beautifully reconstructed by Pólya" (ibid., 73, footnote 3).

But this conflicts with Pólya's own account, because Pólya says: "To begin with, we can scarcely do anything better than examine examples, particular polyhedra" (Pólya 1954, I, 35). Examining them, we observe that $V - E + F = 2$. Moreover, "this relation is verified in all" the polyhedra examined, and "it seems unlikely that such a persistent regularity should be mere coincidence" (ibid., I, 37). So "we are led to the conjecture that, not only in the observed cases, but in any polyhedron the number of faces increased by the number of vertices is equal to the number of edges increased by two" (ibid., I, 38). Therefore, like Euler, Pólya says that we are led to the conjecture $V - E + F = 2$, not by trial and error, but by induction from observed cases.

Lakatos even claims that "it took in this case nearly 2000 years to reach" Euler's "naive conjecture" by "'naive trial and error'. This 'naive' period, the first stage of mathematical discovery, lasted in this particular case from Euclid to Descartes" (Lakatos 1978, II, 96). But this claim seems far-fetched, Euler arrived at his conjecture far more quickly than that, by induction from observed cases. By claiming that it took nearly 2000 years to reach Euler's conjecture by trial and error, Lakatos admits that trial and error is very inefficient, so inefficient that it cannot account for the successes of mathematics. Indeed, the number of trials a mathematician can make is very small with respect to all possible ones, so the probability that he can reach a valuable conjecture by trial and error is very low. This is contradicted by the fact that over 100,000 research papers in mathematics are published every year.

In addition to the shortcomings of Lakatos's account of how Euler's conjecture was reached, Lakatos's assumption that there is a strict relation between the history of mathematics and the logic of mathematical discovery is invalid. For, the history of mathematics is mostly written on the basis of mathematics presented in finished form, and the latter has little or nothing to do with the way it was discovered (see below). Therefore, the history of mathematics does not provide an adequate basis for finding the rules of methodology qua logic of discovery.

Lakatos also claims that the method of proof and refutations is an extension of Pappus's analytic-synthetic method. For, he says that, after we "reach the naive conjecture" by "trial and error," the "naive conjecture is subjected to a sophisticated attempted refutation; analysis and synthesis starts" (Lakatos 1978, II, 96). By 'analysis and synthesis' Lakatos means "Pappusian analysis-synthesis" (ibid., II, 93). Namely, Pappus's analytic-synthetic method. But this does not contribute to the credibility of Lakatos's method of proof and refutations, because Pappus's analytic-synthetic method has serious shortcomings (see Chap. 6).

Therefore, it seems fair to say that Lakatos's method of proof and refutations does not provide a basis for methodology qua logic of discovery.

Lakatos himself ends up admitting it. For, he says that, while "in the seventeenth or even eighteenth century" it "was hoped that methodology would provide scientists" with "rules for solving problems," this hope "has now been given up: modern methodologies or 'logics of discovery' consist merely of a set" of "rules for the appraisal of ready, articulated theories" (ibid., I, 103). In particular, Lakatos's own "'methodology', older connotations of this term notwithstanding," does not presume "to give advice to the scientist" about "how to arrive at good theories," it "only appraises fully articulated theories" (Lakatos 1971, 174). The methodological rules are normative rules, where, however, "'normative' no longer means rules for arriving at solutions, but merely directions for the appraisal of solutions already there" (Lakatos 1978, I, 103, footnote 1). All we can have are rules for the appraisal of solutions already there.

As Nickles observes, it "is astonishing" that "Lakatos's methodology provides ways to appraise" solutions already there, "but stops short of giving advice" (Nickles 1987, 119). For, "the very idea of a method is the idea of something that guides inquiry, however fallibly; and the very idea of methodology is that of something that endorses specific method as preferred directives for future behavior" (ibid., 119–120). So, "the idea of a heuristic methodology which gives no advice is a contradiction in terms. Bluntly stated, Lakatos has no methodology" (ibid., 120).

At least, Lakatos has no methodology qua logic of discovery. This does not invalidate the claim that the original formulation of heuristic philosophy of mathematics can be credited to Lakatos. But it means that, with respect to heuristic philosophy of mathematics, Lakatos is a sort of 'non-playing captain', namely a captain who is not in the field when the game takes place.

3.4 Difference from Practical Heuristics

Heuristic philosophy of mathematics must not be confused with other approaches to mathematics that might seem similar to it.

Thus, heuristic philosophy of mathematics must not be confused with practical heuristics, as formulated by Pólya.

Pólya says that "the greater part of our conscious thinking is concerned with problems" (Pólya 1981, I, 117). And "the most characteristically human activity is solving problems, thinking for a purpose, devising means to some desired end" (ibid., I, 118). This applies also to mathematics, because "mathematics in the making resembles any other human knowledge in the making" (Pólya 1954, I, vi).

But, according to Pólya, there are no general rules for solving problems. Finding "rules applicable to all sorts of problems is an old philosophical dream," rules of that kind "would work magic; but there is no such thing as magic," such rules are like "the philosophers' stone, vainly sought by the alchemists," they are a dream which "will never be more than a dream" (Pólya 2004, 172).

Refusing to admit that there are general rules for solving problems, Pólya ironically says that there are only two rules of discovery: "The first rule of discovery is to have brains and good luck. The second rule of discovery is to sit tight and wait till you get a bright idea" (ibid.). This reminds "somewhat rudely that certain aspirations," such as that of finding rules of discovery, "are hopeless" (ibid.).

According to Pólya, mathematical discovery involves guessing. Indeed, "you have to guess a mathematical theorem before you prove it; you have to guess the idea of the proof before you carry through the details. You have to combine observations and follow analogies; you have to try and try again" (Pólya 1954, I, vi).

But there is no "foolproof method to learn guessing" (ibid.). Guessing "is a practical skill and it is learned, as any other practical skill, by imitation and practice" (ibid.). Dealing with it is the object of heuristics, which "has 'practical' aims" (Pólya 2004, 130). Heuristics only "offers you good examples for imitation and many opportunities for practice: if you wish to learn swimming you have to go into the water, and if you wish to become a problem solver you have to solve problems" (Pólya 1981, ix). So, heuristics is "a practical art, like swimming, or skiing, or playing the piano: you can learn it only by imitation and practice" (ibid.).

Being a practical art, heuristics can only give some practical guidelines to solve problems in mathematics, such as the following. Try to understand the problem, because "incomplete understanding of the problem" is "perhaps the most widespread deficiency in solving problems" (Pólya 2004, 95). If you have difficulty in understanding the problem, "try to draw a figure," even when "your problem is not a problem of geometry," because "to find a lucid geometric representation for your nongeometrical problem could be an important step toward the solution" (ibid., 108). If "you cannot solve the proposed problem, try to solve first some related problem" (ibid., 101). Try to solve some more ambitious problem, because "the new, more ambitious problem" can be "easier to handle than the original problem" (ibid., 121). If you are unable to solve a problem, "leave the problem alone for a while. 'Take counsel of your pillow' is an old piece of advice. Allowing an interval of rest to the problem and to ourselves, we may obtain more tomorrow with less effort" (ibid., 198).

But these practical guidelines are so general that no one is likely to make any mathematical discovery through them. This is consistent with Pólya's view that there is no method of mathematical discovery.

However, inconsistently enough, Pólya also says that there are procedures that are "typically useful in solving problems" and are "practiced by every sane person sufficiently interested in his problems" (ibid., 172). The best of such procedures is "the method of analysis, or method of 'working backwards'" (ibid., 225).

Now, the method of analysis, namely the analytic method, is a general method of solving problems. This contradicts Pólya's claim that there are no general rules for solving problems. The inconsistency depends on the fact that Pólya tacitly assumes that general rules for solving problems ought to be algorithmic, while the method of analysis is not an algorithmic method. But the assumption that general rules for solving problems ought to be algorithmic is invalid (see Chap. 4).

3.5 Difference from Deductive Logic as Logic of Discovery

Heuristic philosophy of mathematics must also not be confused with the view that deductive logic is a logic of discovery, as formulated by Musgrave.

Musgrave says that "there is a logic of invention (discovery). And it is deductive logic" (Musgrave 2011, 222). Indeed, there is no justification in "the argument that logic of discovery must be non-deductive or ampliative because discovery is by definition coming up with something new" (ibid., 221). For, "the conclusion of a valid deduction can have interesting new properties not possessed by any of the premises taken singly" (ibid., 221–222). Deduction does not "take all the inventiveness out" of arguments and renders them "a matter of dull routine," because obtaining the conclusion from the premises "in interesting cases is no trivial or routine task" (ibid., 222). So, a logic of discovery need not be non-deductive or ampliative, on the contrary, "the logic of invention is best regarded as deductive logic" (Musgrave 1989, 32). Indeed, "deductive logic is the only logic that we have or need" (ibid., 16).

But this argument is invalid. Contrary to Musgrave's claim, deduction is non-ampliative (see Chap. 5). Moreover, deduction takes all the inventiveness out of arguments and renders them a matter of dull routine, because there is an algorithm for enumerating all deductions from given premisses. As Weyl says, the algorithm is supposed "to proceed like Swift's scholar, whom Gulliver visits in Balnibarbi, namely, to develop in systematic order, say according to the required number of inferential steps, all consequences and discard the 'uninteresting' ones" (Weyl 1949, 24). Given enough time and space, the algorithm will enumerate all deductions from given premisses. So, discovering a deduction of a desired conclusion from given premisses requires no inventiveness, it is a purely mechanical business. Moreover, to make a deduction, you need premisses, and premisses are not discovered by deduction but by non-deductive inferences. Therefore, deductive logic is not the only logic that we have or need.

3.6 Difference from the Philosophy of Mathematical Practice

Heuristic philosophy of mathematics must also not be confused with the philosophy of mathematical practice. There is no general formulation of the philosophy of mathematical practice, that which comes nearest to it is Mancosu's formulation. Therefore, in what follows, the philosophy of mathematical practice will be considered in Mancosu's formulation.

Mancosu says that the philosophers of mathematical practice "do not engage in polemic with the foundationalist tradition" (Mancosu 2008, 18). In particular, they reject "the polemic against the ambitions of mathematical logic as a canon for philosophy of mathematics," and do not consider "mathematical logic, which had been essential in the development of the foundationalist programs," to be "ineffective in dealing with the questions concerning the dynamics of mathematical discovery" (ibid., 4). Indeed, "many of them work, or have worked, also as mathematical logicians" (ibid.).

The philosophers of mathematical practice are only "calling for an extension" of the foundationalist tradition to "topics that the foundationalist tradition has ignored," namely some "aspects of mathematical practice" (ibid., 18).

This does not mean that the three big foundationalist schools "were removed from such concerns" (ibid., 6–7). Indeed, Frege's development of a formal language "which aimed at capturing formally all valid forms of reasoning occurring in mathematics, required a keen understanding of the reasoning patterns to be found in mathematical practice" (ibid., 7). Hilbert's "distinction between real and ideal elements" also "originates in mathematical practice" (ibid.). Brouwer's intuitionism "takes its origin from the distinction between constructive vs. non-constructive procedures" which was prominent in, "just to name one area, the debates in algebraic number theory in the late nineteenth century (Kronecker vs. Dedekind)" (ibid.). The direct and indirect descendants of the three big foundationalist schools "are also, to various extents, concerned with certain aspects of mathematical practice" (ibid.).

The only difference is that the philosophers of mathematical practice propose to investigate a broader range of aspects of mathematical practice. But this is only a difference in quantity, not in quality.

From this it is clear that, in Mancosu's formulation, there is no conflict between the philosophy of mathematical practice and mainstream philosophy of mathematics. This view is shared by Carter: "I do not intend to claim that there is a necessary tension or conflict between 'philosophy of mathematical practice' and" mainstream "'philosophy of mathematics'" (Carter 2019, 2). Rather, the philosophy of mathematical practice is continuous with mainstream philosophy of mathematics. Therefore, the philosophy of mathematical practice shares the shortcomings of the latter.

Another shortcoming arises from the fact that, according to the philosophers of mathematical practice, "mathematical practice is embodied in the concrete work of mathematicians and that work has taken place in history" (ibid., 13). Therefore, a main concern of the philosophers of mathematical practice is to "cover a broad spectrum" of "case studies arising from mathematical practice" (ibid., 18).

But case studies in the history of mathematics are usually carried out on the basis of finished mathematics, namely mathematics presented in finished form, and the latter has little or nothing to do with the way it was discovered (see below). Therefore, historical case studies can teach us about the sequence of published results and theories, not about discovery. This makes it clear that the aim of the philosophy of mathematical practice is not to account for the making of mathematics, in particular discovery, but only for finished mathematics.

3.7 Finished Mathematics and the Mathematical Process

It has been said above that finished mathematics has little or nothing to do with the way it was discovered. This has been pointed out by several mathematicians.

Thus, Descartes says that "the ancient geometers used a certain analysis," namely the analytic method, "which they extended to the solution of all problems," but "they begrudged revealing it to posterity" (Descartes 1996, X, 373). Therefore, "in their writings, the ancient geometers generally used" synthesis, namely axiomatic demonstration, "not because they were utterly ignorant of analysis," which was, in fact, the method by which they discovered their results, "but because they had such a high regard for it that they kept it to themselves like a sacred mystery" (ibid., VII, 156). They did so "with a kind of pernicious cunning; for, as notoriously many inventors are known to have done where their own discoveries were concerned, they have perhaps feared" that their method, "just because it was so easy and simple, would be depreciated if it were divulged" (ibid., X, 376).

Newton says that his propositions "were invented by analysis" (Newton 1971, 294). Namely by the analytic method. But, "considering that the ancient" mathematicians "admitted nothing into geometry before it was demonstrated by composition," namely by axiomatic demonstration, "I composed what I invented by analysis to make it geometrically authentic and fit for the publick" (ibid.). Admittedly, "if any man who understands analysis will reduce the demonstrations of the propositions from their composition back into analysis," he "will see by what method of analysis they were invented" (ibid.). But this would require considerable skill, because axiomatic demonstration hides analysis, and hence "makes it now difficult for unskilful men to see the analysis by which those propositions were found out" (ibid., 295).

Grothendieck says that the making of mathematics "is not reflected virtually to any extent in the texts or talks that are intended to present such work," whether "textbooks and other didactic texts, or articles and original memoirs, or oral courses and seminar presentations, etc." (Grothendieck 1985, 84). In fact, "from the very beginning of mathematics," there has been "a sort of 'conspiracy of silence' about these 'ineffable works' that prelude to the hatching of any new idea, great or small, which comes to renew our knowledge of a part of this world, in perpetual creation, in which we live" (ibid.).

Henkin says that "mathematical papers are written in a fashion that tends to obscure the process of discovery" (Henkin 1996, 155). Historians of mathematics "describe the development of some subject in a way that can be schematized by the diagram of a directed graph" (ibid.). Specifically, "the nodes, or vertices, of this graph are the publications mentioned," and an edge "leads from vertex u to vertex w in case the author of publication w is thought to have been influenced by publication u" (ibid.). But "the way in which the author of publication w arrived at its ideas" – namely "the process of discovery of the ideas of w – is not treated" (ibid., 156).

That finished mathematics has little or nothing to do with the way it was discovered, has an important consequence. By assuming that the philosophy of mathematics cannot concern itself with the making of mathematics, in particular discovery, as Davies says, mainstream philosophy of mathematics is "left with an impoverished subject that has only a remote connection with what mathematicians do" (Davies 2008, 97).

3.8 Objections to Heuristic Philosophy of Mathematics

Mainstream philosophers of mathematics have raised some objections against Lakatos's original formulation of heuristic philosophy of mathematics. Since similar objections could be raised against heuristic philosophy of mathematics as formulated at the beginning of this chapter, such objections will be discussed here. The main ones are the following.

(1) Feferman says: "Lakatos' fireworks briefly illuminate limited portions of mathematics," but only deductive "logic gives us a coherent picture of mathematics," it "alone throws light on what is distinctive about mathematics, its concepts and methods" (Feferman 1998, 93). The "logical analysis of the structure of mathematics has been especially successful," and one can use formal systems also "to model growth" (ibid., 92).

But this objection is invalid. By the strong incompleteness theorem for second-order logic, there is no consistent formal system for second-order logic capable of deducing all second-order logical consequences of any given set of propositions. But much of mathematics requires second-order logic or beyond. Therefore, deductive logic cannot be said to give us a coherent picture of mathematics, nor to throw light on what is distinctive about mathematics, its concepts and methods. Moreover, as argued in Chap. 2, deductive logic cannot explain why demonstration may yield something new, so formal systems cannot be used to model the growth of mathematical knowledge.

(2) Smoryński says: "Lakatos firmly denies the distinction between mathematics on the one hand and the (rest of the) sciences on the other. His book *Proofs and Refutations*" attempted "to show just this. I cannot accept such a denial: Mathematics" is "a cumulative body of knowledge" (Smoryński 1983, 11).

But this objection is invalid. Mathematics is not a cumulative body of knowledge because there are revolutions in mathematics (see Chap. 10). The claim that mathematics is a cumulative body of knowledge depends on the assumption that the method of mathematics is radically different from the method of natural sciences. Thus, Duhem says that, "underneath their entirely external resemblance, which is due to the borrowing of mathematical language by physics," the method of mathematics and the method of physics "reveal themselves to be profoundly different"

(Duhem 1954, 265). But the assumption is wrong, because the method of mathematics and the method of natural sciences are essentially the same method: the analytic method. Indeed, in both mathematics and the natural sciences we start from problems, we formulate hypotheses for their solution by non-deductive inferences, and we establish the plausibility of hypotheses through a comparison with experience.

(3) Mancosu says: “The ‘maverick tradition’,” namely the tradition ensuing from Lakatos, “has not managed to substantially redirect the course of philosophy of mathematics. If anything, the predominance” of mainstream “approaches to the philosophy of mathematics in the last twenty years proves that the maverick camp did not manage to bring about a major reorientation of the field” (Mancosu 2008, 5). Indeed, “the iconoclastic attitude of the ‘mavericks’ vis-à-vis what had been done in foundations of mathematics had as a consequence a reduction of their sphere of influence. Logically trained” mainstream philosophers of mathematics “felt that the ‘mavericks’ were throwing away the baby with the bathwater” (ibid., 5–6).

But this objection is invalid. The ‘baby’ thrown away is the assumption that mathematical reasoning consists of deductive reasoning. This assumption is very limiting, because it implies that the philosophy of mathematics must be restricted to finished mathematics, which conflicts with Gödel’s incompleteness theorems. Also, the predominance of mainstream approaches to the philosophy of mathematics in the last 20 years does not prove that the maverick camp has failed. Old ideas give way slowly, and yet eventually they are abandoned when their vitality declines. This is apparently the case with the ideas of mainstream philosophy of mathematics.

3.9 The Heuristic View of Mathematics

Heuristic philosophy of mathematics puts forward the heuristic view of mathematics, which is based on the following assumptions.

(I) Mathematical problems are solved by the analytic method. Therefore, mathematics is problem solving by the analytic method.

(II) Solutions to mathematical problems are never certain but only plausible. Indeed, mathematics is part of human knowledge, and human knowledge is never certain but only plausible.

(III) There is no part of mathematics such that all other parts of mathematics can be reduced to it. Specifically, there is no mathematical theory such that all other mathematical theories can be reduced to it. Mathematics consists of a variety of mathematical theories, related in multiple ways.

3.10 Original Formulation of the Heuristic View

Plato gave the first formulation of the heuristic view of mathematics.

Indeed, Plato says that mathematics proceeds "by working through problems" (Plato, *Respublica*, VII 530 b 6). And it proceeds by the analytic method (see Chap. 5).

Plato also says that solutions to mathematical problems are never certain but only plausible, because "certain knowledge is either impossible or extremely difficult to come by in this life," we can only "adopt the best and least refutable of human hypotheses, and embarking upon it as upon a raft, run the risk of sailing the sea of life" (Plato, *Phaedo*, 85 c 3–d 2).

Plato also says that there is no part of mathematics such that all other parts of mathematics can be reduced to it, specifically, there is no mathematical theory such that all other mathematical theories can be reduced to it, because the hypotheses of such a theory would be "the principle of everything" (Plato, *Respublica*, VI 511 b 7). But "it is impossible to have pure knowledge of anything" until our soul is "in the company of the body," so, while we are alive, "knowledge" of the principle of everything "cannot be acquired at all" (Plato, *Phaedo*, 66 e 4–6).

3.11 A Confusion About the Original Formulation of the Heuristic View

That Plato gave the first formulation of the heuristic view of mathematics conflicts with the widespread opinion that, for Plato, the appropriate method for mathematics is the axiomatic method, so mathematics is theorem proving by the axiomatic method.

Thus, Lucas says that "it is Plato who put forward the ideal of axiomatization as something to be pursued fairly consciously" (Lucas 1967, 12). According to this ideal, "we should try and develop the whole of our mathematics by deductive reasoning" from "principles which" could "be established beyond all possible question. Plato put forward this programme. His pupils largely carried it out" (ibid., 13).

Bedürftig and Murawski say that "for Plato" the "appropriate method for mathematics is the axiomatic method: axioms describe the basic properties of mathematical basic concepts. Plato was probably the first who introduced this method and propagated it" (Bedürftig and Murawski 2018, 33).

Karasmanis says that "in the *Republic* Plato" asserts that mathematics is "a body consisting of first principles, conclusions and intermediate propositions woven together" (Karasmanis 2018, 320). It "proceeds downwards from the hypotheses (first principles) to the end," where "first principles are viewed by the mathematicians as self-evident or 'plain to all'. All these features are characteristics of a science already organized in an axiomatic system" (ibid.).

This opinion, however, is unfounded. For, as we will see below, Plato harshly criticizes the view that mathematics is theorem proving by the axiomatic method, to which he opposes the view that mathematics is problem solving by the analytic method. The reason why Plato harshly criticizes the view that mathematics is theorem proving by the axiomatic method is that some member of Plato's Academy, possibly Eudoxus, had introduced the axiomatic method as the method of mathematics. So, Plato felt even more strongly the need to counter an error that lurked in his school.

3.12 The Bottom-Up Approach to Mathematics

Assumption (I) of the heuristic view, that mathematics is problem solving by the analytic method, is an expression of the bottom-up approach to mathematics, according to which every part of mathematics must be developed from below, namely starting from problems, by the analytic method.

An example of the bottom-up approach in the early modern period is Descartes's geometry. In the Latin edition of *La Géométrie*, van Schooten presents a solution to the following problem as representative of Descartes's analytic method or method of analysis: "Given a straight line *AB*, cut at any point *C*, produce it to *D*, so that the rectangle comprised by *AD*, *DB* shall be equal to the square on *CD*" (van Schooten 1649, 167).

The following figure illustrates the problem, where the rectangle comprised by *AD*, *DB* is *ADEF* because *DE* is equal to *DB*, and the square on *CD* is *CDGH*. We must find *DB* such that the rectangle *ADEF* is equal to the square *CDGH*.

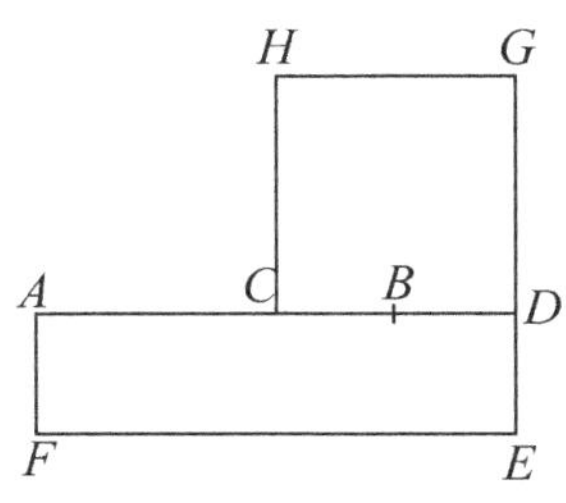

To solve the problem, by analyzing the conditions under which the problem would be solved, van Schooten non-deductively arrives at the following hypothesis, formulated by Descartes:

(I) Solving a geometrical problem reduces to solving the algebraic problem arising as follow: (i) We "give names to all the lines that seem necessary to construct" the geometrical problem, "both the lines which are unknown and those which are known;" (ii) We find equations expressing the conditions for the solvability of the problem, specifically, "as many such equations as there are supposed to be unknown lines" (Descartes 1996, VI, 372).

Hypothesis (I) is a sufficient condition for solving the problem. For, the lines which are known are AC and CB, while the line which is unknown is BD. Let $AC = a$, $CB = b$, and $BD = x$. Then $AD = a + b + x$ and $CD = b + x$. So, the rectangle $ADEF$ is $(a + b + x) \cdot x = ax + bx + x^2$, and the square $CDGH$ is $(b + x)^2 = b^2 + 2bx + x^2$. Thus, by hypothesis (I), the condition that the rectangle $ADEF$ be equal to the square $CDGH$, reduces to the condition that the equation $ax + bx + x^2 = b^2 + 2bx + x^2$ must hold. The equation, simplified, is $ax - bx = b^2$. Therefore $x = \frac{b^2}{a-b}$ gives "the required line BD" (van Schooten 1649, 168). This solves the problem.

Hypothesis (I) is plausible, because the arguments for it are stronger than those against it. But hypothesis (I) is itself a problem that must be solved. And so on.

3.13 The Heuristic View and Open Systems

Assumption (I) of the heuristic view, that mathematics is problem solving by the analytic method, implies that mathematical theories are open systems.

An open system is a system whose development takes place by interacting with the environment, receiving inputs from, and delivering outputs to the outside. Its development does not remain internal to the system, so the system is not a self-sufficient totality.

Assumption (I) implies that mathematical theories are open systems, because a mathematical theory initially consists only of problems and possibly other data already available, and its development consists in obtaining more and more hypotheses to solve the problems by non-deductive rules, and in checking that the hypotheses are plausible (see Chap. 8).

Since the hypotheses are obtained by non-deductive rules and non-deductive rules are ampliative, the hypotheses are not implicitly contained in the problems and the other data already available. Moreover, the hypotheses need not belong to the same part of mathematics as the problems and the other data already available, but may belong to other parts of mathematics. So, the development of a mathematical theory involves interactions with other mathematical theories.

3.14 Problem Solving vs. Theorem Proving

Assumption (I) of the heuristic view, that mathematics is problem solving by the analytic method, is opposed to assumption (I) of the foundationalist view, that mathematics is theorem proving by the axiomatic method. The opposition is already apparent from Plato and Aristotle.

On the one hand, Plato harshly criticizes the view that mathematics is theorem proving by the axiomatic method. Indeed, he says that the mathematicians who demonstrate theorems by the axiomatic method take for granted certain propositions, they "assume them as hypotheses, and do not feel any further need to give a justification for them, either to themselves or to anyone else, as if they were axiomatic principles; then, starting from these, they draw consequences from them, and finally reach the conclusions that they had set out to reach in their research" (Plato, *Respublica*, VI 510 c 6–d 3). But those mathematicians "are only dreaming about what is, and cannot possibly have any awaking awareness of it, as long as they leave untouched the hypotheses they use and remain incapable of accounting for them" (ibid., VII 533 b 8–c 3). For, "if your starting-point" is unjustified and hence "is unknown, and the conclusion and intermediate steps are also constructed out of unknown material," then the starting-point, the intermediate steps and the conclusion will be mere conventions; and "how can you imagine that such a fabric of convention can ever become science?" (ibid., VII 533 c 3–6).

To the view that mathematics is theorem proving by the axiomatic method, Plato opposes the view that mathematics is problem solving by the analytic method. The analytic method "does not treat hypotheses as axiomatic principles" for which no justification is given, but "truly as hypotheses" with a justification, and "as stepping stones from which to take off and proceed" (ibid., VI 511 b 4–5). Only the analytic method, "doing away with the hypotheses, proceeds this way up," it "gently pulls the eye of the soul from the barbarian mud in which it is buried, and leads it upwards" (ibid., VII 533 c 8–d 3). It proceeds "passing through all attempts to disprove" the hypotheses, "and trying to disprove them not according to opinion," namely not on the basis of other hypotheses, "but according to the reality of things" (ibid., VII 534 c 1–3). This is a formidable task, but is an essential one because, "without passing this exploration of all possibilities and in all directions, it is impossible for the mind to attain the truth" (Plato, *Parmenides*, 136 e 1–3).

On the other hand, Aristotle criticizes the view that mathematics is problem solving by the analytic method. Indeed, he says that the analytic method "does not permit one to know anything in an absolute way, but only on the basis of a hypothesis" (Aristotle, *Analytica Posteriora*, A 22, 84 a 5–6). But one cannot have scientific knowledge on the basis of a hypothesis, because scientific knowledge "proceeds from necessary principles (since that of which we have scientific knowledge cannot be otherwise)" (ibid., A 6, 74 b 5–6). And "that which is in itself necessary and must be thought to be so is not a hypothesis" (ibid., A 10, 76 b 23–24).

To the view that mathematics is problem solving by the analytic method, Aristotle opposes the view that mathematics, meaning by this finished mathematics, is theorem proving by the axiomatic method. Indeed, he says that scientific knowledge is necessary, and only "that which is known by demonstrative knowledge," namely by "knowledge we have by having a demonstration" from axioms, "is necessary" (ibid., A 4, 73 a 22–23).

3.15 Problems vs. Theorems

The opposition between assumption (I) of the heuristic view, that mathematics is problem solving by the analytic method, and assumption (I) of the foundationalist view, that mathematics is theorem proving by the axiomatic method, is in particular an opposition between problems and theorems.

Pappus says that, "among the ancients, some say that all things are problems, others say that all things are theorems" (Pappus, *Collectio*, III, 30.7–8, ed. Hultsch). This difference of opinion first arose within Plato's Academy in which, "the mathematicians of the school of Menaechmus thought it correct to call all inquiries problems" (Proclus, *In primum Euclidis Elementorum librum commentarii*, 78.8–9, ed. Friedlein). For, all inquiries involve "acts of production" (ibid., 78.21). Conversely others, "such as the followers of Speusippus," thought it correct "to call all propositions 'theorems', holding the designation 'theorems' more appropriate than the designation 'problems' for the theoretical sciences" (ibid., 77.15–19). For, a problem proposes "to bring into being or to produce something not previously existing" (ibid., 77.22–78.1). But the theoretical sciences "deal with eternal things; and there is no coming to be among eternal things, so a problem has no place here" (ibid., 77.19–22).

That Menaechmus was a mathematician and Speusippus a philosopher is significant, because mathematicians generally endorsed the view that mathematics is problem solving by the analytic method, while philosophers generally endorsed the view that mathematics is theorem proving by the axiomatic method.

The reason why mathematicians generally endorsed the view that mathematics is problem solving by the analytic method, is that they were primarily interested in promoting the growth of mathematics, and thought that it could be promoted only through the analytic method, which led to mathematical discovery.

Thus, Carpus of Antioch says that "problems are prior in rank to theorems, because the subjects about which properties are sought are discovered by means of problems" (ibid., 242.1–4). And, "in the case of problems, one general procedure has been invented, namely the method of analysis, by following which we can always hope to find a solution. Thus it is that even the most obscure problems can be pursued" (ibid., 242.14–17). On the contrary, in the case of theorems, "no one to this day has been able to give us a general method of approaching them" (ibid., 242.19–20).

On the other hand, the reason why philosophers generally endorsed the view that mathematics is theorem proving by the axiomatic method, is that they were primarily interested in showing that mathematical knowledge is firmly grounded, and thought that this could be achieved only through the axiomatic method.

Thus, Aristotle says that problems are "questions concerning which there are conflicting reasonings, the difficulty being whether something is so-and-so or not, there being convincing arguments for both views" (Aristotle, *Topica*, A 11, 104 b

12–14). For "them we have no conclusive argument, because they are so vast, and we find it difficult to give our reasons" (ibid., A 11, 104 b 14–16). Therefore, we cannot achieve firmly grounded knowledge by the analytic method. We can achieve it only by the axiomatic method, because knowledge is firmly grounded only "when we possess it in virtue of having" an axiomatic "demonstration" (Aristotle, *Analytica Posteriora*, A 4, 73 a 23).

However, that philosophers generally endorsed the view that mathematics is theorem proving by the axiomatic method, has at least one exception, namely Plato. Indeed, as we have seen, he criticizes this view, to which he opposes the view that mathematics is problem solving by the analytic method.

3.16 Mathematics as Problem Solving and Mathematicians

Although, as said in Chap. 2, assumption (I) of the foundationalist view, that mathematics is theorem proving by the axiomatic method, is shared by the majority of mathematicians, some mathematicians have remained sceptical. Their position, at least to some extent, is related to assumption (I) of the heuristic view, that mathematics is problem solving by the analytic method.

Thus, Halmos says: "Mathematics" is "never deductive in its creation. The mathematician at work makes vague guesses, visualizes broad generalizations, and jumps to unwarranted conclusions" (Halmos 1968, 380–381). Although axioms, proofs, and theorems are all essential ingredients of mathematics, "none of them is at the heart of the subject," indeed, "the mathematician's main reason for existence is to solve problems," and "therefore, what mathematics 'really' consists of is problems and solutions" (Halmos 1980, 519).

Thom says: "During the past few years the importance of axiomatization as an instrument of systematization and discovery has been much emphasized. As a method of systematizing, it is certainly effective; as for discovery, the matter is more doubtful" (Thom 1971,697). Indeed, "no new theorem of any importance came out of the immense effort at systematization of Nicolas Bourbaki" (ibid., 697–698). Discovery calls for other kinds of processes, "such as analogy" (ibid., 699).

Hamming says: "If the Pythagorean theorem were found to not follow from" Euclid's "postulates, we would again search for a way to alter the postulates until it was true. Euclid's postulates came from the Pythagorean theorem, not the other way" (Hamming 1980, 87). Indeed, "mathematics is not simply laying down some arbitrary postulates and then making deductions, it is much more; you start with some of the things you want and you try to find the postulates to support them! Bourbaki to the contrary, notwithstanding!" (Hamming 1998, 645). The "idea that theorems follow from the postulates does not correspond to simple observation" (Hamming 1980, 87).

Hersh says: "A naive non-mathematician" who "looks into Euclid" and "observes that axioms come first," justifiably "concludes that in mathematics, axioms come first. First your assumptions, then your conclusions, no? But anyone who has done mathematics knows what comes first – a problem" (Hersh 1997, 6). Indeed, "mathematics is a vast network of interconnected problems and solutions," in developing and understanding a subject, "problems, and solutions come first. Later come axiom sets," therefore "the view that mathematics is in essence derivations from axioms is backward. In fact, it's wrong" (ibid.).

Giusti says: "The posing of axioms is never the starting point, but rather the arrival point of a theory; better yet, an intermediate point, which occurs mostly when a certain number of results, which will turn out to be founding, have already been obtained, but the theory has not yet been fully developed" (Giusti 1999, 20). In fact, "the occasions in which one started from the axioms are more the exception than the rule" (ibid.).

3.17 The Heuristic View and Gödel's Incompleteness Theorems

As argued in Chap. 2, assumption (I) of the foundationalist view, that mathematics is theorem proving by the axiomatic method, is refuted by Gödel's incompleteness theorems. On the contrary, assumption (I) of the heuristic view, that mathematics is problem solving by the analytic method, is unaffected and even confirmed by Gödel's results.

Assumption (I) is unaffected and even confirmed by Gödel's first incompleteness theorem, because the analytic method does not confine mathematics within the closed space of an axiomatic system, it lets mathematics develop in an open space, making use of interactions with other systems of knowledge. Indeed, in the analytic method, the solution to a problem is obtained from the problem, and possibly other data already available, by means of hypotheses not necessarily belonging to the same part of mathematics as the problem. Since Gödel's first incompleteness theorem implies that solving a problem of a given part of mathematics may require hypotheses from other parts, Gödel's result provides evidence for the assumption that mathematics is problem solving by the analytic method.

Assumption (I) is also unaffected and even confirmed by Gödel's second incompleteness theorem, because the analytic method does not assume that the solution to a problem is certain. Indeed, in the analytic method, the hypotheses for the solution to a problem are always provisional and only plausible, therefore no solution to a problem can be certain. Since Gödel's second incompleteness theorem implies that no solution to a problem can be said to be certain, Gödel's result provides evidence for the assumption that mathematics is problem solving by the analytic method.

3.18 Other Advantages of the Heuristic View

In addition to being unaffected and even confirmed by Gödel's incompleteness theorems, assumption (I) of the heuristic view, that mathematics is problem solving by the analytic method, has other advantages.

(1) Assumption (I) accounts for how mathematicians solve problems. For, according to it, to solve a problem, one finds some hypothesis capable of solving it, then, to solve the problem posed by that hypothesis, one finds some other hypothesis capable of solving it, and so on. For example, to solve the problem posed by Fermat's conjecture, Ribet used the Taniyama-Shimura conjecture, to solve the problem posed by the Taniyama-Shimura conjecture, Wiles and Taylor used other hypotheses, and so on.

(2) Assumption (I) accounts for the fact that solving a problem of a given part of mathematics may require hypotheses from other parts of mathematics. For, according to the analytic method, the hypotheses to solve a problem need not belong to the same part of mathematics as the problem, they may belong to other parts of mathematics.

(3) Assumption (I) accounts for the fact that a demonstration may yield something new. Peirce says that "it has long been a puzzle how it could be that, on the one hand, mathematics is purely deductive in its nature," and, "on the other hand, it presents as rich and apparently unending a series of surprising discoveries as any observational science" (Peirce 1931–1958, 3.363). The puzzle, however, arises because Peirce assumes that mathematics is purely deductive in its nature, and hence is based on non-ampliative rules. This assumption is refuted by Gödel's first incompleteness theorem, by which mathematics cannot consist in the deduction of propositions from given axioms. Assumption (I) solves the puzzle because, in the analytic method, the hypotheses for the solution to a problem are obtained from the problem, and possibly other data, by means of some non-deductive rule, and hence contain something essentially new with respect to them, because non-deductive rules are ampliative.

(4) Assumption (I) accounts for the fact that new solutions, even hundreds of them, are often sought for problems for which a solution is already known. Wittgenstein says: "Every proof, even of a proposition which has already been proved, is a contribution to mathematics," but "why is it a contribution if its only point was to prove the proposition?" (Wittgenstein 1978, III, § 60). Assumption (I) of the heuristic view gives an answer to this question. A mathematical problem can be seen from different perspectives, each of which may suggest different hypotheses that may lead to different solutions to the problem. Each solution establishes new relations between the problem and other parts of mathematics, showing the problem in a new light. This is an important implication of assumption (I). As Atiyah says, "any good theorem should have several proofs, the more the better," indeed, "if you cannot look at a problem from different directions, it is probably not very interesting; the more perspectives, the better" (Atiyah 2004, 24).

(5) Assumption (I) accounts for the fact that different demonstrations of the same proposition may have different degrees of reliability. For, they may be based on hypotheses with a different degree of plausibility, so the demonstrations need not be equally reliable.

3.19 The Heuristic View and Mathematical Creativity

Assumption (I) of the heuristic view, that mathematics is problem solving by the analytic method, also accounts for mathematical creativity.

According to the prevailing view, mathematical creativity is based on illumination, not on logic, because logic does not generate new ideas, it only organizes and communicates already existing ideas. Therefore, mathematics transcends logic.

Thus, Byers says that there is an episode of mathematical creativity when "a light has suddenly illuminated something that was formerly obscure" and "the individual has a 'Eureka' moment," so "creativity is connected to, even based upon, illumination" (Byers 2011, 41). Illumination "is not a logical process" (Byers 2017, 54). For, "logical arguments do not generate ideas," indeed "logic organizes, stabilizes and communicates ideas but the idea exists prior to the logical formulation" (Byers 2007, 259). Therefore, "mathematics transcends logic" (ibid., 26).

The view that mathematical creativity is based on illumination, not on logic, relegates mathematical creativity to the sphere of irrationality. For, no rational account can be given of illumination. This view makes a pair with other irrational views of creativity, such as the one that "creativity might be the result of random firing of neurons," giving "rise to a new idea" (De Cruz 2006, 187). Illumination is as random as a random firing of neurons.

The view that mathematical creativity is based on illumination depends on the assumption that logic is deductive logic. For, deduction is non-ampliative, namely the conclusion is implicitly contained in the premisses, so logical arguments do not generate new ideas (see Chap. 5). On the contrary, creativity involves novelty. So, if logic is deductive logic, it only remains that mathematical creativity be based upon illumination.

But the assumption that logic is deductive logic is invalid. From antiquity, it has been recognized that, since deductive logic is non-ampliative, and hence cannot contribute to the advancement of knowledge, the latter requires a different kind of logic, non-deductive logic, which is ampliative (see Chap. 5).

Admittedly, there is widespread distrust in non-deductive logic. This is well exemplified by Calvino's Mr. Palomar, for whom "deduction, in any case, was one of his favorite activities," while "induction, on the contrary, was something he did not really trust" (Calvino 1985, 109). For, according to Mr. Palomar, "you have to start with something; that is, you have to have principles, from which, by deduction, you develop your own line of reasoning;" indeed, "if you did not have them, you could not even begin thinking" (ibid., 108–109).

The distrust in non-deductive logic leads to not giving a rational explanation of mathematical creativity, and to believing that mathematical creativity is based on illumination.

This belief, as Russell says, is "the belief in insight as against discursive analytic knowledge" (Russell 1994, 26). It is the "belief in the possibility of a way of knowledge which may be called revelation or insight or intuition, as contrasted with sense, reason, and analysis, which are regarded as blind guides leading to the morass of illusion" (ibid., 27).

But the belief is unfounded, in fact quite the opposite is the case. Rather than sense, reason, and analysis, it is intuition that is a blind guide leading to the morass of illusion. For, as argued in Chap. 2, intuition is unreliable and inadequate as a basis for mathematics.

To give a rational explanation of mathematical creativity, non-deductive logic is needed. Heuristic philosophy of mathematics starts from here. Instead of assuming that logic is deductive logic, it assumes that logic consists of both deductive logic and non-deductive logic, which is as genuinely logic as deductive logic.

Then, assumption (I) of the heuristic view, that mathematics is problem solving by the analytic method, accounts for mathematical creativity. For, in the analytic method on which heuristic philosophy of mathematics is based, non-deductive arguments generate the hypotheses for solving problems. While deductive arguments do not generate new ideas, non-deductive arguments generate new ideas by providing hypotheses for solving problems, whose plausibility is then investigated. Mathematical creativity consists in this. Thus, mathematical creativity is not based on illumination, of which no rational account can be given, but on non-deductive logic, which is genuinely logic. Therefore, mathematics does not transcend logic, but is a wholly rational process.

3.20 Mathematics and Plausibility

Assumption (II) of the heuristic view, that solutions to problems are never certain but only plausible and therefore mathematics is only plausible, implies that mathematics is neither true nor certain.

Hume says that "there is no algebraist nor mathematician so expert in his science, as to place entire confidence in any truth immediately upon his discovery of it, or regard it as any thing, but a mere probability" (Hume 2007, 121). To be sure, "every time he runs over his proofs, his confidence encreases; but still more by the approbation of his friends; and is rais'd to its utmost perfection by the universal assent and applauses of the learned world" (ibid.). However, "this gradual encrease of assurance is nothing but the addition of new probabilities" (ibid.). Therefore, mathematics "resolves itself into probability, and becomes at last of the same nature with that evidence, which we employ in common life" (ibid., 122).

Hume's conclusion is a perfectly reasonable one if it is interpreted as follows. First, the mathematician's confidence, the approbation of his friends, and the universal assent and applauses of the learned world, are not based on mere opinion, but on a careful evaluation of the arguments for and against the proposition in question. Second, 'probability' means 'plausibility'. This interpretation is not far fetched, because by 'probability' Hume does not mean the mathematical concept of probability, but "that evidence, which is still attended with uncertainty" (Hume 2007, 86). He says that "one single contrariety of experiment" does not "entirely destroy all our reasoning. The mind ballances the contrary experiments, and deducting the inferior from the superior, proceeds with that degree of assurance or evidence, which remains" (ibid., 259). So, Hume's concept of probability is related to that of plausibility. With this interpretation, Hume's conclusion becomes the perfectly reasonable one that mathematics resolves itself into plausibility, and becomes at last of the same nature with that evidence, which we employ in common life.

That mathematics resolves itself into plausibility means that mathematics is neither true nor certain. This is no limitation, because there is no source of knowledge capable of guaranteeing truth or certainty. Plausible knowledge is the most we can achieve in all fields, including mathematics, where absolutely certain knowledge cannot be reached by Gödel's second incompleteness theorem, and for the other reasons already considered in Chap. 2.

As Xenophanes says, "as for certain truth, no man has known it, nor will he know it," but "on every thing one can only make hypotheses" (Xenophanes 21 B 34, ed. Diels-Kranz). And yet "in time, as they search, men discover better and better things" (ibid., 21 B 18).

3.21 Mathematics and Non-Finality of Solutions to Problems

Assumption (II) of the heuristic view, that solutions to problems are never certain but only plausible and therefore mathematics is only plausible, also implies that solutions to problems are not final, they are always revisable, and hence are not permanent.

This contradicts the belief of several mathematicians, that solutions to mathematical problems are permanent.

Thus, Franks says that "mathematical truths established at the time of Euclid are still held valid today," and "a theorem proven today may be forgotten in the future because it is uninteresting but it will never cease to be true" (Franks 1989, 68).

Mac Lane says that a result of mathematics, when proved, lasts forever, because "mathematics rests on proof – and proof is eternal" (Mac Lane 1994, 193).

Krantz says that, "once a theorem is proved, and its proof checked and validated, then the theorem stands forever. It is just as true, and just as useful, today as when it was proved. Thus we use the Pythagorean Theorem with confidence today, because Pythagoras proved it 2500 years ago" (Krantz 2011, 169).

This belief is invalid. As Goodman says, "the results of mathematics are no more certain or everlasting than the results of any other science, even though, for sociological reasons, our histories of mathematics tend to disguise that fact" (Goodman 1991, 125). For example, "the eighteenth century mathematicians had a theory of infinite series," on the basis of which "they freely manipulated divergent series and came to conclusions which we find absurd" (ibid., 124). When "this theory was replaced by the modern one at the beginning of the nineteenth century, essentially by the work of Cauchy, many of Euler's results were simply discarded. Almost all of his arguments" had "to be replaced by new arguments" (ibid., 124–125). But "the textbooks do not describe this as a scientific revolution," they merely "say that Euler made some mistakes which were later corrected by Cauchy" (ibid., 125). This contrasts with the fact that Cauchy's change in the foundations of the calculus was really a scientific revolution by which "Euler's theory of infinite series" is "completely dead," just as "the phlogiston theory," and "the true parts of Euler's theory were incorporated into the new function theory of Gauss and Cauchy," just as "the true parts of the phlogiston theory were incorporated into the new chemistry of Lavoisier" (ibid.).

Thus, the textbooks misrepresent reality. This is due to the fact that several mathematicians and historians of mathematics deny that there are revolutions in mathematics. But, as it will be argued in Chap. 10, revolutions in mathematics do exist, and Cauchy's change in the foundations of the calculus is one of them. As Grabiner says, "mathematics may often grow smoothly by the addition of methods, but it did not do so in this case," because "the conceptual difference between the eighteenth-century way of looking at and doing the calculus and nineteenth-century views was simply too great. It is this difference which justifies our claim that the change was a true scientific revolution," and "the most important figure in the initiation" of this revolution "was Augustin-Louis Cauchy" (Grabiner 2005, 2).

3.22 Mathematics as Interaction Between Open Systems

Assumption (III) of the heuristic view, that there is no part of mathematics such that all other parts of mathematics can be reduced to it, specifically, there is no mathematical theory such that all other mathematical theories can be reduced to it, corresponds to the fact that mathematics is interaction between open systems.

Indeed, by what has been said above, an open system does not contain a complete representation of the knowledge of a given mathematical field, only a partial one, so, to acquire knowledge not available to it, the system must interact with other open systems. The interaction does not merely result in a cumulative increase in knowledge, but rather in a restructuring of knowledge which causes changes that are not only local but global.

Through the interaction with other open systems, new relations are established between the problem to be solved and other problems, so the problem acquires new significance and meanings. In general, a problem is all the more significant and important as it can enter into relation with other problems. As Hardy says, "a mathematical idea is 'significant' if it can be connected, in a natural and illuminating way, with a large complex of other mathematical ideas" (Hardy 1992, 89). Also, a problem need not have a single solution but may have multiple solutions, and each new solution establishes new relations between the problem and other problems.

The picture of mathematics that emerges from this is not the image, supported by mainstream philosophy of mathematics, of an apparent variety of mathematical theories that can ultimately be reduced to a single mathematical theory. It is rather the image of a genuine variety of mathematical theories connected in a multitude of ways, in which one mathematical theory draws resources from other mathematical theories that enable it to solve problems it could not otherwise solve.

The picture of mathematics supported by mainstream philosophy of mathematics overlooks that, as Grosholz says, "certain important areas of mathematical activity do not, and may never, lend themselves" to be formulated as axiomatic theories, but "exist rather as a collection of solved and unsolved" mathematical "problems" (Grosholz 2000, 82). And yet, "they can come to stand in significant relations to other areas of mathematical activity that contribute to the growth of mathematical knowledge" (ibid.). Such relations "may themselves be powerful vehicles, not for providing mathematics with foundations" but "for solving problems and constructing novel objects. The vision of a unified mathematics" through a reduction to a single mathematical theory "then seems not only actually unattainable but even misleading as a regulative ideal" (ibid.).

3.23 Heuristic vs. Mainstream Philosophy of Mathematics

Heuristic philosophy of mathematics is not subject to the limitations of mainstream philosophy of mathematics pointed out in Chap. 2.

(1) According to mainstream philosophy of mathematics, the philosophy of mathematics cannot concern itself with the making of mathematics, in particular discovery. But this conflicts with the fact that the philosophy of mathematics has concerned itself with the making of mathematics from its very beginning. For example, the Pythagoreans analyzed the concept of number and concluded that "number is a bounded multitude or composition of monads" (Nichomachus, *Introductio arithmetica*, I, 7, 1). On the basis of that analysis, they demonstrated several properties of numbers. On the contrary, according to heuristic philosophy of mathematics, the philosophy of mathematics can concern itself with the making of mathematics.

(2) According to mainstream philosophy of mathematics, the philosophy of mathematics can concern itself only with finished mathematics, because only finished mathematics is objective, so it can be completely justified. But this conflicts with Gödel's second incompleteness theorem, by which finished mathematics cannot be completely justified. On the contrary, according to heuristic philosophy of mathematics, the philosophy of mathematics can concern itself with finished

mathematics, but the latter is never really finished as such, because every mathematical concept or hypothesis can always be called into question, modified, or reinterpreted. This is unaffected and even confirmed by Gödel's second incompleteness theorem.

(3) According to mainstream philosophy of mathematics, since the philosophy of mathematics cannot concern itself with the making of mathematics, it cannot contribute to the advancement of mathematics. But this conflicts with the fact that, as argued in the Introduction, philosophy has contributed to the advancement of mathematics. On the contrary, according to heuristic philosophy of mathematics, the philosophy of mathematics can possibly contribute to the advancement of mathematics.

(4) According to mainstream philosophy of mathematics, the task of the philosophy of mathematics is primarily to give an answer to the question: How do mathematical propositions come to be completely justified? But this question is empty because, by Gödel's second incompleteness theorem, mathematical propositions cannot be completely justified. On the contrary, according to heuristic philosophy of mathematics, the task of the philosophy of mathematics is primarily to give an answer to the question: How is mathematics made? The philosophy of mathematics can give an answer to this question because it can account for the mathematical process, in particular discovery.

(5) According to mainstream philosophy of mathematics, the method of mathematics is the axiomatic method. But this conflicts with Gödel's incompleteness theorems, by which, for any choice of axioms for a given part of mathematics, there will always be true propositions of that part which cannot be deduced from the axioms, and there is no guarantee that the axioms are justified. On the contrary, according to heuristic philosophy of mathematics, the method of mathematics is the analytic method. This is unaffected and even confirmed by Gödel's incompleteness theorems.

(6) According to mainstream philosophy of mathematics, the role of axiomatic demonstration, namely demonstration based on the axiomatic method, is to guarantee the truth of a proposition. But this conflicts with Gödel's second incompleteness theorem, by which the truth of a proposition cannot be guaranteed. On the contrary, according to heuristic philosophy of mathematics, the role of analytic demonstration, namely demonstration based on the analytic method, is to discover a solution to a problem. This is unaffected and even confirmed by Gödel's second incompleteness theorem.

(7) According to mainstream philosophy of mathematics, since the method of mathematics is the axiomatic method, mathematics is a body of truths, and indeed of truths that are certain. But this conflicts with Gödel's second incompleteness theorem, by which mathematics cannot be said to be a body of truths that are certain. On the contrary, according to heuristic philosophy of mathematics, since the method of mathematics is the analytic method, mathematics is a body of problems and solutions to them that are plausible. This is unaffected and even confirmed by Gödel's second incompleteness theorem.

(8) According to mainstream philosophy of mathematics, since the method of mathematics is the axiomatic method, mathematical reasoning consists of deductive reasoning. But this conflicts with the strong incompleteness theorem for second-order logic. On the contrary, according to heuristic philosophy of mathematics, since the method of mathematics is the analytic method, mathematical reasoning consists of both deductive reasoning and non-deductive reasoning. Deductive reasoning plays an important role also in the analytic method, but 'deductive reasoning' means 'informal deductive reasoning', not 'formal deductive reasoning'. This is unaffected and even confirmed by the strong incompleteness theorem for second-order logic.

(9) According to mainstream philosophy of mathematics, the philosophy of mathematics is a new independent subject introduced by Frege. But this conflicts with historical reality. On the contrary, according to heuristic philosophy of mathematics, the philosophy of mathematics goes back to the beginning of philosophy, many major philosophers have made substantial contributions to it, and their work remains important even today. This agrees with historical reality.

(10) According to mainstream philosophy of mathematics, the philosophy of mathematics can be developed independently of experience, because mathematics is an armchair subject that is the product of thought alone, so it does not depend on experience. But this conflicts with the fact that, by Gödel's second incompleteness theorem, the solutions to mathematical problems can only be plausible, so they depend on experience. On the contrary, according to heuristic philosophy of mathematics, the philosophy of mathematics cannot be developed independently of experience, because several mathematical problems have an extra-mathematical origin, and the solutions to mathematical problems are only plausible, so they are evaluated in terms of their compatibility with experience. This is unaffected and even confirmed by Gödel's second incompleteness theorem.

Therefore, it seems fair to say that heuristic philosophy of mathematics provides an adequate account of mathematics.

3.24 Other Features of Heuristic Philosophy of Mathematics

Heuristic philosophy of mathematics also differs from mainstream philosophy of mathematics as regards the three further questions considered in Chap. 2: mathematical genius, mathematical logic, and attitude towards philosophy.

(1) Mainstream philosophy of mathematics subscribes to the myth of mathematical genius because it assumes that the making of mathematics, in particular discovery, is based on intuition, and hence is incapable of rational account. On the contrary, heuristic philosophy of mathematics has no need of the myth of mathematical genius because, according to it, the making of mathematics, in particular discovery, is a rational processes that can be analyzed in terms of rules, and is the result of ordinary thought processes that produce an extraordinary outcome.

(2) Mainstream philosophy of mathematics considers mathematical logic a proper and adequate tool in the philosophy of mathematics. But this conflicts with the strong incompleteness theorem for second-order logic. On the contrary, heuristic philosophy of mathematics finds mathematical logic inadequate to account for mathematical reasoning.

(3) The attitude of mainstream philosophy of mathematics towards philosophy is either anti-philosophical, or consigns philosophy to irrelevance. On the contrary, heuristic philosophy of mathematics views the philosophy of mathematics as embedded in the philosophical tradition, and is ready to make use of all the means provided by the latter, together with new means.

References

Atiyah, Michael. 2004. Interview. In Martin Raussen, and Christian Skau, Interview with Michael Atiyah and Isadore Singer. *Newsletter of the European Mathematical Society* 53: 24–30.

Bedürftig, Thomas, and Roman Murawski. 2018. *Philosophy of mathematics*. Berlin: de Gruyter.

Byers, William. 2007. *How mathematicians think: Using ambiguity, contradiction, and paradox to create mathematics*. Princeton: Princeton University Press.

———. 2011. *The blind spot: Science and the crisis of uncertainty*. Princeton: Princeton University Press.

———. 2017. Can you say what mathematics is? In *Humanizing mathematics and its philosophy: Essays celebrating the 90th birthday of Reuben Hersh*, ed. Bharath Sriraman, 45–60. Cham: Springer.

Calvino, Italo. 1985. *Mr Palomar*. New York: Harcourt Brace Jovanovich.

Carter, Jessica. 2019. Philosophy of mathematical pratice – Motivations, themes and prospects. *Philosophia Mathematica* 27: 1–32.

Davies, Edward Brian. 2008. Interview. In *Philosophy of mathematics: 5 questions*, ed. Vincent Hendricks and Hannes Leitgeb, 87–99. New York: Automatic Press/VIP.

Davis, Philip J., and Reuben Hersh. 1981. *The mathematical experience*. Cham: Birkhäuser.

De Cruz, Helen. 2006. Toward a Darwinian approach to mathematics. *Foundations of Science* 11: 157–196.

Descartes, René. 1996. *Oeuvres*. Paris: Vrin.

Duhem, Pierre. 1954. *The aim and structure of physical theory*. Princeton: Princeton University Press.

Euler, Leonhard. 1758. Demonstratio nonnullarum insignium proprietatum, quibus solida hedris planis inclusa sunt praedita. *Novi Commentarii Academiae Scientiarum Petropolitanae* 4: 140–160.

Feferman, Solomon. 1998. *In the light of logic*. Oxford: Oxford University Press.

Franks, John. 1989. Review of James Gleick, *Chaos: Making a new science*. *The Mathematical Intelligencer* 11: 65–69.

Giusti, Enrico. 1999. *Ipotesi sulla natura degli oggetti matematici*. Torino: Bollati Boringhieri.

Goodman, Nicolas D. 1991. Modernizing the philosophy of mathematics. *Synthese* 88: 119–126.

Grabiner, Judith V. 2005. *The origins of Cauchy's rigorous calculus*. Mineola: Dover.

Grosholz, Emily. 2000. The partial unification of domains, hybrids, and the growth of mathematical knowledge. In *The growth of mathematical knowledge*, ed. Emily Grosholz and Herbert Breger, 81–91. Dordrecht: Springer.

Grothendieck, Alexander. 1985. *Récoltes et semailles: Réflexions et témoignage sur un passé de mathématicien*. Montpellier: Université des Sciences et Techniques du Languedoc.

Halmos, Paul Richard. 1968. Mathematics as a creative art. *American Scientist* 56: 375–389.

———. 1980. The heart of mathematics. *The American Mathematical Monthly* 87: 519–524.
Hamming, Richard Wesley. 1980. The unreasonable effectiveness of mathematics. *The American Mathematical Monthly* 87: 81–90.
———. 1998. Mathematics on a distant planet. *The American Mathematical Monthly* 105: 640–650.
Hardy, Godfrey Harold. 1992. *A mathematician's apology*. Cambridge: Cambridge University Press.
Henkin, Leon. 1996. The discovery of my completeness proofs. *The Bulletin of Symbolic Logic* 2: 127–158.
Hersh, Reuben. 1997. *What is mathematics, really?* Oxford: Oxford University Press.
———. 2014. *Experiencing mathematics: What do we do, when we do mathematics?* Providence: American Mathematical Society.
Hume, David. 2007. *A treatise of human nature*. Vol. 1. Oxford: Oxford University Press.
Karasmanis, Vassilis. 2018. The axiomatization of mathematics and Plato's conception of knowledge in the *Meno* and the *Republic*. In *Revolutions and continuity in Greek mathematics*, ed. Michalis Sialaros, 319–333. Berlin: de Gruyter.
Krantz, Steven G. 2011. *The proof is in the pudding: The changing nature of mathematical proof.* New York: Springer.
Lakatos, Imre. 1961. *Essays in the logic of mathematical discovery*. Cambridge: PhD Dissertation.
———. 1963–1964. Proofs and refutations. *The British Journal for the Philosophy of Science* 14: 1–25, 120–139, 221–245, 296–342.
———. 1971. Replies to critics. In *PSA 1970: In memory of Rudolf Carnap*, ed. Roger C. Buck and Robert S. Cohen, 174–182. Dordrecht: Reidel.
———. 1976. *Proofs and refutations: The logic of mathematical discovery*. Cambridge: Cambridge University Press.
———. 1978. *Philosophical papers*. Cambridge: Cambridge University Press.
Lucas, John Randolph. 1967. Plato and the axiomatic method. In *Problems in the philosophy of mathematics*, ed. Imre Lakatos, 11–14. Amsterdam: North-Holland.
Mac Lane, Saunders. 1994. Response to Arthur Jaffe and Frank Quinn, Theoretical mathematics: Toward a cultural synthesis of mathematics and theoretical physics. *Bulletin of the American Mathematical Society* 30: 190–193.
Mancosu, Paolo. 2008. Introduction. In *The philosophy of mathematical practice*, ed. Paolo Mancosu, 1–21. Oxford: Oxford University Press.
Musgrave, Alan. 1989. Deductive heuristics. In *Imre Lakatos and theories of scientific change*, ed. Kostas Gavroglu, Yorgos Goudaroulis, and Pantelis Nicolacopoulos, 15–32. Cham: Springer.
———. 2011. Popper and hypothetico-deductivism. In *Handbook of the history of logic*, ed. Dov Gabbay, Stephan Hartmann, and John Woods, vol. 1, 205–234. Amsterdam: North-Holland.
Newton, Isaac. 1971. MS Add. 3968, f. 101. In I. Bernard Cohen, *Introduction to Newton's 'Principia'*, 292–294. Cambridge: Cambridge University Press.
Nickles, Thomas. 1987. Methodology, heuristics, and rationality. In *Rational changes in science*, ed. Joseph P. Pitts and Marcello Pera, 103–132. Springer: Dordrecht.
———. 2000. Lakatos. In *A companion to the philosophy of science*, ed. William Herbert Newton-Smith, 207–212. Malden: Blackwell.
Peirce, Charles Sanders. 1931–1958. *Collected papers*. Cambridge: Harvard University Press.
Pólya, George. 1954. *Mathematics and plausible reasoning*. Princeton: Princeton University Press.
———. 1981. *Mathematical discovery: On understanding, learning, and teaching problem solving*. New York: Wiley.
———. 2004. *How to solve it: A new aspect of mathematical method*. Princeton: Princeton University Press.
Rota, Gian-Carlo. 1997. *Indiscrete thoughts*. Cham: Birkhäuser.
Russell, Bertrand. 1994. *Mysticism and logic*. London: Routledge.

Smoryński, Craig. 1983. Mathematics as a cultural system. *The Mathematical Intelligencer* 5 (1): 9–15.
Thom, René. 1971. 'Modern' mathematics: An educational and philosophic error? *American Scientist* 59: 695–699.
van Schooten, Frans. 1649. In geometriam Renati Des Cartes commentarii. In René Descartes, *Geometria*, 162–336. Leiden: Maire.
Weyl, Hermann. 1949. *Philosophy of mathematics and natural science*. Princeton: Princeton University Press.
Wittgenstein, Ludwig. 1978. *Remarks on the foundations of mathematics*. Oxford: Blackwell.
———. 2005. *The big typescript TS 213*. Oxford: Blackwell.
———. 2016. *Lectures, Cambridge 1930–1933: From the notes of G. E. Moore*. Cambridge: Cambridge University Press.

Part II
Discourse on Method

Chapter 4
The Question of Method

Abstract An essential aspect of the contrast between mainstream philosophy of mathematics and heuristic philosophy of mathematics concerns the method of mathematics. According to the former, the method of mathematics is the axiomatic method, according to the latter, the method of mathematics is the analytic method. So the question of method is a central element of distinction in the philosophy of mathematics. The question of method is important, not only for the philosophy of mathematics, but for mathematics itself. Despite this, attention to the question of method has declined in the last three centuries. In particular, in the last century, several arguments have been used to deny that there is a method of discovery, and even that there is a method at all. The chapter examines these arguments and finds them wanting, then it reasserts the importance of method.

Keywords Origin of method · Oblivion of method · Modern science and analytic-synthetic method · Attempts to improve the analytic-synthetic method · End of method · Arguments against method · Discovery and invention

4.1 The Centrality of Method

As we have seen in Chaps. 2 and 3, an essential aspect of the contrast between mainstream philosophy of mathematics and heuristic philosophy of mathematics concerns the method of mathematics. According to the former, the method of mathematics is the axiomatic method, according to the latter, the method of mathematics is the analytic method. So, the question of method is a central element of distinction in the philosophy of mathematics. The question of method is important not only for the philosophy of mathematics but for mathematics itself, because it essentially affects how mathematics is made.

Despite this, attention to the question of method has declined in the last three centuries. In particular, in the last century, several arguments have been used to deny that there is a method of discovery, or even that there is a method at all. The chapter examines these arguments and finds them wanting, then it reasserts the importance of method.

C. Cellucci, *The Making of Mathematics*, Synthese Library 448,
https://doi.org/10.1007/978-3-030-89731-4_4

4.2 The Origin of Method

The idea of method as a way of discovery and justification is implicit in the method used by Hippocrates of Chios to solve problems in mathematics, and by Hippocrates of Cos to solve problems in medicine. But the Greek term 'methodos' first occurs as a technical term in Plato.

Plato introduces the term 'methodos' in the context of a comparison between medicine and rhetoric. This may sound strange because, in the last centuries, rhetoric has been identified with "the art of using the weakness of people for one's own purposes," an art that "is not worthy of any respect at all" (Kant 2000, 205). But by 'rhetoric' Plato does not mean the bad art capable of producing "persuasion which will make one appear to know better than those who know, to those who do not know" (Plato, *Gorgias*, 459 c 1–2). Namely the bad art "that provides belief without knowledge" (ibid., 454 e 7). Instead, by 'rhetoric' Plato means the good art "that provides knowledge" (ibid., 454 e 7–8). Providing knowledge, rhetoric is virtually the same as philosophy, because "philosophy is acquisition of knowledge [ktesis epistemes]" (Plato, *Euthydemus*, 288 d 8).

Plato establishes a comparison between medicine and rhetoric by saying that "the manner of the medical art is somehow the same as that of rhetoric" because "in both cases one has to analyze a nature, in one the nature of the body, in the other the nature of the soul" (Plato, *Phaedrus*, 270 b 1–2, 4–5). Just as one must analyze the nature of the body if one wants "to give the body health and strength," so one must analyze the nature of the soul if one wants "to give the soul the desired conviction and virtue" (ibid., 270 b 7–9). And just as it is not "possible to know the nature of soul satisfactorily without knowing the nature in general," so, "if one has to believe Hippocrates" of Cos, "it is not possible to know the nature of the body either without this method [methodos]" (ibid., 270 c 1–5). For, in both cases a method is required. Not using a method "would be like walking with the blind. But someone who goes about his subject skillfully must not be like the blind" (ibid., 270 d 9–e 2). Only a method can indicate the road that "would bring us to the place where we may, as it were, rest from the road and come to the end of our journeying" (Plato, *Respublica*, VII 532 e 1–3).

This shows the idea of method at its very formation. For, it suggests that there is a way which is common to different arts, medicine and rhetoric, and hence is independent of the particular art. The method used by Hippocrates of Chios to solve problems in mathematics, and by Hippocrates of Cos to solve problems in medicine, acts as a powerful stimulus for Plato to formulate the general idea of method.

But there is more than that. The method acts as a powerful stimulus for Plato to formulate not only the general idea of method, but also the idea that there is a general method of discovery and justification, and that such method is the analytic method. This leads Plato to give the first formulation of the analytic method (see Chap. 5).

After Plato, 'methodos' as a technical term is taken up by Aristotle, together with the idea that there is a way that is common to different arts, and hence is independent of the particular art to which it is applied. Like Plato, Aristotle refers to medicine and rhetoric. Indeed, he says that "we shall possess the method completely when we shall be in the same condition as in the case of rhetoric, medicine, and other such abilities" (Aristotle, *Topica*, A 3, 101 b 5–6).

Aristotle also takes up Plato's idea that there is a general method of discovery and justification. For, he says that "the goal of this investigation is to find a method with which we shall be able to construct syllogisms concerning any problem proposed" (ibid., A 1, 100 a 18).

However, Aristotle rejects Plato's idea that the general method of discovery and justification is the analytic method. To this, he opposes the idea that the general method of discovery and justification is the analytic-synthetic method. Indeed, Aristotle gives the first formulation of the analytic-synthetic method, and, as a byproduct, the first formulation of the synthetic method or axiomatic method, which is the synthetic part of the analytic-synthetic method (see Chap. 6).

4.3 The Oblivion of Method

Method plays a central role in both Plato and Aristotle because, according to them, the analytic method and the analytic-synthetic method, respectively, are the method of all arts, including philosophy.

In contrast, from the late Hellenistic period to the early modern period, there is an oblivion of the term 'method', and of the idea itself of method.

In particular, 'methodus' – the Latinized form of the Greek 'methodos' – never becomes established in classical Latin, especially because of the influence of Cicero, who deliberately avoids it, using instead the phrase 'via et ratio', or 'ratio et via'. Thus, Cicero says that "in inquiry, every discourse which is to proceed by way and manner ['via et ratio'] must begin with" a declaration such as "'The matter at hand is as follows', so that the debaters may agree as to what is the subject of the debate. Plato set out this procedure in the *Phaedrus*" (Cicero, *De finibus bonorum et malorum*, II.3–4). On the other hand, Aristotle's *Topica* contain "the discipline invented by Aristotle for discovering arguments, so that we arrive at them by manner and way ['ratio et via'] without any error" (Cicero, *Topica*, I.2).

Even when, from the late Hellenistic period to the early modern period, the term 'methodus' is used, it is not intended as meaning a general method of discovery and justification, but only a method of teaching.

Thus, Ramus says that "method [methodus] is disposition by which, out of many homogeneous propositions, either known by themselves," namely axioms, "or known by the judgment of syllogism," namely deduced from axioms, "the proposition which is first in the absolute order of knowledge is placed first, the second is placed second, the third is placed third, and so on" (Ramus 1585, 455). By such "disposition, the whole matter can be more easily taught and perceived" (Ramus 1552, 259).

A change in attitude towards method, in the sense of a return to Plato's and Aristotle's idea that there is a general method of discovery and justification, takes place in the sixteenth century, within the Aristotelian School of Padua.

Thus, Zabarella says that method "is the way leading to the acquirement of science" (Zabarella 1578, 134). Contrary to Ramus's view, method is different from disposition or order, because "order, qua order, has no power of producing" new knowledge, "but only of ordering" knowledge already produced, while, "on the contrary, method has power to infer" new knowledge, and actually "produces it" (ibid., 150). Order is only "an instrumental habitus, by which we are able to so organize the parts of each discipline that the discipline in question is learned as best and easiest as possible" (ibid., 103). Instead, "method is an intellectual instrument producing knowledge of the unknown from the known" (ibid., 150).

Zabarella also says that the general method of discovery and justification is Aristotle's analytic-synthetic method, because "all scientific progress from the known to the unknown is either from cause to effect or from effect to cause; the former is the demonstrative method," or synthesis, "the latter is the resolutive method," or analysis, and, apart from these two processes, "there is no other process" (ibid., 178). The "demonstrative method is a syllogism" which deduces the effect "from propositions that are necessary, with no middle term," namely immediate, "better known, and the causes of the conclusion" (ibid., 179). Conversely, the "resolutive method is a syllogism" which "leads from things that are posterior and effects better known to the discovery of things that are prior and causes" (ibid.). The demonstrative method proceeds "through a priori demonstration," namely demonstration which starts from causes, and "at the same time states that a thing is so and so and why it is so and so" (Zabarella 1587, 130). Conversely, the resolutive method proceeds "through a posteriori demonstration," namely demonstration which starts from the effect, and "only declares that a thing is so and so" (ibid.).

Thus, the Aristotelian School of Padua of the sixteenth century takes up where the ancient tradition of method of Plato and Aristotle left off. It provides an intellectual climate congenial to scientific research, which bears its fruit with Galileo, who acknowledges it by saying that Padua is the place "where I spent the best eighteen years of all my life" (Galilei 1872, II, 210).

4.4 Aristotle's Object of Science

That the Aristotelian School of Padua provides an intellectual climate congenial to scientific research is not enough, however, to explain the rise of the Scientific Revolution of the seventeenth century. The decisive factor is Galileo's philosophical revolution: the introduction of a basic change of the object of science with respect to Aristotle. This sharply distinguishes Galileo from Aristotle and his Paduan followers.

According to Aristotle and his Paduan followers, the object of science is to know the essence of things.

Thus, Aristotle says that "we have knowledge of a thing when we know its essence" (Aristotle, *Metaphysica*, Z 6, 1031 b 6–7). Indeed, "to know a thing is to know its essence," because "any single thing and its pure essence coincide" (ibid., Z 6, 1031 b 19–21).

Zabarella says that we have knowledge of a thing when we "understand the essence of the thing" (Zabarella 1602, 1046).

This involves that, for science to be possible, human beings should have a faculty capable of grasping the essence of things. According to Aristotle and his Paduan followers, this faculty is intuition.

Indeed, Aristotle says that only "intuition" can know "what a thing is according to its essence" (Aristotle, *De Anima*, Γ 6, 430 b 28). For, "nothing apart from intuition can be truer than scientific knowledge" (Aristotle, *Analytica Posteriora*, B 19, 100 b 11–12).

Zabarella says that the "intuiting intellect" is "capable of apprehending not only the images" of sensible things, but also "the essences and quiddities of things, which are represented through the images" (Zabarella 1602, 972–973).

This, however, conflicts with the fact that human beings do not have any faculty capable of grasping the essence of things. In particular, this faculty cannot consist in intuition, because intuition is subjective, and hence is prone to mistakes.

For example, intuition leads Aristotle to believe that it is part of the essence of a body that heavier bodies fall faster than light ones. Indeed, Aristotle claims that, "if a given weight moves a given distance in a given time, a weight which is as great and more moves the same distance in a less time, the times being in inverse proportion to the weights" (Aristotle, *De Coelo*, A 6, 273 b 30–274 a 1). In particular, "if one weight is twice another, it will take half as long over a given movement" (ibid., A 6, 274 a 1–2).

But Aristotle's claim is invalid. As Galileo points out, "a cannon ball weighing one, two hundred pounds or even more, will not arrive at the ground by as much as a span ahead of a musket ball weighing only half a pound, even if they are dropped from a height of two hundred cubits" (Galilei 1968, VIII, 107). Thus, a simple experience shows that Aristotle's claim is invalid.

4.5 Galileo's Object of Science

Since human beings do not actually have any faculty capable of grasping the essence of things, the view that the object of science is to know the essence of things becomes a hindrance to the development of science.

Therefore, Galileo formulates the alternative: "Either, by speculating, we want to try to penetrate the true and intrinsic essence of natural substances; or we want to content ourselves with a knowledge of some of their phenomenal properties" (Galilei 1968, V, 187). Now, "trying for the essence" is "a not less impossible undertaking and a not less vain effort with regard to the closest elemental substances than with those most remote and celestial" (ibid.). Thus, we will content ourselves with coming to know certain phenomenal "properties of them, such as location, motion, shape, size, opacity, mutability, generation, and dissolution" (ibid., V, 188). Such phenomenal properties are mathematical in kind, including mutability, generation, and dissolution, because they are changes in size and shape. They are mathematical in kind in the sense that they can be dealt with by means of mathematics. While we are unable to know the essence of natural substances, with the help of mathematics "we need not despair of our ability" to come to know such phenomenal properties "even with respect to the remotest bodies, just as those close at hand" (ibid.).

Thus Galileo changes the object of science with respect to Aristotle. The object of science is no longer to know the essence of things, but only to know certain phenomenal properties of things, mathematical in kind.

This change is subscribed by Newton, who says that "the moderns – rejecting substantial forms and occult qualities – have undertaken to reduce the phenomena of nature to mathematical laws" (Newton 1999, 27).

Galileo's change of the object of science is the core of the Scientific Revolution of the seventeenth century, and the distinctive mark of modern science. Without it, modern science simply would not exist.

4.6 Aristotle's Science and Mathematics

Since Galileo assumes that the object of science is to know certain phenomenal properties of things, mathematical in kind, with Galileo's change of the object of science, the application of mathematics to the world becomes possible.

On the contrary, since Aristotle assumes that the object of science is to know the essence of things, with him the application of mathematics to the world is impossible. One cannot demonstrate natural things by mathematical means.

This follows from Aristotle's theory of genera. Aristotle says that "a genus is that which, in the essence, is predicated of many things which are different in species" (Aristotle, *Topica*, A 5, 102 a 31–32). Here, "by 'that which, in the essence, is predicated'," Aristotle means "that which it would be appropriate to answer when one is asked what the thing in question is, just as, when one is asked what a man is, it is appropriate to answer that it is an animal" (ibid., A 5, 102 a 32–35).

Each science is concerned with a single genus, "because a science is one if it is of one genus" (Aristotle, *Analytica Posteriora*, A 28, 87 a 38). Different genera are treated by different sciences, "because the underlying genus is different" (ibid., A 9, 76 a 12). Now, the principles of a science "must be in the same genus as the things demonstrated" (ibid., A 28, 87 b 2–3). Since different genera are treated by different sciences, from this it follows that "a proposition of one science cannot be demonstrated by another science" (ibid., A 7, 75 b 14). For example, it is not possible to demonstrate "a geometrical proposition by arithmetic" (ibid., A 7, 75 a 38–39). For, in the case of "arithmetic and geometry," the "genus is different" (ibid., A 8, 75 b 3–4).

Natural science and mathematics are concerned with different genera, because "natural science treats of certain movable things," while mathematics treats of "things which are immovable" (Aristotle, *Metaphysica*, E 1, 1026 a 12, 14–15). Indeed, if you add motion to an object, you get a physical object, so "physical objects" are got "by addition" (Aristotle, *De Coelo*, Γ 1, 299 a 16–17). On the other hand, if you subtract motion from it, you get a mathematical object, so "mathematical objects" are "got by subtraction" (ibid., Γ 1, 299 a 15–16). Since natural science and mathematics are concerned with different genera, "the method of mathematics is not that of natural science" (ibid., α 3, 995 a 16–17). Therefore, one cannot demonstrate natural things by mathematical means.

The only exception to this are natural things dealt with by a science which is subordinate to mathematics. Such is the case of optics, which is subordinate to geometry, and harmonics, which is subordinate to arithmetic. But neither optics nor harmonics "considers its objects qua sight or qua voice, but qua straight lines and numbers; the latter, in fact, are properties of the former" (ibid., M 3, 1078 a 15–16). Namely, optics and harmonics consider their objects as a part of geometry or arithmetic. Therefore, they have nothing to say about their objects qua sight or qua voice.

Aristotle's view that one cannot demonstrate natural things by mathematical means is reaffirmed by the Aristotelian opponents of Galileo.

Thus, the Aristotelian philosopher di Grazia says that those, like Galileo, who "want to demonstrate natural things by mathematical means" are "far from truth" (di Grazia 1968, 385). For, "all sciences and all arts have their proper principles and proper causes, by means of which they demonstrate the proper accidents of their proper object; so, it is not permissible to use the principles of one science to demonstrate the effects of another science" (ibid.). Then, "anyone who thinks he can demonstrate natural accidents by mathematical means is greatly demented, because these two sciences are most different from each other. Indeed, the natural scientist studies natural bodies that have motion as their natural and proper state, while the subject matter of the mathematician abstracts from all motion" (ibid.).

4.7 Galileo's Science and Mathematics

In particular, the Aristotelian opponents of Galileo hold that one cannot demonstrate natural things by mathematical means, because the propositions of mathematics are true in abstract but false in concrete.

Indeed, they object to Galileo that "mathematicians may very well demonstrate by means of their principles, for example, that 'sphaera tangit planum in puncto' [a sphere touches a plane at a single point]," but "when one comes to matter, things succeed quite another way" (Galilei 1968, VII, 229). For, whenever a bronze sphere is "placed upon a plane, its weight" may "so depress it that the plane" may "yield somewhat, or the sphere itself" may "be mashed at the contact" (ibid., VII, 233). So, the sphere will not touch the plane at a single point, but with part of its surface. Thus, "because of the imperfection of matter, that body which ought to be perfectly spherical, and that plane which ought to be perfectly flat, do not turn out to be such in the concrete, as one imagines them to be in the abstract" (ibid.). Generally, "the imperfection of matter makes things, taken in the concrete, to disagree with those taken in the abstract" (ibid.). Therefore, mathematical propositions "are true in the abstract, but applied to sensible and physical matter, they do not work out" (ibid., VII, 229).

Galileo answers this objection by observing that, "to show that a material sphere does not touch a material plane at a single point," the Aristotelians "make use of a sphere that is not a sphere, and of a plane that is not a plane" (ibid., VII, 233). Indeed, they do not apply a sphere to a plane, but "an imperfect sphere to an imperfect plane," and "say that these do not touch at a single point" (ibid.). Now, "even in the abstract, an immaterial sphere that is not a perfect sphere may touch an immaterial plane that is not a perfect plane, not at a single point, but with part of its surface," so, "that which happens in the concrete, does in like manner happen in the abstract" (ibid.). Therefore, the objection is "very much out of place" (ibid.).

According to Galileo, it "would be novel indeed if computations and ratios made in abstract numbers should not thereafter correspond to concrete gold and silver coins and merchandise" (ibid., VII, 233–234). When a tradesman in the marketplace wants to calculate the weight of the goods sold, he must subtract the weight of "containers, straps, and other packing items" (ibid., VII, 234). In the same way, "when the geometrical philosopher wants to find in the concrete the effects demonstrated in the abstract, he must deduct the impediments of matter" (ibid.). Namely, he must apply correction factors. And, "if he knows how to do that, rest assured things will agree no less exactly than arithmetical computations" (ibid.). So, "errors lay neither in the abstract nor in the concrete, neither in geometry nor in physics, but in the calculator, who does not know how to do the accounting correctly" (ibid.). Indeed, "if one had a perfect sphere and a perfect plane, though they were material, there is no doubt that they would touch at a single point" (ibid.). And, if such a sphere or plane "was and is impossible to be procured, it was very much out of place to say that 'sphaera aenea non tangit planum in puncto' [a bronze sphere does not touch a plane at a single point]" (ibid.).

Thus Galileo answers the objection of the Aristotelian opponents by observing that, to understand the concrete, one must use the abstract, deducting the impediments of matter, namely applying the necessary correction factors. Only so one can reach the true nature of the laws governing the concrete.

4.8 A Misinterpretation of Galileo's Book of the Universe

It has been said above that, with Galileo's change of the object of science, the application of mathematics to the world becomes possible.

Actually, with Galileo's change of the object of science, the application of mathematics to the world becomes not only possible but also necessary, because treating phenomenal properties of things, mathematical in kind, requires mathematics. For this reason, Galileo says that "trying to deal with natural questions without geometry is attempting to do what is impossible to be done" (Galilei 1968, VII, 229).

However, that, with Galileo's change of the object of science, the application of mathematics to the world becomes not only possible but also necessary, does not mean, as Husserl claims, that, according to Galileo, the world "is, in its true being-in-itself, mathematical" (Husserl 1970, 54).

To be sure, Galileo says that natural "philosophy is written in this very great book that is continually open to us before our eyes (I mean the universe)" (Galilei 1968, VI, 232). It "is written in mathematical language, and the characters are triangles, circles, and other geometrical figures, without the use of which it is impossible to humanly understand any word of it" (ibid.).

But, by these expressions, Galileo does not mean to say that the world is, in its true being-in-itself, mathematical. For, this would contradict Galileo's change of the object of science. Indeed, if the world were, in its true being-in-itself, mathematical, then, by coming to know properties of natural substances mathematical in kind, we would come to know the essence of natural substances. But this is incompatible with Galileo's fundamental assertion, that trying for the essence is a not less impossible undertaking and a not less vain effort with regard to the closest elemental substances than with those most remote and celestial.

Rather, Galileo means to be polemical with contemporary Aristotelians, who claim that natural philosophy is written in Aristotle's books. According to them, one can "draw from his books demonstrations of all that can be known, because there is every thing in them" (ibid., VII, 134). Therefore, "to philosophize is, and cannot be other than, to make great practice on Aristotle's texts" (ibid., V, 190).

Against this claim, Galileo says that natural philosophy is written in the book of the universe, it is written in mathematical language, and anyone who knows that language can read it, not only Aristotle, whose eyes, then, do not have "to see for all his posterity" (ibid.). But, by this, he only means to say that the phenomenal properties of natural substances, mathematical in kind, which are the object of science, can be treated by means of mathematics, and anyone who knows mathematics can have knowledge of them.

4.9 Galileo and Aristotle's Analytic-Synthetic Method

That, with Galileo, the object of science changes with respect to Aristotle, does not mean that, with him, the method of science also changes.

According to a widespread view, the creators of the Scientific Revolution abandoned Aristotle's analytic-synthetic method and regarded themselves as the founders of the true scientific method, in the modern sense.

Thus, Kline says that "modern science owes its origin and present flourishing state to a new scientific method which was fashioned almost entirely by Galileo" (Kline 1985, 284).

Cohen says that Galileo's approach to science "earned him a place as a founder of the scientific method of inquiry," which permitted "predictions capable of test, without concern over causal considerations in the old Aristotelian sense of recourse to metaphysical principles" (Cohen 1985, 142).

Martins says that "the Scientific Revolution, that had its culmination in the period from Galileo to Newton, was both a revolution of concepts and of methods," and "the creators of the Scientific Revolution regarded themselves as the founders of the 'true' scientific method (as against the Aristotelian method)" (Martins 1993, 210).

But this view is unfounded. For, the creators of the Scientific Revolution did not abandon Aristotle's analytic-synthetic method, nor regarded themselves as the founders of the true scientific method. On the contrary, they declared to use, and actually used, Aristotle's analytic-synthetic method as the true scientific method.

This is clear from Galileo. Indeed, Galileo criticizes the Aristotelians for saying that "Aristotle first laid his main foundation on a priori discourse, demonstrating the necessity" of the conclusion "by means of natural, manifest, and clear principles," and only "afterwards established the same conclusion a posteriori, by means of the senses" (Galilei 1968, VII, 75). This, Galileo says, "is the method by which" Aristotle "has written his doctrine," namely the method by which he gave a presentation of it, not "the method by which he investigated it" (ibid.).

In fact, Aristotle "first procured, by way of senses, experiments and observations," so by analysis, "to assure himself as much as possible of the conclusion," and only "afterwards he sought means to demonstrate it," so by synthesis, "because this is what is done for the most part in the demonstrative sciences" (ibid.). This "happens because, when the conclusion is true, by making use of the resolutive method," namely analysis, "one may easily encounter some proposition which is already demonstrated, or arrive at some principle which is known by itself" (ibid.). From the certainty of these propositions, "the truth of our" conclusion, which is derived from them by synthesis, "will acquire strength and certainty" (ibid., VII, 435). Conversely, "if the conclusion is false, one can go on endlessly without ever hitting upon any known truth, if indeed one does not encounter some impossibility or manifest absurdity" (ibid., VII, 75).

This is the method that "Aristotle teaches us in his Dialectics" (ibid., XVIII, 248). And is also the method by which Galileo proceeds. Indeed, Galileo professes to "observe more religiously the Peripatetic, or I should rather say, Aristotelian teaching than do many" Aristotelians, and hence to be "a better Peripatetic" and to "more dextrously use that doctrine than" many of them, who make "use of it clumsily" (ibid., XVIII, 234). For, "to be truly Peripatetic, namely an Aristotelian philosopher, consists principally in philosophizing conformably with the Aristotelian teachings, proceeding by those methods and with those true assumptions and principles on which scientific reasoning is founded" (ibid., XVIII, 248). This is how Galileo proceeds. So, "if Aristotle should return to earth, he would accept" Galileo "among his followers" much more than he would accept "a great many others who, in order to sustain his every saying as true, go pulling out from his texts concepts that would have never crossed his mind" (ibid., XVIII, 249).

4.10 Newton and Aristotle's Analytic-Synthetic Method

That the creators of the Scientific Revolution declared to use, and actually used, Aristotle's analytic-synthetic method as the true scientific method, is also clear from Newton.

Indeed, Newton says that, "as in mathematicks, so in natural philosophy, the inquiry of difficult things by the method of analysis, ought ever to precede the method" of synthesis, or "composition. This analysis consists in making experiments and observations, and in drawing general conclusions from them by induction" (Newton 1952, 404). Then, "if no exception occur from phaenomena, the conclusion may be pronounced generally," while, if "any exception shall occur from experiments," the conclusion will "then begin to be pronounced with such exceptions as occur" (ibid.). For, in natural philosophy "propositions gathered from phenomena by induction should be considered either exactly or very nearly true" until "yet other phenomena make such propositions either more exact or liable to exceptions" (Newton 1999, 442).

By this way of analysis we proceed "from effects to their causes, and from particular causes to more general ones, till the argument end in the most general. This is the method of analysis" (Newton 1952, 404). On the other hand, "synthesis consists in assuming the causes discovered, and established as principles, and by them explaining the phaenomena proceeding from them, and proving the explanations" (ibid., 404–405). The making of science is based on this way of analysis. On the other hand, finished science is based on synthesis, since it is only concerned with showing how the causes, discovered and established as principles, "may be assumed in the method of composition," namely in synthesis, "for explaining the phaenomena arising from them" (ibid., 405).

This is Aristotle's analytic-synthetic method, and is also the method by which Newton proceeds. Indeed, as we have seen in Chap. 3, Newton says that the propositions of his *Principia Mathematica* were invented by analysis, then he demonstrated by synthesis what he had invented by analysis. This is confirmed by Cotes, the editor of the second edition of Newton's *Principia*, who says that "those whose natural philosophy is based on experiment" proceed "by a twofold method, analytic and synthetic. From certain selected phenomena they deduce by analysis the forces of nature and the simpler laws of those forces, from which they then give the constitution of the rest of the phenomena by synthesis" (Cotes 1999, 32). This is the method "which our most celebrated author," Newton, "thought should be justly embraced," this "alone he judged worthy of being cultivated" (ibid.).

It might be objected that Newton says that "the main business of natural philosophy is to argue from phaenomena" and "to deduce causes from effects" (Newton 1952, 369). Thus, he seems to say that hypotheses are not obtained from phenomena by induction, but are deduced from them. But this objection is invalid. For, Newton specifies that, in natural philosophy, "propositions are deduced from the phenomena and are made general by induction," in particular "the laws of motion and the laws of gravity have been found by this method" (Newton 1999, 589). Then, what are deduced from the phenomena are not the hypotheses but singular propositions, namely the descriptions of observations, which are then made general by induction. Therefore, it is not deduction but induction that yields the hypotheses.

4.11 Attempts to Improve Aristotle's Analytic-Synthetic Method

Although, according to both Galileo and Newton, the true scientific method is Aristotle's analytic-synthetic method, there is a feature of Aristotle's analytic-synthetic method which conflicts with Galileo's change of the object of science.

According to Aristotle's analytic-synthetic method, in analysis the hypotheses are found either by induction or by syllogism from effects. The latter is Aristotle's procedure for finding the premisses of a syllogism given the conclusion. (For a detailed description of Aristotle's procedure, see Cellucci 2013, Section 7.4).

Indeed, Aristotle says that analysis "proceeds either by induction or by syllogism" (Aristotle, *Ethica Nicomachea*, Z 3, 1139 b 27–28). So, in analysis, one gets "the necessary premisses either by syllogism or by induction" (Aristotle, *Topica*, Θ 1, 155 b 36). By 'syllogism' Aristotle means syllogism from effects.

Similarly, Zabarella says that the resolutive method, namely analysis, "divides into two kinds," the one kind is syllogism "from effects," the "other is induction" (Zabarella 1578, 180).

Now, while induction causes no problem, the syllogism from effects conflicts with Galileo's change of the object of science. For, Aristotle says that, to apply syllogism from effects, one must "first set down the thing itself, its definition, and whatever properties are peculiar to it," namely the essential properties of the thing; then one must set down "whatever follows the thing, what is followed by it, and whatever cannot belong to the thing" (Aristotle, *Analytica Priora*, A 27, 43 b 2–5). Thus, to apply syllogism from effects, one must first set down the essence of the thing. This conflicts with Galileo's change of the object of science because, as we have seen, Galileo says that 'trying for the essence' is an impossible undertaking.

Because of this, in the seventeenth and eighteenth centuries, attempts were made to improve Aristotle's analytic-synthetic method, freeing it from the need to set down the essence of the thing. These attempts, however, failed. We will consider two assumptions that contributed to their failure.

4.12 The Assumption that a Method Must Be Algorithmic

An assumption that contributed to the failure of the attempts to improve Aristotle's analytic-synthetic method, is that a method must be algorithmic.

Thus, Leibniz says that a method must be "a sensible and mechanical direction for the mind, as it were, that even the most stupid will acknowledge" (Leibniz 1965, VII, 14). This is necessary because "our intellect, due to its weakness, must be directed by a certain mechanical thread" (Leibniz 1903, 351). An algorithmic method provides such a mechanical thread.

To comply with the assumption that a method must be algorithmic, Leibniz maintains, on the one hand, that the method of science is Aristotle's analytic-synthetic method, but, on the other hand, that the latter must be an algorithmic method.

Indeed, Leibniz says that method is twofold, one by analysis, the other by synthesis, where "analysis is when, whenever a conclusion is given or a problem is proposed, we seek the principles by which we may demonstrate the conclusion or solve the problem," while "synthesis is when, starting from the principles, we compound theorems and problems" (Leibniz 1965, I, 194–195). Analysis "is destined to advance the sciences, and to discover the things that we want," while synthesis "pertains to the instauration of sciences, and to judging the things already discovered" (Leibniz 1923–, VI.4, 354).

Now, Aristotle's analytic-synthetic method cannot be based on intuition, because for intuition there are no rules. Indeed, if the method is based on intuition, it becomes "nearly like the precept of I do not know which chemist: Take what you should and do what you should, and you will get what you want" (Leibniz 1965, IV, 329).

Instead of being based on intuition, Aristotle's analytic-synthetic method must be an algorithmic one. For, if Aristotle's analytic-synthetic method is an algorithmic method, then it will be a "calculus of reasoning" by means of which one will be able to "reason easily and infallibly" (Leibniz 1923–, VI.4, 274). The calculus of reasoning will permit to "discover and judge everything from the data" (Leibniz 1903, 219). By means of it, "everything that can be obtained from the data through reasoning, even by a very great and highly trained intelligence," will "eventually be found by anyone and with a sure method" (Leibniz 1965, VII, 202).

This, however, is faced with the difficulty that there cannot be any calculus of reasoning that will allow to discover and judge everything from the data. For, on the one hand, by the theorem of undecidability of provability, there is no calculus of reasoning that will allow to discover everything from the data. On the other hand, by the strong incompleteness theorem for second-order logic, there is no consistent formal system for second-order logic capable of deducing all second-order logical consequences of any given set of propositions. Thus, there is no calculus of reasoning that will allow to judge everything from the data.

This prevents Leibniz from actually improving Aristotle's analytic-synthetic method.

4.13 The Assumption that Discovery is the Work of Mathematical Genius

Another assumption that contributed to the failure of the attempts to improve Aristotle's analytic-synthetic method is that discovery is the work of mathematical genius. This assumption is based on the myth of mathematical genius, put forward by Novalis.

Novalis says that mathematical creations, in particular discoveries, "are leaps – (intuitions, resolutions)" which are a prerogative "of the genius – of the leaper par excellence" (Novalis 2007, 28). The mathematician, when creating, "is en état de Createur absolu" (ibid., 140). And "whoever is able to bring forth numerous" creations "is called a genius" (ibid., 194). Therefore, mathematics is the work of mathematical genius.

To comply with the assumption that mathematics is the work of mathematical genius, Novalis maintains, on the one hand, that the method of mathematics is Aristotle's analytic-synthetic method, but, on the other hand, that the latter is ultimately based on genius.

Indeed, Novalis says that the method of mathematics is "analysis and synthesis" (ibid., 195). Namely, Aristotle's analytic-synthetic method. For, "solution and proof" are based, respectively, on "the analytic and synthetic ability" (ibid., 174).

But the essential part of Aristotle's analytic-synthetic method is the synthetic method. Admittedly, "the synthetic method" is "the freezing – wilting, crystallizing, structuring and successive method. The analytic method, in contrast, is a warming, dissolving and liquefying method. The former seeks the whole, the latter the parts" (ibid., 175). But the synthetic method plays the crucial role in solution and proof, because mathematics "originates from a flash of insight" (ibid., 111). And "a flash of insight is a synthetic thought" (ibid., 84). Therefore, in mathematics "a true method of progressing synthetically is the main thing" (ibid., 100).

Now, the synthetic method is the "method of the divinatory genius" (ibid., 100). For, "genius is the synthesizing principle" (ibid., 215). Indeed, "a truly synthetic person" is "a genius" (ibid., 10).

Since the essential part of Aristotle's analytic-synthetic method is the synthetic method, and the synthetic method is the method of the divinatory genius, Aristotle's analytic-synthetic method is ultimately based on genius.

This, however, is faced with the difficulty that for genius there are no rules, since the genius proceeds by leaps of intuition, and for intuition there are no rules.

This prevents Novalis from actually improving Aristotle's analytic-synthetic method.

4.14 The Decline of Method

As a result of the failure of the attempts to improve Aristotle's analytic-synthetic method, in the eighteenth century method declines. The decline progresses further in the nineteenth century, which endorses the Romantic assumption that discovery is the work of the mathematical genius.

Thus, Whewell says that "an art of discovery is not possible. At each step" of the investigation "are needed invention, sagacity, genius – elements which no art can give. We may hope in vain, as Bacon hoped, for an Organon which shall enable all men to construct scientific truths" (Whewell 1847, I, viii). Scientific discovery "must ever depend upon some happy thought, of which we cannot trace the origin – some fortunate cast of intellect, rising above all rules. No maxims can be given which inevitably lead to discovery. No precepts will elevate a man of ordinary endowments to the level a man of genius" (ibid., II, 20–21). The task "of the philosophy of science" must be "rather classification and analysis" of what has been done "than precept and method" (ibid., I, viii).

Science "begins with common observation of facts" (ibid., II, 36). But "the first and great instrument by which facts, so observed with a view to the formation of exact knowledge, are combined into important and permanent truths, is that peculiar sagacity which belongs to the genius of a discoverer," that "cannot be limited by rules" (ibid., II, 40).

It would be "impossible to describe in words the habits of thought which led Archimedes to refer the conditions of equilibrium on the lever to the conception of pressure" (ibid., II, 40). Or those "which impelled Pascal to explain, by means of the conception of the weight of air, the facts which his predecessors had connected by the notion of nature's horror of a vacuum" (ibid.). These are "felicitous and inexplicable strokes of inventive talent," and "no rules" can "enable men who do not possess similar endowments, to make like advances in knowledge" (ibid., II, 41). They must be "described as happy guesses" (ibid.).

4.15 The End of Method

After declining in the eighteenth and nineteenth century, method comes to an end in the twentieth century.

As Nickles says, "in the twentieth century, many logical positivists and Popperians not only upheld the divorce of discovery from justification but also expelled the topic of discovery from epistemology" (Nickles 2000, 87).

In fact, they also expelled other topics concerning method from epistemology. As a result, today it is widely believed that there is no method of discovery, or that there is no method of justification, or that there is no method at all, or that method has become obsolete.

Thus, in the preface to one of the very rare histories of method written in the last few decades, Gower says: "Those of my friends and colleagues who knew that I was writing a book about scientific method often expressed their surprise. Why, they said, should anyone wish to revive such a long-expired steed? People do not now believe in scientific method" (Gower 1997, vi).

The end of method is a particularly serious problem in the case of mathematics, because mathematics is problem solving, and problem solving is empty without a method, so the method of mathematics is the core of mathematics.

Several arguments have been put forward for the view that there is no method of discovery, or there is no method of justification, or there is no method at all, or that method has become obsolete. In what follows these arguments will be discussed and will be found wanting.

4.16 The Argument of Subjectivity

An argument that has been put forward for the view that there is no method of discovery is the argument of subjectivity, according to which the question of discovery is subjective and psychological. Only the question of justification is objective, because there is a method of justification, which is given by the axiomatic method.

Thus, Frege says that "we can inquire, on the one hand, how we have gradually arrived at a given proposition and, on the other, how we can finally provide it with the most secure foundation" (Frege 1967, 5). The first question, namely the question of discovery, "may have to be answered differently for different persons" because it is subjective and psychological, so the question of discovery may concern only its "psychological genesis" (ibid.). To study the psychological genesis of discovery "is certainly a feasible undertaking, but it is not a logical one" (Frege 1979, 146). So, there cannot be a method of discovery.

On the contrary, the second question, namely the question of justification, is objective, because "the answer to it is connected with the inner nature of the proposition considered" (Frege 1967, 5). For, it concerns the ultimate ground upon which the justification of the proposition rests, not "the conditions, psychological, physiological and physical, which have made it possible to form the content of the proposition in our consciousness" (Frege 1960, 3). This requires determining "upon what primitive laws" the proposition "is based" (Frege 1984, 235). The primitive laws must be true, because a justification is given only by "those grounds of judgment which are truths" (Frege 1979, 3). Once ascertained that the primitive laws are true, we must deduce the proposition from them. This will provide the required justification for holding the proposition to be true. For, to deduce is "to make a judgment because we are cognisant of other truths as providing a justification for it" (ibid.). There are "laws governing this kind of justification," namely the laws of deduction, and logic studies these laws, since they are the "laws of valid inference" (ibid.). Then, there is a method of justification, given by the axiomatic method. In mathematics we start from axioms, namely "propositions appealed to without proof" which are "explicitly declared as such, so that it can be clearly recognised on what the whole structure rests," and we proceed from them to the theorem by rules of deduction "listed in advance" (Frege 2013, I, VI).

But the claim that the question of discovery is subjective and psychological, conflicts with the fact that the methods of discovery that have been used from antiquity, namely the analytic method and Aristotle's analytic-synthetic method, have nothing subjective and psychological about them. The non-deductive rules by which hypotheses are obtained are as objective as the deductive rules of Frege's method of justification.

On the other hand, the claim that the question of justification is objective since there is a method of justification, given by the axiomatic method, conflicts with the fact that the axiomatic method is subject to Gödel incompleteness theorems. By Gödel's first incompleteness theorem, for any consistent, sufficiently strong, formal system, there are propositions of the system that are true but cannot be deduced from the axioms of the system. So, mathematics cannot consist in the deduction of propositions from given axioms. And, by Gödel's second incompleteness theorem, for any consistent, sufficiently strong, formal system, it is impossible to prove, by absolutely reliable means, that the axioms of the system are consistent. So, the mathematical knowledge resulting from the deduction of propositions from given axioms cannot be said to be really justified.

4.17 The Argument of Non-Algorithmicity

Another argument that has been put forward for the view that there is no method of discovery is the argument of non-algorithmicity, according to which there are no algorithmic methods to devise new theories, there are only algorithmic methods to justify theories already devised.

Thus, Carnap says that there cannot be a "machine – a computer into which we can put all the relevant observational sentences and get, as an output, a neat system of laws that will explain the observed phenomena" (Carnap 1966, 33). Indeed, "one cannot simply follow a mechanical procedure based on fixed rules to devise a new system of theoretical concepts, and with its help a theory," to this purpose "creative ingenuity is required" (ibid.). There are only mechanical procedures to justify theories already devised. Indeed, "it is in many cases possible to determine, by mechanical procedures," the "degree of confirmation" of a hypothesis "*h* on the basis of" evidence "*e*" (ibid., 34). Therefore, "the aim of epistemology" can only be "the formulation of a method for the justification of cognitions" (Carnap 2003, 305).

But, that there are no algorithmic methods to devise new theories, does not mean that there are no methods to devise them. A method for solving problems need not be algorithmic, it can be heuristic. Between the dullness of algorithmic methods and the inscrutability of creative ingenuity, there is an intermediate region inhabited by heuristic methods, which are neither algorithmic nor require creative ingenuity. While an algorithmic method guarantees to always produce a correct solution to a problem, a heuristic method does not guarantee that. And yet it can greatly reduce the search space, namely, the domain within which the solution is sought, making a solution feasible where no algorithmic method would do.

That a method for solving problems need not be algorithmic and can be heuristic, has been recognized from antiquity. Indeed, the method used by Hippocrates of Chios to solve problems in mathematics, and by Hippocrates of Cos to solve problems in medicine, was a heuristic method.

Lakatos mockingly says that "primitive men worship algorithms," because they feel unsafe if they move "beyond the bounds of ritual," so "their concept of rationality, like that of Leibniz, of Wittgenstein and of modern formalists, is essentially algorithmic" (Lakatos 1978, II, 72). In particular, "primitive men prefer decision-procedures," since "with the help of a decision-procedure one can decide mechanically whether a conjecture is true or false" (ibid.). But "the Greeks did not find a decision-procedure for their geometry," they "did, however, find a compromise solution: a heuristic procedure" which "does not always yield the desired result, but which is still a heuristic rule, a standard pattern of the logic of discovery" (ibid.).

Also, the claim that, while there are no algorithmic methods to devise new theories, there are algorithmic methods to justify theories already devised, is invalid. For, as Putnam observes, if "there is no logic of discovery," then "there is no logic of testing, either" (Putnam 1975–1983, I, 268). The belief that "correct ideas just come from the sky, while the methods for testing them are highly rigid and predetermined, is one of the worst legacies of the Vienna Circle" (ibid.). Indeed, "all the formal algorithms proposed for testing, by Carnap, by Popper, by Chomsky, etc., are, to speak impolitely, ridiculous; if you don't believe this, program a computer to employ one of these algorithms and see how well it does at testing theories!" (ibid.). By the theorem of undecidability of validity, there is not even an algorithmic method for testing whether a formula is logically valid.

4.18 The Argument of Creative Intuition

Another argument that has been put forward for the view that there is no method of discovery is the argument of creative intuition, according to which there is no method of discovering a scientific theory, discovery is the result of creative intuition, by which the basic laws of the theory are obtained. Then, from the basic laws, conclusions are deduced and compared with experience, so the scientific method is the method of deductive testing.

Thus, Popper says that "there is no method of discovering a scientific theory" (Popper 2000, 6). Indeed, "there is no such thing as a logical method of having new ideas, or a logical reconstruction of this process," every "discovery contains 'an irrational element', or 'a creative intuition', in Bergson's sense" (Popper 2005, 8). And creative intuition "is a gift of the gods" (Popper 1974,48). From a new idea obtained by creative intuition, "conclusions are drawn by means of logical deduction. These conclusions are then compared with one another" to test "the internal consistency of the system," and compared with experience, to test the system "by way of empirical applications of the conclusions" (Popper 2005, 9). This is the scientific method, which is then the method of "deductive testing" (ibid.).

But the claim that discovery is the result of a creative intuition, conflicts with the fact that an unrestrained intuition might generate so many candidate basic laws that it would be impossible to test all of them. There must be some criterion for picking up some basic law among all the candidate basic laws, and for judging it worth testing. This criterion would act as a rule of discovery, so, at least in this sense, there must be a method of discovery.

On the other hand, the claim that the scientific method is the method of deductive testing conflicts with the strong incompleteness theorem for second-order logic, by which there is no consistent formal system for second-order logic capable of deducing all second-order logical consequences of any given set of propositions.

4.19 The Argument of Luck

Another argument that has been put forward for the view that there is no method of discovery is the argument of luck, according to which discovery, in particular discovery of proof, is just a matter of luck.

Thus, Quine says that a proof is "discoverable in general only by luck" (Quine 1951, 291). The "mathematician hits upon his proof by unregimented insight and good fortune" (ibid., 87). Actually, "a proof once discovered can be mechanically checked, but the actual discovery of the proof is a hit and miss matter" (ibid., 6).

But the claim that discovery, in particular the discovery of a proof, is just a matter of luck, is unfounded. For, Quine assumes that a proof is a formal deduction from axioms, as it is clear from the fact that he says that a proof once discovered can be mechanically checked. Then, there is no justification for the claim that a proof is discoverable in general only by luck. For, there is an algorithm for enumerating all formal deductions from given axioms, so if a proof is a formal deduction from axioms, then it is discoverable by a computer.

4.20 The Argument of Serendipity

Another argument that has been put forward for the view that there is no method of discovery is the argument of serendipity, according to which all discoveries are serendipitous, namely such that one fails to discover what one is looking for, but by chance discovers something else unintended and unexpected. If all discoveries are serendipitous, this means that there is no method of discovery.

Thus, Kantorovich says that "science can make significant progress only by serendipity" (Kantorovich 1993, 157). Only "serendipity enables science to break through the boundaries of the prevailing paradigm" (ibid., 203). Therefore, "serendipity is essential for generating ideas which deviate from tradition" (ibid., 216).

Yewdell says that all "discoveries are serendipitous" (Yewdell 2008, 494). They do not require "having a good idea (or even a bad idea, sometimes any idea will do, as they can all lead to serendipity)" (ibid., 493).

An often quoted example of serendipitous discovery is the theory of "chaos as serendipitously discovered by Poincaré" (Krishnaswami and Senapati 2019, 87). The king of Sweden had established a prize for a solution to the problem of predicting the motion of n celestial bodies interacting with each other gravitationally. Newton had solved the case $n = 2$, Poincaré presented a solution for the case $n = 3$ and won the prize. To his dismay, however, he soon realized that there was a major mistake in his argument, his alleged solution did not indicate a stable orbit. In a feverish attempt to correct his work, Poincaré discovered that even the smallest variation in the initial conditions – mass, location, speed, and direction of motion of the bodies – would lead to vastly different orbits. Therefore, one cannot predict the motion of the three celestial bodies over time, because it is impossible to know the initial conditions with an infinite degree of accuracy. So, while failing to discover what he was looking for, namely a solution to the problem of predicting the motion of the n celestial bodies for the case $n = 3$, Poincaré discovered something else unintended and unexpected, namely the foundations of chaos theory.

But the claim that all discoveries are serendipitous conflicts with the fact that, in reality, the so-called serendipitous discoveries are not made by chance but require a prepared mind.

This is supported by the responses of readers of *L'Enseignement Mathématique* to a questionnaire that asked "what is the part of chance" in "mathematical discoveries," and whether the "main discoveries" are "the result of deliberate work, directed in a specific direction," or come to "mind spontaneously" (Fehr 1908: 31). According to the responses of readers, "the part of chance" in mathematical research is "infinitely small" (ibid., 45). And "mathematical discoveries" are "never born of spontaneous generation. They always presuppose a soil seeded with prior knowledge, and well prepared by labor, both conscious and unconscious" (ibid., 47).

In fact, Poincaré did not discover that even the smallest variation in the initial conditions would lead to vastly different orbits by chance, but by analyzing the mistake of his argument, which required a prepared mind.

Poincaré's discovery can be accounted for by the analytic method. For, in the analytic method if, in the course of solving a problem, a hypothesis turns out to be implausible, one analyses the reasons why the hypothesis is implausible and arrives at some alternative hypothesis (see Chap. 5). This is exactly what Poincaré did.

Generally, the phenomenon of serendipity can be accounted for in terms of the analytic method. For, one of the distinguishing features of the analytic method is that the hypotheses intended to solve a problem may turn out to be unable to solve that problem, but able to solve some other problem (see Chap. 5).

Since serendipity can be accounted for in terms of the analytic method, which is a method of discovery, the claim that all discoveries are serendipitous, hence by chance, is unfounded.

4.21 The Argument of the Criterion of Truth

An argument that has been put forward for the view that there is no method of justification is the argument of the criterion of truth, according to which there is no criterion of truth, namely no criterion by which we can recognize whether a theory is true.

Thus, Popper says that "science is the search for truth," and "truth is therefore the aim of science" (Popper 1996, 39). But "we have no criterion of truth," namely no criterion "by which we can recognize truth" (Popper 2002, 306).

Now, Popper is right in saying that we have no criterion of truth. This can be shown by modifying an argument used by Frege to show that truth is undefinable. To have a criterion of truth, "certain characteristics" of truth "would have to be specified. And, in application to any particular case, the question would always arise whether it were true that the characteristics were present" (Frege 1984, 353). But, to answer this question, we would need a criterion of truth. Then "we should be going round in a circle" (ibid.). Therefore, we have no criterion by which we can recognize whether a theory is true.

But, that we have no criterion of truth, does not mean that there is no method of justification, it only means that the assumption that the aim of science is truth is problematic. Indeed, rather than truth, the aim of science is plausibility and, while we have no criterion of truth, we have a criterion of plausibility, namely the plausibility test procedure (see Chap. 5). By means of that criterion, we can recognize whether a theory is plausible, so there is a method of justification.

4.22 The Argument of Zero Probability

Another argument that has been put forward for the view that there is no method of justification is the argument of zero probability, according to which there is no way of justifying a theory in terms of probability, in the sense of the probability calculus, because every theory has zero probability.

Thus, Popper says that there is no way of justifying a theory in terms of probability, because a theory involves universal hypotheses, and every "universal hypothesis makes assertions about an infinite number of cases, while the number of observed cases can only be finite," so "every universal hypothesis h goes so far beyond any empirical evidence e that its probability $p(h,e)$ will always remain zero" (Popper 2000, 219).

But, that there is no way of justifying a theory in terms of probability, in the sense of the probability calculus, does not mean that there is no method of justification, it only means that the aim of science is not probability. As already said above, the aim of science is plausibility, and we have a criterion of plausibility by which we can recognize whether a theory is plausible, so there is a method of justification.

4.23 The Argument of No Scientific Method

An argument that has been put forward for the view that there is no method at all is the argument of no scientific method, according to which scientific method does not exist.

Thus, Popper says that "Plato, Aristotle, Bacon and Descartes, as well as most of their successors" until the nineteenth century, "believed that there existed a method of finding scientific truth" (Popper 2000, 6). Afterwards, there were people "who believed that there existed a method, if not of finding a true theory, then at least of ascertaining whether or not some given hypothesis was true," or "whether some given hypothesis was at least 'probable' to some ascertainable degree" (ibid.). But in fact "there is no method of discovering a scientific theory," and "there is no method of ascertaining the truth of a scientific hypothesis, that is, no method of verification" and "there is no method of ascertaining whether a hypothesis is 'probable', in the sense of the probability calculus" (ibid.). Therefore, "scientific method does not exist," and "I ought to know, having been, for a time at least, the one and only professor of this non-existent subject within the British-Commonwealth" (ibid., 5).

This argument, however, is invalid because it depends on the invalid assumption that the aim of science is truth or probability. As argued above, rather than truth or probability, the aim of science is plausibility; there is a method of discovering a scientific theory, namely the analytic method; and there is a method to ascertain the plausibility of a scientific hypothesis, namely the plausibility test procedure.

4.24 The Argument of Anything Goes

Another argument that has been put forward for the view that there is no method at all is the argument of anything goes, according to which progress in science is not due to method but to violations of method, and even to violations of reason. An example of this is how Galileo leads the Copernican theory to victory. Therefore, the only rule compatible with progress in science is: there are no rules, anything goes.

Thus, Feyerabend says that violations of method "are necessary for progress" (Feyerabend 1993, 14). What is more, "without a frequent dismissal of reason, no progress" (ibid., 158). For example, "Galileo violates important rules of scientific method which were invented by Aristotle," and makes progress "because he does not invariably follow these rules" (Feyerabend 1974, 297). Indeed, at the time of Galileo, the Copernican theory is inconsistent with facts, but Galileo leads it to victory by means of "psychological tricks," which are "arguments in appearance only" (Feyerabend 1993, 65). Therefore, "those who admire science and are also slaves of reason" must "make a choice. They can keep science; they can keep reason; they cannot keep both" (ibid., 214). The "only principle that does not inhibit progress is: anything goes" (ibid., 14).

However, the claim that progress in science is not due to method but to violations of method, and even to violations of reason, is by no means confirmed by how Galileo leads the Copernican theory to victory. Galileo gives arguments by which the Copernican theory, until then less plausible than the Ptolemaic theory, becomes more plausible than the latter, so Galileo leads the Copernican theory to victory by observing reason.

Therefore, it is unjustified to say that the only rule compatible with progress in science is: there are no rules, anything goes. A theory that is implausible at a certain stage may become plausible at a later stage, when the arguments for the theory become stronger than those against it. So, progress in science is possible because in science there are rules.

4.25 The Argument of Big Data

An argument that has been produced for the view that method has become obsolete is the argument of Big Data, according to which, since today we are confronted with petabytes of data, the traditional scientific method, based on hypotheses or theories, is totally inadequate. It must be replaced with the Big Data approach, according to which we can analyze the data without hypotheses or theories, using statistical algorithms, which can explore huge databases and find correlations and regularities therein. This will permit to make predictions without any need for hypotheses or theories.

Thus, Anderson says that, since today we are confronted with data "at the petabyte scale," the traditional scientific method, based on hypotheses or theories, is inadequate, science "calls for an entirely different approach" (Anderson 2008). We can throw the data "into the biggest computer clusters the world has ever seen and let statistical algorithms find patterns where" the traditional scientific method "cannot" (ibid.). While the traditional scientific method seeks causal explanations, with the new approach "correlation supersedes causation," and "science can advance" without "theories, or really any mechanistic explanation at all" (ibid.). Indeed, "petabytes allow us to say: 'Correlation is enough'," we "can analyze the data without hypotheses about what it might show" (ibid.).

But the claim that we can analyze the data without hypotheses or theories, conflicts with the fact that data do not speak for themselves, they acquire meaning only when interpreted, and interpreting them requires a hypothesis or theory, on the basis of which to observe them and extract information from them. There is always a 'viewpoint' preceding observation and experiment, namely a theory or hypothesis which guides observation and experiment, and generally data-finding. As Berman acknowledges, "for Big Data projects, holding a prior theory or model is almost always necessary; otherwise, the scientist is overwhelmed by the options" (Berman 2013, 147).

Also, the claim that statistical algorithms can explore huge databases and find correlations and regularities therein, permitting to make predictions without any need for hypotheses or theories, conflicts with the nature of statistical algorithms. For, statistical algorithms are found making use of some theory, so they are based on that theory. In addition, as Calude and Longo say, "the overwhelming majority of correlations are spurious. In other words, there will be regularities, but, by construction, most of the time (almost always, in the mathematical sense), these regularities cannot be used to reliably predict and act" (Calude and Longo 2017, 609). Therefore, "the more data, the more arbitrary, meaningless and useless (for future action) correlations will be found in them. Thus, paradoxically, the more information we have, the more difficult is to extract meaning from it" (ibid., 600). This shows that "big data analytics cannot replace science" (ibid., 610).

4.26 The Separation Between Discovery and Invention

A consequence of the decline and end of method has been the separation between discovery and invention.

According to the current view, discovery and invention are different, because "to dis-cover is to uncover, as we might reveal what lies on a table by removing the cloth covering it. Such an act of uncovering involves an existing entity being first hidden to us but then being made evident to us," so discovery "is not itself a human creation" (Miller 2013, 138). On the contrary, invention is the "bringing something new into existence," so invention is a human "creation" (ibid.).

This view has an illustrious pedigree. Indeed, Kant says that "inventing and discovering are different. The former is to first produce something that did not yet exist, whereas the latter is to first find something that already existed" (Kant 2012, 398).

This, however, conflicts with the fact that, if to dis-cover is to uncover an existing entity being first hidden to us, then, once a discovery is made, it should last forever. But the history of science and the history of mathematics show that no concept, theory or discovery lasts forever, it is always called into question, modified, or reinterpreted, and this could not be the case if discoveries were simply uncoverings.

4.27 The Separation Between Discovery and Invention in Mathematics

The view that discovery and invention are different also concerns mathematics.

One version of the view that discovery and invention are different in mathematics is that a concept is invented, a theorem is discovered.

Thus, Rav says that "concepts can be defined explicitly, as in the case of prime numbers, or implicitly, by a system of axioms, like the concept of group. In either case it is an inventive act" (Rav 1989, 67). On the contrary, theorems "have more the character of a discovery, in the sense that one discovers a road linking different localities" because, "once certain concepts have been introduced, and so to speak are already there, it is a matter of discovering their connection, and this is the function of proofs" (ibid.). For example, "the concept of a prime number is an invention," while "the theorem that there are infinitely many prime numbers is a discovery" (ibid., 66). For, through the theorem, one discovers a road map "linking 'set of primes', 'number of elements', etc., to yield a path to the conclusion" (ibid., 67). This is the function of the proof of the theorem.

But this conflicts with the fact that a concept and a theorem are part of the same process of solving a problem. The process involves introducing hypotheses for solving a problem, and concepts are among such hypotheses. The process yields a solution to the problem which consists of both concepts and a theorem. So it does not make sense to say that a concept is invented, a theorem is discovered.

Another version of the view that discovery and invention are different in mathematics is that a proof is invented, a theorem is discovered.

Thus, De Giorgi says that "a theorem is something discovered; its proof something invented" (Emmer 1997, 1101). The "discovery of a theorem can be made by different people, as if it were there waiting for someone to uncover it" in the sense of "pulling the cover off," but "the statement of the theorem is always the same" (ibid.). On the other hand, "proof is an invention – a construction of a road leading to the theorem," and "it happens sufficiently frequently that two mathematicians prove in independent way the same theorem as stated," where, however, unlike the statement of a theorem, "the invented proof can vary greatly according to the mathematician who finds it" (ibid.).

But this conflicts with the fact that a proof and a theorem are part of the same process of solving a problem. The process yields a solution to the problem which consists of both a proof and a theorem. So it does not make sense to say that a proof is invented, a theorem is discovered.

4.28 Discovery and Invention Before the Separation

While a consequence of the decline and end of method has been the separation between discovery and invention, in the centuries before the decline and end of method there is no such separation.

In ancient Greek and Latin, there are even no distinct terms for discovery and invention. The Greek term 'heuresis' and the Latin term 'inventio' mean both discovery and invention.

In the medieval period, there is still no separation between discovery and invention. Thus, in that period, Aristotle's discovery procedure for finding the premisses of a syllogism, given the conclusion, is called "inventio medii [invention of the middle term]" (Albertus Magnus 1890, 170).

In the early modern period, there is no separation between discovery and invention until the early eighteenth century.

Thus, Zabarella says that "invention," and specifically "the invention of causes," is "the aim of the resolutive method," namely analysis, because "through resolution we investigate the causes from the effects, in order, then, to know the effects from the causes" (Zabarella 1578, 179). Indeed, "the resolutive method" leads "to the invention of principles, so that, once they have been invented, from them we can then demonstrate natural effects" (ibid.). Now, the resolutive method is a method of discovery.

Arnauld and Nicole say that invention is the aim of "analysis or method of resolution" or "method of invention," which is a method "for discovering the truth when we do not know it" (Arnauld and Nicole 1992, 281–282).

Leibniz says that invention is "the art of making hypotheses, or the art of conjecturing" (Leibniz 1903, 174). Specifically, invention is "the art of discovering the causes of phenomena, or genuine hypotheses" (Leibniz 1965, V, 436).

Afterwards, however, the decline and end of method determines the separation between discovery and invention.

4.29 Negative Effects of the End of Method

In addition to causing the separation between discovery and invention, the decline and end of method has had very negative effects on the methodology of science and on philosophy.

On the one hand, the decline and end of method has led to abandoning any attempt to investigate how new knowledge is acquired and to find new procedures of knowledge acquirement. Knowledge acquirement has been relegated to the realm of irrationality, by considering it as the exclusive product of intuition and genius.

On the other hand, the decline and end of method has led to abandoning the view that philosophy aims at knowledge and methods to acquire knowledge (see Cellucci 2018, 2021). Thus the vigorous and fruitful discussions about the scientific method as a method to acquire knowledge in the seventeenth century, have given way to the feeble and sterile discussions on the definition of the concept of knowledge in the twentieth century.

For example, Ryle claims that philosophy is "intended not to increase what we know" (Ryle 2009a, lix). It can only deal with questions such as: What is knowledge? And the answer is that "'know' is a capacity verb" signifying the ability to "get things right" (ibid., 117). Namely, signifying the ability to get the right information in the right way. Conversely, philosophy should not concern itself with method, because "preoccupation with questions about methods tends to distract us from prosecuting the methods themselves. We run, as a rule, worse, not better, if we think a lot about our feet" (Ryle 2009b, II, 331).

These claims sharply contrast with Descartes, who says that philosophy aims to acquire "a perfect knowledge of all things that mankind is capable of knowing" (Descartes 1996, IX–2, 2). To this purpose, there is need for a method. Indeed, "it is far better to never think of investigating the truth of anything than to do so without a method" (ibid., X, 371). For, to have no method is to be like a person who "is burning with such a stupid desire to find a treasure, that he constantly roams about the streets to see if by chance he might find one lost by a passer-by" (ibid.).

Descartes is quite right in saying so. If philosophy does not want to be like that person, it must take up where the ancient tradition of method of Plato and Aristotle, and the modern tradition of method of the seventeenth century, left off, investigating how new knowledge is acquired and finding new procedures of knowledge acquirement.

References

Albertus Magnus. 1890. *Opera omnia*, vol. 2. Paris: Vives.

Anderson, Chris. 2008. The end of theory: The data deluge makes the scientific method obsolete. *Wired Magazine*, 23 June.

Arnauld, Antoine, and Pierre Nicole. 1992. *La logique ou l'art de penser*. Paris: Gallimard.

Berman, Jules. 2013. *Principles of big data: Preparing, sharing, and analyzing complex information*. Amsterdam: Elsevier.

Calude, Cristian S., and Giuseppe Longo. 2017. The deluge of spurious correlations in big data. *Foundations of Science* 22: 595–612.

Carnap, Rudolf. 1966. *Philosophical foundations of physics: An introduction to the philosophy of science*. New York: Basic Books.

———. 2003. *The logical structure of the world and pseudoproblems in philosophy*. La Salle: Open Court.

Cellucci, Carlo. 2013. *Rethinking logic: Logic in relation to mathematics, evolution, and method*. Cham: Springer.

———. 2018. Philosophy at a crossroads: Escaping from irrelevance. *Syzetesis* 5 (1): 13–53.

———. 2021. Philosophy, discovery, and advancement of knowledge. *Syzetesis* 8: 9–32.

Cohen, I. Bernard. 1985. *Revolution in science*. Cambridge: Harvard University Press.

Cotes, Roger. 1999. Editor's preface of the second edition. In Isaac Newton, *The Principia: Mathematical principles of natural philosophy*, 31–54. Oakland: University of California Press.

Descartes, René. 1996. *Œuvres*. Paris: Vrin.

di Grazia, Vincenzio. 1968. Considerazioni sopra al discorso di Galileo Galilei intorno alle cose che stanno in su l'acqua o in quella si muovono. In Galileo Galilei, *Opere*, vol. 4, 371–440. Florence: Barbera.

Emmer, Michele. 1997. Interview with Ennio De Giorgi. *Notices of the American Mathematical Society* 44: 1097–1101.

Fehr, Henri. 1908. *Enquête de 'L'Enseignement Mathématique' sur la méthode de travail des mathématiciens*. Paris: Gauthier-Villars.

Feyerabend, Paul. 1974. Machamer on Galileo. *Studies in History and Philosophy of Science Part A* 5: 297–304.
———. 1993. *Against method*. London: Verso.
Frege, Gottlob. 1960. *The foundations of arithmetic*. New York: Harper.
———. 1967. *Begriffsschrift*, a formula language, modeled upon that of arithmetic, for pure thought. In *From Frege to Gödel*, ed. Jean van Heijenoort, 1–82. Cambridge: Harvard University Press.
———. 1979. *Posthumous writings*. Oxford: Blackwell.
———. 1984. *Collected papers on mathematics, logic, and philosophy*. Oxford: Blackwell.
———. 2013. *Basic laws of arithmetic*. Oxford: Oxford University Press.
Galilei, Galileo. 1872. *Epistolario*. Livorno: Vigo.
———. 1968. *Opere*. Barbera: Firenze.
Gower, Barry. 1997. *Scientific method: An historical and philosophical introduction*. London: Routledge.
Husserl, Edmund. 1970. *The crisis of European sciences and transcendental phenomenology*. Evanston: Northwestern University Press.
Kant, Immanuel. 2000. *Critique of the power of judgment*. Cambridge: Cambridge University Press.
———. 2012. *Lectures on anthropology*. Cambridge: Cambridge University Press.
Kantorovich, Aharon. 1993. *Scientific discovery: Logic and tinkering*. Albany: State University of New York Press.
Kline, Morris. 1985. *Mathematics for the nonmathematician*. Mineola: Dover.
Krishnaswami, Govind S., and Himalaya Senapati. 2019. An introduction to the classical three-body problem: From periodic solutions to instabilities and chaos. *Resonance* 24: 87–114.
Lakatos, Imre. 1978. *Philosophical papers*. Cambridge: Cambridge University Press.
Leibniz, Gottfried Wilhelm. 1903. *Opuscules et fragments inédits*. Paris: Alcan.
———. 1923. *Sämtliche Schriften und Briefe*. Berlin: Academie Verlag.
———. 1965. *Die Philosophischen Schriften*. Hildesheim: Olms.
Martins, Roberto de Andrade. 1993. *Huygens's reaction to Newton's gravitational theory*. In *Renaissance and revolution: Humanists, scholars, craftsmen and natural philosophers in early modern Europe*, ed. Judith Veronica Field and Frank A. James, 203–213. Cambridge: Cambridge University Press.
Miller, David Philip. 2013. Attributing creativity in science and engineering: The discourses of discovery, invention and breakthrough. In *Handbook of research and creativity*, ed. Kerry Thomas and Janet Chan, 138–149. Cheltenham: Edgar Elgar.
Newton, Isaac. 1952. *Opticks, or a treatise of the reflections, refractions, inflections and colours of light*. Mineola: Dover.
———. 1999. *The Principia: Mathematical principles of natural philosophy*. Oakland: University of California Press.
Nickles, Thomas. 2000. Discovery. In *A companion to the philosophy of science*, ed. William Herbert Newton-Smith, 85–96. Malden: Blackwell.
Novalis. 2007. *Notes for a romantic encyclopedia: Das Allgemeine Brouillon*. Albany: State University of New York Press.
Popper, Karl Raimund. 1974. *Autobiography*. In *The philosophy of Karl Popper*, ed. Paul Arthur Schilpp, vol. 1, 3–181. La Salle: Open Court.
———. 1996. *In search of a better world*. London: Routledge.
———. 2000. *Realism and the aim of science*. London: Routledge.
———. 2002. *Conjectures and refutations: The growth of scientific knowledge*. London: Routledge.
———. 2005. *The logic of scientific discovery*. London: Routledge.
Putnam, Hilary. 1975–1983. *Philosophical papers*. Cambridge: Cambridge University Press.
Quine, Willard Van Orman. 1951. *Mathematical logic*. Cambridge: Harvard University Press.

Ramus, Petrus. 1552. *Institutionum dialecticarum libri tres*. Paris: David.
———. 1585. *Dialectica Audomari Talaei praelectionibus illustrata*. Basel: Eusebius & Nicolaus Episcopius.
Rav, Yehuda. 1989. Philosophical problems of mathematics in the light of evolutionary epistemology. *Philosophica* 43: 49–78.
Ryle, Gilbert. 2009a. *The concept of mind*. Chicago: University of Chicago Press.
———. 2009b. *Collected papers*. London: Routledge.
Whewell, William. 1847. *The philosophy of the inductive sciences*. London: Parker.
Yewdell, Jonathan W. 2008. How to succeed in science: a concise guide for young biomedical scientists. Part II: making discoveries. *Nature Reviews Molecular Cell Biology* 9: 491–494.
Zabarella, Jacopo. 1578. *Opera logica*. Venice: Meietto.
———. 1587. *In duos Aristotelis libros Posteriores Analyticos commentarii*. Venice: Meietto.
———. 1602. *De rebus naturalibus libri XXX*. Köln: Zetzner.

Chapter 5
Analytic Method

Abstract According to heuristic philosophy of mathematics, one of the tasks of the philosophy of mathematics is to give an answer to the question: What is the method of mathematics? And its answer is that the method of mathematics is the analytic method. The chapter examines the analytic method, its mathematical origin and medical origin, its original formulation, features, fortune, and its relation to abduction and reductio ad absurdum.

Keywords Analytic method · Non-ampliativity of deductive rules · Ampliativity of non-deductive rules · Plausibility · Mathematical origin of the analytic method · Medical origin of the analytic method · Analytic method and abduction · Analytic method and reductio ad absurdum

5.1 Statement of the Analytic Method

As we have seen in Chap. 3, according to heuristic philosophy of mathematics, one of the tasks of the philosophy of mathematics is to give an answer to the question: What is the method of mathematics? And its answer is that the method of mathematics is the analytic method. A brief description of the analytic method has already been given there, a fuller description will be given here.

The analytic method or method of analysis is the method according to which, to solve a problem, one looks for some hypothesis that is a sufficient condition for solving the problem, namely such that a solution to the problem can be deduced from the hypothesis. The hypothesis is obtained from the problem, and possibly other data including data acquired from mathematical diagrams, by some non-deductive rule – induction, analogy, metaphor, etc. – and must be plausible, namely such that the arguments for the hypothesis are stronger than the arguments against it, on the basis of experience. But the hypothesis is in turn a problem that must be solved, and is solved in the same way. Namely, one looks for another hypothesis that is a sufficient condition for solving the problem posed by the previous hypothesis, it is obtained

C. Cellucci, *The Making of Mathematics*, Synthese Library 448,
https://doi.org/10.1007/978-3-030-89731-4_5

from the latter, and possibly other data including data acquired from mathematical diagrams, by some non-deductive rule, and must be plausible. And so on. Thus, solving a problem is a potentially infinite process.

If, in the course of this process, a hypothesis turns out to be implausible, one analyses the reasons why the hypothesis is implausible, and arrives at some alternative hypothesis.

In the analytic method, there are no principles, everything is a hypothesis. The problem, and the other data already available, are the only basis for solving a problem. The aim of the problem solving process is to discover hypotheses that are sufficient conditions for solving the problem and are plausible. The hypotheses are conditions of solvability of the problem.

In the analytic method, logic plays a double role, as a means to obtain hypotheses and as a means to deduce conclusions from them. The former role is played by non-deductive rules, the latter by deductive rules. Thus the analytic method comprises a double movement, an upward movement from problems to hypotheses, by non-deductive rules, and a downward movement, from hypotheses to problems, by deductive rules.

The analytic method can be schematically represented as follows.

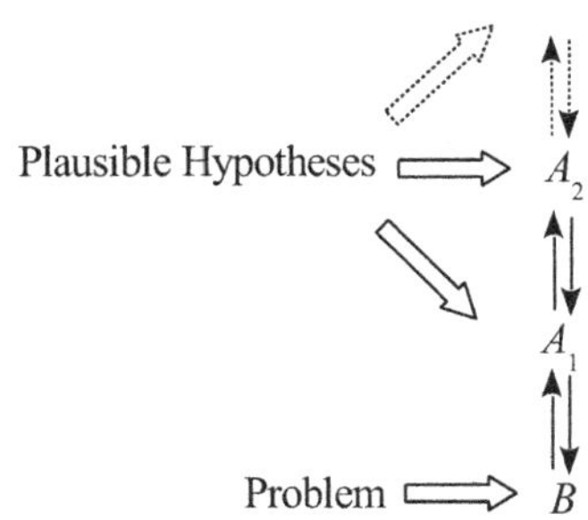

5.2 Open-Ended Character of Hypotheses and Inference Rules

In the analytic method, the hypotheses are provisional arrival points. At any stage, the inquiry may bring up new data, incompatible with some of the hypotheses on which the solution is based, showing that the hypothesis is no longer plausible. Then the hypothesis is modified or replaced with a new one, obtained from the problem, and possibly other data already available including data acquired from mathematical diagrams, by some non-deductive rule, and plausible.

Even when the inquiry does not bring up incompatible data, a hypothesis remains a problem to be solved, it is solved by looking for another hypothesis, and so on. Therefore, any solution to a problem is always provisional.

The non-deductive rules by which the hypotheses are obtained are not given once for all, but can always be extended when they turn out to be insufficient to obtain some hypothesis.

Even the deductive rules by which solutions to problems are deduced from the hypotheses are not given once for all. For, by the strong incompleteness theorem for second-order logic, they may turn out to be insufficient to deduce some solution. When this occurs, they can be extended. For this reason, even the deductive rules are not given once for all, but can always be extended when they turn out to be insufficient to deduce some solution.

Each extension of the analytic method by adding new rules, non-deductive or deductive, is a development of the analytic method, which grows as new rules are added.

5.3 Ampliativity and Non-Ampliativity of Inference Rules

That, in the analytic method, the hypotheses are obtained by non-deductive rules rather than by deductive rules, it is because non-deductive rules are ampliative, namely their conclusion is not already implicitly contained in the premisses. On the contrary, deductive rules are non-ampliative, namely their conclusion is already implicitly contained in the premisses. Therefore, the conclusion of a non-deductive rule says more than was already said in the premisses, while the conclusion of a deductive rule says nothing more than was already said in the premisses.

In addition to being ampliative, non-deductive rules are not truth-preserving, namely their conclusion can be false even if the premisses are true. They are not truth-preserving exactly because they are ampliative. Indeed, the conclusion of a non-deductive rule can be false even if the premisses are true, because the conclusion says more than was said in the premisses. So the conclusion is not a mere reformulation of the content or part of the content of the premisses, it contains something more. Therefore, non-deductive rules can extend our knowledge.

Conversely, in addition to being non-ampliative, deductive rules are truth-preserving, namely their conclusion cannot be false if the premisses are true. They are truth-preserving exactly because they are non-ampliative. The conclusion of a deductive rule cannot be false if the premisses are true, exactly because the conclusion says nothing more that was already said in the premisses. Indeed, the conclusion is only a reformulation of the content or part of the content of the premisses. The conclusion may be psychologically surprising, but cannot augment the content of the premisses. Therefore, deductive rules cannot extend our knowledge.

5.4 Ampliativity and Non-Ampliativity in Antiquity

The ampliativity of non-deductive rules and the non-ampliativity of deductive rules have been recognized already in antiquity.

Thus, the Epicureans say that non-deductive rules are ampliative, because the conclusion apprehended by a non-deductive inference will essentially differ from the premisses from which we infer. For example, suppose we infer, from the premiss "Men among us are mortal," the conclusion "Men everywhere are also mortal" (Philodemus, *De signis*, Col. XVI, 6–10, ed. De Lacy). Then the conclusion essentially differs from the premiss. For, by this non-deductive inference we "pass from appearances to the non-apparent" (ibid., Col. XIX, 12–13). And "the non-apparent differs from the apparent" (ibid., Col. XX, 22–23). Indeed, the conclusion says more than was said in the premiss.

On the other hand, deductive rules are non-ampliative, because "the conclusion apprehended by" a deductive inference "will not differ from the" premisses "from which we infer" (ibid., Col. III, 5–8). So, the conclusion will say nothing more than was already said in the premisses. Therefore, "we will not say that syllogism can make us know anything" (Epicurus, *De natura*, XXVIII, 17, I, ed. Sedley).

Since non-deductive rules are ampliative, they are essential to the advancement of knowledge. Indeed, to advance knowledge, "we should use all the things that have been obtained through experience, in serious inductive reasoning and not apart from inductive reasoning" (Philodemus, *De signis*, Frag. 4, 2–5, ed. De Lacy).

Against the Epicureans, the Stoics object that, while deductive inference is cogent, non-deductive inference "is not cogent" (ibid., Col. IV, 31–32). Thus, the inference from 'Men among us are mortal' to 'Men everywhere are mortal', is not cogent because "it is not necessary that," because certain things "exist in our experience, they exist also in places that are unperceived" (ibid., Col. Ia, 5–7). For, we cannot "presuppose that the men about whom we infer are like those among us even in being mortal" (ibid., Col. XVI, 10–14).

But the Epicureans answer this objection by saying that, if non-deductive inference is not cogent, this does not mean that it cannot yield knowledge. Thus, the inference from 'Men among us are mortal' to 'Men everywhere are mortal' yields knowledge. For, by means of it, "we infer, from the fact that all men among us are alike even in being mortal, that universally all men are liable to death, because nothing opposes the inference or draws us even a step toward the view that they do not admit of death" (ibid., Col, XVI, 16–25). The inference is sound because it does not conflict with experience. Generally a non-deductive inference "is sound with this condition, that no appearance or previously demonstrated fact conflicts with the inference" (ibid., Col. XXXII, 23–27).

Non-conflict with experience is essential because, according to the Epicureans, all knowledge depends on experience. Indeed, Epicurus says that, "of opinions," some "are true and some are false. True ones are those that are testified in favour of, and not testified against, by experience, while false ones are those that are testified against, and not testified in favour of, by experience" (Sextus Empiricus, *Adversus Logicos*, I.211, ed. Mutschmann).

Thus, the Epicureans clearly distinguish between non-deductive rules and deductive rules, they recognize that non-deductive rules are essential to the advancement of knowledge, and give a criterion for the soundness of non-deductive inferences, namely compatibility with experience.

5.5 Non-Ampliativity of Deductive Rules Since Antiquity

The non-ampliativity of deductive rules has been repeatedly reaffirmed since antiquity.

Thus, Descartes says that logicians cannot "form any syllogism leading to a true conclusion unless" they "already knew the very same truth which is deduced in the syllogism," so they "can learn nothing new from such form of reasoning" (Descartes 1996, X, 406).

Kant says that deductive logic "teaches us nothing at all about the content of cognition," so "using it as a tool (organon) for an expansion and extension of its information, or at least the pretension of so doing, comes down to nothing but idle chatter" (Kant 1998, A61/B86).

De Morgan says that deduction is subject "to the great rule of all search after truth, that nothing is to be asserted as a conclusion, more than is actually contained in the premises" (De Morgan 1835, 99).

Mill says that, in deductive inference, the conclusion is "merely a repetition of the same, or part of the same, assertion, which was contained in the first" (Mill 1963–1986, VII, 158). Thus, "there is in the conclusion no new truth, nothing but what was already asserted in the premisses, and obvious to whoever apprehends them" (ibid., VII, 160).

Peirce says that, in deductive inference, "certain facts are first laid down in the premisses," and "part or all of them" will be thrown "into a new statement" that "will be the conclusion of" the deductive "inference" (Peirce 1931–1958, 2.680). Therefore, deductive inference "is evidently entirely inadequate to the representation" of inference "which goes out beyond the facts given in the premisses" (ibid., 2.681).

Poincaré says that deduction is "incapable of adding anything to the data" which are "given it; these data reduce themselves" to "axioms, and we should find nothing else in the conclusions" (Poincaré 2015, 31).

Wittgenstein says that, "if one proposition follows from another, then the latter says more than the former, and the former less than the latter" (Wittgenstein 2002, 5.14). Hence, "there can never be surprises in logic" (ibid., 6.1251).

5.6 Objections to the Non-Ampliativity of Deductive Rules

Although the non-ampliativity of deductive rules has been widely reaffirmed since antiquity, in the past century some arguments have been put forward in favour of their ampliativity. These arguments, however, are invalid. Here are the main ones.

(1) Deductive rules are ampliative because, even if the conclusion of a deduction is contained in the premisses, extracting it from the premisses requires labour.

Thus, Frege says that, even if the conclusions of deductions "are in a way contained covertly in the whole set" of premisses "taken together," this "does not absolve us from the labour of actually extracting them and setting them out in their own right" (Frege 1960, 23). For, "what we shall be able to infer from" the premisses, "cannot be inspected in advance; here, we are not simply taking out of the box again what we have just put into it," so "the conclusions we draw" from the premisses may "extend our knowledge, and ought therefore, on Kant's view, to be regarded as synthetic" (ibid., 100–101). The conclusions are indeed contained covertly in the premisses, but only "as plants are contained in their seeds, not as beams are contained in a house" (ibid., 101).

This argument, however, is invalid because there is an algorithm for enumerating all deductions from given premisses. Given enough time and space, a computer will grind out all the deductions from the premisses one by one. If a conclusion can be deduced from them, then the computer, operating in this blind fashion, will sooner or later find a deduction of it. Thus, extracting the conclusions from the premisses is a purely mechanical task, it can be performed by a computer, and hence requires no labour.

Moreover, it is misleading to say that the conclusions are contained in the premisses as plants are contained in their seeds. Plants can develop from seeds only by absorbing water from the soil and harvesting energy from the sun, hence using resources from the environment not contained in the seeds. On the contrary, conclusions are deduced from the premisses without using anything not contained in the premisses.

Frege's claim that the conclusions we draw from the premisses extend our knowledge mistakes psychological novelty for logical novelty. The conclusions may be psychologically surprising, not because they are not contained in the premisses, but because we are incapable of making even comparatively short deductions without some external aid. For that reason, extracting conclusions from the premisses seems to require labour.

(2) Deductive rules are ampliative because the conclusion of a deductive inference, even when it only involves individuals mentioned in the premisses, tells us something about those individuals that the premisses did not tell us.

Thus, Russell says that "it is an old debate among philosophers whether deduction ever gives new knowledge. We can now see that in certain cases, at least, it does do so" (Russell 1998, 44). For example, "if we already know that two and two always make four, and we know that Brown and Jones are two, and so are Robinson and Smith, we can deduce that Brown and Jones and Robinson and Smith are four" (ibid.). This "is new knowledge, not contained in our premisses, because the general proposition, 'two and two are four', never told us there were such people as Brown and Jones and Robinson and Smith, and the particular premises do not tell us that there were four of them" (ibid.).

This argument, however, is invalid because the premiss, 'Brown and Jones are two', told us that there were such people as Brown and Jones and that there were two of them. On the other hand, the premiss, 'Robinson and Smith are two', told us that there were such people as Robinson and Smith and that there were two of them. Then this, together with the premiss that two and two always make four, already told us that there were such people as Brown and Jones and Robinson and Smith, and that there were four of them.

In fact, Russell himself eventually acknowledges that his argument is invalid, because he says that "deduction has turned out to be much less powerful than was formerly supposed; it does not give new knowledge, except as to new forms of words for stating truths in some sense already known" (Russell 2009, 140). Thus, "all mathematical proof consists merely in saying in other words part or the whole of what is said in the premisses. If, from a theorem *A*, you deduce a theorem *B*, it must be the case that *B* repeats *A* (or part of it) in other words" (Russell 1983–, XI, 360).

(3) Deductive rules are ampliative, otherwise mathematics would be trivial. For, as soon as we would know the axioms of a mathematical theory, we would thereby know all the theorems.

Thus, Dummett says that, if deductive rules were non-ampliative, then, "as soon as we had acknowledged the truth of the axioms of a mathematical theory, we should thereby know all the theorems. Obviously, this is nonsense" (Dummett 1991, 195). Therefore, deduction must have the power "to yield knowledge that we did not previously possess" (ibid.).

This argument, however, is invalid because, as already said, we are incapable of making even comparatively short deductions without some external aid.

Thus, Hadamard says: "I have several times happened to overlook results which" were "immediate consequences of other ones which I had obtained" (Hadamard 1954, 50). For example, "two theorems, important to the subject" of my doctoral thesis, "were such obvious and immediate consequences of the ideas contained therein that, years later, other authors imputed them to me, and I was obliged to confess that, evident as they were, I had not perceived them" (ibid., 51).

Since we are incapable of making even comparatively short deductions without some external aid, we do not know all the theorems that can be deduced from the axioms of a mathematical theory as soon as we acknowledge that the axioms are true. This has nothing to do with ampliativity.

(4) Deductive rules are ampliative because they are truth-preserving, so if we know that the premisses of a valid deductive inference are true, then the argument gives us knowledge of the conclusion. Conversely, non-deductive rules are non-ampliative because they are not truth-preserving, so even if we know that the premisses of a correct non-deductive inference are true, the argument can only yield reasonable belief, not knowledge of the conclusion.

Thus, Swinburne says that "deductive inferences extend knowledge" (Swinburne 1974, 7). For, "if an argument is a valid deductive one, then, if the premisses are true, the conclusion must be," so, "if we know the premisses to be true, and see that the argument is a valid deductive one, then the argument gives us knowledge of the conclusion" (ibid., 6). Conversely, non-deductive inferences do not extend knowledge, they only yield "reasonable belief" (ibid.). For, "if we know the premisses to be true and see that the argument is a correct" non-deductive "one, all that we can conclude is something about the probability of the conclusion" (ibid.).

This argument, however, is invalid because it depends on the assumption that knowledge must be true. This assumption is invalid even for mathematical knowledge because, by Gödel's second incompleteness theorem, mathematical knowledge cannot be said to be true. But, if knowledge cannot be said to be true, then the premisses of a deductive argument cannot be said to be true, so the condition 'if we know that the premisses of a valid deductive inference are true' cannot be satisfied. Therefore, if knowledge must be true, then the argument cannot be said to give us knowledge of the conclusion.

5.7 The Paradox of Inference

That, in the analytic method, the hypotheses are obtained by non-deductive rules, rather than by deductive rules, is necessary. For, the hypotheses must contain something that is not already contained in the problem or in the data already available. Therefore, the hypotheses must be obtained by ampliative rules, and ampliative rules cannot be valid.

This is confirmed by Cohen and Nagel's "paradox of inference" (Cohen and Nagel 1964, 173). According to it, if in an inference the conclusion is not contained in the premisses, then the inference cannot be valid; and if the conclusion is contained in the premisses and hence has no novelty with respect to them, then the inference cannot be ampliative; but the conclusion cannot be contained in the premisses and also have novelty with respect to them; therefore, an inference cannot be both valid and ampliative.

From the paradox of inference it follows that, since deductive rules are valid, they cannot be ampliative, and since non-deductive rules are not valid, they can be ampliative.

The paradox of inference, however, does not mean that an inference rule cannot be both non-valid and non-ampliative, but only that it cannot be both valid and ampliative. In fact, there are inference rules that are both non-valid and non-ampliative. As we will see below, abduction is an example.

5.8 Plausibility and Novelty

As already said, in the analytic method a hypothesis must be plausible, namely such that the arguments for the hypothesis are stronger than the arguments against it, on the basis of experience.

That a hypothesis is plausible implies that it agrees with experience. But this does not mean that the hypothesis has no novelty with respect to experience.

For example, let us consider hyperbolic geometry, the non-Euclidean geometry obtained by replacing Euclid's fifth postulate with the hyperbolic postulate: Given any straight line and any point not on it, there are at least two distinct straight lines through that point which do not intersect that straight line.

Hyperbolic geometry was not well received at first. In a review of Lobachevski's work *On the principles of geometry*, it was even said that Lobachevski had "failed to give his book an appropriate title," a more appropriate title would have been "*A satire on geometry* or *A caricature of geometry* or a similar thing" (Rosenfeld 1988, 209).

The decisive step for the acceptance was the construction of models of hyperbolic geometry in Euclidean geometry, such as the Beltrami-Klein model, which can be described as follows. Take a circle in the Euclidean plane. Interpret points as the points inside the circle. Interpret straight lines as chords of the circle, without their endpoints.

All Euclid's postulates except the fifth one hold in the Beltrami-Klein model. For example, the first postulate, that given any two points, there is exactly one straight line passing through them, holds. For, given any two points A and B inside the circle, there is exactly one chord of the circle passing through A and B.

On the other hand, the hyperbolic postulate also holds. For, as the following figure suggests, given any chord c of the circle and any point A inside the circle not on c, there are at least two distinct chords of the circle through A which do not intersect c.

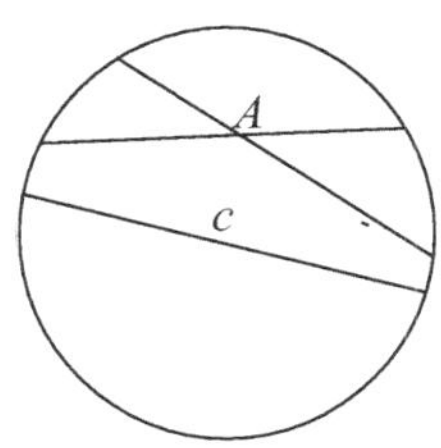

Thus, all the postulates of hyperbolic geometry hold in the Beltrami-Klein model. Since the Beltrami-Klein model is a circle in the Euclidean plane, it follows that, if the Euclidean geometry is consistent, then so is the hyperbolic geometry.

Therefore, the hypotheses of hyperbolic geometry agree with Euclidean experience. But this does not mean that they have no novelty respect to Euclidean experience. Indeed, the hyperbolic postulate cannot be deduced from Euclid's postulates.

5.9 Plausibility and Truth

Plausibility is sometimes confused with other concepts, such as truth, probability, and persuasiveness.

Plausibility is sometimes confused with truth. Thus, Peirce says that "what we mean by the truth" is "the opinion which is fated to be ultimately agreed to by all who investigate" (Peirce 1931–1958, 5.407).

This amounts to confusing plausibility with truth. For, the opinion which is fated to be ultimately agreed to by all who investigate is plausible, but not true.

Indeed, plausibility is different from truth. For, truth is an absolute concept, since a true proposition is true forever. On the contrary, plausibility is a relative concept. A hypothesis that is implausible at a certain time may become plausible at a later time, if the arguments for it become stronger than those against it, on the basis of the experience available at that time. And, conversely, a hypothesis that is plausible at a certain time may become implausible at a later time, if the arguments against it become stronger than those for it, on the basis of the experience available at that time.

Also, as argued in Chap. 4, there is no criterion of truth. On the contrary, there is a criterion of plausibility, namely the plausibility test procedure (see below).

Moreover, a proposition that is true is certain, because a true proposition is true forever. On the contrary, a hypothesis that is plausible is not certain, because there is no guarantee that no counterexample will ever be found.

Not only plausibility is different from truth, but is alternative to truth. For, by Gödel's second incompleteness theorem, mathematics cannot be concerned with truth, it can only be concerned with plausibility.

5.10 Plausibility and Probability

Plausibility is sometimes confused with probability. Thus, Pólya says that one can "use the calculus of probability to render more precise our views on plausible reasoning" (Pólya 1954, II, 116). For, "the calculus of plausibilities obeys the same rules as the calculus of probabilities" (Pólya 1941, 457).

This amounts to confusing plausibility with probability. For, evaluating the evidence of a hypothesis by the rules of the calculus of probability establishes its probability, not its plausibility. Indeed, plausibility is different from probability. For, probability is a mathematical concept. On the contrary, plausibility concerns whether the arguments for a hypothesis are stronger than the arguments against it, so it is not a mathematical concept.

In the case of classical probability, that plausibility is different from probability is made clear already by Kant. Indeed, Kant says that "probability is a fraction, whose denominator is the number of all the possible cases, and whose numerator contains the number of winning cases" (Kant 1992, 328). Mathematics "determines certain rules, in accordance with which the object can be cognized probably" (ibid., 331). So probability is a mathematical concept. On the contrary, "plausibility is concerned with whether, in the cognition, there are more grounds for the thing than against it" (ibid.). Thus, plausibility is "the relation of the grounds for the truth to the grounds of the opposite" (ibid., 477). Plausibility is "a holding-to-be-true based on insufficient grounds insofar as these are greater than the grounds of the opposite" (ibid., 583). So plausibility is not a mathematical concept. In probability, cases "must be enumerated," in plausibility, grounds "must be weighed" (ibid., 584). Therefore, plausibility is different from classical probability. (On Kant's distinction between plausibility and probability, see Capozzi 2013, 2019, 2020).

However, plausibility is different from probability not only in the case of classical probability, but also in that of all the other kinds of probability, including subjective probability, to which it is often equated.

In particular, as regards subjective probability, de Finetti says that "subjective probability is one's degree of belief in an outcome, based on an evaluation making the best use of all the information available to him and his own skill" (de Finetti 1974, 16). As Galavotti points out, by asking to make the best use of all the information available, de Finetti does not overlook "objective information" and does not "consider equally acceptable whatever evaluation is made" (Galavotti 2018, 154). But de Finetti makes it clear that, although "every evaluation of a probability is based on all the available information, including objective data," only "our subjective judgment can guide our selection of what information to consider as relevant," and "the subjective decision to admit only such information as relevant and to make use of it in the ordinary ways is what transforms objective data into a probability" (de Finetti 1974, 16).

On the contrary, the evaluation of the plausibility of a hypothesis is not based only on our subjective judgment, but on an assessment of the arguments for and against the hypothesis that are made by the mathematical community as a whole. Therefore, plausibility is different from subjective probability.

5.11 Plausibility and Persuasiveness

Plausibility is sometimes confused with persuasiveness. Thus, Walton says that "the very best definition of plausibility was given by Carneades," whose "most important legacy to philosophy was his famous theory of plausibility" (Walton 2001, 152). According to the latter, "something is plausible if it appears to be true, or (is even more plausible) if it appears to be true and is consistent with other things that appear to be true. Or thirdly, it is even more plausible if it is stable (consistent with other things that appear to be true), and is tested" (ibid.).

This amounts to confusing plausibility with persuasiveness. For, Carneades's theory is not a theory of plausibility but a theory of persuasiveness [pithanotes], since it concerns the conduct of life, while plausibility concerns knowledge.

Indeed, Carneades says that "there is absolutely no criterion of truth" for knowledge, "neither reason, nor sense perception, nor appearance, nor anything else that there is; for all of these, singularly or as a group, deceive us" (Sextus Empiricus, *Adversus Logicos*, I.159, ed. Mutschmann). Nevertheless, there is "some criterion for the conduct of life and the attainment of happiness," and this criterion is "the persuasive appearance [pithane phantasia]" (ibid., I.166). For, "it is contrary to nature that there is nothing persuasive," therefore "the wise man will make use of whatever persuasive appearance he encounters" (Cicero, *Academica*, II.99).

The persuasive appearance is "the appearance that seems to be true" (Sextus Empiricus, *Adversus Logicos*, I.169, ed. Mutschmann). Then, to say that the criterion is the persuasive appearance is to say that "the criterion is the appearance that seems to be true" (ibid., I.174). Admittedly, the appearance that seems to be true "sometimes happens actually to be false" but, "since this rarely occurs," one should not "distrust the one that for the most part abides by truth. For, our judgments and our actions are regulated by what applies for the most part" (ibid., I.175).

The persuasive appearance, however, is only the first level of persuasiveness. In addition to it, there are two other levels. The second level is the appearance that is "persuasive and at the same time unreversed" (ibid., I.176). Namely the appearance that is a concurrence of several appearances, all of which are persuasive and hence "in unison seem to be true" (ibid., I.177). The third level is the appearance that is "persuasive, unreversed and thoroughly examined" (ibid., I.184).

For the conduct of life, "on ordinary matters we use the persuasive appearance as a criterion," while "on more important matters we use the appearance that is" also "unreversed, and on matters that contribute to happiness we use the appearance that is" also "thoroughly examined" (ibid.).

All examples of persuasive appearance, or appearance with a higher level of persuasiveness, considered by Carneades concern the conduct of life, not knowledge. Thus, Carneades considers the case of "the wise man" who "embarks on board of a ship" for the purpose of going "from this place to Puteoli" (Cicero, *Academica*, II.100). He cannot be sure that "he is definitely going to arrive," but he has "a persuasive appearance" that, if the ship is "a seaworthy vessel, with a good pilot, and in fine weather," he "will arrive there safely" (ibid.). On the basis of this persuasive appearance, the wise man deliberates to embark on board of the ship. Generally, one "will deliberate about what to do or not to do on the basis of appearances of this kind" (ibid.).

From the fact that all examples of persuasive appearance, or appearance with a higher level of persuasiveness, considered by Carneades, concern the conduct of life, not knowledge, it is clear that Carneades's theory is a theory of persuasiveness, not of plausibility. This confirms that saying, as Walton does, that Carneades's most important legacy to philosophy was his famous theory of plausibility, amounts to confusing plausibility with persuasiveness.

But plausibility is different from persuasiveness. First, as already said, plausibility concerns knowledge while, according to Carneades's theory, persuasiveness concerns the conduct of life. Of course, even the conduct of life requires knowledge, but the knowledge it requires is practical knowledge, not the kind of theoretical knowledge that is involved in mathematics and science, to which plausibility refers. Walton himself admits that Carneades's theory "is not a theory of knowledge or belief. It is a guide to rational acceptance or commitment, a guide to action" (Walton 2001, 153).

Also, plausibility concerns judgments. On the contrary, Carneades's persuasiveness concerns appearances, and appearances are not judgments because, as Carneades says, "there is no appearance capable of judging" (Sextus Empiricus, *Adversus Logicos*, I.165, ed. Mutschmann).

Moreover, as already said, plausibility is alternative to truth. On the contrary, Carneades's persuasiveness concerns truth, because a persuasive appearance is an appearance that seems to be true. To be sure, Carneades maintains that truth "cannot be apprehended" (Sextus Empiricus, *Pyrrhoniae Hypotyposes*, I.3, ed. Mutschmann). However, to say that an appearance seems to be true does not require apprehending the truth, but only knowing what means to be true.

5.12 Plausibility and Endoxa

While plausibility is different from truth, probability, and persuasiveness, plausibility is somewhat related to Aristotle's 'endoxa'.

Aristotle says that endoxa are opinions accepted "by everyone, or by the great majority, or by the wise, and among them either by all of them, or by the great majority, or by the most famous and esteemed" (Aristotle, *Topica*, A 1, 100 b 21–23).

This is usually supposed to be how "in the *Topics*, Aristotle defines endoxa" (Haskins 2004, 23). But, if it were so, then the superficial or biased opinions of the great majority would be endoxa. This is not what Aristotle means.

Rather, how Aristotle defines endoxa, is implicit in his assertion that, for each thesis concerning some given subject, we must "examine the arguments for it and the arguments against it" (ibid., Θ 14, 163 a 37–b 1). For, "going through the difficulties on either side, we shall more readily discern the true as well as the false in any subject" (ibid., A 2, 101 a 35–36). Indeed, "if the objections are answered and the endoxa remain, we shall have proved the case sufficiently" (Aristotle, *Ethica Nicomachea*, Z 1, 1145 b 6–7).

Thus, endoxa are theses that are accepted on the basis of an examination of the arguments for and against them, from which the former turn out to be stronger than the latter. This justifies the claim that plausibility is somewhat related to Aristotle's endoxa. Striker even uses 'plausible' "to translate the Greek endoxon" (Striker 2009, 77).

However, there is a basic difference between plausibility and endoxa. As already said above, plausibility is alternative to truth. On the contrary, endoxa are not alternatives to truth but means to truth.

This is clear from the fact that, on the one hand, according to Aristotle, problems are solved by means of Aristotle's analytic-synthetic method and, in the analytic part of Aristotle's analytic-synthetic method, until one arrives at some principle, one proceeds through hypotheses which, for the moment, are only plausible, or endoxa (see in Chap. 6). On the other hand, according to Aristotle, "the aim of theoretical science is truth" (Aristotle, *Metaphysica*, α 1, 993 b 20–21). And "mathematics is a theoretical science" (ibid., E 1, 1025 a 7–8). Therefore, the aim of mathematics is truth. This explains why endoxa are not alternatives to truth, but means to truth.

5.13 The Plausibility Test Procedure

To determine whether a hypothesis is plausible, one will generally need to consider its consequences. This is already made clear by Plato, who says that, for any hypothesis, you wouldn't go on "until you had examined its consequences" (Plato, *Phaedo*, 101 d 4–5).

Specifically, to determine whether a hypothesis is plausible, the following plausibility test procedure can be used.

(1) Deduce conclusions from the hypothesis.

(2) Compare the conclusions with each other, to see that the hypothesis does not lead to contradictions.

(3) Compare the conclusions with other hypotheses already known to be plausible, and with results of observations or experiments, to see that the hypothesis is compatible with them.

If the hypothesis passes the plausibility test procedure, then for the moment it is approved. Indeed, 'plausible' comes from the Latin 'plausibilis', which derives from 'plaudere', which means 'to applaud', 'to approve'. The hypothesis, however, is approved only for the moment, because new data can always emerge with which the hypothesis may turn out to be incompatible.

On the other hand, if the hypothesis does not pass the plausibility test procedure, it is not rejected outright. Rather, it is put on a waiting list, subject to further investigation, since nothing is definitively plausible or definitively implausible.

5.14 Inference Rules and Plausibility Preservation

As argued above, inference rules differ with respect to truth-preservation, because deductive rules are truth-preserving, while non-deductive rules are not truth-preserving. The situation is different with respect to plausibility-preservation, because non-deductive rules and deductive rules are both not plausibility-preserving.

Non-deductive rules are not plausibility-preserving. For example, let us call a side of a triangle a bounding element of the triangle, and a face of a tetrahedron a bounding element of the tetrahedron. Now, a triangle is analogous to a tetrahedron in so far as the triangle is the figure in a plane bounded by the minimum number of bounding element, namely three, and the tetrahedron is the figures in space bounded by the minimum number of bounding element, namely four. Let us say that lines in a plane or in space are concurrent if they intersect at a single point. Then, the three altitudes of a triangle are concurrent.

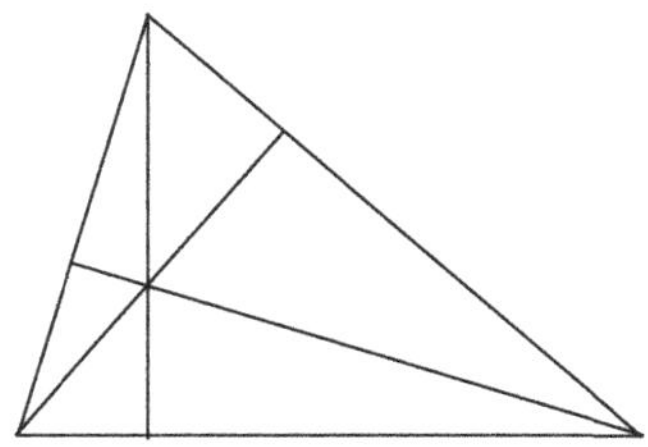

From this, by analogy – and specifically, by analogy by agreement, see Chap. 7 – we may infer that the four altitudes of a tetrahedron are concurrent. But, while the premiss 'The three altitudes of a triangle are concurrent' is plausible, the conclusion 'The four altitudes of a tetrahedron are concurrent' is not plausible, because the four altitudes of a tetrahedron are not generally concurrent.

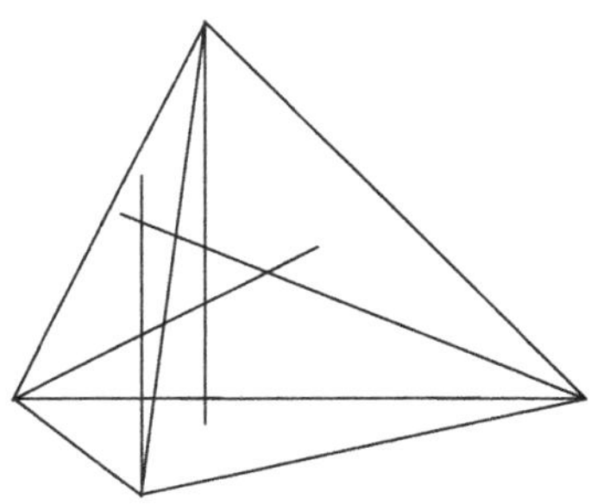

This example shows that non-deductive rules are not plausibility-preserving.

But deductive rules are not plausibility-preserving either. As Prawitz says, "a deductive inference does not in general preserve the level" of "plausibility," so "there is no guarantee that a person who has good but non-conclusive grounds for the premisses of a deductive inference gets a good ground for the conclusion by performing the inference" (Prawitz 2014, 83).

For example, suppose that, at some time, two mutually incompatible hypotheses, *A* and *B*, are both plausible. This is possible because the arguments for and against *A* may be different from the arguments for and against *B*. Then, for the moment, *A* and *B* may be considered to be both plausible. But their conjunction, $A \wedge B$, cannot be considered to be plausible, because *A* and *B* are mutually incompatible. Therefore, the conjunction introduction rule, 'From *A* and *B* infer $A \wedge B$', is not plausibility-preserving.

5.15 Analytic Method and Doubling the Cube

The first uses of the analytic method of which we have textual evidence are those of Hippocrates of Chios to solve problems in mathematics, and of Hippocrates of Cos to solve problems in medicine.

For example, Hippocrates of Chios uses the analytic method to solve the following problem.

Doubling the cube: Find the side of the cube that shall be double a given cube.

To solve the problem, by analyzing the conditions under which the problem would be solved, Hippocrates of Chios non-deductively arrives at the following hypothesis:

(I) Given any two straight lines, a and b, we can always find two other straight lines, x and y, which are the mean proportionals in continued proportion between a and b, namely such that $a : x = x : y = y : b$.

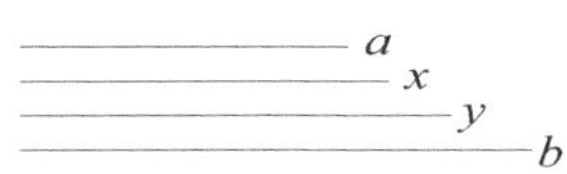

Hypothesis (I) is a sufficient condition for solving the problem. For, by hypothesis (I), for any given straight lines a and b, we can find two other straight lines, x and y, such that $a : x = x : y = y : b$. From this it follows $(a : x)^3 = (a : x)(x : y)(y : b) = (a : b)$. For $b = 2a$, this yields $(a : x)^3 = (1 : 2)$, hence $x^3 = 2a^3$, namely a cube double of a given cube of side a. Therefore, $x = a \cdot \sqrt[3]{2}$ Thus Hippocrates of Chios shows that, "if two mean proportionals could be found in continued proportion between two straight lines, of which the greater was double the lesser, the cube would be doubled" (Eutocius, *Commentarii in libros de sphaera et cylindro. Ut Eratosthenes*. In Archimedes, *Opera Omnia*, III, 104.11–15, ed. Heiberg). This solves the problem.

Hypothesis (I) is plausible, because the arguments for it are stronger than those against it. But hypothesis (I) is itself a problem that must be solved. To solve it, by analyzing the conditions under which the problem would be solved, Menaechmus non-deductively arrives at the following hypothesis:

(II) Given any two straight lines, a and b, the desired straight lines, x and y, are the coordinates of the meeting point of the parabola with vertex O, axis OY and parameter a, and the parabola with vertex O, axis OX and parameter b.

Hypothesis (II) is a sufficient condition for solving problem (I). For, let a and b be any two straight lines. Draw two straight lines, $AO = a$ and $BO = b$, perpendicular at O, and produce AO to Y and BO to X. Draw a parabola with vertex O, axis OY, and parameter AO, and a parabola with vertex O, axis OX, and parameter BO. Let P be the meeting point of the two parabolas. Draw PM, PN perpendicular to OY, OX respectively.

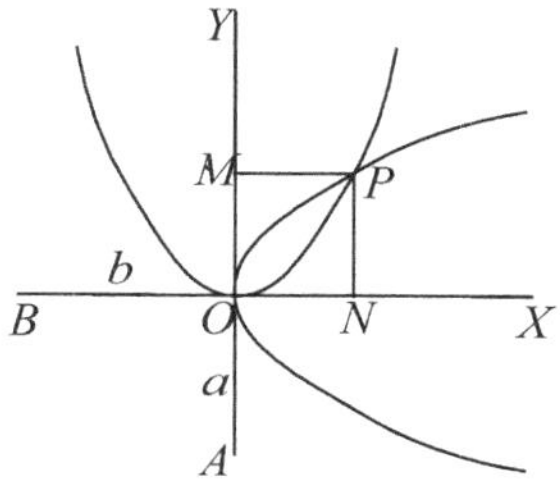

By the property of parabolas, $PM^2 = AO \cdot OM$ and $PN^2 = BO \cdot ON$, so $ON^2 = a \cdot OM$ and $OM^2 = b \cdot ON$. Since ON and OM are the coordinates of the point P, by hypothesis (II), $x = ON$ and $y = OM$. So, from $ON^2 = a \cdot OM$ and $OM^2 = b \cdot ON$, it follows $x^2 = ay$ and $y^2 = bx$. Hence $a : x = x : y$ and $x : y = y : b$, and therefore $a : x = x : y = y : b$. Namely, hypothesis (I). This is that "which it was required to find" (Eutocius, *Commentarii in libros de sphaera et cylindro. Ut Menechmus*. In Archimedes, *Opera Omnia*, III, 98.14, ed. Heiberg). This solves the problem. The solution increases the plausibility of hypothesis (I).

Hypothesis (II) is plausible, because the arguments for it are stronger than those against it. But hypothesis (II) is itself a problem that must be solved. To solve it, it is necessary to formulate another hypothesis. And so on.

The process through which the problem of doubling the cube is solved can be schematically represented as follows.

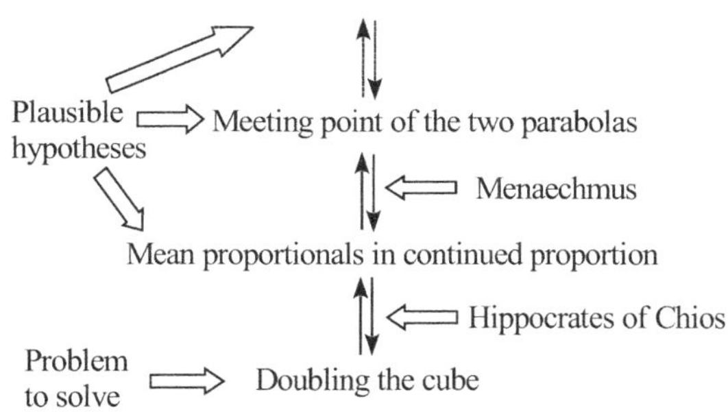

5.16 Analytic Method and Quadrature of the Lunule

Hippocrates of Chios also uses the analytic method to solve the problem of the quadrature of certain lunules, such as the following.

> *Quadrature of the lunule whose outer circumference is a semicircle*: Find the area of "the lunule whose outer circumference was a semicircle," namely the lunule obtained "by circumscribing about a right-angled isosceles triangle a semicircle, and about the base a segment of a circle similar to those cut off by the sides" (Simplicius, *In Aristotelis Physicorum Libros Quattuor Priores Commentaria*, I 2, 61.20–27, ed. Diels). Here, "similar segments of circles" are those which "have equal angles" (ibid., I 2, 61.32–34).

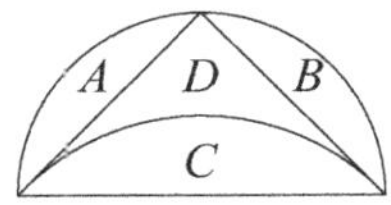

To solve the problem, by analyzing the conditions under which the problem would be solved, Hippocrates of Chios non-deductively arrives at the following hypothesis:

(I) "Similar segments of circles are to each other as the squares on their bases" (ibid., I 2, 61.6–7).

Hypothesis (I) is a sufficient condition for solving the problem. For, by the Pythagorean theorem, the square on the base of the triangle is equal to the sum of the squares on the sides. From this, by hypothesis (I), it follows that the segment on the base of the triangle is equal to the sum of the segments on the sides, namely $C = A + B$. Now, the lunule consists of the sum of the segments on the sides plus the part of the triangle above the segment about the base, thus lunule $=A + B + D$. On the other hand, the triangle consists of the segment about the base plus the part of the triangle above the segment about the base, thus triangle $=C + D$. Since $C = A + B$, it follows lunule $=A + B + D = C + D=$ triangle. Thus "the lunule is equal to the triangle. Therefore the lunule, having been demonstrated equal to the triangle, can be squared" (ibid., I 2, 62.6–8). This solves the problem.

Hypothesis (I) is plausible, because the arguments for it are stronger than those against it. But hypothesis (I) is itself a problem that must be solved. To solve it, by analyzing the conditions under which the problem would be solved, Hippocrates of Chios non-deductively arrives at the following hypothesis:

(II) "Circles are to each other as the squares on their diameters" (ibid., 61.11).

Hypothesis (II) is a sufficient condition for solving problem (I). Indeed, "as circles are to each other, so too are similar segments. For, similar segments are the same part of the circle," hence, "similar segments have equal angles" (ibid., I 2, 61.11–15). Then, from hypothesis (II), it follows that similar segments of circles are to each other as the squares on their bases. Namely hypothesis (I). This solves the problem. The solution increases the plausibility of hypothesis (I).

Hypothesis (II) is plausible, because the arguments for it are stronger than those against it. But hypothesis (II) is itself a problem that must be solved. To solve it, it is necessary to formulate another hypothesis. And so on.

The process through which the problem of the quadrature of a certain lunule is solved can be schematically represented as follows.

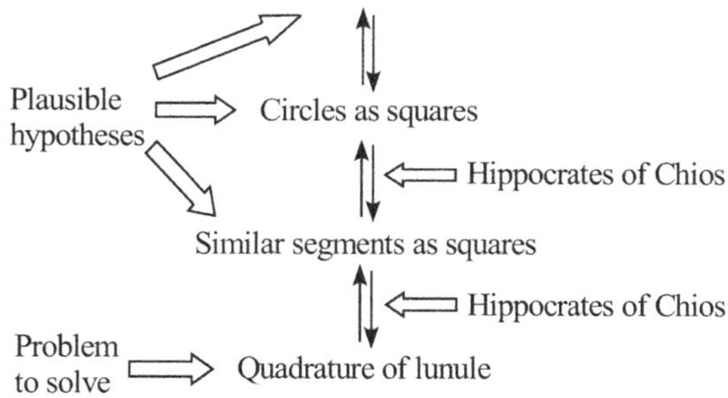

5.17 Analytic Method and Impact of Food on Health

Hippocrates of Cos uses the analytic method to solve problems in medicine, such as the following.

> *Impact of food on health*: Explain why "the human being is affected and altered" by each one of different foods "in one way or another" (Hippocrates, *De vetere medicina*, I, 600.14–16, ed. Littré).

To solve the problem, by analyzing the conditions under which the problem would be solved, Hippocrates of Cos non-deductively arrives at the following hypothesis:

(I) The substances which are in the food "are also in the human being" (ibid., I, 602.8–9).

Hypothesis (I) is a sufficient condition for solving the problem. For, when the substances which are in the human being are well "mixed and blended with one another," they do not "cause the human being pain," while, when one of the substances which are in the human being "separates off and comes to be on its own," it "causes the human being pain" (ibid., I, 602.12–14). So, the human being is affected and altered by the substances which are in him. From this, by the hypothesis (I), it follows that the human being is affected and altered by the substances which are in the food. Since different foods contain different substances, one may then conclude that the human being is affected and altered by different foods in one way or another. This solves the problem.

Hypothesis (I) is plausible, because the arguments for it are stronger than those against it. But hypothesis (I) is itself a problem that must be solved. To solve it, by analyzing the conditions under which the problem would be solved, Hippocrates of Cos non-deductively arrives at the following hypothesis:

(II) Each humour in the body draws from food and drink that which is homogeneous to it, "each humour attracting its like through the veins" (Hippocrates, *De morbis*, VI, 548.9–10, ed. Littré).

Hypothesis (II) is a sufficient condition for solving problem (I). For, from hypothesis (II) it follows that, since each humour in the body draws from food and drink that which is homogeneous to it, the substances which are in the food are also in the human being. Namely hypothesis (I). This solves the problem. The solution increases the plausibility of hypothesis (I).

Hypothesis (II) is plausible, because the arguments for it are stronger than those against it. But hypothesis (II) is itself a problem that must be solved. To solve it, it is necessary to formulate another hypothesis. And so on.

The process through which the problem of impact of food on health is solved can be schematically represented as follows.

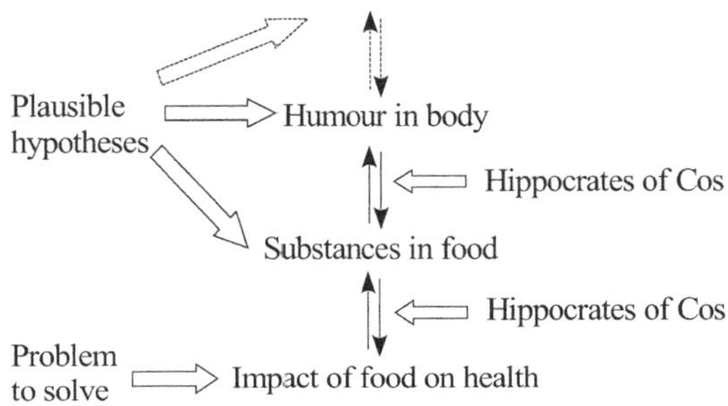

5.18 Original Formulation of the Analytic Method

As already mentioned in the Introduction, although the analytic method was used by Hippocrates of Chios to solve problems in mathematics, and by Hippocrates of Cos to solve problems in medicine, neither of them gave a formulation of the method, they simply used it, Plato gave the first formulation.

Sometimes it is said that "Plato discovered the method of analysis" (Doxiadis and Sialaros 2013, 373). This, however, is due to a misinterpretation of Diogenes Laertius and Proclus. For, Diogenes Laertius says that Plato "was the first to explain the method of solving problems by analysis to Leodamas of Thasos" (Diogenes Laertius, *Vitae Philosophorum*, III.24, 270–272, ed. Dorandi). And Proclus says that Plato "taught this method to Leodamas, who is also reported to have made many discoveries in geometry by means of it" (Proclus, *In primum Euclidis Elementorum librum commentarii*, 211.21–23, ed. Friedlein). Then, clearly, neither Diogenes Laertius nor Proclus says that Plato discovered the method of analysis, but only that Plato explained or taught it to Leodamas. Indeed, Plato did not discover the analytic method, he only gave the first formulation of it.

Plato begins his formulation of the analytic method by saying that it proceeds "from a hypothesis," by which "I mean the sort of thing geometers often use in their inquiries" (Plato, *Meno*, 86 e 4–5).

To solve a problem, "I thought I should take refuge in certain hypotheses, and consider in them the truth of things" (Plato, *Phaedo*, 99 e 5–6).

Specifically, "on each occasion, I posit the hypothesis which I judge to be the strongest, and I lay down as true whatever seems to me to agree with it", while "I put down as not true whatever does not seem to me to agree with it" (ibid., 100 a 3–7).

Once you had assumed a hypothesis, "if anyone attacked the hypothesis, you would not dismiss him and would not avoid answering to him until you had investigated its consequences, to see whether they are in accord, or are not in accord, with each other" (ibid., 101 d 3–5).

Even when the consequences of the hypothesis are in accord with each other, this does not mean that the hypothesis is correct and hence conclusively justified. For, if in formulating the hypothesis "a small imperceptible error has been made," it may well be that all the propositions "that follow, however numerous, are consistent with each other" (Plato, *Cratylus*, 436 d 2–4). Therefore, "it is necessary to carefully examine and investigate the starting point of each question, to see if the hypothesis is correct or not; after carefully examining it, the subsequent steps will appear to be consequent to it" (ibid., 436 d 4–7).

Thus, even if the hypotheses appear "convincing, you will have to re-examine them more carefully" (Plato, *Phaedo*, 107 b 5–6). This means that each hypothesis must be accounted for. And "you would account for it in the same way, positing another hypothesis, whichever should seem best of the higher ones, and so on" (ibid., 101 d 6–7). For, all hypotheses are "stepping stones from which to take off and proceed upward" (Plato, *Respublica*, VI 511 b 5).

You should go on positing hypotheses "until you came to something adequate" (Plato, *Phaedo*, 101 e 1). But the hypothesis so reached will in turn be a problem that eventually will have to be solved, it will be solved by positing a new hypothesis. And so on. Thus solving a problem is "an infinite task" (Plato, *Parmenides*, 136 c 7).

This is because mathematics is "knowledge of what eternally is" (Plato, *Respublica*, VII 527 b 4). And getting that kind of knowledge is an infinite undertaking.

5.19 Plato's Dependence Upon the Two Hippocrates

As we have seen above, Plato begins his formulation of the analytic method by saying that it proceeds from a hypothesis, by which he means the sort of thing that geometers often use in their inquiries. Now hypotheses, as used by Hippocrates of Chios, are conditions of solvability of a problem, and Plato means hypotheses in that very sense. So, Plato draws the view that hypotheses are conditions of solvability of a problem from Hippocrates of Chios.

In fact, from Hippocrates of Chios, Plato draws not only this view but also all the other features of the analytic method.

However, Plato does not draw everything from Hippocrates of Chios, he draws something also from Hippocrates of Cos. As we have seen in Chap. 4, in *Phaedrus*, Plato mentions Hippocrates of Cos in relation to the necessity of a research method. More importantly, as we have seen in Chap. 3, Plato criticizes the use of the axiomatic method in mathematics because it bases demonstrations upon unjustified hypotheses. Now, his criticism of the use of the axiomatic method in mathematics is akin to Hippocrates of Cos's criticism of the use of the axiomatic method in medicine.

Indeed, Hippocrates of Cos says that "all those who have undertaken to speak or write about medicine" clearly "go wrong in much that they say" by "basing their discourse on" an unjustified "hypothesis, hot or cold or wet or dry or anything else they want," so "laying down the same one or two things as the cause in all cases" (*De vetere medicina*, I, 570.1–6, ed. Littré). For, medicine must not use an unjustified hypothesis. Anyone who tries to investigate on the basis of an unjustified hypothesis "and says that he has discovered something, has been deceived and continues to deceive himself" (ibid., I, 572.14–15).

Thus, Hippocrates of Cos criticizes those who carry out an inquiry in medicine by the axiomatic method, because they take their hypotheses for granted, and do not give any justification for them. Now, Plato criticizes those who carry out an inquiry in mathematics by the axiomatic method for that very same reason.

5.20 Analytic Method and Teachability of Virtue

Plato uses the analytic method to solve problems in philosophy, such as the following.

Teachability of virtue: "Is virtue teachable?" (Plato, *Meno*, 70 a 1–2).

To solve the problem, by analyzing the conditions under which the problem would be solved, Plato non-deductively arrives at the following hypothesis:

(I) "Virtue is knowledge" (ibid., 87 c 5).

Hypothesis (I) is a sufficient condition for solving the problem. For, it is "plain to everyone that a man cannot be taught anything but knowledge" (ibid., 87 c 2–3). From this, by hypothesis (I), it follows that "virtue is teachable" (ibid., 87 c 5–6). This solves the problem.

Hypothesis (I) is plausible, because the arguments for it are stronger than those against it. But hypothesis (I) is itself a problem that must be solved. To solve it, by analyzing the conditions under which the problem would be solved, Plato non-deductively arrives at the following hypothesis:

(II) "Virtue is something good" (ibid., 87 d 2–3).

Hypothesis (II) is a sufficient condition for solving the problem. For, "knowledge embraces everything that is good" (ibid., 87 d 6–7). From this, by (II), it follows that virtue is knowledge. Namely hypothesis (I). This solves the problem. The solution increases the plausibility of hypothesis (I).

Hypothesis (II) is plausible, because the arguments for it are stronger than those against it. But hypothesis (II) is itself a problem that must be solved. To solve it, it is necessary to formulate another hypothesis. And so on.

The process through which the problem of the teachability of virtue is been solved can be schematically represented as follows.

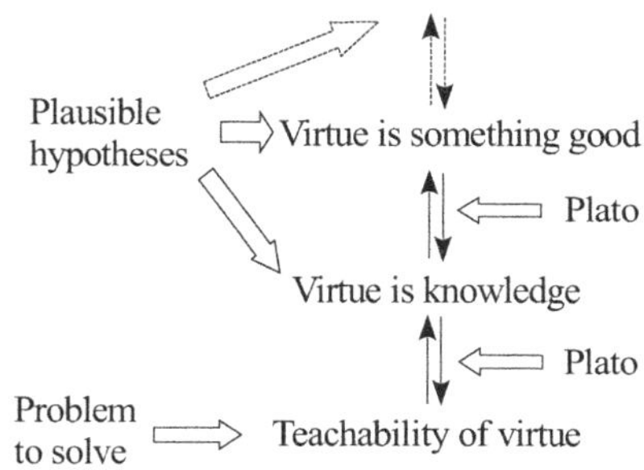

5.21 Analytic Method and Doubling the Square

Plato, however, does not use the analytic method only to solve problems in philosophy, but also to solve problems in mathematics. Eudemus even says that "Plato, who appeared after" Hippocrates of Chios, "greatly advanced mathematics in general and especially in geometry because of his commitment in this field" (Proclus, *In primum Euclidis Elementorum librum commentarii*, 66.8–11, ed. Friedlein).

In particular, in *Meno* Plato considers the following problem in mathematics.

Doubling the square: For any given square, find "the side of the square double in size" (Plato, *Meno*, 82 e 2).

To solve the problem, by analyzing the conditions under which the problem would be solved, Plato non-deductively arrives at the following hypothesis:

(I) A diagonal "cuts each square in half" (ibid., 85 a 1).

Hypothesis (I) is a sufficient condition for solving the problem. For, let *ABCD* be the given square. Then, by hypothesis (I), the diagonal *DB* cuts *ABCD* in half, so *ABCD* consists of two halves. Four such halves, arranged into a square, will make a square double in size, because "the relation of four to two" is "double" (ibid., 85 a 8–9). Then, the question is only how to arrange the four halves into a square. Now, clearly, the only possible way of doing so is to arrange them into the square *DBFH*.

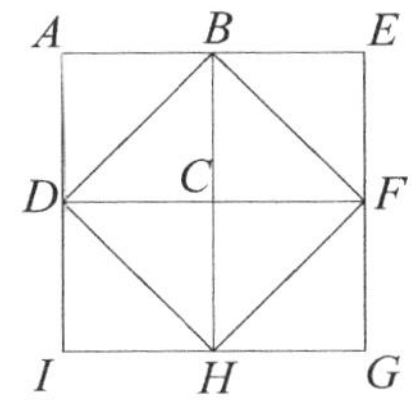

The square *DBFH* is surrounded by four lines, each of which "cuts each of these four" equal squares *ABCD*, *BEFC*, *CFGH*, and *DCHI*, "in half" (ibid., 85 a 5–6). So, each of the four lines is a diagonal of one of the squares in question. Hence, "it is the diagonal that will produce the figure double in size" (ibid., 85 b 5–6). Therefore, for any given square, the side of the square double in size is a diagonal of the given square. This solves the problem.

Hypothesis (I) is plausible, because the arguments for it are stronger than those against it. But hypothesis (I) is itself a problem that must be solved. To solve it, by analyzing the diagram and the conditions under which the problem would be solved, Plato non-deductively arrives at the following hypothesis:

(II) If two triangles have their corresponding sides equal, then they are equal, because "a thing to which nothing is added and from which nothing is taken away is neither increased nor diminished, but always remains the same in amount" (Plato, *Theaetetus*, 155 a 7–9).

Hypothesis (II) is a sufficient condition for solving the problem. For, the two right-angled isosceles triangles into which a diagonal cuts a square have their corresponding sides equal, hence by hypothesis (II) they are equal. Therefore, a diagonal cuts each square in half. Namely hypothesis (I). This solves the problem. The solution increases the plausibility of hypothesis (I).

Hypothesis (II) is plausible, because the arguments for it are stronger than those against it. But hypothesis (II) is itself a problem that must be solved. To solve it, it is necessary to formulate another hypothesis. And so on.

The process through which the problem of doubling the square is solved can be schematically represented as follows.

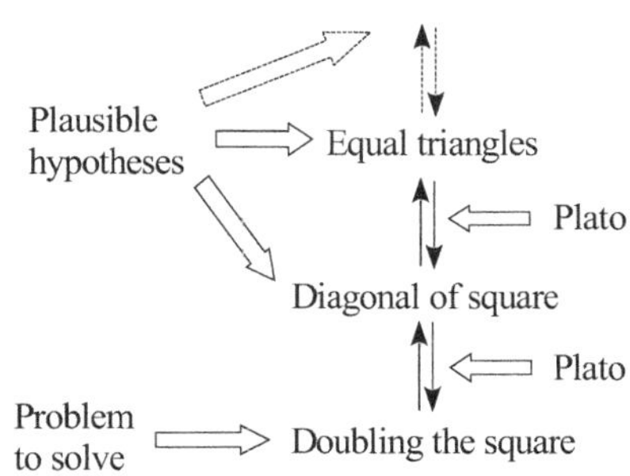

5.22 Analytic Method and Inscription of Square as Triangle in Circle

In *Meno*, immediately after the problem of doubling the square, Plato considers another problem in mathematics, namely the problem "whether this area can be inscribed as a triangle in a given circle" (Plato, *Meno*, 86 e 6–87 a 1).

Plato says that, if this problem is posed to a geometer, he might reply: "I think I have a hypothesis useful for the problem: If this area is such that, when placed alongside the given line, it falls short by an area such as the one which has been placed alongside the given line, I think one result follows, while if it is impossible for it to fall so, another result follows" (ibid., 87 a 2–7).

This passage is obscure, several conjectures have been made as to its meaning, but, as Lloyd says, each of them "falls foul of one or more difficulties" (Lloyd 1992, 167). However, the more plausible conjecture seems to be that of Benecke 1867. According to him, 'this area' is the given square, *ABCD*, in Plato's solution to the problem of doubling the square; 'the given line' is a diameter of the circle; and 'an area such as the one which has been placed alongside the given line' is a square equal to the square *ABCD*. Then the problem is the following.

Inscription of square as triangle in circle: Can the square *ABCD* be inscribed as a triangle in a given circle, and if so, under what condition?

To solve the problem, by analyzing the conditions under which the problem would be solved, the geometer non-deductively arrives at the following hypothesis (the same as hypothesis (I) in Plato's solution to the problem of doubling the square):

(I) A diagonal cuts each square in half.

Hypothesis (I) is a sufficient condition for solving the problem. For, let *DF* be a circle with radius equal to the side of the square *ABCD*. Let *BEFC* be a square equal to *ABCD*. Let *DB* be a diagonal of *ABCD*, and *BF* a diagonal of *BEFC*. Then, by hypothesis (I), the square *ABCD* is equal in area to the triangle *DBF*.

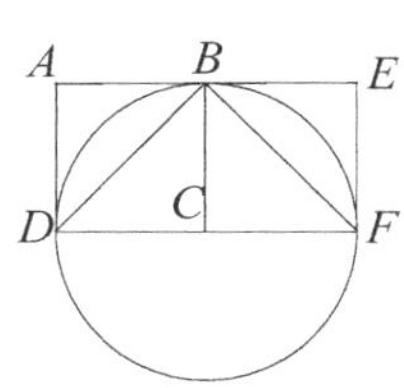

So, if the square *ABCD* is such that, when placed alongside the diameter of the circle *DF*, it falls short by a square *BEFC* equal to *ABCD*, then the square *ABCD* can be inscribed as a triangle in the circle *DF*. This condition is satisfied if the side of *ABCD* is equal to the radius of the circle. Conversely, if the condition is not satisfied, the square *ABCD* cannot be so inscribed. Thus, the square *ABCD* can be inscribed as a triangle in a given circle if and only if the side of *ABCD* is equal to the radius of the circle. This solves the problem.

As already said in Plato's solution to the problem of doubling the square, hypothesis (I) is plausible, but is itself a problem that must be solved. It can be solved by means of the hypothesis (II) considered there, namely, if two triangles have their corresponding sides equal, then they are equal. The solution increases the plausibility of hypothesis (I). And so on.

The process through which the problem of the inscription of square as triangle in circle is solved can be schematically represented as follows.

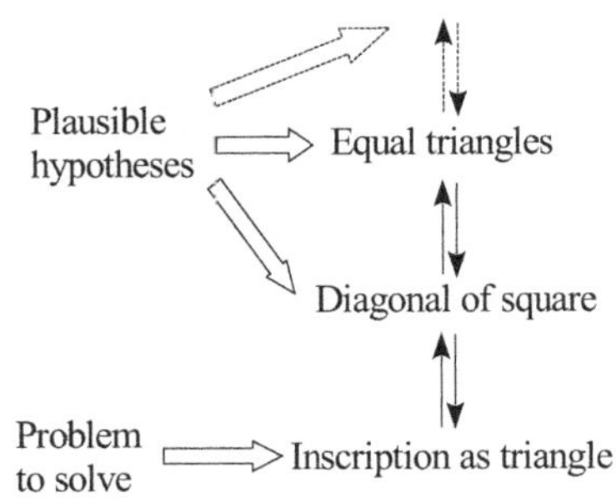

5.23 Analytic Method and the Beginnings of Greek Mathematics

Although the first uses of the analytic method of which we have textual evidence are those of Hippocrates of Chios to solve problems in mathematics, and of Hippocrates of Cos to solve problems in medicine, it seems likely that the analytic method was used by other Greek mathematicians before Hippocrates of Chios.

An example may be Thales's solution to the following problem.

Inscription of triangle in semicircle: Show that any triangle inscribed in a circle, with one side being a diameter of the circle, is a right-angled triangle.

Its first solution is attributed to Thales because "Pamphile says that" Thales "was the first to inscribe a right-angled triangle in a circle, after which he sacrificed an ox" (Diogenes Laertius, *Lives of eminent philosophers*, I.24, 36–37, ed. Dorandi).

The problem captured the poetic imagination of Dante, who described it as the problem "se del mezzo cerchio far si pote | triangol sì ch'un retto non avesse [if in a semicircle there can be made | a triangle so that it would not have a right angle]" (Dante, *La Divina Commedia*, Canto XIII, 101–102).

The problem could be easily solved making use of the fact that, in any triangle, the sum of the internal angles is equal to two right angles. But "Eudemus the Peripatetic attributes to the Pythagoreans the discovery of this theorem, that every triangle has internal angles equal to two right angles" (Proclus, *In primum Euclidis*

Elementorum librum commentarii, 379.2–5, ed. Friedlein). (For a description of the demonstration of the Pythagoreans, see Chap. 7). Therefore, the theorem was not known at Thales's time.

Alternatively, on the basis of what was known at Thales's time, we can imagine that, to solve the problem, Thales inscribes a triangle *ABC* in a circle with *AB* a diameter of the circle, then he constructs another triangle *ABD* by rotating triangle *ABC* by 180° about the centre of the circle *O*.

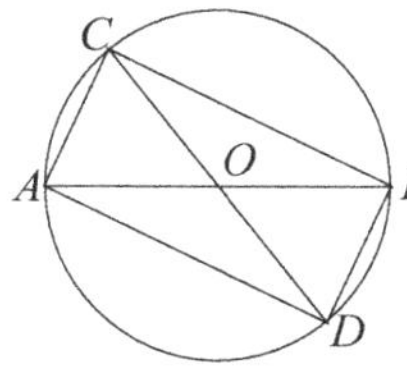

By analyzing the diagram and the conditions under which the problem would be solved, Thales non-deductively arrives at the following hypothesis:

(I) The quadrilateral *ACBD* is a rectangle.

Hypothesis (I) is a sufficient condition for solving the problem. For, by hypothesis (I), the quadrilateral *ACBD* is a rectangle, so all angles of *ACBD* are right angles, hence in particular so is the angle *ACB*. Therefore, the triangle *ABC* is a right-angled triangle. This solves the problem.

Hypothesis (I) is plausible, because the arguments for it are stronger than those against it. But hypothesis (I) is itself a problem that must be solved. To this aim, we can imagine that Thales draws a line *CD* through the centre of the circle *O*. By analyzing the diagram and the conditions under which the problem would be solved, Thales non-deductively arrives at the following hypothesis:

(II) If the diagonals of a parallelogram are equal, then the parallelogram is a rectangle.

Hypothesis (II) is a sufficient condition for solving the problem. For, the triangle *ABD* has been constructed by rotating triangle *ABC* by 180° about the centre of the circle *O*, and a rotation by 180° maps lines onto parallel lines, hence lines *AC* and *BD* are parallel, and so are lines *AD* and *BC*. Therefore, the quadrilateral *ACBD* is a parallelogram. On the other hand, since the diagonals *AB* and *CD* of the parallelogram *ACBD* are both diameters of the circle, they are equal. Therefore, by hypothesis (II), the parallelogram *ACBD* is a rectangle. Namely hypothesis (I). This solves the problem. The solution increases the plausibility of hypothesis (I).

Hypothesis (II) is plausible, because the arguments for it are stronger than those against it. But hypothesis (II) is itself a problem that must be solved. To solve it, it is necessary to formulate another hypothesis. And so on.

The process through which the problem of the inscription of triangle in semicircle is solved can be schematically represented as follows.

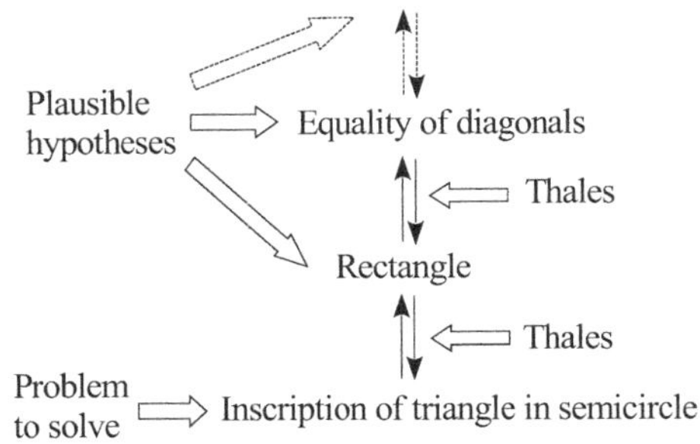

5.24 The Analytic Method Before the Greeks

Not only it seems likely that the analytic method was used by other Greek mathematicians before Hippocrates of Chios, but it also seems likely that the analytic method was used in pre-Greek mathematics. Indeed, some solutions to problems in pre-Greek mathematics seem almost to suggest it.

For example, let us consider the solution by an unknown Egyptian mathematician to Problem 51 of the Rhind Mathematical Papyrus: "If it is said to you: What is the area" of an isosceles "triangle of 10 khet on the height of it and 4 khet on the base of it?" then, to give an answer to this question, "take 1/2 of 4, namely, 2, in order to get" the base of "its rectangle. Multiply 10 times 2," therefore "its area is 20" (Clagett 1999, 163).

Thus the Egyptian mathematician reduces the problem of the area of an isosceles triangle of base 4 and height 10 to the problem of the area of a rectangle of base 2 and height 10. This solution holds not only for an isosceles triangle of base 4 and height 10, but for any isosceles triangle. Then, the problem is the following.

Area of isosceles triangle: Find the area of an isosceles triangle with given base and height.

On the basis of the description given in the Rhind Mathematical Papyrus, we can imagine that, to solve the problem, the Egyptian mathematician draws an isosceles triangle with base b and height h and draws the height of the triangle.

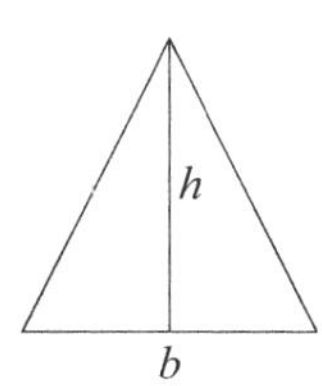

By analyzing the conditions under which the problem would be solved, the Egyptian mathematician non-deductively arrives at the following hypothesis:

(I) The height cuts an isosceles triangle in half.

Hypothesis (I) is a sufficient condition for solving the problem. For, by hypothesis (I), the height cuts the isosceles triangle in half. Then, cut the isosceles triangle along its height, reverse one piece, turn it upside down, and glue both pieces together, obtaining a rectangle with base $b/2$ and height h.

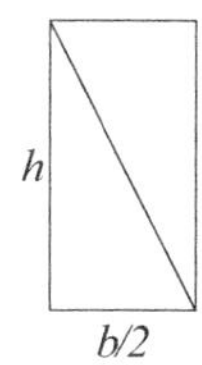

The area of the isosceles triangle is equal to the area of the rectangle, namely $(b/2)h$. Therefore, the area of the triangle is $(b/2)h$. This solves the problem.

Hypothesis (I) is plausible, because the arguments for it are stronger than those against it. But hypothesis (I) is itself a problem that must be solved. To solve it, by analyzing the conditions under which the problem would be solved, the Egyptian mathematician non-deductively arrives at the following hypothesis:

(II) If two triangles have their corresponding sides equal, then they are equal.

Hypothesis (II) is a sufficient condition for solving the problem. For, the two triangles into which the height cuts the isosceles triangle have their corresponding sides equal, so by hypothesis (II) they are equal. Hence the height cuts an isosceles triangle in half. Namely, hypothesis (I). This solves the problem. The solution increases the plausibility of hypothesis (I).

Hypothesis (II) is plausible, because the arguments for it are stronger than those against it. But hypothesis (II) is itself a problem that must be solved. To solve it, it is necessary to formulate another hypothesis. And so on.

The process through which the problem of the area of the isosceles triangle is solved can be schematically represented as follows.

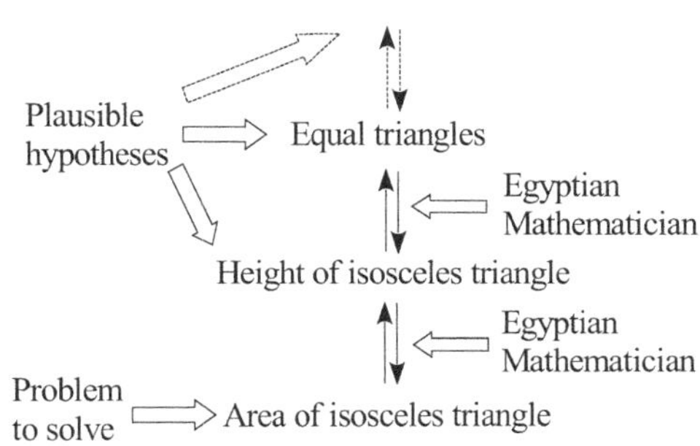

5.25 Shortcomings of the Original Formulation of the Analytic Method

With respect to the analytic method as formulated at the beginning of the chapter, Plato's original formulation of the analytic method has some significant shortcomings.

(1) Plato gives no indication as to how to find hypotheses to solve problems. As it is clear from his uses of the analytic method described above, Plato arrives at hypotheses non-deductively, but he does not explicitly indicate by which non-deductive rules the hypotheses are obtained.

(2) Plato only requires that the consequences of hypotheses agree with each other. But this does not guarantee that hypotheses agree with experience.

(3) Plato assumes that mathematics is knowledge that is absolutely true and certain. But this conflicts with Gödel's second incompleteness theorem, by which mathematics cannot be said to be knowledge that is absolutely true and certain.

The analytic method as formulated at the beginning of the chapter is not subject to these shortcomings.

It is not subject to shortcoming (1), because it specifies that hypotheses are obtained by non-deductive rules. (Several non-deductive rules for obtaining hypotheses will be described in Chap. 7).

It is not subject to shortcoming (2), because it requires that hypotheses be plausible, which guarantees that they agree with experience.

It is not subject to shortcoming (3), because it does not assume that mathematics is knowledge that is absolutely true and certain. Being based on plausible hypotheses, mathematics is knowledge that can only be plausible.

5.26 The Hindrance of the Body

It has been said above that Plato assumes that mathematics is knowledge that is absolutely true and certain. This, however, needs clarification. According to Plato, knowledge that is absolutely true and certain can be acquired only after death.

Indeed, Plato says that "in our investigation, the body turns up everywhere, causes confusion and turmoil, and overwhelms us, so as to prevent us from being able to see the truth" (Plato, *Phaedo*, 66 d 5–7). Therefore, "as long as we have the body and our soul is contaminated by such an evil, we will never properly acquire what we desire," namely "truth" (ibid., 66 b 5–7).

Thus, "if we are ever to have full knowledge of something, we must be separated from the body and contemplate the things by themselves with the soul by itself," because "only then, namely when we have died, in other words when we are outside life, we will get what we desire, wisdom" (ibid., 66 d 8–e 4). Only after death we will see things "in pure light, being ourselves pure and untainted by this which now we carry around with us and call the body, in which we are imprisoned like an oyster in its shell" (Plato, *Phaedrus*, 250 c 3–6).

Indeed, only after death the process of passing from hypothesis to hypothesis will come to an end, and we will reach "what is unhypothetical, the principle of the whole" (Plato, *Respublica*, VI 511 b 5–6). We will grasp it by "intuition," which is "the highest faculty" (ibid., VI 511 d 8). Indeed, what is unhypothetical, the principle of the whole, is "visible only by the pilot of the soul, intuition" (Plato, *Phaedrus*, 247 c 7–8).

Then, "having grasped this principle," we will reverse the process and, "going through the consequences that follow from" the principle, we will "go down to the conclusion" (Plato, *Respublica*, VI 511 b 6–c 1). So we will acquire knowledge that is absolutely true and certain, being based on intuition.

Since this is possible only after death, somewhat paradoxically Plato says that the best preparation to acquire knowledge that is absolutely true and certain is to "practise nothing other than dying and being dead" (Plato, *Phaedo*, 64 a 5–6).

5.27 Characters of the Analytic Method

The analytic method as formulated at the beginning of the chapter has the following characters.

(1) The hypotheses formulated to solve a problem need not belong to the field of the problem, they may belong to other fields. So, the search for a solution to a problem is not carried out in a closed, predetermined space.

(2) The hypotheses formulated to solve a problem are not global but local, they are not general principles, good for all problems, but are aimed at a specific problem.

(3) Being local, the hypotheses formulated to solve a problem can be efficient. For, being designed to meet a specific problem, they can take care of the peculiarities of the problem. This can be essential for the feasibility of a solution.

(4) Different problems will generally require different hypotheses. This follows from the fact that the hypotheses are aimed at a specific problem, so they depend on the peculiarities of the problem.

(5) The same problem can be solved using different hypotheses. A hypothesis is a window through which one looks upon a problem. But a problem can be looked at through different windows from different perspectives, which may reveal heretofore unsuspected aspects, leading to different solutions to the problem. When a problem seems to be solvable only by a single hypothesis, one should worry because it might mean that the solution is wrong, or that the problem is ill-posed. Indeed, if one cannot view a problem from different perspectives, the problem may be not very interesting.

(6) The hypotheses formulated to solve a problem may turn out to be unable to solve that problem, but may be able to solve some other problem. Such other problem may be an insignificant one, but sometimes may have an interest of its own.

(7) The hypotheses formulated to solve a problem may turn out to be able to solve also some other problem, in addition to the intended one. For, the hypotheses

formulated to solve a problem may involve concepts not occurring in that problem, so they may have implications also for other problems.

(8) Solving a problem is both a process of discovery and a process of justification. It is a process of discovery, because it involves finding hypotheses by non-deductive rules. It is a process of justification, because it involves establishing that the hypotheses are plausible.

(9) Solving a problem requires a comparison with experience. For, that hypotheses must be plausible, involves that the arguments for them must be stronger than the arguments against them, on the basis of experience.

(10) Solving a problem is a dynamic process. First, the problem is posed. Then a hypothesis to solve it is obtained and shown to be plausible. Afterwards, a new hypothesis to solve the problem posed by the previous hypothesis is obtained and shown to be plausible. And so on.

(11) Solving a problem yields something new. Being obtained from the problem, and possibly other data already available, by some non-deductive rules, a solution is not contained in the problem or in the other data already available. It has novelty with respect to them, because non-deductive rules are ampliative. This explains why solving a problem yields something new.

(12) Solving a problem does not involve intuition. For, intuition has no role in the formulation of hypotheses, because the latter are obtained from the problem, and possibly other data already available, by some non-deductive rule, so not by intuition but by inference. It has no role in the justification of hypotheses, because their plausibility is established by the plausibility test procedure, so not by intuition but by inference. Therefore, intuition has no role in solving a problem.

(13) Solving a problem is not subjective or psychological. For, hypotheses are obtained by some non-deductive rule, so by inference, and their plausibility is established by the plausibility test procedure, so by inference.

(14) Hypotheses are neither true nor certain, but this does not diminish the value of the analytic method. Hypotheses are neither true nor certain, because the analytic method is a heuristic method, not an algorithmic one, so it cannot guarantee to yield true and certain knowledge, only plausible one. But this does not diminish the value of the analytic method, because there is no source of knowledge capable of guaranteeing truth and certainty, plausible knowledge is the most we can achieve, and, on the other hand, without plausible knowledge there is no knowledge at all.

(15) Solving a problem is a potentially infinite process, so no solution to a problem is final. For, any hypothesis which provides a solution to a problem can at most be plausible, and it is itself a problem that must be solved, so no hypothesis is final.

5.28 Knowledge as an Infinite Process

Since, with the analytic method, solving a problem is a potentially infinite process, the analytic method is based on a view of knowledge, essentially going back to Plato, according to which, as Natorp puts it, knowledge is a potentially "infinite process" because, "beyond every (relative) beginning, we must look for a prior beginning" (Natorp 1910, 13–14). Every such new "beginning leads to wider and deeper developments" (ibid., 15). This opens the way to an "unlimited deepening of the problem" (ibid., 16). So, as regards knowledge, "we can speak of an infinite backward path from hypotheses to ever more fundamental hypotheses" (ibid., 15).

This view of knowledge was behind the philosophical practice of Plato and the Academy, which was based on the analytic method. As Berti says, "it seems that, at the beginning of each discussion, Plato posed a problem and invited his friends to propose some solutions ('hypotheses'); then each proposed solution was submitted to severe scrutiny, through questions, answers and, in fact, attempts of refutation," so that "the solution capable of proving more resistant to 'all the refutations' could be, at least provisionally, regarded as true" (Berti 2010, ix). But, being regarded as true only provisionally, the solution (hypothesis) was itself a problem to be solved, and so on. For this reason often, at the end of several Plato's Dialogues, speaking to the respondent, the questioner says something of the kind: "I would like us, having come this far, to continue" (Plato, *Protagoras*, 361 c 4–5). To which the respondent replies: "We will examine these things later, whenever you like" (ibid., 361 e 5–6). (For more on this, see Cellucci 2013, Section 7.2).

By being a potentially infinite process, knowledge is never definitive, it is an ongoing process that leads to ever deeper hypotheses.

This agrees with the views of Hume and Kant on solutions to problems. Indeed, Hume says that "each solution" to a problem "still gives rise to a new question as difficult as the foregoing, and leads us on to farther enquiries" (Hume 2007, 14). Kant says that "any answer given according to principles of experience always begets a new question which also requires an answer" (Kant 2002, 141–142).

5.29 Analytic Method and the Inexhaustibility of Mathematics

The analytic method accounts for what Gödel calls "the phenomenon of the inexhaustibility of mathematics" (Gödel 1986–2002, III, 305).

Gödel says that the incompleteness theorems establish "the incompletability or inexhaustibility of mathematics" (ibid.). Specifically, it is the second incompleteness theorem that "makes the incompletability of mathematics particularly evident" (ibid., III, 309). For, "it makes it impossible that someone should to set up a certain well-defined system of axioms and rules and consistently make the following assertion about it: All of these axioms and rules I perceive (with mathematical certitude) to be correct, and moreover I believe that they contain all of mathematics" (ibid.). Indeed, "if someone makes such a statement he contradicts himself" because, "if he perceives the axioms in consideration to be correct, he also perceives" that "they are consistent. Hence he has a mathematical insight not derivable from his axioms" (ibid.). This means that "no well-defined system of correct axioms can contain all of mathematics," in the sense of "the system all true mathematical propositions," because "the proposition which states the consistency of the system is true, but not demonstrable in the system" (ibid.).

Thus, however, Gödel does not account for the inexhaustibility of mathematics. For, to say that no well-defined system of correct axioms can contain all of mathematics expresses a limitation of such systems, it does not explain the reason of the inexhaustibility of mathematics.

Conversely, the analytic method accounts for the inexhaustibility of mathematics because it explains such reason. For, according to it, any hypothesis which provides a solution to a problem can at most be plausible, and is itself a problem that must be solved, so no hypothesis is final. Therefore, solving a problem is a potentially infinite process.

5.30 Analytic Method and Infinite Regress

Since, with the analytic method, there are no primitive premisses and hence the series of the premisses will be infinite, it might be objected that one cannot acquire knowledge by means of it.

This is the substance of Aristotle's infinite regress argument, already discussed in Chap. 2. Gödel uses essentially the same argument when he says that one cannot acquire knowledge "by trying to give explicit definitions for concepts and proofs for axioms, since for that one obviously needs other undefinable abstract concepts and axioms holding for them. Otherwise one would have nothing from which one could define or prove" (Gödel 1986–2002, III, 383).

This objection, however, is unjustified. For, the fact that, with the analytic method, there are no primitive premisses, does not mean that by such method one cannot acquire knowledge. One could not acquire knowledge only if the hypotheses were arbitrary. But they are not arbitrary, they must be plausible. So, by the analytic method, one can acquire knowledge. On the other hand, the knowledge one can acquire by means of it is fallible, in the sense that it cannot be ruled out that it can lead to error. New data can always emerge with which the hypotheses on which knowledge is based may turn out to be incompatible.

Lewis says that "to speak of fallible knowledge, of knowledge despite uneliminated possibility of error, just sounds contradictory" (Lewis 1996, 549). But even mathematical knowledge is fallible because, by Gödel's second incompleteness theorem, it cannot be ruled out that it can lead to error. A fortiori, non-mathematical knowledge is fallible. So, if Lewis were right, no knowledge would be possible.

5.31 Analytic Method and Non-Finality of Solutions to Problems

Since, with the analytic method, no solution to a problem is final, it might be objected that the analytic method contrasts with the fact that the things of mathematics are determinate and unchanging.

Thus, Grosholz says: "I do not think that the naturalist view that we should focus on plausibility instead of truth, and construe justification as always somewhat empirical and revisable, captures the peculiar nature of mathematical reasoning" (Grosholz 2016, 49). For, "the things of mathematics are determinate and unchanging as the things of nature are not: why would we ever revise our belief that the sum of the squares of the two oppositive sides of a Euclidean right triangle is equal to the square of its hypotenuse, or that 2 + 3 = 5?" (ibid.).

This objection, however, is unjustified. As we have seen in Chap. 3, several mathematicians believe that solutions to problems are permanent, but this belief is invalid. As other mathematicians have pointed out, solutions to problems are not permanent, they are always revisable.

Thus, Poincaré says that, in mathematics, there are not "solved problems and others which are not; there are only problems more or less solved," although "it often happens" that "an imperfect solution guides us toward a better one" (Poincaré 2015, 377–378).

Davis says that, in mathematics, "there is no finality in the creation, formulation and solutions of problems," we can only "settle for provisional, 'good enough' solutions" (Davis 2006, 164). In fact, "discovering a sense in which a solved problem is still not completely solved but leads to new and profound challenges, is one important direction that mathematical research takes" (ibid., 177).

The analytic method provides such sense. According to it, a solved problem is still not completely solved because any hypothesis which yields a solution to the problem can at most be plausible, and moreover, it is itself a problem that must be solved, so no hypothesis is final. Therefore, a solved problem leads to new and profound challenges.

Contrary to what the objection assumes, if there is no reason why we would ever revise our belief that the sum of the squares of the two legs of a Euclidean right-angled triangle is equal to the square of its hypotenuse, or that 2 + 3 = 5, it is not because the things of mathematics are determinate and unchanging. It is rather because, at our scale, the world is mostly made up of discrete objects that combine according to equations such as 2 + 3 = 5, and of continuous objects such that the sum of the squares of the two legs of a Euclidean right-angled triangle is equal to the square of its hypotenuse, and the beliefs in question have been embodied in human beings by natural selection. They would have been different if human beings had evolved in a different environment, so they are not absolute truths but only plausible beliefs.

As Dehaene says, "perhaps our arithmetic would have been radically different," in particular we would not have held the belief that 1 + 1 = 2, "if, like cherubs, we had evolved in the heavens where one cloud plus another cloud was still one cloud" (Dehaene 2011, 231).

5.32 Fortune of the Analytic Method

The analytic method is not confined to antiquity. Although this is not generally realized, it has continued to be used, even unknowingly, until the present day.

For example, let us consider the solution to the problem posed by Fermat's conjecture: Show that there are no positive integers x, y, z such that $x^n + y^n = z^n$ for $n > 2$. According to the current prevailing view, the problem was solved by Wiles 1995, Taylor and Wiles 1995. But this is not literally correct. Rather, the problem was solved by Ribet 1990 using the Taniyama-Shimura conjecture: Every elliptic curve over the rational numbers is modular. For, Ribet showed: Taniyama-Shimura conjecture $\Rightarrow$ Fermat's conjecture. Thus Ribet showed that the Taniyama-Shimura conjecture is a sufficient condition for solving the problem posed by Fermat's conjecture. Specifically, Ribet showed that, if there existed positive integers x, y, z such that $x^n + y^n = z^n$ for $n > 2$, they would yield an elliptic curve over the rational numbers which would not be modular, contradicting the Taniyama-Shimura conjecture.

But the Taniyama-Shimura conjecture was in its turn a problem that had to be solved. It was solved by Wiles and Taylor using hypotheses from various fields of mathematics, from differential geometry to complex analysis. Thus what Wiles and Taylor solved was not the problem posed by Fermat's conjecture, but the problem posed by the Taniyama-Shimura conjecture.

The process through which the problem posed by Fermat's conjecture was solved can then be schematically represented as follows.

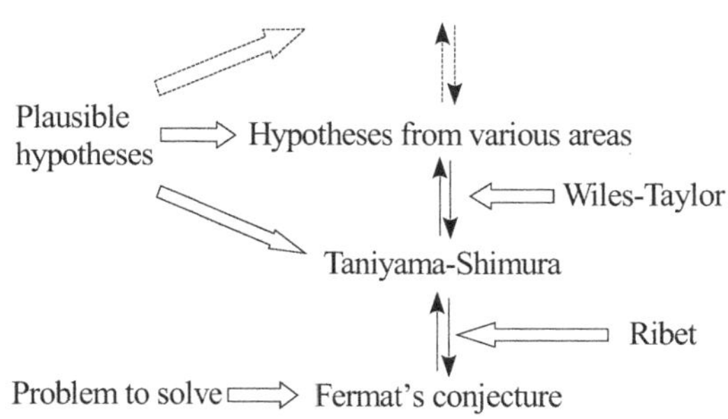

More precisely, what Wiles and Taylor solved was not the problem posed by the Taniyama-Shimura conjecture, but a special case of it: Every semistable elliptic curve over the rational numbers is modular, which was enough to solve the problem posed by Fermat's conjecture. The problem posed by the full Taniyama-Shimura conjecture was solved by Breuil et al. 2001, building on the earlier work of Wiles and Taylor. But this is inessential here.

Most mathematicians would object that the problem posed by Fermat's conjecture was not solved by Ribet, because Ribet used a hypothesis, the Taniyama-Shimura conjecture, that at the time had not been demonstrated yet. Since the problem posed by the Taniyama-Shimura conjecture was solved by Wiles and Taylor, the solution to the problem posed by Fermat's conjecture should be credited to them.

But this objection is invalid. For, Wiles and Taylor solved the problem posed by the Taniyama-Shimura conjecture by making use of hypotheses from various mathematics fields which depended on other hypotheses, and ultimately on the axioms of set theory. Now, the axioms of set theory are a hypothesis that has not been demonstrated yet. So, if Ribet did not solve the problem posed by Fermat's conjecture because he used a hypothesis, the Taniyama-Shimura conjecture, that at the time had not been demonstrated yet, then Wiles and Taylor did not solve the problem posed by Fermat's conjecture because they used hypotheses from various mathematics fields which ultimately depended on a hypothesis, the axioms of set theory, that to this very day has not been demonstrated yet.

This by no means intends to diminish the value of the result of Wiles and Taylor. Indeed, Lang says that the Taniyama-Shimura "conjecture is one of the most important of the century" (Lange Lang 1995, 1301). In contrast, Rav says that Fermat's Last Theorem "is insignificant (historical interest apart)" (Rav 1999, 25). To say that what Wiles and Taylor solved was not the problem posed by Fermat's conjecture, but the problem posed by the Taniyama-Shimura conjecture, is only meant to clarify the nature of the process involved.

5.33 Analytic Method and Abduction

Some people have claimed that there is a relation between the analytic method and Peirce's abduction.

Thus, Niiniluoto says that "Peirce's description of" abduction, "as a retroductive inference of a cause from its effect, is an instance" of "analysis" (Niiniluoto 2018, 19).

Koskela, Paavola, and Kroll say that "one interesting parallel to Peirce's abduction (especially when interpreted as a regressive inference of a cause from its effect) is the method of analysis (where regressive inferences also play a central part)" (Koskela, Paavola, and Kroll 2018, 157).

But this is invalid. The analytic method must not be confused with abduction, because it is a method of discovery, while abduction is not a method of discovery.

Indeed, according to Peirce's formulation, abduction is an inference of the form: P, if Q then P, therefore Q.

For, Peirce says that abduction is: "The surprising fact, *C*, is observed. But if *A* were true, *C* would be a matter of course. Hence, there is reason to suspect that *A* is true" (Peirce 1931–1958, 5.189). For example, Galileo observes the surprising fact *C*: 'Four bodies change their position around Jupiter'. But if *A*: 'Jupiter has satellites', were true, *C* would be a matter of course. 'Hence', Galileo concludes, 'there is reason to suspect that *A* is true'.

The question is what value is to be assigned to this kind of inference. In this regard, Peirce's views fall into two periods.

In a first period, Peirce claims that abduction introduces new ideas, and indeed is "the only logical operation which introduces any new idea" (ibid., 5.172). So "all ideas of science come to it by way of abduction" (ibid., 5.145). Therefore, "abduction must cover all operations by which theories and conceptions are engendered" (ibid., 5.590).

But these claims are invalid. Abduction is non-ampliative because the conclusion *Q* already appears in the premiss 'if *Q* then *P*'. Now, as Frankfurt observes, "if the new idea, or hypothesis, must appear in one of the premisses of the abduction, it cannot be the case that it originates as the conclusion of such an inference; it must have been invented before the conclusion was drawn" (Frankfurt 1958, 594). Then, what generates a new idea, or hypothesis, is not abduction, but the process that yields the premiss 'if *Q* then *P*', so a process prior to abduction. Therefore, "the idea that hypotheses originate as the conclusions of abductions, or that new ideas result from abductive inferences, cannot be accepted" (ibid., 595).

In a later period, however, Peirce admits that abduction is non-ampliative. He recognizes that a conclusion "cannot be abductively inferred, or if you prefer the expression, cannot be abductively conjectured until its entire content is already present in the premiss" (Peirce 1931–1958, 5.189). Therefore, "quite new conceptions cannot be obtained from abduction" (ibid., 5.190).

Since abduction is non-ampliative, it cannot be a method of discovery. On the other hand, the analytic method is a method of discovery. Therefore, the analytic method must not be confused with abduction.

An example of this confusion is Niiniluoto's claim that "many of the inferential steps and 'guesses' that" Sherlock Holmes performs during his "stage of investigation are abductive" in "Peirce's sense" (Niiniluoto 2018, 28).

This claim is invalid. The inferential steps and guesses in question are not abductive in Peirce's sense, they are made by the analytic method. As Franchella says, the "long tradition of analysis is the method which Sherlock Holmes referred to" (Franchella 2011, 233).

Indeed, Sherlock Holmes says that "in solving a problem of this sort, the grand thing is to be able to reason backward," namely "analytically" (Doyle 2020, 115). Most people can only "reason forward," namely "synthetically," but "there are few people" who, "if you told them a result, would be able to evolve from their own inner consciousness what the steps were which led up to that result. This power is what I mean when I talk of reasoning backwards, or analytically" (ibid., 115–116).

From this, it is clear that the science of deduction and analysis to which Sherlock Holmes refers is the analytic method, where analysis is an upward movement from problems to hypotheses, and deduction is a downward movement from hypotheses to problems.

In addition to being non-ampliative, abduction is also non-valid because, with *P* true and *Q* false, the premisses *P* and 'if *Q* then *P*' are both true, but the conclusion *Q* is false. Earlier in the chapter, it has been mentioned that there are inference rules that are both non-valid and non-ampliative. Then, abduction is an example of such rules.

5.34 Analytic Method and Reductio ad Absurdum

While there is no relation between the analytic method and abduction, there is a relation between the analytic method and reductio ad absurdum.

Reductio ad absurdum is an inference of the form: if *P*, then *Q* and not-*Q*, therefore not-*P*. The reason why there is a relation between the analytic method and reductio ad absurdum is that, in the latter, to solve a problem of the form not-*P*, one assumes a hypothesis *P* that is a sufficient condition for solving the problem. So, one does not start from already given principles, but from something assumed as a hypothesis, hence, like the analytic method, reductio ad absurdum is a procedure by hypothesis.

Aristotle underlines this by saying that in reductio ad absurdum "the conclusion is reached from a hypothesis" (Aristotle, *Analytica Priora*, A 44, 50 a 32). All those who reach a conclusion by absurdum "deduce the falsehood by a syllogism, but demonstrate" the "conclusion from a hypothesis, when something impossible results from the assumption of the contradictory" (ibid., A 23, 41 a 23–26).

5.35 Example of Reductio ad Absurdum

As an illustration of reductio ad absurdum, let us consider the following problem:

> *Incommensurability of diagonal and side of square*: For any given square, show that the diagonal of the square is incommensurable with the side.

The first solution to the problem is credited to the Pythagoreans. However, there is no textual evidence regarding their solution.

Aristotle describes a solution to the problem saying that "one demonstrates that the diagonal is incommensurable because, if one assumes that it is commensurable, odd numbers turn out to be equal to even ones. That odd numbers turn out to be equal to even ones is deduced by a syllogism, but that the diagonal is incommensurable is demonstrated from a hypothesis, since a falsehood results from the assumption of the contradictory" (Aristotle, *Analytica Priora*, A 23, 41 a 26–30).

Heath claims that the demonstration "referred to by Aristotle" is that "by which the Pythagoreans proved the incommensurability of $\sqrt{2}$ with unity" and is the demonstration which appears "in the text of Euclid" but "is undoubtedly an interpolation," and accordingly is relegated "to an Appendix" (Heath 1956, III, 2). (For the demonstration which appears in the text of Euclid, see Chap. 14).

But these claims do not seem to be valid because, as Corry says, "the extant Greek sources do not show any evidence of symbolic manipulation of the kind assumed by Heath" (Corry 2015, 54). Also, "there is a clear historical difference" between "the claim that the diagonal is incommensurable with the side and the claim that $\sqrt{2}$ is an irrational number" because, "in the Greek mathematical culture, the ratio between diagonal and side was by no means a number but was just that, a ratio" (ibid.).

A possible alternative solution to the problem, which relies only on ideas used by the Pythagoreans in their work, is the following (see Knorr 1975, 26–27). Let us consider the diagram that appears in Plato's solution to the problem of doubling the square.

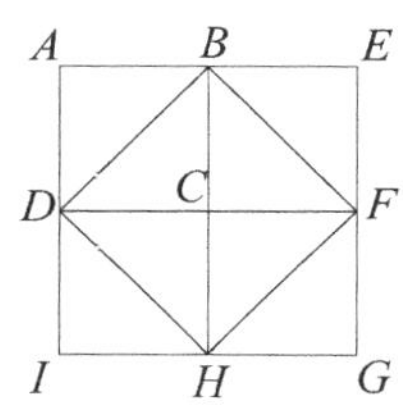

To solve the problem, the Pythagoreans assume *P*: The diagonal *DF* and the side *DB* of the square *DBFH* are commensurable, namely they have a common measure. Then, each of *DF* and *DB* can be viewed as representing a number, namely the number of times that each is measured by their common measure. It can be assumed that *DF* and *DB* do not both represent even numbers (for, otherwise there would be a larger common measure). Then *Q*: At least one of *DF* and *DB* represents an odd number.

From the diagram it is clear that the squares *AEGI* and *DBFH* represent square numbers (in the sense of the Pythagorean figurate numbers). Moreover, *AEGI* represents a square number that is double the square number represented by *DBFH* (counting the number of equal triangles on each), so *AEGI* represents an even square number. Hence, its side *AE* represents an even number (for, if a square represents an even square number, its side represents an even number, as it can be seen from the Pythagorean figurate numbers). But then *AEGI* represents a multiple of four. Since *AEGI* represents a square number which is double the square number represented by *DBFH*, from this it follows that *DBFH* represents a square number which is a multiple of two, hence an even square number. Hence, its side *DB* represents an even number (for, if a square represents an even square number, its side represents an even number). On the other hand, since *DF* is equal to *AE*, *DF* represents an even number. So *DF* and *DB* both represent even numbers. Then not-*Q*: None of *DF* and *DB* represents an odd number.

Thus, from the assumption *P*, the Pythagoreans deduce a contradiction, *Q* and not-*Q*. Therefore, by reductio ad absurdum, they conclude not-*P*: The diagonal *DF* and the side *DB* of the square *DBFH* are incommensurable.

5.36 Original Reason of Reductio ad Absurdum

Reductio ad absurdum was first used to demonstrate conclusions contrary to immediate experience. Such is the case of the Pythagoreans, who used it to demonstrate the incommensurability of diagonal and side of square. This is contrary to immediate experience, which would suggest that two lines could always be made commensurable by taking a sufficiently small unit.

Such is also the case of the Eleats. For example, Zeno used reductio ad absurdum to demonstrate that in a race, if given a head start, the slowest runner will never be overtaken by the fastest runner. This is contrary to immediate experience, which would suggest that the fastest runner would always overtake the slowest runner. The problem considered by Zeno is the following.

> *Achilles and the tortoise*: Show that, in a race, if given a head start, the slowest runner will never be overtaken by the fastest runner.

To solve the problem, Zeno assumes *P*: In a race, if given a head start, the slowest runner will be overtaken by the fastest runner. Now, in the race between Achilles and the tortoise in which the tortoise is given a head start, the tortoise is the slowest runner, and Achilles the fastest runner. Then *Q*: The tortoise will be overtaken by Achilles. On the other hand, in a race, if given a head start, "the slowest runner will never be overtaken by the fastest runner, because it is necessary that the pursuer should first reach the point from which the pursued started, so that necessarily the slowest runner is always somewhat in advance" (Zeno 29 A 26, ed. Diels-Kranz). Then, $\neg Q$: The tortoise will never be overtaken by Achilles. Thus, from the assumption *P*, Zeno deduces a contradiction, *Q* and not-*Q*. Therefore, by reductio ad absurdum, he concludes not-*P*: In a race, if given a head start, the slowest runner will not be overtaken by the fastest runner.

5.37 Differences Between Analytic Method and Reductio ad Absurdum

Even if there is a relation between the analytic method and reductio ad absurdum, there are important differences between them.

(1) The analytic method is an upward path from the problem to hypotheses. Conversely, reductio ad absurdum is a downward path because, to demonstrate not-*P*, one assumes *P* and deduces a contradiction, *Q* and not-*Q*, from it.

(2) In the analytic method, hypotheses are not known from the beginning, to find them is the goal of the research. Conversely, in reductio ad absurdum, the hypothesis *P* is known from the beginning, since it is the opposite of the desired conclusion not-*P*, which of course is known from the beginning.

(3) In the analytic method, hypotheses must be plausible. As Aristotle says, in it "a preliminary agreement is needed" about the plausibility of the hypothesis, "if the other people is to consent" (Aristotle, *Analytica Priora*, A 44, 50 a 33–34). Conversely, in reductio ad absurdum, the hypothesis *P* cannot be plausible, because it is the opposite of the desired conclusion not-*P*, which is meant to be plausible. In it, "people concede the point even with no preliminary agreement, because the falsehood is evident; for example, when it is assumed that the diagonal" of the square" is commensurable "with the side," we immediately see the falsehood of the ensuing proposition "that odd numbers will be equal to even numbers" (ibid., A 44, 50 a 35–38).

References

Benecke, Adolph. 1867. *Ueber die geometrische Hypothesis in Platons Menon*. Elbing: Kafemann.

Berti, Enrico. 2010. *Sumphilosophein: La vita nell'Accademia di Platone*. Rome: Laterza.

Breuil, Christophe, Brian Conrad, Fred Diamond, and Richard Taylor. 2001. On the modularity of elliptic curves over **Q**: Wild 3-adic exercises. *Journal of the American Mathematical Society* 14: 843–939.

Capozzi, Mirella. 2013. *Kant e la logica I*. Naples: Bibliopolis.

———. 2019. Le ipotesi secondo Kant: Requisiti, giustificazione, status epistemico e euristica. *Syzetesis* 6 (1): 153–189. 183–184.

———. 2020. *Wahrscheinlichkeit* and *Scheinbarkeit*: A key issue in Kant's logic and philosophy. In *Kants Schriften in Übersetzungen, Sonderheft 15 of Archiv für Begriffsgeschichte*, ed. Gisela Schlüter, 513–533. Hamburg: Meiner.

Cellucci, Carlo. 2013. *Rethinking logic: Logic in relation to mathematics, evolution, and method*. Cham: Springer.

Clagett, Marshall. 1999. *Ancient Egyptian science: A source book*, vol. 3: *Ancient Egyptian mathematics*. Philadelphia: American Philosophical Society.

Cohen, Morris Raphael, and Ernest Nagel. 1964. *An introduction to logic and scientific method*. London: Routledge.

Corry, Leo. 2015. *A brief history of numbers*. Oxford: Oxford University Press.

Davis, Philip J. 2006. *Mathematics and common sense: A case of creative tension*. Natick: A K Peters.

de Finetti, Bruno. 1974. The true subjective probability problem. In *The concept of probability in psychological experiments*, ed. Carl-Axel S. Staël von Holstein, 15–23. Dordrecht: Springer.

De Morgan, Augustus. 1835. Review of Georg Peacock, *A treatise on algebra*. *Quarterly Journal of Education* 9 (91–110): 293–311.

Deahene, Stanislas. 2011. *The number sense: How the mind creates mathematics*. Oxford: Oxford University Press.

Descartes, René. 1996. *Oeuvres*. Paris: Vrin.

Doxiadis, Apostolos, and Michalis Sialaros. 2013. Sing, muse, of the hypotenuse: Influences of poetry and rhetoric on the formation of Greek mathematics. In *Writing science: Medical and mathematical authorship in ancient Greece*, ed. Markus Asper, 367–409. Berlin: de Gruyter.

Doyle, Arthur Conan. 2020. A study in scarlet. In Arthur Conan Doyle, *Sherlock Holmes: The complete novels and stories*, vol. 1, 1–120. New York: Vintage.
Dummett, Michael. 1991. *The logical basis of metaphysics*. Cambridge: Harvard University Press.
Franchella, Miriam. 2011. Where the most logical mind may be at fault. In *Sherlock Holmes and philosophy: The footprints of a gigantic mind*, ed. Josef Steiff, 229–236. Chicago: Open Court.
Frankfurt, Harry. 1958. Peirce's notion of abduction. *The Journal of Philosophy* 55: 593–597.
Frege, Gottlob. 1960. *The foundations of arithmetic: A logico-mathematical enquiry into the concept of number*. New York: Harper.
Galavotti, Maria Carla. 2018. Who is afraid of subjective probability? In *From arithmetic to metaphysics: A path through philosophical logic*, ed. Ciro de Florio and Alessandro Giordani, 151–157. Berlin: de Gruyter.
Gödel, Kurt. 1986–2002. *Collected works*. Oxford: Oxford University Press.
Grosholz, Emily. 2016. *Starry reckoning: Reference and analysis in mathematics and cosmology*. Cham: Springer.
Hadamard, Jacques. 1954. *The psychology of invention in the mathematical field*. Mineola: Dover.
Haskins, Ekaterina V. 2004. *Logos and power in Isocrates and Aristotle*. Columbia: University of South Carolina Press.
Heath, Thomas L. 1956. *The thirteen books of Euclid's elements*. New York: Dover.
Hume, David. 2007. *An enquiry concerning human understanding and other writings*. Cambridge: Cambridge University Press.
Kant, Immanuel. 1992. *Lectures on logic*. Cambridge: Cambridge University Press.
———. 1998. *Critique of pure reason*. Cambridge: Cambridge University Press.
———. 2002. *Theoretical philosophy after 1781*. Cambridge: Cambridge University Press.
Knorr, Wilbur Richard. 1975. *The evolution of Euclidean elements: A study of the theory of incommensurable magnitudes and its significance for early Greek mathematics*. Dordrecht: Reidel.
Koskela, Lauri, Sami Paavola, and Ehud Kroll. 2018. The role of abduction in production of new ideas in design. In *Advancement in the philosophy of design*, ed. Pieter E. Vermaas and Stéphane Vial, 153–183. Cham: Springer.
Lang, Serge. 1995. Some history of the Shimura-Taniyama conjecture. *Notices of the American Mathematical Society* 42: 1301–1309.
Lewis, David Kellogg. 1996. Elusive knowledge. *Australasian Journal of Philosophy* 74: 549–567.
Lloyd, Geoffrey Ernest Richard. 1992. The *Meno* and the mysteries of mathematics. *Phronesis* 37 (2): 166–183.
Mill, John Stuart. 1963–1986. *Collected works*. Toronto: University of Toronto Press.
Natorp, Paul. 1910. *Die logischen Grundlagen der exakten Wissenschaften*. Leipzig: Teubner.
Niiniluoto, Ilkka. 2018. *Truth-seeking by abduction*. Cham: Springer.
Peirce, Charles Sanders. 1931–1958. *Collected papers*. Cambridge: Harvard University Press.
Poincaré, Henri. 2015. *The foundations of science: Science and hypothesis, The value of science, Science and method*. Cambridge: Cambridge University Press.
Pólya, George. 1941. Heuristic reasoning and the theory of probability. *The American Mathematical Monthly* 48: 450–465.
———. 1954. *Mathematics and plausible reasoning*. Princeton: Princeton University Press.
Prawitz, Dag. 2014. The status of mathematical knowledge. In *From a heuristic point of view*, ed. Cesare Cozzo and Emiliano Ippoliti, 73–90. Newcastle upon Tyne: Cambridge Scholars Publishing.
Rav, Yehuda. 1999. Why do we prove theorems? *Philosophia Mathematica* 7: 5–41.
Ribet, Kenneth A. 1990. From the Taniyama-Shimura conjecture to Fermat's last theorem. *Annales de la Faculté Mathematiques et des Sciences de Toulouse* 11: 116–139.
Rosenfeld, Boris Abramovich. 1988. *A history of non-Euclidean geometry: Evolution of the concept of a geometric space*. New York: Springer.
Russell, Bertrand. 1983. *Collected papers*. London: Routledge.
———. 1998. *The problems of philosophy*. Oxford: Oxford University Press.
———. 2009. *Human knowledge: Its scope and limits*. London: Routledge.

Striker, Gisela. 2009. Commentary. In Aristotle, *Prior Analytics*, Book I, 67–246. Oxford: Oxford University Press.

Swinburne, Richard. 1974. Introduction. In *The justification of induction*, ed. Richard Swinburne, 1–17. Oxford: Oxford University Press.

Taylor, Richard, and Andrew Wiles. 1995. Ring-theoretic properties of certain Hecke algebras. *Annals of Mathematics* 141: 553–572.

Walton, Douglas. 2001. Abductive, presumptive and plausible arguments. *Informal Logic* 21 (2): 141–169.

Wiles, Andrew. 1995. Modular elliptic curves and Fermat's last theorem. *Annals of Mathematics* 141: 443–551.

Wittgenstein, Ludwig. 2002. *Tractatus Logico-Philosophicus*. London: Routledge.

Chapter 6
Analytic-Synthetic Method and Axiomatic Method

Abstract From antiquity, a number of methods have been put forward as an alternative to the analytic method. The chapter examines Aristotle's analytic-synthetic method and its relation to the analytic method. Then, it examines Pappus's analytic-synthetic method and its relation to reductio ad absurdum. Finally, it examines the material axiomatic method, a byproduct of Aristotle's analytic-synthetic method, and the formal axiomatic method, the form of the axiomatic method that has prevailed for the past century.

Keywords Aristotle's analytic-synthetic method · A priori demonstration · A posteriori demonstration · Pappus's analytic-synthetic method · Material axiomatic method · Formal axiomatic method

6.1 Aristotle vs. Analytic Method

In Chap. 5 it has been argued that the analytic method can yield knowledge, though only plausible knowledge. This contrasts with the fact that, from antiquity, it has been assumed that knowledge must be true, since knowledge that is not true cannot be knowledge.

This assumption is made by Aristotle, who says that "knowledge and intuition are always true" (Aristotle, *Analytica Posteriora*, B 19, 100 b 7–8).

For this reason, as we have seen in Chap. 3, Aristotle opposes the analytic method, arguing that it does not proceed on the basis of a principle but only of a hypothesis, and one cannot have scientific knowledge on the basis of a hypothesis. For, a hypothesis is not necessarily true, while "knowledge must proceed from premisses that are true" (ibid., A 2, 71 b 20–21). Therefore, the analytic method does not yield knowledge.

Actually, Aristotle does not completely dismiss the analytic method, but makes some substantial changes to it. This gives rise to a substantially different method, Aristotle's analytic-synthetic method.

C. Cellucci, *The Making of Mathematics*, Synthese Library 448,
https://doi.org/10.1007/978-3-030-89731-4_6

6.2 Statement of Aristotle's Analytic-Synthetic Method

A brief description of Aristotle's analytic-synthetic method has already been given in Chap. 2, a fuller description will be given here.

Aristotle's analytic-synthetic method or method of analysis and synthesis is the method according to which, to solve a problem, one looks for some hypothesis that is a sufficient condition for solving the problem, namely such that a solution to the problem can be deduced from the hypothesis. The hypothesis is obtained from the problem, and possibly other data already available, by some non-deductive rule, and must be plausible, namely such that the arguments for the hypothesis are stronger than the arguments against it, on the basis of experience.

If the hypothesis so obtained is not a principle (or a proposition deduced from principles), one looks for another hypothesis that is a sufficient condition for solving the problem posed by the previous hypothesis, it is obtained from the latter, and possibly other data already available, by some non-deductive rule, and must be plausible. And so on, until one arrives at a principle (or a proposition deduced from principles). The principles must be true. When one arrives at a principle (or a proposition deduced from principles), the process terminates. This is analysis.

At this point, one tries to see whether, inverting the order of the steps followed in analysis, one obtains a deduction of the solution of the problem from the principle (or proposition already deduced from principles) arrived at in analysis. This is synthesis.

Aristotle's analytic-synthetic method can be schematically represented as follows.

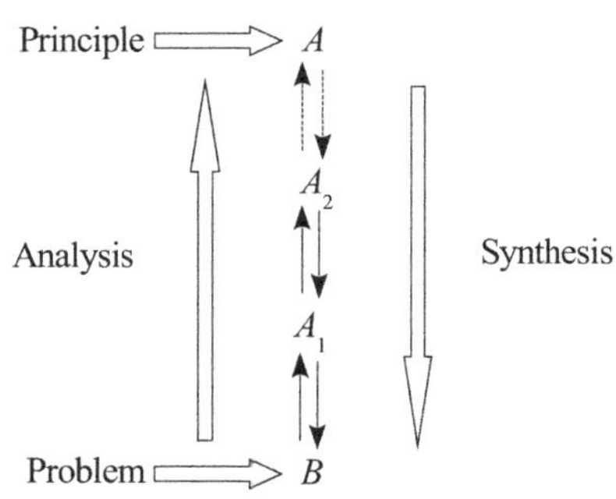

6.3 Original Formulation of Aristotle's Analytic-Synthetic Method

Hintikka and Remes claim that "the only extensive general description of the method of analysis and synthesis to be found in the surviving ancient literature is given by Pappus of Alexandria" (Hintikka and Remes 1976, 254). But this claim is invalid, because Aristotle gives an extensive general description of the method of analysis and synthesis.

Indeed, Aristotle formulates the analytic-synthetic method by saying that "the process of knowledge proceeds from what is more knowable and clearer to us to what is clearer and more knowable by nature" (Aristotle, *Physica*, A 1, 184 a 16–18). Therefore, "we must necessarily proceed in this way, namely, starting from the things that are less known by nature but clearer to us, towards those which are clearer and more knowable by nature" (ibid., A 1, 184 a 18–21). Now, what is more knowable and clearer to us is the problem to solve, while what is clearer and more knowable by nature are the principles. They are "most knowable" by nature because, "by reason of these, and from these, all other things are known, but these are not known by means of the things subordinate to them" (Aristotle, *Metaphysica*, A 2, 982 b 2–4). So, by saying that we must proceed from what is more knowable and clearer to us to what is clearer and more knowable by nature, Aristotle means to say that we must proceed from the problem to principles.

Specifically, given the problem to solve, we must find "the necessary hypotheses through which the syllogisms come about" (Aristotle, *Topica*, Θ 1, 155 b 29). As we have seen in Chap. 4, according to Aristotle, we will find them either by syllogism from effects or by induction. When the hypotheses are found, we "should not put forward these" hypotheses "right away, but rather should stand off as far above them as possible" (ibid., Θ 1, 155 b 29–30). Namely, we should find other hypotheses from which the previous hypotheses can be deduced. And so on, until we arrive at a principle. The principles must be true, "because it is not possible to have scientific knowledge of what is not the case, for example, that the diagonal" and side of a square are "commensurate" (Aristotle, *Analytica Posteriora*, A 2, 71 b 25–26).

When we arrive at a principle, the process terminates. This is analysis.

But analysis is not always successful and, "if we meet with an impossibility, we give up" (Aristotle, *Ethica Nicomachea*, Γ 3, 1112 b 24–25). Conversely, "if the thing appears possible, we try to do it" (ibid., Γ 3, 1112 b 26–27). Namely, we try to see whether, by inverting the order of the steps followed in analysis, we may obtain a deduction of the solution of the problem from the principle arrived at in analysis. This is synthesis.

This method serves not only to acquire scientific or mathematical knowledge, but also to solve problems in practical fields, such as medicine, which require deliberation to reach the intended end. In such fields "one deliberates not about ends, but about means to ends" (ibid., Γ 3, 1112 b 11–12). For example, doctors do not deliberate whether they are to heal, but take the end that they want to achieve, namely healing, and "consider how and by what means it is to be attained" (ibid., Γ 3, 1112 b 15–16). Then they consider "by what other means" these "means will be achieved," and so on, "until they arrive at the first cause, which is last in the order of discovery" (ibid., Γ 3, 1112 b 18–20). This is analysis. At that point, they try to see whether, by inverting the order of the steps followed in analysis, starting from the first cause arrived at in analysis, they can achieve the intended end, namely healing. This is synthesis or construction. Thus, "he who deliberates seems to carry out an inquiry and an analysis in the way described as though he were carrying out an analysis in a geometrical construction" (ibid., Γ 3, 1112 b 20–21). Since the synthesis, or construction, starts from the first cause, which is last in the order of discovery, "what is last in the order of analysis is first in the order of construction" (ibid., Γ 3, 1112 b 23–24).

From this, it is clear that there is no justification for Hintikka and Remes's claim that the only extensive general description of the method of analysis and synthesis to be found in the surviving ancient literature is given by Pappus of Alexandria. Moreover, as we will see below, unlike Aristotle's analytic-synthetic method, the description of the method of analysis and synthesis given by Pappus is problematic, because Pappus makes some apparently incompatible assertions about the direction of analysis.

6.4 Example of Aristotle's Analytic-Synthetic Method

As an example of Aristotle's analytic-synthetic method, let us consider a solution to the following problem.

> *Vertically opposite angles* (Euclid, *Elementa*, I, Proposition 15). Show that, if two straight lines cut one another, then they make the vertically opposite angles equal to one another. (The vertically opposite angles are the angles opposite each other at the vertex).

Let two straight lines *AB* and *CD* cut one another at the point *E*. We must show that the angles *CEA* and *DEB* are equal to one another, and the angles *CEB* and *AED* are equal to one another.

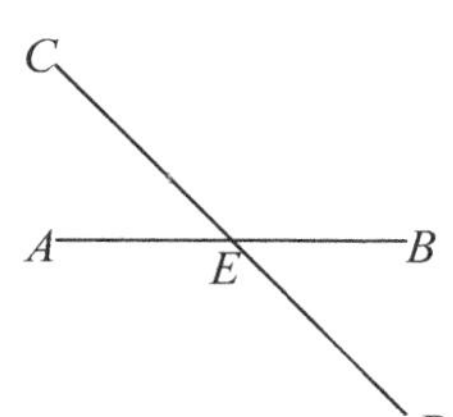

By analyzing the conditions under which the problem would be solved, we non-deductively arrive at the following hypothesis:

(I) If a straight line stands on another straight line, then it makes angles whose sum is equal to two right angles.

Hypothesis (I) is a sufficient condition for solving the problem. For, by hypothesis (I), since the straight line *AE* stands on the straight line *CD*, it makes angles *CEA* and *AED* whose sum is equal to two right angles. Again, by hypothesis (I), since the straight line *DE* stands on the straight line *AB*, it makes angles *AED* and *DEB* whose sum is equal to two right angles. Being both equal to two right angles, the sum of the angles *CEA* and *AED* and the sum of the angles *AED* and *DEB* are equal to one another. Therefore, the angles *CEA* and *DEB* are equal to one another. Similarly, it can be shown that the angles *CEB* and *AED* are equal to one another. Therefore, the vertically opposite angles are equal to one another. This solves the problem.

But hypothesis (I) is Euclid, *Elementa*, I, Proposition 13. Thus, we have arrived at a proposition deduced from principles, so the process terminates.

This is analysis.

Now we invert the process. We start from Euclid, *Elementa*, I, Proposition 13, and we deduce the solution to the problem by the same procedure by which we showed that hypothesis (I) is a sufficient condition for solving the problem.

This is synthesis.

6.5 The Direction of Analysis in Aristotle's Analytic-Synthetic Method

In Aristotle's analytic-synthetic method, analysis is an upward path. Some people, however, claim that analysis is a downward path. They argue that analysis must be a deductive procedure, otherwise it would always be trivially convertible into synthesis.

Thus, Knorr claims that Aristotle "portrays analysis as a deductive procedure" (Knorr 1993, 75). And rightly so, because the attempt to portray analysis "as a search for appropriate antecedents" would "render superfluous any concern over convertibility" of analysis into synthesis, since "the analysis (of antecedents) would of itself have produced the deductive sequence of the synthesis" (ibid., 95, footnote 65).

But this claim is invalid. For, Aristotle gives two good reasons why analysis as an upward path may not always be convertible into synthesis.

(1) Aristotle says that, "if it were impossible to prove something true from something false, it would be easy to make analyses," namely to discover premisses from which to deduce the conclusion, "because then the propositions would convert of necessity" (Aristotle, *Analytica Posteriora*, A 12, 78 a 6–8). But "it is possible to deduce a true conclusion from false premisses" (ibid., B 2, 53 b 8). Therefore, "sometimes it happens, as in the case of geometrical demonstrations, that, after making the analysis, we are unable to make the synthesis" (Aristotle, *Sophistici Elenchi*, 16, 175 a 27–28).

(2) Aristotle points out that the principle obtained by analysis may not be the cause of the conclusion. When this happens, the principle does not produce scientific knowledge, because "we have scientific knowledge of something only when we know its cause" (Aristotle, *Analytica Posteriora*, A 2, 71 b 30–31).

Reasons (1) and (2), which explain why analysis is not always convertible into synthesis, support the interpretation of analysis as an upward path.

That, in Aristotle's analytic-synthetic method, analysis is an upward path, is also confirmed by Alexander of Aphrodisias, who says that, for Aristotle, "analysing is a route from the end up to the principles" (Alexander of Aphrodisias, *In Aristotelis Analyticorum Priorum Librum Primum Commentarium*, 7.14–15, ed. Wallies).

6.6 Basic Changes with Respect to the Analytic Method

The analytic method and Aristotle's analytic-synthetic method are often confused with each other. For example, Menn says that both "Plato and Aristotle" hold that "there are two stages of argument," namely, "first to the archai," that is, to the principles, "(contrary to the 'natural' order of things) and then from the archai (following the 'natural' order)" (Menn 2002, 193). Thus, Menn claims that Plato and Aristotle both hold that analysis is an upward path to principles.

But this is invalid. According to Plato, analysis is an upward path not to principles but to plausible hypotheses. Therefore, the analytic method and Aristotle's analytic-synthetic method are essentially different.

Admittedly, as we have seen in Chap. 5, according to Plato, the process of passing from hypothesis to hypothesis will eventually come to an end, and we will reach what is unhypothetical, the principle of the whole. However, this cannot happen while we are alive, but only after death. While we are alive, there remain essential differences between analysis in the analytic method and analysis in Aristotle's analytic-synthetic method.

With respect to the analytic method, Aristotle's analytic-synthetic method involves three basic changes.

(1) The search for a solution to a problem is a finite process, so the ascending sequence of the premisses must terminate. For, if it "does not terminate and there is always something above whatever premiss has been taken, then there will be demonstrations of all things" (Aristotle, *Analytica Posteriora*, A 22, 84 a 1–2). Including falsehoods. Thus, there would be no scientific knowledge.

(2) Analysis is ultimately a means for finding deductions of given conclusions from already given principles, so for finding demonstrations in given axiomatic systems. As Geminus says, for Aristotle "analysis is the discovery of demonstration" (Ammonius, *In Aristotelis Analyticorum Priorum Librum I Commentarium*, 5.28, ed. Wallies). Specifically, demonstration in a given axiomatic system.

(3) Since analysis is ultimately a means for finding demonstrations in given axiomatic systems, analysis is subordinated to the axiomatic method. As a means for discovering unknown hypotheses, it loses its function. Thus, as Lakatos says,

analysis is no longer "a venture into the unknown," but only "an exercise in mobilizing and ingeniously connecting the relevant parts of the known" (Lakatos 1978, II, 100).

Aristotle's analytic-synthetic method can be viewed as what results from the analytic method when the process of finding deeper and deeper hypotheses is arbitrarily stopped at a certain stage, and the hypotheses reached at that stage are assumed as principles.

6.7 Aristotle's Analytic-Synthetic Method and Intuition

In Aristotle's analytic-synthetic method, intuition plays an essential role. As already mentioned, in that method the principles must be true. Then the question arises of how the principles "become known" to be true, and "what is the state which gets to know them" (Aristotle, *Analytica Posteriora*, B 19, 99 b 17–18).

Now, according to Aristotle, the principles cannot become known to be true by demonstration, nor can the state which gets to know them be discursive thinking, otherwise the principles would be demonstrable. But the principles "are indemonstrable," or else there would be an infinite regress, "therefore, it will not be scientific knowledge but intuition that is concerned with the principles" (Aristotle, *Magna Moralia*, A 34, 1197 a 22–23). Then, the principles must become known to be true by intuition, and the state which gets to know them must be intuitive thinking. So, although "the hypotheses are the end" of analysis, it is not "argument that teaches us the principles" (Aristotle, *Ethica Nicomachea*, Z 8, 1151 a 16–18).

Analysis and intuition are essentially different. While analysis belongs to discursive thinking, intuition belongs to intuitive thinking. Moreover, while analysis is fallible since the hypotheses found by means of it need not be true, intuition is supposed to be infallible.

6.8 A Priori Demonstration and A Posteriori Demonstration

With reference to Aristotle's distinction between analysis and synthesis, in the medieval period a distinction is made between two kinds of demonstration, 'a priori demonstration' and 'a posteriori demonstration'. The former is demonstration which proceeds from causes to effect, the latter is demonstration which proceeds from effect to causes.

More precisely, in the medieval period, first, a distinction is made between two kinds of demonstration, 'demonstration propter quid' and 'demonstration quia'.

Thus, Thomas Aquinas says that "demonstration is twofold. One is through the cause, and is called propter quid, and this is to argue from what is prior absolutely; the other is through the effect, and is called a demonstration quia, and this is to argue from what is prior relatively to us" (Thomas Aquinas, *Summa Theologica*, Part I, Question 2, Article 2).

By the expressions 'what is prior absolutely' and 'what is prior relatively only to us', Thomas Aquinas means the same as that which Aristotle means by the expressions 'what is clearer and more knowable by nature' and 'what is more knowable and clearer to us'. So, his use of the expressions 'demonstration propter quid' and 'demonstration quia' reflects Aristotle's text.

Later, 'demonstration propter quid' and 'demonstration quia' are also called 'a priori demonstration' and 'a posteriori demonstration', respectively.

Thus, Albert of Saxony says that demonstration is twofold, one "proceeds from causes to effect, and is called a priori demonstration and demonstration propter quid," and the other "proceeds from effects to causes," and "is called a posteriori demonstration and demonstration quia" (Albert of Saxony, *Questiones Subtilissime in Libros Aristotelis de Caelo et Mundo – Questiones Subtilissime super Libros Posteriorum*, Book I, Question IX, f. 8 r).

The expressions 'a priori demonstration' and 'a posteriori demonstration' were still in use in the early modern period. Indeed, as we have seen in Chap. 4, Zabarella used them. They will continue to be used until the second half of the seventeenth century, when Arnauld and Nicole say that "proving the effects by the causes" is "called demonstrating a priori," while "proving the causes by the effects" is "called proving a posteriori" (Arnauld and Nicole 1992, 281).

6.9 Pappus's Analytic-Synthetic Method

In late antiquity, Pappus formulated a different form of the analytic-synthetic method, Pappus's analytic-synthetic method.

Pappus's analytic-synthetic method is the method according to which, to solve a problem, one assumes the problem as if it were established, and deduces a conclusion from it. The conclusion must be plausible. If the conclusion is not something established, namely a principle or something already known, one deduces a conclusion from it. The conclusion must be plausible. And so on, until one arrives at something established. Then the process terminates. This is analysis.

At this point, one tries to see whether, inverting the order of the steps followed in analysis, one obtains a deduction of the solution of the problem from the something established arrived at in analysis. This is synthesis.

While, in Aristotle's analytic-synthetic method, analysis is an upward path and synthesis a downward path, in Pappus's analytic-synthetic method, analysis and synthesis are both downward paths, being both deductive processes.

Pappus's analytic-synthetic method can be schematically represented as follows.

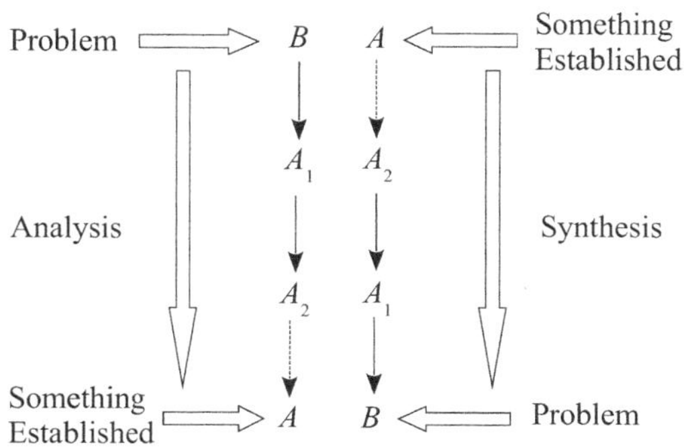

6.10 Clarifying Some Confusions

Aristotle's analytic-synthetic method is often confused with Pappus's analytic-synthetic method.

Thus, Hintikka and Remes claim that "Pappus's description of analysis and synthesis and the Aristotelian passage" in which Aristotle says that he who deliberates seems to carry out an analysis in a geometrical construction, "are closely similar" (Hintikka and Remes 1974, 86).

But this claim is invalid. For, as already pointed out above, while in Aristotle's analytic-synthetic method, analysis is an upward path and synthesis a downward path, in Pappus's analytic-synthetic method, analysis and synthesis are both downward paths, since they are deductive processes.

Even the analytic method or method of analysis is often confused with Pappus's analytic-synthetic method.

Thus, Sayre claims that a method "well known among mathematicians of Plato's day was the so-called 'method of analysis', in which proof of a given proposition is sought by deducing consequences from it until one is reached which is known independently to be true" (Sayre 1969, 22). Since geometers "typically are concerned with deductions in which both premises and conclusions" are "mutually convertible, a demonstration of the proposition in question often could be produced subsequently by a deductive movement in the opposite direction," which is "the method of synthesis" (ibid.). The method described by Plato in *Phaedo* is the one "familiar to the mathematicians of his day" (ibid.).

But this claim is invalid. For, while in the analytic method analysis is an upward path, as already said, in Pappus's analytic-synthetic method, analysis and synthesis are both downward paths, being deductive processes.

6.11 Original Formulation of Pappus's Analytic-Synthetic Method

In his original formulation of Pappus's analytic-synthetic method, Pappus first gives a definition of analysis and synthesis.

Indeed, Pappus says that "analysis is the way from what is sought, as if it were established, through its consequences, to something that is established in synthesis" (Pappus, *Collectio*, II, 634.11–13, ed. Hultsch). For, "in analysis we assume what is sought as if it had been achieved, and look for the thing from which it follows, and again what comes before that, until, by regressing in this way, we come upon something already known, or ranking as a principle. We call this way of proceeding analysis, as being a reduction backward" (ibid., II, 634.13–18). On the other hand, "in synthesis we assume what was reached last in analysis to have been achieved already, and, setting now in natural order, as antecedents, what before were following, and fitting them one with another, we in the end arrive at the construction of what was sought. This we call synthesis" (ibid., II, 634.18–23).

After giving a definition of analysis and synthesis, Pappus distinguishes between two different kinds of analysis, theorematic analysis, and problematic analysis.

Indeed, Pappus says that there are "two kinds of analysis; one seeks after the truth, and is called 'theorematic', the other tries to find what was demanded, and is called 'problematic'" (ibid., II, 634.24–26).

In theorematic analysis, "we assume what is sought as a fact and true, and proceed through its consequences, as if they were true facts according to the hypothesis, to something established" (ibid., II, 636.1–4). Then, "if the thing that has been established is true, the thing that was sought too will be true, and its demonstration will be the reverse of the analysis; but if we come upon something established to be false, the thing that was sought too will be false" (ibid., II, 636.5–7).

In problematic analysis, "we assume the required thing to be known, and proceed through its consequences, as if true, to something established" (ibid., II, 636.8–10). Then, "if the thing established is possible and obtainable," the "required thing too will be possible, and again the demonstration will be the reverse of the analysis; but if we come upon something established to be impossible, the problem too will be impossible" (ibid., II, 636.10–14).

In fact, Pappus's distinction between two different kinds of analysis, theorematic analysis and problematic analysis, is inessential, because theorematic analysis can be considered to be the special case of problematic analysis in which what is sought is the problem of whether what is sought is true.

6.12 Example of Pappus's Analytic-Synthetic Method

As an example of Pappus' analytic-synthetic method, let us consider a solution to the following problem, interpolated by an unknown scholiast to Euclid's *Elementa* XIII.

Extreme and mean ratio (Euclid, *Opera Omnia*, IV, Appendix I to Book XIII, Proposition 4, ed. Heiberg). Show that, if a straight line AB is cut in extreme and mean ratio at the point C, and AC is the greater segment, then $AB^2 + BC^2 = 3AC^2$

Let AB be cut in extreme and mean ratio at the point C, AC being the greater segment, namely $AB : AC = AC : BC$ (Euclid, *Elementa*, VI, Definition 3).

We assume what is sought, namely $AB^2 + BC^2 = 3AC^2$, as if it were established. Now, $AB^2 + BC^2 = 2AB \cdot BC + AC^2$ (Euclid, *Elementa*, II, Proposition 7). From the latter and $AB^2 + BC^2 = 3AC^2$ it follows $2AB \cdot BC + AC^2 = 3AC^2$, so $2AB \cdot BC = 2AC^2$, hence $AB \cdot BC = AC^2$. Therefore $AB : AC = AC : BC$. Thus, we have arrived at something established.

This is analysis.

Now we try to reverse the analysis in a synthesis.

Since $AB : AC = AC : BC$, we have $AB \cdot BC = AC^2$. So, $2AB \cdot BC = 2AC^2$, hence $2AB \cdot BC + AC^2 = 3AC^2$. But $AB^2 + BC^2 = 2AB \cdot BC + AC^2$ (Euclid, *Elementa*, II, Proposition 7). From the latter and $2AB \cdot BC + AC^2 = 3AC^2$ it follows $AB^2 + BC^2 = 3AC^2$, that is what was to be demonstrated.

This is synthesis.

6.13 The Direction of Analysis in Pappus's Analytic-Synthetic Method

The original formulation of Pappus's analytic-synthetic method, however, is puzzling, because it contains incompatible assertions about the direction of analysis.

Indeed, in his definition of analysis and synthesis, on the one hand, Pappus says that analysis is the way from what is sought, as if it were established, through its consequences. This suggests that analysis is a downward way. On the other hand, Pappus says that, in analysis, we assume what is sought as if it had been achieved, and look for the thing from which it follows. This suggests that analysis is an upward way.

In his definition of theorematic analysis, on the one hand, Pappus says that, in theorematic analysis, we assume what is sought as a fact and true, and proceed through its consequences, to something established. This suggests that analysis is a downward way. On the other hand, Pappus says that, if the thing that has been established is true, the thing that was sought will be true. But, if analysis were a downward way, this would be in conflict with the fact that, if something true follows from a given thing, the latter will not necessarily be true because, from a false premiss, a true conclusion may follow.

In his definition of problematic analysis, on the one hand, Pappus says that, in problematic analysis, we assume the required thing to be known, and proceed through its consequences, to something established. This suggests that analysis is a downward path. On the other hand, Pappus says that, if the thing that has been established is possible and obtainable, the required thing will be possible. But, if analysis were a downward way, this would be in conflict with the fact that, if something possible follows from a given thing, the latter will not necessarily be possible because, from an impossible premiss, a possible conclusion may follow.

A possible explanation for the fact that the formulation of Pappus's analytic-synthetic method contains incompatible assertions about the direction of analysis, is that Pappus draws the formulation from earlier texts that are mutually incompatible. Indeed, Gulley says that Pappus "is repeating two different formulations" of analysis, where according to one of them "analysis is a resolution upwards through propositions antecedent to an initial assumption until something already known or ranking as a first principle" is reached, while according to the other one "analysis is deductive" (Gulley 1958, 13).

There is evidence for this. Indeed, on the one hand, Heron describes analysis as an upward way. For, he says that "analysis" is "when some question or other is posed to us," and "we resolve it to something whose proof is already had," while "synthesis" is "when one begins with the known things; then one combines them until the unknown is found" (Al-Nayrizi 2009, 22–23).

On the other hand, a scholium to Euclid describes analysis as a downward way. For, it says that "analysis is a taking of that which is sought, as if it were established, through its consequences, up to something admitted as true. Synthesis is a taking of that which is admitted through its consequences up to that which is sought" (Euclid, *Opera Omnia*, IV, Appendix I to Book XIII, Propositions 1–5, ed. Heiberg).

In any case, Pappus's analytic-synthetic method has been widely interpreted as meaning that analysis and synthesis are both downward paths, being both deductive processes. In particular, the downward interpretation coheres with most of the mathematical practice of Pappus himself and other Greek geometers.

6.14 Pappus's Analytic-Synthetic Method and Reductio ad Absurdum

In Chap. 5 it has been said that there is a relation between the analytic method and reductio ad absurdum. Now, there is an even stricter relation between Pappus's analytic-synthetic method and reductio ad absurdum.

As Szabó says, "the 'analysis' of Pappus that leads to something known being false is just a special case of the so-called reductio ad absurdum; because in such a case our starting theorem is, beyond any doubt, false" (Szabó 1974, 126).

Indeed, in Pappus's theorematic analysis, if we come upon something established being false, the thing sought will be false. Thus, reductio ad absurdum corresponds to the case of Pappus's theorematic analysis when analysis leads to something established being false.

Similarly, in Pappus's problematic analysis, if we come upon something established being impossible, the problem will be impossible. Thus, reductio ad absurdum corresponds to the case of Pappus's problematic analysis when analysis leads to something established being impossible.

On the other hand, in Pappus's theorematic analysis, if we come upon something established being true, we have to complete our 'analysis' with the reverse process ('synthesis'), because a true conclusion can follow from a false premiss.

Similarly, in Pappus's problematic analysis, if we come upon something established being possible, we have to complete our 'analysis' with the reverse process ('synthesis'), because a possible conclusion can follow from an impossible premiss.

6.15 Fortune of Pappus's Analytic-Synthetic Method

Pappus's analytic-synthetic method, in its downward interpretation, has been very influential in the early modern period. Its influence was helped by the publication, in 1588, of Commandino's Latin translation of Pappus's *Sunagoge*, which made Pappus's work widely accessible.

Thus, Viète says that "analysis" is "the assumption of that which is sought as if it were admitted, through the consequences, to what is admitted as true," while "synthesis is the assumption of what is admitted, through the consequences, to the end and comprehension of that which is sought" (Viète 1646, 1).

Arnauld and Nicole say that "the analysis of geometers" consists in this, that, "when a question is proposed to them, of which they ignore the truth or falsity if it is a theorem, the possibility or impossibility if it is a problem, they assume that it is as proposed" (Arnauld and Nicole 1992, 287). Namely they assume that the question is true if it is a theorem, or that it is possible if it is a problem. Then, "examining what follows from it, if they arrive, in this examination, at some clear truth from which what is proposed to them follows necessarily, they conclude from this that what is proposed to them is true; and then starting over from where they had ended, they demonstrate it by the other method which is called composition" (ibid.). Conversely "if, as a necessary consequence of what is proposed to them, they fall into some absurdity or impossibility, they conclude from this that what had been proposed to them is false and impossible" (ibid.).

The influence of Pappus's analytic-synthetic method extends to the contemporary period. Thus, Pólya says that "in analysis, we start from what is required, we take it for granted, and we draw consequences from it, and consequences from the consequences," till "we come eventually upon something already known or admittedly true" (Pólya 2004, 142). On the other hand, "in synthesis, reversing the process, we start from the point which we reached last of all in the analysis, from the thing already known or admittedly true" (ibid.). Then "we derive from it what preceded it in the analysis, and go on making derivations until, retracing our steps, we finally succeed in arriving at what is required" (ibid.).

6.16 Analytic Method vs. Analytic-Synthetic Method

With respect to the analytic method, the analytic-synthetic method, both in Aristotle's and Pappus's form, has serious limitations.

(1) The analytic method, as argued in Chap. 3, is compatible with Gödel's incompleteness theorems. Conversely, the synthetic part of the analytic-synthetic method is the axiomatic method (see below), thus, since the axiomatic method is incompatible with Gödel's incompleteness theorems, so is the analytic-synthetic method.

(2) The analytic method accounts for the fact that a solution to a problem should consist of steps that are all explicable. Conversely, the analytic-synthetic method does not account for this fact, because it involves principles that are supposed to be known to be true by a faculty, intuition, which is inexplicable.

(3) The analytic method accounts for the fact that a solution to a problem may require hypotheses from parts of mathematics other than the one to which the problem belongs, because a solution to a problem is sought in an open space. Conversely, the analytic-synthetic method does not account for this fact, because a solution to a problem is sought in a closed space. The principles are supposed to contain all the knowledge concerning the part of mathematics in question, since the solution must be deduced from them.

(4) The analytic method accounts for the fact that the solution to a problem yields something new, because it involves finding hypotheses by means of non-deductive rules, which are ampliative. Conversely, the analytic-synthetic method does not account for this fact, because the solution to a problem involves deducing a solution from the principles, and deductive rules are non-ampliative, so the solution is already contained in the principles and hence has no novelty with respect to them.

(5) The analytic method accounts for the fact that often multiple solutions are found for the same problem. For, any problem has several sides, each of which can lead to a distinct hypothesis and hence to a distinct solution to the problem. Each solution establishes new connections between the problem and the existing knowledge, so each solution may lead to progress even when other solutions are already known. Conversely, the analytic-synthetic method does not account for this fact, because the purpose of demonstration is to give a foundation or justification of a proposition. Once a demonstration has been found, and hence a foundation or justification of the proposition has been given, there is no point in giving other demonstrations of that proposition, even four hundred of them, as in the case of the Pythagorean theorem.

(6) The analytic method accounts for the fact that different solutions to a problem may have different degrees of reliability. For, different solutions can be based on hypotheses with different degrees of plausibility. Conversely, the analytic-synthetic method does not account for this fact, because a solution consists in a deduction from principles, and deduction is either correct or incorrect, so different degrees of reliability are not possible.

6.17 The Trivialization of Analysis

In addition to the limitations with respect to the analytic method described above, Pappus's analytic-synthetic method has a limitation also with respect to Aristotle's analytic-synthetic method: it trivializes analysis.

This is because, while in Aristotle's analytic-synthetic method analysis is non-deductive, in Pappus's analytic-synthetic method analysis is deductive. As we have seen, in theorematic analysis, one assumes the theorem as true and deduces conclusions from it, until one ultimately arrives at something established. In problematic analysis, one assumes the problem as solved and deduces conclusions from it, until one ultimately arrives at something established.

This trivializes analysis because, as argued in Chap. 5, deductive rules are non-ampliative, so the conclusion of a deduction is already contained in the premisses. Thus, in Pappus's analytic-synthetic method, through analysis, one can only arrive at something established that is contained in the theorem or problem. This strongly contrasts with the fact that, generally, demonstrating a theorem or solving a problem requires something that is not contained in the theorem or problem. Moreover, there is an algorithm for enumerating all deductions from the theorem or problem. Therefore, in Pappus's analytic-synthetic method, analysis becomes trivial as a method of discovery.

6.18 The Material Axiomatic Method

A byproduct of Aristotle's analytic-synthetic method is the material axiomatic method. The latter is what results from Aristotle's analytic-synthetic method when the analytic part of the method is omitted and only the synthetic part is retained. Thus, the material axiomatic method is the synthetic part of Aristotle's analytic-synthetic method.

Indeed, the material axiomatic method is the method according to which, to present, justify, and teach an already acquired proposition, one starts from given axioms and deduces the proposition from them.

The axioms must be true, in the sense that there must be things, specified in advance, of which the axioms are true.

Then the purpose of the material axiomatic method is not to obtain new knowledge. For, deduction yields presentation, justification, and teaching through what is already known.

The material axiomatic method can be schematically represented as follows.

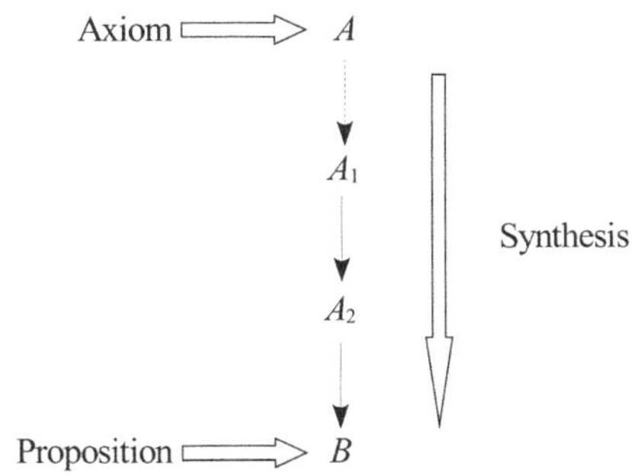

6.19 Original Formulation of the Material Axiomatic Method

The material axiomatic method was already used by some mathematicians of Plato's Academy, possibly Eudoxus, but they did not give a formulation of the method, they simply used it. Aristotle gave the first formulation of the method.

Indeed, Aristotle formulates the material axiomatic method by saying that, to present, justify, and teach an already acquired proposition, we start from the principles proper to the subject matter of that proposition, and deduce the proposition from them. For, "didactic arguments are those that deduce" propositions "from the principles proper to each discipline" (Aristotle, *De Sophisticis Elenchis*, 2, 165 b 1–2). Thus, didactic arguments are the "demonstrative arguments" which are "treated in the *Analytics*" (ibid., 2, 165 b 9).

Since didactic arguments are demonstrative arguments, the principles from which they deduce propositions must be true, because "a thing is demonstrated from what is true" (Aristotle, *Analytica Posteriora*, A 9, 75 b 39).

The principles must be true in the sense that there must be things, specified in advance, of which the principles are true. For, each science is concerned with a single "genus, of which the science considers the attributes which hold of it in itself" (ibid., A 10, 76 b 13). The genus includes all proper objects of that science, "about which the science considers what holds of them in themselves, as for example arithmetic is about units, and geometry is about points and lines" (ibid., A 10, 76 b 4–5).

The purpose of the material axiomatic method is not to acquire new knowledge, but only to present, justify, and teach already acquired propositions. Indeed, didactic arguments do not yield new knowledge, but "produce their teaching through what we already know" (ibid., A 2, 71 a 6–7). For, "all teaching and all intellectual learning come about from already existing knowledge. This is clear if we consider all the branches of intellectual learning. In fact, the mathematical sciences are learned in this way and so is each of the other arts" (ibid., A 1, 71 a 1–4).

6.20 Difference in Purpose from Aristotle's Analytic-Synthetic Method

Aristotle assigns the material axiomatic method a purpose different from that of Aristotle's analytic-synthetic method. While the purpose of Aristotle's analytic-synthetic method is to discover demonstrations in an axiomatic system, the purpose of the material axiomatic method is to present, justify, and teach already acquired propositions.

In fact, in his scientific research work, Aristotle does not use the material axiomatic method, but rather Aristotle's analytic-synthetic method.

Cicero underlines the difference in purpose between Aristotle's analytic-synthetic method and the material axiomatic method, by saying that "all methodical treatment of rational discourse has two parts, one of discovering, the other of judging" (Cicero, *Topica*, II.6). Aristotle "came first in both," while "the Stoics concerned themselves" only "with the latter: indeed, they diligently pursued ways of judging" but "completely neglected the art of discovering," which "was both more powerful in use and certainly prior in the order of nature" (ibid.).

In fact, as we have seen in Chap. 5, in opposition to the Epicureans who acknowledged the importance of non-deductive reasoning, the Stoics restricted logic to deductive reasoning. A similar restriction was reintroduced by Frege in the nineteenth century, and has been endorsed by all of mathematical logic.

6.21 The Formal Axiomatic Method

As already said above, in the material axiomatic method the axioms must be true, in the sense that there must be things, specified in advance, of which the axioms are true.

Another form of the axiomatic method is the formal axiomatic method, also called modern axiomatic method. The formal axiomatic method is like the material axiomatic method, except that the axioms are not required to be true in the sense that there must be things, specified in advance, of which the axioms are true.

The choice of the axioms is arbitrary, subject only to the condition that the axioms must be consistent. The primitive terms occurring in the axioms can be thought of in any way one likes.

Apart from this difference, the purpose of the formal axiomatic method remains the same as that of the material axiomatic method. Namely, its purpose is not to obtain new knowledge, but only to present, justify, and teach already acquired knowledge.

6.22 Original Formulation of the Formal Axiomatic Method

The formal axiomatic method arose gradually from the second half of the nineteenth century to the early decades of the twentieth century, but is best known from Hilbert's formulation.

Hilbert formulates the formal axiomatic method by saying that the mathematician "must of course be free to do as" he pleases "in giving characteristic marks" (Hilbert 1980, 39). Namely, in choosing the axioms. The choice of the axioms is only subject to the condition that the axioms must not contradict one another. For, "if the arbitrarily given axioms do not contradict one another with all their consequences, then they are true and the things defined by the axioms exist. This" is "the criterion of truth and existence" (ibid., 39–40).

The "basic elements," namely the primitive terms occurring in the axioms, "can be thought of in any way one likes" (ibid., 40). So, "any theory can always be applied to infinitely many systems of basic elements" (ibid., 40–41). Thus, "if in speaking" of the primitive terms occurring in my axioms of geometry, e.g. "of my points, I think of some system of things, e.g. the system: love, law, chimney-sweep," and I "assume all my axioms as relations between these things, then my propositions, e.g. Pythagoras' theorem, are also valid for these things" (ibid., 40).

The purpose of "the axiomatic exploration of a mathematical truth" is not to find "new or more general propositions connected to that truth" (Hilbert 1902–1903, 50). It is, instead, "to determine the position of that" truth "within the system of known truths," in "such a way that it can be clearly said which conditions are necessary and sufficient for giving a foundation of that truth" (ibid.). So, the purpose of the formal axiomatic method is not to obtain new knowledge, but only to present and justify already acquired knowledge, and to teach it through textbooks such as Hilbert's *Grundlagen der Geometrie*.

Hilbert's formulation of the formal axiomatic method, however, is not wholly coherent. For, on the one hand, as we have just seen, Hilbert says that the purpose of the axiomatic exploration of a mathematical truth is not to obtain new knowledge. But, on the other hand, he also says that the formal "axiomatic method" is "the general method of research which seems to be coming more and more into its own in modern mathematics" (Hilbert 1996b, 1107). Such method "is logically incontestable and at the same time fruitful; it thereby guarantees the maximum flexibility in research" (Hilbert 1996c, 1120). Now, if the formal axiomatic method is the general method of research in modern mathematics, then its purpose must be to obtain new knowledge.

6.23 Formal Axiomatic Method and Mathematicians

The rise of the formal axiomatic method was not without opposition. A significant example is Frege.

As we have seen, Hilbert says that, if the arbitrarily given axioms do not contradict one another with all their consequences, then they are true and the things defined by the axioms exist. To this, Frege opposes that the axioms are "propositions that are true but are not proved because our knowledge of them flows" from "spatial intuition," and only "from the truth of the axioms it follows that they do not contradict one another" (Frege 1980, 37). Hilbert also says that the primitive terms occurring in the axioms can be thought of in any way one likes, so any theory can always be applied to infinitely many systems of basic elements. To this, Frege opposes that the axioms "must not contain a word or sign whose sense and meaning" was "not already completely laid down," the sense and meaning of each "sign or word that occurs in them" must "already be laid down" so that "there is no doubt about the sense of the proposition and the thought it expresses" (ibid., 36). On this basis, Frege concludes that the formal axiomatic method "is on the whole a failure" (ibid., 90).

But the opposition to the formal axiomatic method by Frege and other people was unsuccessful. As a matter of fact, the majority of mathematicians hold that the method of mathematics is the formal axiomatic method.

Thus, Bourbaki says that "mathematical truth" resides "uniquely in logical deduction starting from premises arbitrarily set by axioms" (Bourbaki 1994, 17). Proofs can be expressed as formal proofs, because "it matters little whether this or that meaning is attached to the words, or signs in the text, or indeed whether any meaning at all is attached to them; the only important point is the correct observance of" the deduction "rules" (Bourbaki 2004, 8).

Sentilles says that a theory consists of "undefined" terms "which are admitted initially as simply undefined" (Sentilles 2011, 89). And of "some facts about these terms. These are the axioms," they "are statements about (that is, relating) the undefined terms" which "need not be true about any particular thing that exists anywhere in this universe, they are only taken as true" (ibid., 90). Along "with the axioms" there is a "set of stated rules" for "manipulating the given statements of the theory" (ibid.).

Vialar says that "mathematics base theories on propositions postulated as true, which are called axioms and use only demonstrations deriving from these axioms" (Vialar 2017, 6). The only requirement on the axioms that "the construction of axiomatic systems imposes" is "consistency," namely "non-contradiction" (ibid.). The "mathematician starts from axioms and definitions, and has also at disposal theorems already demonstrated;" then he "obtains new theorems by means of demonstrations," namely "chains of deduction that obey logical rules" (ibid., 7).

6.24 The Axiomatic Ideology

Hilbert not only formulates the formal axiomatic method, but makes it the basis of a general ideology of science, the axiomatic ideology, whose main characters are the following.

(1) The formal axiomatic method is the method of research of all scientific knowledge. Indeed, the formal "axiomatic method is and remains the indispensable tool, appropriate to our minds, for all exact research in any field whatsoever" (Hilbert 1996c, 1120).

(2) All scientific knowledge is dependent on the formal axiomatic method. Indeed, "anything at all that can be the object of scientific thought," as soon as "it is ripe for the formation of a theory," becomes "dependent on the" formal "axiomatic method" (Hilbert 1996b, 1115).

(3) Through the formal axiomatic method, all scientific knowledge is ultimately dependent on mathematics. Indeed, anything at all that can be the object of scientific thought, insofar as it is dependent on the formal axiomatic method, is "thereby indirectly" dependent "on mathematics, as soon as it is ripe for the formation of a theory" (ibid.).

(4) Through the formal axiomatic method, mathematics has a leadership role over all the sciences. Indeed, "in the sign of the" formal "axiomatic method, mathematics is summoned to a leading role in science" (ibid.).

(5) Through the formal axiomatic method, mathematics is the foundation of the entire modern culture. Indeed, through it, "our entire modern culture, in so far as it rests on the penetration and utilization of nature, has its foundation in mathematics" (Hilbert 1996d, 1163).

(6) Through the formal axiomatic method, mathematics becomes the supreme court that will decide all assertions. Indeed, "mathematics in a certain sense develops into a tribunal of arbitration, a supreme court that will decide questions of principle – and on such a concrete basis that universal agreement must be attainable and all assertions can be verified" (Hilbert 1967, 384).

The axiomatic ideology does not have a scientific purpose, its goal is only to affirm the hegemony of mathematics over all the sciences, and even over the entire modern culture. But, in view of the limitations of mathematics that will be discussed in Chap. 16, this goal seems unrealistic.

6.25 Original Reason of the Formal Axiomatic Method

To explain the passage from the material axiomatic method to the formal axiomatic method, several people appeal to the discovery of non-Euclidean geometry and non-commutative algebra.

Thus, Eves says that "the discovery of a non-Euclidean geometry and, not long after, of a non-commutative algebra led to a deeper study and refinement of axiomatic procedure; thus from the material axiomatics of the ancient Greeks evolved the formal axiomatics of the twentieth century" (Eves 1997, 147).

Koh and Tay say that the conception of "material axiomatics," that "an axiom expresses a property of some basic objects," was "shaken in the nineteenth century by the 'discovery' of a non-Euclidean geometry" and "of a non-commutative algebra," which led to the development of "the so-called formal axiomatics," in which an axiom is "a basic assumption about some undefined primitive term" (Koh and Tay 2019, 23–24).

But this is invalid. For, the founders of non-Euclidean geometries, such as Lobachevski and Riemann, and the founders of non-commutative algebra, such as Hamilton and Grassmann, did not place their work in the realm of the formal axiomatic method, but in that of the material axiomatic method. For, they thought that there were things, specified in advance, of which the axioms were true.

Thus, as regards non-Euclidean geometry, according to Lobachevski, the object of geometry is the study of bodily movements, because "we know nothing in nature but movement" (Lobachevsky Lobachevski 1898–1899, I, 76). Movements are movements of bodies, because "in nature there are neither straight nor curved lines, neither plane nor curved surface; we find in it only bodies" (ibid., I, 82). All other "geometrical concepts" are generated "by our understanding, which derives them from the properties of movement" (ibid., I, 76).

According to Riemann, the object of geometry is the study of "the metric relations of space in the infinitely small," because "the empirical notions on which the metrical determinations of space are founded, the notion of a solid body and of a ray of light, cease to be valid for the infinitely small" (Riemann 2016, 40). Thus, we must establish what are the metric relations of space in the infinitely small, and these relations "are only to be deduced from experience" (ibid., 31).

On the other hand, as regards non-commutative algebra, according to Hamilton, algebra is not merely "a language," a "system of expressions," it is a "system of truths" (Hamilton 1837, 295). Indeed, algebra is "a science properly so called; strict, pure, and independent; deduced by valid reasonings from its own intuitive principles," and hence distinct "from the signs by which it may express its meaning" (ibid.). For, algebra is based on "the intuition of time," and in fact it is "the science of pure time" (ibid.).

According to Grassmann, algebra is not merely "a formal development," it is "a conceptual development" and "progresses along a conceptual development," the "formulas written down at each step" only "symbolically represent the conceptual progress" (Grassmann 1861, VI). Algebra is a conceptual development because it consists of truths about thought forms. Indeed, "pure mathematics is the science of the particular existent that has come to be by thought. The particular existent, understood in this sense, we call a thought form" (Grassmann 1844, XX).

6.26 Romanticism and Mathematics

Rather than being due to the discovery of non-Euclidean geometry and non-commutative algebra, the passage from the material axiomatic method to the formal axiomatic method was due to a somewhat tardy impact of Romanticism on mathematics. Indeed, a basic characteristic of Romanticism is the claim of an absolutely free creativity of the mind. This implies that the mathematician is absolutely free in choosing his axioms, there need not be things, specified in advance, of which the axioms are true.

Thus, Novalis says that the mathematician is absolutely free in his creations, because "the foundation of" mathematical "creation lies in the will" (Novalis 2007, 92).

But, in addition to saying that the mathematician is absolutely free in his creations, Novalis makes two further claims.

The first claim is that mathematics is purely abstract and formal. Indeed, Novalis says that "pure mathematics is not concerned with quantity," it is "purely arbitrary, dogmatically instrumental. Likewise, so it is also with abstract language" (Novalis 1901, II, 282). Both in mathematics and in abstract language we are concerned with arbitrary creations. In abstract language, signs and words are introduced in an arbitrary way. In mathematics, in addition to signs and words, principles are also introduced in an arbitrary way, and all the other propositions are deduced from them, because "to demonstrate something a priori means to deduce something" (ibid., II, 425).

The second claim is that the mathematician can solve any mathematical problem. As we have seen in Chap. 4, for Novalis the true method of mathematics is the synthetic method. The synthetic method "begins with" freedom "and proceeds toward freedom" (Novalis 2003, 171). Only the synthetic method permits the mathematician to construct mathematics in an absolutely free way as "a pure a priori system, without the condition of any external stimulus" (Novalis 1901, II, 109). So, the synthetic method permits the mathematician to deal with any problem and to reach that "perfection of the will that is expressed by saying: he can what it wants" (ibid., II, 453). Such perfection is the expression of "the omnipotence of inner humanity" (Novalis 1996, 74). The freedom that the synthetic method offers the mathematician is the basis for this omnipotence.

6.27 The Impact of Romanticism on Mathematics

The impact of Romanticism on mathematics is clear from the assertion of several mathematicians, in the last decades of the nineteenth century or the first decades of the twentieth century, that the mathematician is absolutely free in choosing his axioms, subject only to the requirement of consistency.

Thus, Cantor says that "the essence of mathematics lies precisely in its freedom" (Cantor 1996, 896). Mathematics is "in its development entirely free and is only bound in the self-evident respect that its concepts must" be "consistent with each other" (ibid.).

Dedekind says that "numbers are free creations of the human mind" (Dedekind 1996a, 791). They are "something new" which "the spirit creates. We are of a divine race and, without any doubt, possess creative power not merely in material things (railroads, telegraphs) but especially in mental things" (Dedekind 1996b, 835). Our creative power is only subject to the condition that, after we have freely created a system of numbers through a system of axioms, the system must not "contain internal contradictions" (Dedekind 1967, 101).

Poincaré says that the facts with which the mathematician is concerned are not independent of him, because it is himself, "almost I had said his caprice, that creates these facts" (Poincaré 2015, 370). The latter are only subject to the condition that they must be consistent with each other, because "in mathematics the word exist can have one meaning, it means free from contradiction" (ibid., 454).

But Cantor, Dedekind, and Poincaré do not endorse the two further Novalis's claims, that mathematics is purely abstract and formal, and that the mathematician can solve any mathematical problem. Conversely, these claims are endorsed by Hilbert.

As we have seen above, Hilbert says that the mathematician must be free to do as he pleases in choosing axioms, only subject to the condition that the axioms must not contradict one another.

But Hilbert also says that mathematics is purely abstract and formal because, through formalization, "the mathematical inferences and definitions become a formal part of the edifice of mathematics" (Hilbert 1996c, 1123). Mathematics comes to be "an inventory of formulas" that "follow each other according to definite rules. Certain of these formulas correspond to the mathematical axioms," and material "inference is replaced by manipulation of signs according to rules" (Hilbert 1967, 381). Thus, "the full transition from a naive to a formal treatment is now accomplished, on the one hand, for the axioms themselves, which originally were naively taken to be fundamental truths but in modern axiomatics had already for a long time been regarded as merely establishing certain interrelations between notions" (ibid.). And, on the other hand, for the logical calculus, which provides "a sign language that is capable of representing mathematical propositions in formulas and of expressing logical inferences through formal processes" (ibid.).

Moreover, Hilbert also says that the mathematician can solve any mathematical problem by the formal axiomatic method, because the formal axiomatic method allows to deploy "the creative power of pure reason" (Hilbert 1996a, 1098). It "brings us the exaltation of the conviction that at least the mathematical understanding encounters no limits" (Hilbert 1998, 233). This makes us confident that, by the formal axiomatic method, "every definite mathematical problem must necessarily be susceptible of an exact settlement" (Hilbert 1996a, 1102). Indeed, when we are confronted with still unsolved mathematical problems, "however unapproachable these problems may seem to us and however helpless we stand before them, we have, nevertheless, the firm conviction that their solution must follow by a finite number of purely logical processes" (ibid.). And that we "can find it by pure reason, for in mathematics there is no 'ignorabimus'" (ibid.). Indeed, in mathematics "there are absolutely no unsolvable problems. Instead of the foolish 'ignorabimus', our answer is on the contrary: We must know, we shall know" (Hilbert 1996d, 1165).

By these claims, Hilbert took side in the 'ignorabimus' controversy initiated by Du Bois-Reymond, who had declared that there are problems with respect to which we cannot simply say 'ignoramus' and "must resign ourselves once and for all to the far more difficult verdict: Ignorabimus" (Du Bois-Reymond 1872, 34). (On the the 'ignorabimus' controversy, see Vidoni 1988; Bayertz et al. 2012).

Hilbert's claim that the mathematician can solve any mathematical problem by the formal axiomatic method was a perfect though somewhat tardy expression of the Romantic spirit. But it was an illusion, that was soon dispelled. By the theorem of undecidability of provability, for any consistent sufficiently strong deductive theory, there is no mechanical procedure for deciding whether or not a mathematical proposition can be demonstrated from the axioms of the theory. So, the question of the decidability of a mathematical question in a finite number of operations has a negative answer. Therefore, the mathematician cannot solve any mathematical problem by the formal axiomatic method.

6.28 Changed Relation Between Mathematics and Physics

The rise of the formal axiomatic method involves a change in the relation between mathematics and physics.

From the seventeenth century until the nineteenth century, there is a close relation between mathematics and physics. Its origin is to be found in Galileo's change of the object of science, which assigns a central role to mathematics in science (see Chap. 4). The effect of Galileo's change is clear, for example, from Aristotle's approach to motion.

Aristotle defines motion by saying that "motion is the actuality of what is potentially, as such" (Aristotle, *Physica*, B 3, 201 a 10–11). But, with this qualitative definition, Aristotle can only establish laws of motion such as: "Everything that is in motion must be moved by something" (ibid., H 1, 242 a 37–38).

Galileo's change permits to replace Aristotle's qualitative definition of motion with a quantitative one. Thus Galilei defines uniform motion as follows: "By steady or uniform motion, I mean one in which the distances traversed by the moving body during any equal intervals of time, are equal to each other" (Galilei 1968, VIII, 191). With this quantitative definition, Galileo can establish laws of motion such as: "If a moving body carried uniformly at a constant speed, traverses two distances, the time-intervals required are to each other in the ratio of these distances" (ibid., VIII, 192).

However, Galileo's approach to motion is limited by the mathematics available to him, essentially Greek geometry and the Eudoxian theory of proportions. That kind of mathematics can adequately deal with static things such as numbers and geometric figures, not with dynamic things such as motion. Thus Galileo's change of the object of science raises two needs: (1) To replace Aristotle's science, which provides a kind of knowledge that is qualitative, concrete, and material, with a science which provides a kind of knowledge that is quantitative, general, and abstract. (2) To develop a new kind of mathematics, capable of dealing with dynamic things such as motion.

Need (1) is satisfied by Galileo himself. Galileo also starts thinking about need (2), as it is clear, for example, from the fact that he says that "the continuum" is "composed of infinitely many indivisibles" (ibid., VIII, 80). But Galileo fails to develop a mathematics capable of dealing with dynamic things such as motion, although he seems to have had in mind to develop it, as it appears from some letters addressed to him by Cavalieri, in which Cavalieri says: "Let me remind you of your work on the indivisibles, that you decided to compose" (ibid., XIII, 309). I "should be grateful if you would apply yourself to it, so that I could publish my own work" on the indivisibles, "which in the meantime I will be polishing" (ibid., XIII, 312).

Need (2) is satisfied by Leibniz and Newton through the development of the calculus of infinitesimals. This would have been unfeasible within Aristotle's science, because Aristotle says that "it is impossible for anything to be in motion at the instant" (Aristotle, *Physica*, Z 3, 234 a 31). By satisfying need (2), Leibniz and Newton lay the foundation for a close relationship between mathematics and physics that lasts until the nineteenth century.

With the rise of the formal axiomatic method, however, mathematicians start introducing concepts unconnected with physical reality and experience generally. This results in a separation between mathematics and physics, a separation that is fully accomplished in the twentieth century.

Thus, Hilbert says that, with the formal axiomatic method, mathematics "becomes completely detached from concrete reality," it "has nothing more to do with real objects," it is "a pure thought construction, of which one can no longer say that it is true or false," and "the task of mathematics is then to develop this framework of concepts in a logical way" (Hilbert 2013, 435).

Stone says that, of all the "important changes that have taken place since 1900 in our conception of mathematics," the "one which truly involves a revolution in ideas is the discovery that mathematics is entirely independent of the physical world" (Stone 1961, 716). Such discovery "may be said without exaggeration to mark one of the most significant intellectual advances in the history of mankind" (ibid.). Mathematics has become "the study of general abstract systems," and "this new orientation, made possible only by the divorce of mathematics from its applications, has been the true source of its tremendous vitality and growth during the present century" (ibid., 717).

Dieudonné says that "recent history has been" very little "willing to conform to the pious platitudes of the prophets of doom, who regularly warn us of the dire consequences mathematics is bound to incur by cutting itself from the applications to other sciences" (Dieudonné 1964, 247–248). Contrary to the proclamations of the prophets of doom, "even if mathematics were to be forcibly separated from" other sciences, "there would remain food for centuries of thought in the big problems we still have to solve within our own science" (ibid., 248).

6.29 Negative Effects of the Formal Axiomatic Method

The separation between mathematics and physics has very negative effects on both mathematics and physics.

(1) The separation makes mathematics an end in itself, with no thought of what objective it might serve. As Kline says, while "mathematics aims to discover something worth knowing," with the separation, "research begets research, which begets research," one "dare not ask for meaning and purpose. Mathematics must not be tainted by reality. The ivy has grown so thick that the researchers within can no longer see the world outside" (Kline 1980, 302). With the separation, mathematicians are "like the mathematicians Gulliver met on his voyage to Laputa," they "live on an island suspended in the air above the earth," and "are doomed ultimately to expire in a vacuum," they "may like to soar into the clouds," but "like birds they must return to earth for food" (ibid., 305).

(2) The separation makes it incomprehensible why many applications of mathematics to the world are successful, and in fact the supporters of the formal axiomatic method are unable to explain it. Thus, Bourbaki asks: "Why do such applications ever succeed?" in particular "Why have some of the most intricate theories in mathematics become an indispensable tool to the modem physicist, to the engineer, and to the manufacturer of atom-bombs?" (Bourbaki 1949, 2). Bourbaki's candid reply is: "Fortunately for us, the mathematician does not feel called upon to answer such questions, nor should he be held responsible for such use or misuse of his work" (ibid.). Indeed, "we are completely ignorant as to the underlying reasons for this fact" and "we shall perhaps always remain ignorant of them" (Bourbaki 1950, 231).

(3) The separation encourages mathematicians to take up quarters in a corner of mathematics, losing sight of the final purpose, or perhaps having no such purpose. As even Bourbaki admits, "many mathematicians take up quarters in a corner of the domain of mathematics, which they do not intend to leave" (ibid., 1266). Indeed, "not only do they ignore almost completely what does not concern their special field, but they are unable to understand the language and the terminology used by colleagues who are working in a corner remote from their own" (ibid.).

(4) The separation leads to a fragmentation of mathematics. Indeed, by permitting to consider arbitrary axioms with the only condition of consistency, it produces an uncontrolled proliferation of theories. As even Bourbaki admits, one may ask "whether this exuberant proliferation makes for the development of a strongly constructed organism, acquiring ever greater cohesion and unity with its new growths, or whether it is the external manifestation of a tendency towards a

progressive splintering, inherent in the very nature of mathematics" (ibid., 1266). Therefore, one may also ask "whether the domain of mathematics is not becoming a tower of Babel, in which autonomous disciplines are being more and more widely separated from one another, not only in their aims, but also in their methods and even in their language" (ibid.).

(5) The separation leads mathematicians to create artificial systems, entirely without applications. As even Bourbaki admits, the separation has favoured the creation of "a whole crop of monster-structures, entirely without applications; their sole merit was that of showing the exact bearing of each axiom, by observing what happened if one omitted or changed it" (ibid., 1275, footnote 9). This seems to justify the "temptation to conclude that these were the only results that could be expected from the axiomatic method" (ibid.).

(6) The separation leads to the insignificance of mathematics. As Courant says, at present, there is an "unfortunate separation between pure mathematics and the vital applications" (Courant and Robbins 1996, n.p.). Specifically, "there seems to be a great danger in the prevailing overemphasis on the deductive-postulational character of mathematics" (ibid). For, "a serious threat to the very life of science is implied in the assertion that mathematics is nothing but a system of conclusions drawn from definitions and postulates that must be consistent but otherwise may be created by the free will of the mathematician" (ibid.). Indeed, "if this description were accurate, mathematics could not attract any intelligent person. It would be a game with definitions, rules, and syllogisms, without motive or goal" (ibid.). In fact, "the notion that the intellect can create meaningful postulational systems at its whim is a deceptive halftruth. Only under the discipline of responsibility to the organic whole, only guided by intrinsic necessity, can the free mind achieve results of scientific value" (ibid.).

(7) The separation leads to the sclerosis of research. This can be seen using a Frege's argument. Frege says that, "where a tree lives and grows, it must be soft and succulent," and, "when all that was green has turned into wood, the tree ceases to grow" (Frege 1980, 33). Similarly, we can say that, where mathematics lives and grows, it must be flexible and fecund, and, when mathematics is organized and presented by the formal axiomatic method, it ceases to grow. To the lignification of the tree, which is the end of its growth, there corresponds the sclerosis of research produced by the formal axiomatic method, which is the end of its flexibility and fecundity.

(8) The separation leads mathematicians to publish mainly for career reasons. As Kline says, "since applied problems require vast knowledge in science as well as in mathematics and since the open ones are more difficult, it is far easier to invent one's own problems and solve what one can" (Kline 1980, 282). Therefore, "professors select problems of pure mathematics that are readily solved," and "assign such problems to their doctoral students," because so doctoral students "can produce theses quickly," and "professors can more readily help them overcome any difficulties they encounter" (ibid.).

6.30 The Axiomatic Method

The material axiomatic method and the formal axiomatic method are essentially the same method, except for the status of the axioms. As we have seen, in the material axiomatic method, the axioms must be true, in the sense that there must be things, specified in advance, of which the principles are true. In the formal axiomatic method, the axioms must only be consistent.

However, 'to be consistent' may be regarded as a weak sense of 'to be true'. Indeed, Hilbert even says that "'consistent' is identical to 'true'" (Hilbert 1931, 122). In this perspective, the axioms must be true also in the formal axiomatic method, though not in the strong sense that there must be things, specified in advance, of which the principles are true, but only in the weak sense that they must be consistent.

On this basis, in what follows 'axiomatic method' will be used to mean either the material axiomatic method or the formal axiomatic method. Which is meant will be clear from context.

Thus, the axiomatic method will be the method according to which, to present, justify, and teach an already acquired proposition, one starts from given axioms which are true, in some sense of 'true', and deduces the proposition from them.

References

Al-Nayrizi, Abu'l Abbas. 2009. *Commentary on Books II-IV of Euclid's 'Elements of Geometry'*. Leiden: Brill.

Arnauld, Antoine, and Pierre Nicole. 1992. *La logique ou l'art de penser*. Paris: Gallimard.

Bayertz, Kurt, Myriam Gerhard, and Walter Jaeschke, eds. 2012. *Der Ignorabimus-Streit*. Hamburg: Meiner.

Bourbaki, Nicholas. 1949. Foundations of mathematics for the working mathematician. *The Journal of Symbolic Logic* 14: 1–8.

———. 1950. The architecture of mathematics. *The American Mathematical Monthly* 57: 221–232.

———. 1994. *Elements of the history of mathematics*. New York: Springer.

———. 2004. *Elements of mathematics: Theory of sets*. Berlin: Springer.

Cantor, Georg. 1996. Foundations of a general theory of manifolds: A mathematico-philosophical investigation into the theory of the infinite. In *From Kant to Hilbert: A source book in the foundations of mathematics*, ed. William Ewald, vol. 2, 881–920. Oxford: Oxford University Press.

Courant, Richard, and Herbert Robbins. 1996. *What is mathematics? An elementary approach to ideas and methods*. Oxford: Oxford University Press.

Dedekind, Richard. 1967. Letter to Keferstein, 27 February 1890. In *From Frege to Gödel: A source book in mathematical logic, 1879–1931*, ed. Jean van Heijenoort, 99–103. Cambridge: Harvard University Press.

———. 1996a. Was sind un was sollen die Zahlen? In *From Kant to Hilbert: A source book in the foundations of mathematics*, ed. William Ewald, vol. 2, 790–833. Oxford: Oxford University Press.

———. 1996b. Letter to Heinrich Weber, 24 January 1888. In *From Kant to Hilbert: A source book in the foundations of mathematics*, ed. William Ewald, vol. 2, 834–835. Oxford: Oxford University Press.

Dieudonné, Jean. 1964. Recent developments in mathematics. *The American Mathematical Monthly* 71: 239–248.
Du Bois-Reymond, Emil. 1872. *Über die Grenzen des Naturerkennens*. Leipzig: Veit.
Eves, Howard. 1997. *Foundations and fundamental concepts of mathematics*. Mineola: Dover.
Frege, Gottlob. 1980. *Philosophical and mathematical correspondence*. Oxford: Blackwell.
Galilei, Galileo. 1968. *Opere*. Barbera: Florence.
Grassmann, Hermann. 1844. *Die lineale Audehnungslehre ein neuer Zweig der Mathematik*. Leipzig: Wigand.
———. 1861. *Lehrbuch der Arithmetik für höhere Lehranstalten*. Berlin: Enslin.
Gulley, Norman. 1958. Greek geometrical analysis. *Phronesis* 33: 1–14.
Hamilton, William Rowan. 1837. Theory of conjugate functions, or algebraic couples; with a preliminary and elementary essay on algebra as the science of pure time. *Transactions of the Royal Irish Academy* 17: 293–423.
Hilbert, David. 1902–1903. Über den Satz von der Gleichheit der Basiswinkel im gleichschenkligen Dreieck. *Proceedings of the London Mathematical Society* 35: 50–68.
———. 1931. Beweis des Tertium non datur. *Nachrichten von der Gesellschaft der Wissenschaften zu Göttingen aus dem Jahre 1931, Mathematisch-Physikalische Klasse*, 120–125. Berlin: Weidmann.
———. 1967. On the infinite. In *From Frege to Gödel: A source book in mathematical logic 1879–1931*, ed. Jean van Heijenoort, 369–392. Cambridge: Harvard University Press.
———. 1980. Letter to Frege, 29 December 1899. In Gottlob Frege, *Philosophical and mathematical correspondence*, 38–41. Oxford: Blackwell.
———. 1996a. From 'mathematical problems'. In *From Kant to Hilbert: A source book in the foundations of mathematics*, ed. William Ewald, vol. 2, 1096–1105. Oxford: Oxford University Press.
———. 1996b. Axiomatic thought. In *From Kant to Hilbert: A source book in the foundations of mathematics*, ed. William Ewald, vol. 2, 1107–1115. Oxford: Oxford University Press.
———. 1996c. The new grounding of mathematics: First report. In *From Brouwer to Hilbert*, ed. William Ewald, vol. 2, 1117–1134. Oxford: Oxford University Press.
———. 1996d. Logic and the knowledge of nature. In *From Kant to Hilbert: A source book in the foundations of mathematics*, ed. William Ewald, vol. 2, 1157–1165. Oxford: Oxford University Press.
———. 1998. Problems of the grounding of mathematics. In *From Brouwer to Hilbert: The debate on the foundations of mathematics in the 1920s*, ed. Paolo Mancosu, 227–233. Oxford: Oxford University Press.
———. 2013. *Lectures on the foundations of arithmetic and logic 1917–1933*. Dordrecht: Springer.
Hintikka, Jaakko, and Unto Remes. 1974. *The method of analysis: Its geometrical origin and its general significance*. Dordrecht: Reidel.
———. 1976. Ancient geometrical analysis and modern logic. In *Essays in memory of Imre Lakatos*, ed. Robert Cohen, Paul Feyerabend, and Max Wartofsky, 253–276. Dordrecht: Springer.
Kline, Morris. 1980. *Mathematics: The loss of certainty*. Oxford: Oxford University Press.
Knorr, Wilbur Richard. 1993. *The ancient tradition of geometric problems*. Mineola: Dover.
Koh, Khee Meng, and Eng Guan Tay. 2019. Some breakthroughs ideas in mathematics. In *Big ideas in mathematics*, ed. Tin Lam Toh and Joseph B.W. Yeo, 11–27. Singapore: World Scientific.
Lakatos, Imre. 1978. *Philosophical papers*. Cambridge: Cambridge University Press.
Lobachevski, Nicolai Ivanovich. 1898–1899. *Zwei geometrische Abhandlungen*. Leipzig: Teubner.
Menn, Stephen. 2002. Plato and the method of analysis. *Phronesis* 47: 193–223.
Novalis. 1901. *Schriften*. Berlin: Reimer.
———. 1996. Christianity or Europe: A fragment. In *The early political writings of the German romantics*, ed. Frederick Beiser, 59–79. Cambridge: Cambridge University Press.
———. 2003. *Fichte studies*. Cambridge: Cambridge University Press.

———. 2007. *Notes for a romantic encyclopedia: Das Allgemeine Brouillon*. Albany: State University of New York Press.
Poincaré, Henri. 2015. *The foundations of science: Science and hypothesis, The value of science, Science and method*. Cambridge: Cambridge University Press.
Pólya, George. 2004. *How to solve it: A new aspect of mathematical method*. Princeton: Princeton University Press.
Riemann, Bernhard. 2016. *On the hypotheses which lie at the bases of geometry*. Cham: Springer.
Sayre, Kenneth M. 1969. *Plato's analytic method*. Chicago: University of Chicago Press.
Sentilles, Dennis. 2011. *A bridge to advanced mathematics*. Mineola: Dover.
Stone, Marshall. 1961. The revolution in mathematics. *The American Mathematical Monthly* 68: 715–734.
Szabó, Árpád. 1974. Working backwards and proving by synthesis. Appendix 1 to Jaakko Hintikka, and Unto Remes. *The method of analysis: Its geometrical origin and its general significance*, 118–130. Dordrecht: Springer.
Vialar, Thierry. 2017. *Handbook of mathematics*. Norderstedt: Books on Demand.
Vidoni, Ferdinando. 1988. *Ignorabimus!* Milan: Marcos y Marcos.
Viète, François. 1646. *Opera mathematica*. Leiden: Ex Officina Bonaventurae & Abrahami Elzeviriorum.

Chapter 7
Rules of Discovery

Abstract In the analytic method, hypotheses are obtained by non-deductive rules. Obtaining them is not a sufficient condition for discovery, because discovery requires showing that the hypotheses are plausible. Nevertheless, obtaining hypotheses is a necessary condition for discovery and, in this sense, non-deductive rules can be said to be rules of discovery. The rules of discovery are not a closed set given once for all, but an open set that can always be extended as research develops. Nevertheless, as a matter of historical fact, some rules of discovery have been widely used in finding hypotheses in mathematics. The chapter considers them, and how they could have been used in finding solutions to certain historically significant mathematical problems.

Keywords Rules and discovery · Induction · Induction from a single case · Induction from multiple cases · Analogy · Analogy by quasi-equality · Analogy by separate indistinguishability · Analogy by agreement · Analogy by agreement and disagreement · Metaphor · Metonymy · Generalization · Specialization

7.1 Non-Deductive Rules as Rules of Discovery

As we have seen in Chaps. 5 and 6, both in the analytic method and in Aristotle's analytic-synthetic method, the hypotheses are obtained by non-deductive rules, because only such rules are ampliative. Of course, finding hypotheses is not a sufficient condition for discovery, because the latter requires to show that the hypotheses are plausible. Nevertheless, finding hypotheses is a necessary condition for discovery and, in this sense, non-deductive rules can be said to be rules of discovery.

The rules of discovery are not a closed set, given once for all, but an open set, which can always be extended as research develops. Indeed, as already said in Chap. 5, the non-deductive rules by which the hypotheses are obtained can always be extended when they turn out to be insufficient to obtain certain hypotheses. Descartes even says that every solution to a problem that is found can be used "as a rule for finding other ones" (Descartes 1996, VI, 20–21).

C. Cellucci, *The Making of Mathematics*, Synthese Library 448,
https://doi.org/10.1007/978-3-030-89731-4_7

On the other hand, although the rules of discovery are not a closed set, given once for all, some rules of discovery have been widely used in mathematics. The chapter formulates such rules.

Against this, Grosholz objects that, rather than on formulating rules of discovery, the effort would be "better spent on studying the history of mathematics and the actual development of important recent problem-solutions" (Grosholz 2016, 49).

But this objection does not take into account that, as already pointed out in Chap. 3, the history of mathematics, including important recent problem-solutions, is mostly written on the basis of mathematics presented in finished form, and the latter has little or nothing to do with the way it was discovered. Therefore, studying the history of mathematics can teach us about the sequence of published mathematical results and theories, not about the real mathematical process.

Besides formulating rules of discovery, the chapter considers how such rules could have been used in finding solutions to certain historically significant mathematical problems.

This is meant in the spirit of Pólya, who says: "I cannot tell the true story how the discovery did happen, because nobody really knows that. Yet I shall try to make up a likely story how the discovery could have happened. I shall try to emphasize" the "inferences that led to it" (Pólya 1954, I, vi–vii).

Since we cannot tell the true story how the discovery did happen, the chapter tries to make up a likely story how it could have happened, trying to emphasize the inferences involved. Namely, the chapter makes conjectures as to the rule of discovery that could have led to the hypothesis on which the solution to the problem considered was based. In some cases it will be clear that other rules of discovery could have led to the hypothesis, but the one considered in the chapter seems to be the most significant for the problem in question.

The rules considered in the chapter are: induction, in particular, induction from a single case, and induction from multiple cases; analogy, in particular, analogy by quasi-equality, analogy by separate indistinguishability, analogy by agreement, and analogy by agreement and disagreement; metaphor; metonymy; generalization; and specialization.

7.2 Induction

Induction is a kind of reasoning which infers, from the fact that some things of a given kind have a certain property, that all things of that kind have that property.

Aristotle formulates induction by saying that "induction is a passage from particulars to the universal" (Aristotle, *Topica*, A 11, 105 a 13–14). The importance of induction derives from the fact that "demonstration proceeds from universals," and induction is essential to arrive at universals, since "it is impossible to consider universals except through induction" (Aristotle, *Analytic Posteriora*, A 18, 81a 41–b 2).

The importance of induction in obtaining hypotheses in mathematics has been highlighted by several mathematicians.

Thus, Laplace says that "in the mathematical sciences themselves, the main means of arriving at truth" are "induction and analogy" (Laplace 1878–1912, VII, v).

Gauss says that "in arithmetic it most frequently happens that, by some unexpected luck, the most elegant new truths spring up by induction" (Gauss 1863–1933, II, 3).

Pólya says that "many mathematical results were found by induction first and proved later. Mathematics presented with rigor is a systematic deductive science but mathematics in the making is an experimental inductive science" (Pólya 2004, 117).

There are several kinds of induction. With respect to mathematics, the main ones are: induction from a single case, and induction from multiple cases.

7.3 Induction from a Single Case

Induction from a single case (ISC) is an inference of the form: a thing a of kind A is B, therefore all things of kind A are B. So, it is an inference by the rule:

$$(\mathrm{ISC})\,\frac{A(a) \qquad B(a)}{\forall x(A(x) \rightarrow B(x))}.$$

For example, let us consider the problem: Is the sum of the first x odd numbers x^2, for arbitrary given x?

To solve the problem, the Pythagoreans consider a diagram such as the following.

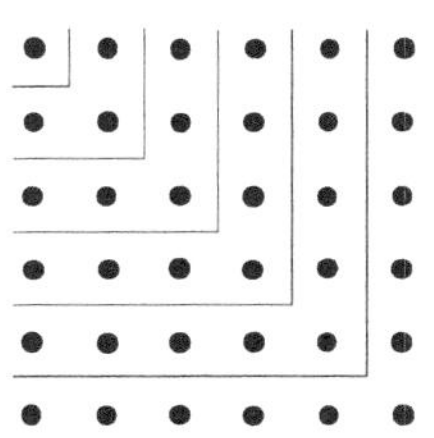

In the diagram, the odd numbers 1, 3, 5, 7, 9, 11 are represented by dots arranged as a carpenter's square, or 'gnomon'. From the diagram it is clear that, if one makes the sum of the number of dots in each gnomon, $1 + 3 + 5 + 7 + 9 + 11$, or one makes the product of the number of rows of dots and the number of columns of dots, 6×6, one obtains the same result, 6^2. Thus $1 + 3 + 5 + 7 + 9 + 11 = 6^2$, namely the sum of the first 6 odd numbers is 6^2. From this, by (ISC), the Pythagoreans arrive at the hypothesis: If the "odd numbers" are "added to the" odd "numbers in order," a square "is produced" (Nicomachus, *Introductionis Arithmeticae libri II*, II.9.3, ed. Hoche). The "side of" the square "consists of as many units as there are numbers taken into the sum to produce it" (ibid., II.9.4). Namely the Pythagoreans arrive at the hypothesis: The sum of the first x odd numbers, for arbitrary given x, is the square of side x, namely $1 + 3 + 5 + \ldots + (2x - 1) = x^2$.

Thus, the argument pattern is as follows. Let $A(x)$: x is a natural number. Let $B(x)$: $1 + 3 + 5 + \ldots + (2x - 1) = x^2$. Then $A(6)$ and $B(6)$. From this, by (ISC), it follows $\forall x(A(x) \rightarrow B(x))$.

Note that, in accordance with what has been said above, the expression 'From this, by (ISC), the Pythagoreans arrive at the hypothesis' does not mean that the Pythagoreans actually arrived at the hypothesis by (ISC), or that they formulated (ISC), or even that they were aware of (ISC). It only means that the use of (ISC) is a hypothetical explanation of the discovery process through which the Pythagoreans arrived at the hypothesis. (The same applies to all other rules of discovery and examples of their use considered in this chapter).

On the other hand, while there is no textual evidence that the Pythagoreans were aware of (ISC), there is textual evidence that Aristotle was aware of it. Indeed, Aristotle considers inferences such as: "The wise are good, since Pittacus is good" (Aristotle, *Analytica Priora*, B 27, 70 a 16–17). Or: "The fact that Socrates is wise and just is a sign that the wise are just" (Aristotle, *Rhetorica*, A 2, 1357 b 12–13).

Induction from a single case (ISC) is more widely used than generally realized. According to Singer, this is because "learning is or can be instantaneous," and "the basic association typically appears full-bloom after a single exposure" (Singer 1971, 1011). Although, "in each case, additional positive instances may be said to have some effect on the organism's behavior with respect to the 'belief'," the "belief itself, the basic formulation of the association or 'induction'" typically "takes place in a single exposure or trial" (ibid.).

The importance of induction from a single case in obtaining hypotheses in mathematics depends on the fact that there are problems which have some representative special case, namely some special case such that a solution of the problem for that case works just as well in the general case.

7.4 Induction from Multiple Cases

Induction from multiple cases (IMC) is an inference of the form: several things $a_1, \ldots, a_n$ of kind A are B, therefore all things of kind A are B. So, it is an inference by the rule:

$$(\text{IMC})\frac{A(a_1) \wedge \ldots \wedge A(a_n) \quad B(a_1) \wedge \ldots \wedge B(a_n)}{\forall x(A(x) \rightarrow B(x))}.$$

For example, let us consider the problem: Is every natural number either a square or the sum of two, three, or four squares?

To solve the problem, Bachet observes that $1 = 1, 2 = 1 + 1, 3 = 1 + 1 + 1, 4 = 4, 5 = 1 + 4, 6 = 1 + 4 + 1, 7 = 1 + 1 + 1 + 4, 8 = 4 + 4, 9 = 9, 10 = 1 + 9, \ldots,$

325 = 1 + 324, so every natural number up to 325 is either a square, or the sum of two, three, or four squares. From this, by (IMC), Bachet arrives at the hypothesis: "Every number is either a square, or the sum of two, three, or four squares" (Bachet 1621, 241).

Thus, the argument pattern is as follows. Let $A(x)$: x is a natural number. Let $B(x)$: x is either a square or the sum of two, three, or four squares. Then $A(1) \wedge \ldots \wedge A(325)$ and $B(1) \wedge \ldots \wedge B(325)$. From this, by (IMC), it follows $\forall x(A(x) \rightarrow B(x))$.

Kant formulates induction from multiple cases by saying that it is the inference "from the particular to the universal" according to "the principle of universalization: What belongs to many things of a genus belongs to the remaining ones too" (Kant 1992, 626). Although induction from multiple cases is non-deductive, "we cannot do without it" because, "along with it most of our cognitions would have to be abolished at the same time" (ibid., 232).

Euler highlights the importance of induction from multiple cases in obtaining hypotheses in mathematics, by saying that, "in investigating the nature of numbers, very much must be attributed to observation and induction" (Euler 1761, 186). Indeed, "many notable properties of numbers" which "up to now have been found and demonstrated," in fact "have been observed" by "their discoverers before they thought of demonstrating them" (ibid., 185). Thus, "Fermat inferred" some "remarkable properties of numbers by induction long before he learned to demonstrate them" (ibid., 186). On the other hand, "many properties of numbers are known to us, which however we are not yet able to demonstrate, so we have been led to their knowledge only by observations" (ibid., 19). So, "in the science of numbers," we "must expect most from observations, indeed we are immediately led by them to new properties of numbers, for which later a demonstration has to be worked out" (ibid.).

7.5 Induction and Probability

According to a widespread opinion, all inductive reasoning is reasoning in terms of probability.

Thus, Reichenbach says that "the inductive inference must be conceived as an operation belonging in the framework of the calculus of probability" (Reichenbach 1951, 233).

This opinion, however, is invalid. For, if all inductive reasoning is reasoning in terms of probability, then one should adopt the hypothesis that is most probable. Indeed, as Reichenbach says, "the degree of probability supplies a rating of the" hypothesis, "it tells us how good the" hypothesis "is" (ibid., 240).

But, if one should adopt the hypothesis that is most probable, then, for example, the Pythagoreans should not have adopted the hypothesis: $1+3+5+\ldots+(2x-1)=x^2$, for arbitrary given x. For, this hypothesis is about infinitely many cases, so, being obtained by induction from a single case (ISC), it has probability zero. Then, that the

Pythagoreans adopted this hypothesis becomes inexplicable if all inductive reasoning is reasoning in terms of probability.

This is just an example of why the opinion that all inductive reasoning is reasoning in terms of probability is invalid.

7.6 Analogy

Analogy is a kind of reasoning which infers, from the fact that two things are similar and one of them has a certain property, that the other one has that property.

Let $a \approx b$: a is similar to b. Analogy is an inference of the form: $a \approx b$, and a is A, therefore b is A. So, it is an inference by the rule:

$$(\mathrm{AN})\,\frac{a \approx b \quad A(a)}{A(b)}.$$

Aristotle formulates (AN) as follows: "Among similars, what is true of one is true of the rest" (Aristotle, *Topica*, A 18, 108 b 13–14).

Weil highlights the importance of analogy in obtaining hypotheses in mathematics by saying that "nothing is more fruitful, all mathematicians know this, than these obscures analogies, these hazy reflections of one theory on another," and "nothing gives more pleasure to the researcher" (Weil 2009, 408).

Dieudonné says that "in many cases analogy has played a role of powerful ferment and driving force of research," especially "in mathematics, where I am sure that at least 50% of the problems and fecund ideas have come out of it" (Dieudonné 1981, 257). In particular, "it can be affirmed without hesitation that the great discoveries in mathematics" in the twentieth century "have arisen from the unexpected bringing together of superficially distinct notions" (ibid., 266).

Banach even characterizes the mathematician in terms of analogy, since he says that "a mathematician is a person who can find analogies between theorems; a better mathematician is one who can see analogies between proofs and the best mathematician can notice analogies between theories" (Kaluza 1995, 92). Then one can suppose that "the ultimate mathematician is one who can see analogies between analogies" (ibid.).

The above formulation of (AN), however, is vacuous if one does not specify what meaning is to be attached to the expression 'a is similar to b'. For, any two things can be viewed as having some property in common, for example, the property of being a thing, and hence can be viewed as being similar with respect to that property.

Depending on the meaning attached to the expression 'a is similar to b', several kinds of analogy can be distinguished. With respect to mathematics, the main ones are: analogy by quasi-equality, analogy by separate indistinguishability, analogy by agreement, and analogy by agreement and disagreement.

7.7 Analogy by Quasi-Equality

Two things are said to be quasi-equal if, while not identical, they are a very close approximation to each other.

Let $a \cong b$: a is quasi-equal to b. Analogy by quasi-equality is an inference of the form: $a \cong b$, and a is A, therefore b is A. So, it is an inference by the rule:

$$(\text{AQE})\,\frac{a \cong b \quad A(a)}{A(b)}.$$

For example, let us consider the problem: What is the area of a circle with circumference c and radius r?

To solve the problem “Antiphon, having drawn a circle, inscribed in it one of the polygons that can be inscribed therein” (Simplicius, *In Aristotelis Physicorum Libros Quattuor Priores Commentaria*, I 2, 54.20–22, ed. Diels). Then, bisecting the sides of the inscribed polygon, he formed a polygon of twice as many sides; and doing the same again and again, he concluded that in this manner “a polygon would be inscribed in the circle whose sides would by and large coincide with the circumference of the circle” (ibid., 55.7–8).

Thus, given a circle with circumference c and radius r, let us inscribe in it a regular n-sided polygon with side length s. We may consider the polygon as consisting of n isosceles triangles with the same base s and the same height h, hence with the same area $(1/2)sh$.

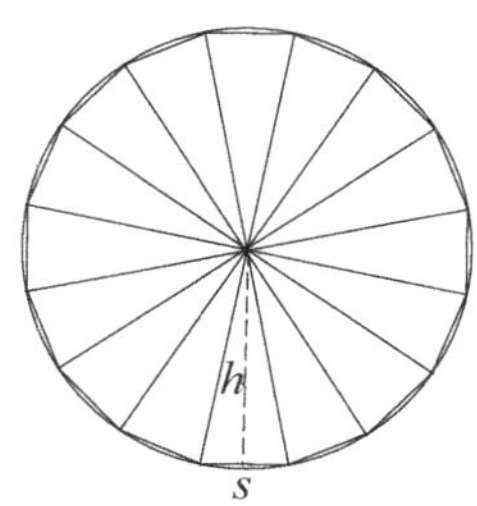

Let p be the perimeter of the polygon. Then $p = ns$, and the area of the polygon will be $n(1/2)sh = (1/2)nsh = (1/2)ph$. As n increases, the perimeter p of the polygon will approximate more and more closely the circumference c of the circle, and the height h of the triangles will approximate more and more closely the radius r of the circle. Therefore, for a sufficiently large n, the polygon and the circle will be quasi-equal. Then, from the fact that the area of the polygon is $(1/2)ph$, by (AQE), Antiphon arrives at the hypothesis: The area of the circle with circumference c and radius r is $(1/2)cr$.

Thus, the argument pattern is as follows. Let a be an n-sided polygon inscribed in a circle b. Let $A(x)$: area of $x = (1/2)\cdot$(perimeter of x)$\cdot$(height of x) where, if x is a circle, ‘perimeter of x’ is to mean ‘circumference of x’, and ‘height of x’ is to mean ‘radius of x’. Then, for a suitably large n, we have $a \cong b$, and $A(a)$. From this, by (AQE), it follows $A(b)$.

Leibniz highlights the importance of analogy by quasi-equality in obtaining hypotheses in the calculus of infinitesimals by saying: "I consider to be equal not only those things whose difference is absolutely nothing, but also those whose difference is incomparably small; and although this difference need not be called absolutely nothing, neither is it a quantity comparable with those whose difference it is" (Leibniz 1971, V, 322). Indeed, "I think that only those homogeneous quantities are comparable, one of which can become larger than the other if multiplied by a finite number. And I judge that those whose difference is not such a quantity are equal" (ibid.). This "is precisely what is meant by saying that the difference is smaller than any given quantity" (ibid.). Since quantities whose difference is incomparably small are considered to be equal, they can be supposed to have the same properties.

Leibniz's statements are the basis for Postulate I in L'Hôpital 2015, considered in Chap. 2.

7.8 Analogy by Separate Indistinguishability

Two things are said to be indistinguishable when observed separately if, as Leibniz says, they "cannot be distinguished when observed in isolations from one another" (Leibniz 1971, V, 180).

Let $a \sim b$: a and b are indistinguishable when observed separately. Analogy by separate indistinguishability is an inference of the form: $a \sim b$ and a is A, therefore b is A. So, it is an inference by the rule:

$$(\text{ASI})\frac{a \sim b \quad A(a)}{A(b)}.$$

For example, let us consider the problem: Are circles to each other as the squares on their diameters? As we have seen in Chap. 5, this is the problem to which Hippocrates of Chios reduces the problem of the quadrature of the lunule whose outer circumference is a semicircle.

To solve the problem, Leibniz proceeds as follows. Let "a circle be described with diameter *AB*," and let "a square *CD* on the diameter be circumscribed to it" (ibid., V, 182). Also, let "a circle be described with diameter *LM*," and let "a square *NO* on the diameter be circumscribed to it" (ibid.). (By a square on the diameter it is meant a square whose sides are equal to the diameter).

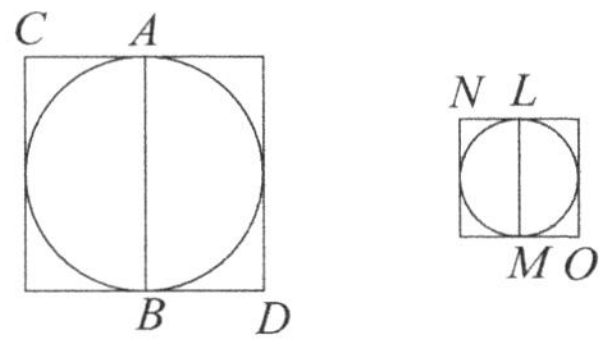

The resulting figures *ABCD* and *LMNO* are indistinguishable when observed separately. Of course, they "can be distinguished by magnitude" (ibid., V, 181). But "magnitude can be grasped only through the compresence of the things or through the application of some intervening thing to both" (ibid., V, 180). Indeed, "magnitude can be known only by co-observation either of both" figures "at the same time, or of each of them with some unit of measure, but in this case they would not be viewed in isolations from one another, as postulated" (ibid., V, 181). Therefore, it is justified to say that the two figures are indistinguishable when observed separately.

From the fact that two figures are indistinguishable when observed separately it follows that there is "the same ratio or proportion among the parts in each figure" (ibid., V, 181). Indeed, from the fact that the two figures *ABCD* and *LMNO* are indistinguishable when observed separately, by two applications of (ASI), it follows that there is the same ratio or proportion among the parts in *ABCD* and *LMNO*. So, "the circle *AB* is to the square *CD* as the circle *LM* is to the square *NO*; hence also, the circle *AB* is to the circle *LM* as the square *CD* is to the square *NO*" (ibid., V, 182). Therefore, "circles are to each other as the squares on their diameters" (ibid.).

Thus the argument pattern is as follows. Let $A(x)$: there is a certain determinate ratio or proportion among the parts in x. Now, $ABCD \sim LMNO$. From this, by two applications of (ASI), it follows that $A(ABCD)$ if and only if $A(LMNO)$, namely there is the same ratio or proportion among the parts in *ABCD* and *LMNO*. As argued above, from this it follows that circles are to each other as the squares on their diameters.

Leibniz highlights the importance of analogy by separate indistinguishability in obtaining hypotheses in geometry, by saying that "besides quantity, figure in general includes also quality or form; and as those figures are equal whose magnitude is the same, so those are similar whose form is the same" (ibid., V, 179). So "a true geometric analysis considers not only equalities" but "also similarities" (ibid.). Now, "the true reason why geometers have not made enough use of consideration of similarity" is that "they did not have any general concept of it sufficiently distinct or adapted to mathematical inquiries" (ibid., V, 179–180). Such a concept is provided by similarity as separate indistinguishability, according to which "we call two presented figures similar if nothing can be observed in one, viewed singly, which cannot be equally observed in the other" (Leibniz 1971, V, 181).

7.9 Analogy by Agreement

Two things are said to be analogous by agreement if they have some properties in common.

Analogy by agreement is an inference of the form: a and b are $A_1, \ldots, A_n$, and a is B, therefore b is B. So, it is an inference by the following rule, where $\widehat{A}(x)$ is short for $A_1(x) \wedge \ldots \wedge A_n(x)$:

$$(\mathrm{AA})\frac{\widehat{A}(a) \quad \widehat{A}(b) \quad B(a)}{B(b)}.$$

For example, let us consider the problem: In any convex polygon, what is the sum of its internal angles?

To solve the problem, Proclus starts from the fact that, in any triangle, the sum of the internal angles is equal to two right angles. Now, a triangle and a convex polygon agree in that they are both convex polygons. Moreover, the fact that, in any triangle, the sum of the internal angles is equal to two right angles, can also be written as follows: In any convex polygon of 3 sides, the sum of the internal angles is equal to $2 \cdot (3 - 2)$ right angles. From this, by (AA), Proclus arrives at the hypothesis that, in any convex polygon of n sides, the sum of the internal angles is equal to $2 \cdot (n - 2)$ right angles. Indeed, Proclus says that any convex polygon "is divisible into triangles two less in number than the number of its sides" (Proclus, *In primum Euclidis Elementorum librum commentarii*, 382.5–6, ed. Friedlein). And "the number which is double the number of the triangles" constituting the convex polygon "will give the number of right angles to which the angles of any" convex polygon "are equal" (ibid., 382.15–18).

Thus, the argument pattern is as follows. Let a be a triangle. Let b be a convex polygon. Let $A(x)$: x is a convex polygon. Let $B(x)$: sum of the internal angles of $x =$ $2 \cdot$ ((number of sides of x) $-$ 2)right angles. Then $A(a)$ and $A(b)$ and $B(a)$. From this, by (AA), it follows $B(b)$.

Kant formulates analogy by agreement saying that it is the inference by which, "if two things agree under as many determinations as I have become acquainted with, then I infer that they agree also in the other determinations. I infer, then, from some determinations, which I cognize, that the others belong to the thing too" (Kant 1992, 503). Thus, analogy by agreement "infers from" the partial "similarity of two things" to their total similarity, "according to the principle of specification: Things of one genus, which we know to agree in much, also agree in what remains, with which we are familiar in some things of this genus but which we do not perceive in others" (ibid., 626).

Analogy by agreement (AA) can be derived from induction from a single case (ISC). This can be seen as follows:

$$
(\rightarrow E)\ \dfrac{\widehat{A}(b) \qquad (\forall E)\ \dfrac{(ISC)\ \dfrac{\widehat{A}(a) \qquad B(a)}{\forall x(\widehat{A}(x) \rightarrow B(x))}}{\widehat{A}(b) \rightarrow B(b)}}{B(b)}
$$

7.10 Analogy by Agreement and Disagreement

Two things are said to be analogous by agreement and disagreement if they have some properties in common and some other properties not in common.

Analogy by agreement and disagreement is an inference of the form: a and b are $A_1, \ldots, A_n$, but a is $B_1, \ldots, B_k$ while b is not all of $B_1, \ldots, B_k$, and a is C, therefore b is C. So, it is an inference by the following rule, where $\widehat{A}(x)$ is short for $A_1(x) \wedge \ldots \wedge A_n(x)$, and $\widehat{B}(x)$ is short for $B_1(x) \wedge \ldots \wedge B_k(x)$:

$$
(AAD)\ \frac{\widehat{A}(a) \quad \widehat{A}(b) \quad \widehat{B}(a) \quad \neg\widehat{B}(b) \quad C(a)}{C(b)},
$$

provided that $\neg\widehat{B}(b)$ is not incompatible with $C(b)$.

For example, let us consider the problem: In any ellipse, if two straight lines, which are not through the centre, cut each other, do they cut each other in half?

To solve the problem, Apollonius starts from Euclid's proposition that, in any circle, if two straight lines, which are not through the centre, cut each other, then they do not cut each other in half (Euclid, *Elementa*, III, Proposition 4).

Now, a circle and an ellipse have in common the property that they are both a closed curved shape on a plane, but do not have in common the property that all points on them are equally far from the centre, because the circle has this property, but not the ellipse. Then, from Euclid's solution, by (AAD), Apollonius arrives at the hypothesis: "In any ellipse," if "two straight lines, which are not through the centre, cut each other, then they do not cut each other in half" (Apollonius, *Conica*, II, Proposition 26, ed. Heiberg).

Thus, the argument pattern is as follows. Let a be a circle. Let b an ellipse. Let $A(x)$: x is a closed curved shape on a plane. Let $B(x)$: all points on x are equally far from the centre. Let $C(x)$: In x, if two straight lines, which are not through the centre, cut each other, they do not cut each other in half. Then $A(a)$, $A(b)$, $B(a)$, $\neg B(b)$, and $C(a)$. From this, by (AAD), it follows $C(b)$. This is justified because $\neg B(b)$ is not incompatible with $C(b)$.

7.11 Induction and Analogy

Hume says that induction is nothing other than an analogy, because it is "founded on the similarity, which we discover among" certain "objects, and by which we are induced to expect effects similar to those, which we have found to follow from such objects" (Hume 2007, 26).

But this is invalid in general, because it conflicts with (ISC). For, (ISC) considers a single case, so it cannot be founded on the similarity which we discover among certain objects.

Conversely, Kant says that "analogy" is "nothing other than an induction, only an induction in respect of the predicate" (Kant 1992, 232). For, when "2 things have come together in respect of all attributes that I have been able to cognize in them, then they will also come together in the remaining attributes, which I have not cognized in them, and thus runs the inference in regard to analogy" (ibid.).

But this is invalid in general. Admittedly, in a sense (AA) can be said to be nothing other than an induction. For, as we have seen above, (AA) can be derived from (ISC). But (AAD) cannot be said to be nothing other than an induction. For, while (AAD) takes account of properties that *a* and *b* do not have in common, (ISC) and (IMC) do not take account of them. This is relevant because inferences that would be legitimate by (AA), which does not take account of properties that *a* and *b* do not have in common, are not legitimate by (AAD), which takes account of them. For example, a circle and an ellipse have in common the property that they are both a closed curved shape on a plane. From this, it would be legitimate to infer by (AA) the implausible hypothesis that a circle and an ellipse have both a constant radius throughout the shape. Conversely, since a circle and an ellipse do not have in common the property that all points on them are equally far from the centre, it would not be legitimate to infer such hypothesis by (AAD).

7.12 Metaphor

Metaphor is asserting that things of one domain, called the 'target domain', are to be considered as if they were things of another domain, called the 'source domain'. The implication is that, if the things of the source domain have a certain property, the things of the target domain will have that very same property. Thus metaphor transfers properties of objects of the source domain to objects of the target domain.

Let S by the source domain and T the target domain. Let $T \rhd S$: The elements of T are to be considered as if they were elements of S. Metaphor is an inference of the form: $T \rhd S$, $a \in T$, and if $x \in S$ then x is A, therefore a is A. Thus it is an inference by the rule:

$$(\text{MTA})\,\frac{T \rhd S \quad a \in T \quad \begin{matrix}[x \in S] \\ A(x)\end{matrix}}{A(a)},$$

where the assumption class $[x \in S]$ written above the premiss $A(x)$, indicates that the premiss in question depends on assumptions belonging to that class, which may be discharged by the rule, in the sense that the conclusion $A(a)$ no longer depends on them, but only on the yet undischarged assumptions.

For example, let us consider the problem: What is the derivative of x^n?

To solve the problem, Newton says: "I consider" mathematical "quantities as though they were generated by continuous increase in the manner of a space which a moving object describes in its course" (Newton 1967–1981, III, 73). So, "mathematical quantities" are to be considered "not as consisting of least possible parts," namely of indivisibles, "but as described by continuous motion" (ibid., VIII, 123). In particular, "lines are described and by describing generated" through "the continuous motion of points; surface-areas are through the motion of lines, solids through the motion of surface-areas, angles through the rotation of sides times through continuous flux, and the like in other cases" (ibid.). These geneses do not take place merely in the mind, they "take place in the reality of physical nature," and "much in this manner the ancients, by 'drawing' mobile straight lines into the length of stationary ones, taught the genesis of rectangles" (ibid.).

On this basis, Newton says: "I was led to seek a method of determining quantities out of the speeds of motion or increment by which they are generated; and naming these speeds of motion or increment 'fluxions' and the quantities so borne 'fluents', I fell" upon "the method of fluxions" (ibid.). Namely, upon Newton's calculus of infinitesimals. In the latter, "fluxions are very closely near as the augments of their fluents begotten in the very smallest equal particles of time: to speak accurately, indeed, they are in the first ratio of the nascent augments" (ibid., VIII, 123, 125).

In terms of Newton's calculus, then, the above problem becomes: What is the fluxion of the quantity x^n?

To solve the problem, Newton observes that, "in the time that the quantity x comes in its flux to be $x + o$, the quantity x^n will come to be $(x + o)^n$" (ibid., VIII, 127). Namely, when $(x + o)^n$ is expanded by the method of infinite series, x^n will come to be

$$x^n + nox^{n-1} + \frac{1}{2}(n^2 - n)o^2x^{n-2} + \ldots.$$

So "the augments o and $nox^{n-1} + \frac{1}{2}(n^2 - n)o^2x^{n-2} + \ldots$ are to each other as 1 and $nx^{n-1} + \frac{1}{2}(n^2 - n)ox^{n-2} + \ldots$" (ibid., VIII, 127, 129). Now "let those augments come to vanish and their last ratio will be 1 to nx^{n-1}; consequently the fluxion of the quantity x is to the fluxion of the quantity x^n as 1 to nx^{n-1}" (ibid., VIII, 129). On the basis of the assumption that mathematical quantities are to be considered as if they were quantities generated by continuous motion, from this, by (MTA), Newton arrives at the hypothesis: The fluxion of the quantity x^n is nx^{n-1}.

Thus, the argument pattern is as follows. Let S be the domain of quantities generated by continuous motion, and T the domain of mathematical quantities. Let $A(x)$: the fluxions of x^n is nx^{n-1}. Let us consider the elements of T as if they were elements of S. Then $T \rhd S$. Let $a \in T$. If $x \in S$ then $A(x)$. From this, by (MTA), it follows $A(a)$.

Lakoff and Núñez highlight the importance of metaphor in obtaining hypotheses in mathematics by saying that "metaphors are an essential part of mathematical thought, not just auxiliary mechanisms used for visualization or ease of understanding" (Lakoff and Núñez 2000, 6). For, they allow us "to bring to one domain of mathematics the ideas and the methods of precise calculation of another domain" (ibid., 351–352). By virtue of the use of metaphors, "present mathematics can be extended to create new forms by importing structure from one branch to another and by fusing mathematical ideas from different branches" (ibid., 379).

7.13 Metaphor and Analogy

Aristotle says that metaphor is nothing other than an analogy, because "metaphor makes what is signified somehow familiar on account of the similarity; for those who make metaphors do so on account of some similarity" (Aristotle, *Topica*, Z 2, 140 a 9–12).

This, however, is invalid because it would make metaphor vacuous. For example, on this view, Newton's metaphor that mathematical quantities are to be considered as if they were quantities generated by continuous motion, means that mathematical quantities are to be considered similar to quantities generated by a continuous motion. But, since this statement does not specify with respect to which properties they are similar, the metaphor is vacuous. This is akin to the problem with analogy (AN) considered above, that any two things can be viewed as having some property in common, and hence as being similar with respect to that property.

But even when one specifies with respect to which properties the two things are similar, such properties may not mean the same thing in the target domain and in the source domain.

For example, in Newton's metaphor that mathematical quantities are to be considered as if they were quantities generated by continuous motion, mathematical quantities and quantities generated by continuous motion have both the property that they can vary. But this does not mean the same thing in the domain of mathematical quantities and in the domain of quantities generated by continuous motion. While mathematical quantities can vary in the sense that they can take different values, quantities generated by continuous motion can vary in the sense that they are physical objects whose physical condition can change in time.

Saying that mathematical quantities are to be considered as if they were quantities generated by continuous motion creates a similarity, because it suggests that mathematical quantities and quantities generated by continuous motion are similar in the sense that both can vary. Thus, it is the metaphor that creates a similarity, rather than being based on a pre-existing similarity.

Moreover, while similarity is symmetrical, metaphor is not symmetrical. For example, 'mathematical quantities are quantities generated by continuous motion' is a metaphor, but 'quantities generated by continuous motion are mathematical quantities' does not make sense. However, if metaphor were nothing other than an analogy, metaphor would have to be symmetrical.

7.14 Metonymy

Metonymy is letting one thing stand for another thing that is associated with it. As Lakoff and Johnson say, "metaphor and metonymy are different kinds of processes. Metaphor is principally a way of conceiving of one thing in terms of another," while metonymy "allows us to use one entity to stand for another" (Lakoff and Johnson 2003, 36).

Metonymy is the basis of all mathematical symbolism, so it is omnipresent in mathematics and is involved in virtually every mathematical discovery.

Let $a \Rightarrow b$: Let a stand for b. Metonymy is an inference of the form: $a \Rightarrow b$, and a is A, therefore b is A. Thus it is an inference by the rule:

$$(\mathrm{MTO})\,\frac{a \Rightarrow b \quad A(a)}{A(b)}.$$

For example, let us consider the problem: Demonstrate that, in any triangle, the sum of the internal angles is equal to two right angles.

Eudemus says that the Pythagoreans "demonstrated it as follows. Let *ABC* be a triangle, and through *A* draw a line *DE* parallel to *BC*" (Proclus, *In primum Euclidis Elementorum librum commentarii*, 379.5–8, ed. Friedlein).

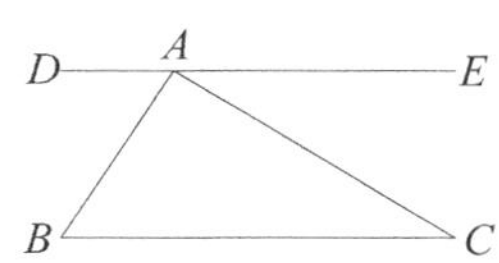

Then, since *DE*, *BC* are parallel, the alternate angles *ABC*, *DAB* are equal, and so are the alternate angles *ACB*, *EAC*. Therefore, the sum of the angles *ABC*, *ACB* is equal to the sum of the angles *DAB*, *EAC*. Add the angle *BAC* to each. Then the sum of the angles *ABC*, *ACB*, *BAC* is equal to the sum of the angles *DAB*, *EAC*, *BAC*, and hence to the sum of the angles *DAB*, *BAC*, *EAC*, so it is equal to two right angles. Therefore, the sum of the angles *ABC*, *ACB*, *BAC* is equal to two right angles. This solves the problem. Such "is the demonstration of the Pythagoreans" (ibid., 379.17–18):

Now, when the Pythagoreans say 'Let *ABC* be a triangle', they mean to say 'Let *ABC* stand for a triangle'. So, they mean to say: $ABC \Rightarrow b$, where b is a triangle. Then, from the fact that, in *ABC*, the sum of the internal angles is equal to two right angles, by (MTO), the Pythagoreans arrive at the hypothesis: In any triangle, the sum of the internal angles is equal to two right angles.

Thus, the argument pattern is as follows. Let $ABC \Rightarrow b$. Let $A(x)$: In the triangle x, the sum of the internal angles is equal to two right angles. Then $A(ABC)$. From $ABC \Rightarrow b$ and $A(ABC)$, by (MTO), it follows $A(b)$.

It might seem improper to include (MTO) among the rules of discovery, but it is not so. The importance of metonymy in obtaining hypotheses in mathematics depends on the fact that it permits to reason on a concrete object and then to transfer properties established on it to an abstract object, which may have great heuristic value. Thus, in the case of the above problem, metonymy permits to reason on *ABC*, which is concrete being a figure, and then to transfer the property established on *ABC* to a triangle b, which is an abstract object since it is everything that satisfies the definition of triangle. This has great heuristic value, because it would be extremely difficult, if not impossible, to discover such property on the basis of the definition of a triangle alone.

7.15 Generalization

Generalization is a kind of reasoning which infers, from the fact that all members of a given set of objects have a certain property, that any member of a larger set, containing the given set as a subset, has that property.

Inference by generalization is an inference of the form: $S \subseteq T$, $a \in T$, and if $x \in S$ then x is A, therefore a is A. Thus it is an inference by the rule:

$$(\text{GEN})\,\frac{S \subseteq T \quad a \in T \quad \begin{matrix}[x \in S] \\ A(x)\end{matrix}}{A(a)},$$

where the assumption class $[x \in S]$ written above the premiss $A(x)$, indicates that the premiss in question depends on assumptions belonging to that class, which may be discharged by the rule, in the sense that the conclusion $A(a)$ no longer depends on them, but only on the yet undischarged assumptions.

For example, let us consider the problem: In any right-angled triangle, is the square on the hypotenuse equal to the sum of the squares on the legs?

To solve the problem, an unknown Egyptian mathematician finds it easier to consider the following special case of the problem: In any right-angled isosceles triangle, is the square on the hypotenuse equal to the sum of the squares on the legs?

The Egyptian mathematician solves this special case by using the following figure which, as Allman says, is inspired to him by "the contemplation of a draught-board, or of a floor covered with square tiles, or of a wall ruled with squares" (Allman 1889, 29).

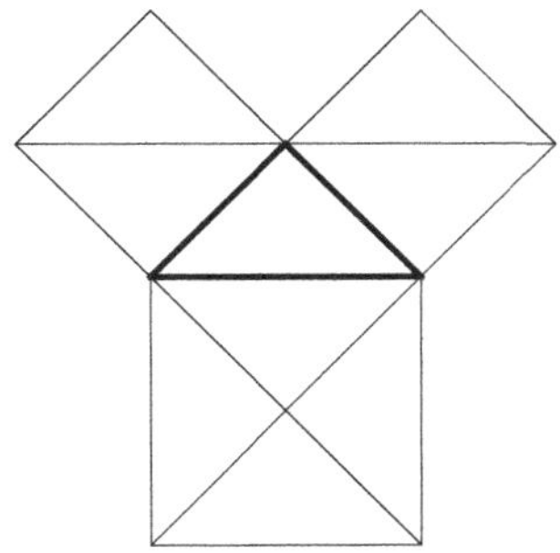

The figure immediately solves the problem, because it shows that, in a right-angled isosceles triangle, the square on the hypotenuse is equal to the sum of the squares on the legs (counting the number of equal triangles on each). From this, by (GEN), the Egyptian mathematician arrives at the hypothesis: In a right-angled triangle, the square on the hypotenuse is equal to the sum of the squares on the legs.

Thus, the argument pattern is as follows. Let S be the domain of right-angled isosceles triangles. Let T be the domain of right-angled triangles. Let $A(x)$: in the right-angled triangle x, the square on the hypotenuse is equal to the sum of the squares on the legs. Then $S \subseteq T$. Let $a \in T$. If $x \in S$ then $A(x)$. From this, by (GEN), it follows $A(a)$.

The importance of generalization in obtaining hypotheses in mathematics depends on the fact that often a special case may be easier to solve than a general problem. Then, one may try to see whether the solution for the special case can be extended to the general problem.

7.16 Specialization

Specialization is a kind of reasoning which infers, from the fact that all members of a given set of objects have a certain property, that any member of a smaller set, contained in the given set as a subset, has that property.

Inference by specialization is an inference of the form: $S \subseteq T$, $a \in S$, and every $x \in T$ is A, therefore a is A. Thus it is an inference by the rule:

$$(\text{SPE})\,\frac{S \subseteq T \quad a \in S \quad \begin{matrix}[x{\in}T]\\ A(x)\end{matrix}}{A(a)},$$

where the assumption class $[x \in T]$ written above the premiss $A(x)$, indicates that the premiss in question depends on assumptions belonging to that class, which may be discharged by the rule, in the sense that the conclusion $A(a)$ no longer depends on them, but only on the yet undischarged assumptions.

For example, let us consider the problem: What is the sum of the numbers 1, 2, 3, . . ., 100?

According to folklore, this is the problem almost instantly solved by the nine year old Gauss.

Sartorius narrates: "Gauss had just entered the class when" the schoolteacher "Büttner gave out for a problem the adding of a series of numbers from 1 to 100. The problem was barely stated before Gauss threw his slate on the table with the words (in the low Braunschweig dialect): There it lies" (Sartorius von Waltershausen 2018, 4). When, "at the end of the hour the slates were turned bottom up" and "Büttner read out the answer, to the surprise of all present that of young Gauss was found to be correct, whereas many of the others were wrong" (ibid.).

Again according to folklore, to solve the problem, Gauss proceeds as follows. Instead of the problem 'What is the sum of the numbers 1, 2, 3, . . ., 100?', Gauss considers the more general problem: What is the sum of the numbers 1, 2, 3, . . ., x, for arbitrary given x?

To solve the problem, Gauss imagines the numbers 1, 2, 3, ..., x to be displayed in two rows in reverse order:

1	2	3	. . .	$x-2$	$x-1$	x
x	$x-1$	$x-2$	. . .	3	2	1

The sum of each column is $x + 1$ and there are x columns, so the total sum is $x(x + 1)$, therefore the sum of the numbers 1, 2, 3, . . ., x, for arbitrary given x, is $x(x + 1)/2$.

Note that Gauss's solution to the problem 'What is the sum of the numbers 1, 2, 3, . . ., x, for arbitrary given x?' does not use mathematical induction. It is a uniform argument that works for any natural number x, the kind of argument that Herbrand calls a "prototype" (Herbrand 1971, 289, footnote 5). For more on this, see Cellucci 2009.

From the fact that the sum of all numbers 1, 2, 3, . . ., x, for arbitrary given x, is $x(x + 1)/2$, by (SPE), Gauss arrives at the hypothesis: The sum of the numbers 1, 2, 3, . . ., 100 is $100 \cdot 101/2 = 5050$,

Thus, the argument pattern is as follows. Let $S = \{1, 2, 3, \ldots, 100\}$. Let $T = \{1, 2, 3, \ldots, x\}$ for $x \geq 100$. Let $A(x)$: $1 + 2 + 3 + \ldots + x = x(x + 1)/2$. Then $S \subseteq T$ and $100 \in S$. If $x \in T$ then $A(x)$. From this, by (SPE), it follows $A(100)$.

The importance of specialization in obtaining hypotheses in mathematics depends on the fact that often the more general problem may be easier to solve than a special case.

7.17 A More Significant Example

A more significant example of the fact that often the more general problem may be easier to solve than a special case is provided by the problem considered earlier in this chapter: Is the sum of the first x odd numbers x^2, for arbitrary given x?

Suppose that, instead of that problem, we consider the problem: Is the sum of the first x odd numbers a square, for arbitrary given x? The former problem is more general than this one because, if the sum of the first x odd numbers is x^2, for arbitrary given x, then the sum of the first x odd numbers is a square. (A problem is more general than another one if every solution to the former is also a solution to the latter).

If one tries to arrive by mathematical induction at the more general hypothesis 'The sum of the first x odd numbers is x^2, namely $1 + 3 + 5 + \ldots + (2x - 1) = x^2$, for arbitrary given x' the induction step works smoothly. Indeed:

Basis. $1 = 1^2$. *Induction Step.* Assume as induction hypothesis that $1 + 3 + 5 + \ldots + (2x - 1) = x^2$. Then $1 + 3 + 5 + \ldots + (2x - 1) + (2(x + 1) - 1) = 1 + 3 + 5 + \ldots + (2x - 1) + (2x + 1) = x^2 + (2x + 1) = (x + 1)^2$.

But, if one tries to arrive by mathematical induction at the hypothesis 'The sum of the first x odd numbers, namely $1 + 3 + 5 + \ldots + (2x - 1)$, is a square, for arbitrary given x', then the induction step breaks down. For, the induction hypothesis, '$1 + 3 + 5 + \ldots + (2x - 1)$ is a square, for arbitrary given x', does not say which square $1 + 3 + 5 + \ldots + (2x - 1)$ actually is. But this would be indispensable to conclude that the square in question, added to $2x + 1$, is also a square, and hence that $1 + 3 + 5 + \ldots + (2x - 1) + (2x + 1)$ is a square, for arbitrary given x.

7.18 Rules of Discovery and Rationality

Pólya says that "we should not ask: Are there rules of discovery?" but rather: "Are there" practical guidelines "of some sort expressing attitudes useful in problem solving?" (Pólya 1981, II, 90). Pólya says so because, as we have seen in Chap. 3, according to him, there is no method of mathematical discovery, there can only be some practical guidelines to solve problems in mathematics. But this view, instead of clarifying the nature of mathematics, confines mathematics to irrationalism.

Indeed, Pólya says that "it seems impossible to evaluate" the chances of an approach to solving a problem "merely on the basis of some clear, rational argument," so "to some extent, in some manner the problem solver must rely on his inarticulate feelings" (ibid.). Even if the problem solver may "carefully consider his articulate reasons for an appropriate time," he must "refer to his feelings afterwards for the final decision" (ibid., II, 91). Therefore, the main practical guideline "of the problem solver should be: Never act against your feelings" (ibid.).

But 'Never act against your feelings' means to rely on intuition and, as argued in Chap. 2, intuition is inadequate as a basis for mathematics.

As an antidote to irrationalism, rules of discovery must be stated and shown to play a role in discovery. Admittedly, as Pólya says, when one considers the history of mathematics, one cannot tell the true story how the discovery did happen, because nobody really knows that. But one can make plausible guesses at the possible rules of discovery involved and, more importantly, the rules of discovery so formulated can be used in future discoveries.

References

Allman, George Johnston. 1889. *Greek geometry from Thales to Euclid*. Dublin: Hodges, Figgis, & Co.

Bachet, Claude Gaspard. 1621. *Diophanti Alexandrini arithmeticorum libri sex, et de numeris multangulis liber unus*. Paris: Drouart.

Cellucci, Carlo. 2009. The universal generalization problem. *Logique & Analyse* 205: 3–20.

Descartes, René. 1996. *Œuvres*. Paris: Vrin.

Dieudonné, Jean. 1981. L'analogie en mathématique. In *Analogie et connaissance*, ed. André Lichnerowicz, François Perroux, and Gilbert Gadoffre, vol. 2, 257–266. Paris: Meloine.

Euler, Leonhard. 1761. Specimen de usu observationum in mathesi pura. *Novi Commentarii Academiae Scientiarum Imperialis Petropolitanae* 6 (1756/1757): 19–21, 185–230.

Gauss, Carl Friedrich. 1863–1933. *Werke*. Göttingen: Königlichen Gesellschaft der Wissenschaften.

Grosholz, Emily. 2016. *Starry reckoning: Reference and analysis in mathematics and cosmology*. Cham: Springer.

Herbrand, Jacques. 1971. *Logical writings*. Dordrecht: Springer.

Hume, David. 2007. *A treatise of human nature*, vol. 1. Oxford: Oxford University Press.

Kaluza, Roman. 1995. *Through a reporter's eyes: The life of Stefan Banach*. Boston: Birkhäuser.

Kant, Immanuel. 1992. *Lectures on logic*. Cambridge: Cambridge University Press.

Lakoff, George, and Mark Johnson. 2003. *Metaphors we live by*. Chicago: The University of Chicago Press.

Lakoff, George, and Rafael E. Núñez. 2000. *Where mathematics comes from*. New York: Basic Books.

Laplace, Pierre Simon, marquis de. 1878–1912. *Oeuvres complètes*. Paris: Gauthier-Villars.

Leibniz, Gottfried Wilhelm. 1971. *Mathematische Schriften*. Hildesheim: Olms.

L'Hôpital, Guillaume. 2015. *Analyse des infiniments petits*. Cham: Springer.

Newton, Isaac. 1967–1981. *The mathematical papers*. Cambridge: Cambridge University Press.

Pólya, George. 1954. *Mathematics and plausible reasoning*. Princeton: Princeton University Press.

———. 1981. *Mathematical discovery: On understanding, learning, and teaching problem solving*. New York: Wiley.

———. 2004. *How to solve it: A new aspect of mathematical method*. Princeton: Princeton University Press.

Reichenbach, Hans. 1951. *The rise of scientific philosophy*. Berkeley: University of California Press.

Sartorius von Waltershausen, Wolfgang. 2018. *Carl Friedrich Gauss: A memorial*. London: Forgotten Books.

Singer, Barry F. 1971. Towards a psychology of science. *American Psychologist* 26: 1010–1015.

Weil, André. 2009. De la métaphysique aux mathématiques. In André Weil, *Œuvres scientifiques – Collected papers*, vol. 2, 408–412. Berlin: Springer.

Chapter 8
Theories

Abstract Mainstream philosophy of mathematics and heuristic philosophy of mathematics yield two different views of theories, the axiomatic view, and the analytic view. The chapter describes them and argues that the axiomatic view is inadequate, only the analytic view is adequate. Then, it discusses some questions raised by the analytic view of theories, namely: What are mathematical problems? How do mathematical problems arise? How are mathematical problems posed? How are mathematical problems solved?

Keywords Axiomatic view of theories · Analytic view of theories · Nature of mathematical problems · Rise of mathematical problems · Mathematical problem posing · Mathematical problem solving

8.1 Different Views of Theories

Mathematics is articulated in theories. But there are different views of theories, because what a theory is depends on what the method of mathematics is supposed to be.

As we have seen in Chaps. 2 and 3, according to mainstream philosophy of mathematics, the method of mathematics is the axiomatic method, while according to heuristic philosophy of mathematics, the method of mathematics is the analytic method. This yields two different views of theories: the axiomatic view, and the analytic view.

These two views are implicit in what has been said about the axiomatic method and the analytic method. The chapter makes the two views explicit, and argues that the axiomatic view is inadequate, only the analytic view is adequate. Then, it discusses some questions raised by the analytic view, namely: What are mathematical problems? How do mathematical problems arise? How are mathematical problems posed? How are mathematical problems solved?

C. Cellucci, *The Making of Mathematics*, Synthese Library 448,
https://doi.org/10.1007/978-3-030-89731-4_8

8.2 The Axiomatic View of Theories

According to the axiomatic view of theories, a theory consists of axioms and deductions from them, and is built demonstrating propositions by the axiomatic method.

Thus, Bourbaki says that "a theory is a section of a mathematical text, consisting of a number of proofs which are grouped together" because "they all have some hypotheses in common; the latter are called the axioms of the theory" (Bourbaki 1949, 7).

A theory can be formalized by expressing its concepts and demonstrations in a suitable formal language.

Thus, Bourbaki says that, by analyzing "proofs in suitably chosen mathematical texts," we see that "a sufficiently explicit mathematical text could be expressed in a conventional language containing only a small number of fixed 'words', assembled according to a syntax consisting of a small number of unbreakable rules: such a text is said to be formalized" (Bourbaki 2004, 7).

The building of a theory is not entirely based on reason, because it is ultimately based on intuition.

Thus, Bourbaki says that the mathematician builds his theories "by a special intuition, which is not the popular sense-intuition, but rather a kind of direct divination (ahead of all reasoning) of the normal behavior, which he seems to have the right to expect of mathematical beings, with whom a long acquaintance has made him as familiar as with the beings of the real world" (Bourbaki 1950, 227).

8.3 Characters of Theories According to the Axiomatic View

According to the axiomatic view, theories have the following basic characters.

(1) A theory is completely determined by its axioms from the beginning. For, deductive rules are non-ampliative, so all the theorems are implicitly contained in the axioms.

(2) The paths from the axioms to the theorems are also completely determined from the beginning. For, there is an algorithm for enumerating all deductions from given axioms.

(3) The purpose of a theory is not to advance mathematical knowledge, but only to present, justify, and teach results already acquired, so a theory belongs to finished mathematics.

8.4 Inadequacy of the Axiomatic View of Theories

The axiomatic view of theories, however, is inadequate.

(1) The axiomatic view is invalid by Gödel's first incompleteness theorem because, by the latter, for any consistent, sufficiently strong, formal system, there are propositions of the system that are true but cannot be demonstrated from the axioms of the system. Thus a mathematical theory cannot consist of axioms and deductions from them. As we have seen in Chap. 2, Dieudonné objects that the undecidable proposition described by Gödel is very artificial, without any connections with any other part of the current theory of numbers. But this objection is invalid because there are propositions of the theory of numbers of the usual kind, such as Goodstein's theorem, which are true but cannot be demonstrated in first-order Peano arithmetic PA.

(2) The axiomatic view is also invalid by Gödel's second incompleteness theorem because, by the latter, if a mathematical theory consists of axioms and deductions from them, then generally a mathematical theory cannot be completely justified, since it is impossible to demonstrate by absolutely reliable means that the axioms are consistent, let alone that they are true. Bourbaki objects that "absence of contradiction" is "an empirical fact," so, if a contradiction arises, let it "be traced back to its cause, and the latter either removed or so surrounded by warning signs as to prevent serious trouble. This, to the mathematician, ought to be sufficient" (Bourkaki 1949, 3). But this objection is invalid, because it assumes that a contradiction sooner or later will be discovered. This conflicts with the fact that the deduction of a contradiction from the axioms may be too long to be feasible, so the contradiction may never be detected. An example of deduction from the axioms that is too long to be feasible is given by Boolos (see Chap. 10 below).

(3) The axiomatic view of theories is also invalid because it is unable to provide any rational account of the building of theories. Saying that the mathematician builds his theories by a special intuition which is a kind of direct divination ahead of all reasoning, does not provide a rational account, because there is no rational justification for divination.

8.5 The Analytic View of Theories

According to the analytic view of theories, a theory consists of problems and solutions to them, and is built discovering solutions by the analytic method.

This means that the building of a theory consists in starting from problems, obtaining hypotheses by non-deductive rules, and deducing solutions from them.

The hypotheses must be plausible, namely the arguments for them must be stronger than the arguments against them, on the basis of experience. But the hypotheses are in turn problems that must be solved, and are solved in the same way. And so on.

Thus the building of a theory does not come to an absolute end, it is an ongoing process.

The analytic view of theories was the original view of theories in Greek mathematics. In the latter, the first theories consisted of problems, such as the doubling of cube or the quadrature of certain lunules, and solutions to them. It is through such theories that mathematics as discipline was born.

8.6 Characters of Theories According to the Analytic View

According to the analytic view, theories have the following basic characters.

(1) A theory is not completely determined by its problems from the beginning. For, non-deductive rules are ampliative, so the hypotheses for solving a problem are not implicitly contained in the problem.

(2) The paths from the problem to the hypotheses are not completely determined from the beginning. For, there is no algorithm for enumerating all the hypotheses for solving a problem. In particular, finding hypotheses may require introducing new concepts. The paths from the problem to the hypotheses are made by the process of finding the hypotheses.

(3) The purpose of a theory is to advance mathematical knowledge by solving problems. This involves finding hypotheses that are the key to the discovery of solutions, and justifying solutions by showing that the hypotheses are plausible. Both these tasks are essential to mathematical knowledge. Finding hypotheses is essential to the growth of mathematical knowledge, justifying solutions is essential to the validation of mathematical knowledge. Neither of these tasks can be performed with certainty. For, on the one hand, the non-deductive rules by which hypotheses are found do not guarantee to yield hypotheses that are the key to the discovery of solutions. On the other hand, no test procedure can guarantee to justify the solutions absolutely, because hypotheses can only be plausible. This, however, is not a limitation peculiar to theories according to the analytic view, it equally applies to theories according to the axiomatic view. For, by Gödel's second incompleteness theorem, it is impossible to demonstrate by absolutely reliable means that theories according to the axiomatic view are true.

8.7 Adequacy of the Analytic View of Theories

Contrary to the axiomatic view of theories, the analytic view of theories is adequate.

(1) The analytic view of theories is unaffected and even confirmed by Gödel's first incompleteness theorem because, by the latter, generally no theory can solve all the problems of a given part of mathematics. Every theory is incomplete and must appeal to other systems to bridge the gaps.

(2) The analytic view of theories is also unaffected and even confirmed by Gödel's second incompleteness theorem because, by the latter, no theory can give solutions to problems that are absolutely certain.

(3) The analytic view of theories can provide a rational account of the building of theories, in terms of hypotheses obtained by non-deductive rules and validated by their plausibility.

This justifies the claim that the analytic view of theories is adequate. As Goodman says, "mathematical theories" are "constructed to solve particular problems," and "are not originally deductive structure based on axioms but, rather, informal bodies of reasoning based on conjectures and bold extrapolation" (Goodman 1991, 124). Conjectures and bold extrapolations are the result of non-deductive inferences.

8.8 The Nature of Mathematical Problems

The analytic view of theories raises the questions: What are mathematical problems? How do mathematical problems arise? How are mathematical problems posed? How are mathematical problems solved? We will now consider these questions.

What are mathematical problems? The term 'problem' comes from the Greek 'problema', which literally means 'anything put forward'. More specifically, Plato uses 'problem' in the sense of a question for which a solution is sought. For, he says: "We must take over the question and try to solve it as one does with a problem [problema]" (Plato, *Theaetetus* 180 c 5–6).

The assumption of the analytic view of theories, that a theory consists of problems and solutions to them, and is built discovering solutions by the analytic method, is related to this meaning of 'problem'.

That a theory consists of problems and solutions to them means that problems are the starting points of mathematical knowledge

8.9 The Rise of Mathematical Problems

How do mathematical problems arise? They arise from situations that are of concern to someone. Nothing is a mathematical problem unless it arises from a situation that is of concern to someone and makes him wish to obtain a solution to it.

The situations from which mathematical problems arise are multifarious. Mathematical problems arise from the needs of social or economic life, such as commercial or financial transactions, the construction of edifices, bridges and dams, ships and aircrafts; from the need to investigate the physical world; from the solution of some other mathematical problem that gives origin to new problems for which one feels the desire to obtain a solution; even from the sense of mathematical beauty, which may lead mathematicians to deal with problems and obtain solutions to them which give great aesthetic satisfaction.

That the situations from which mathematical problems arise are multifarious has been the case from the very origin of mathematics. Thus, problems of similarity of triangles arose from the need to measure lengths that were not directly accessible by touch, but only indirectly by sight. Solving a problem of similarity of triangles enabled Thales to satisfy this need.

Indeed, to calculate the height of the pyramid, Thales "set a stick upright at the end of the shadow cast by the pyramid and, since two triangles were formed by the sun's rays," he "demonstrated that the height of the pyramid had the same ratio to the length of the stick as the one shadow had to the other" (Plutarch, *Septem sapientium convivium*, 147 a 7–11, ed. Hercher).

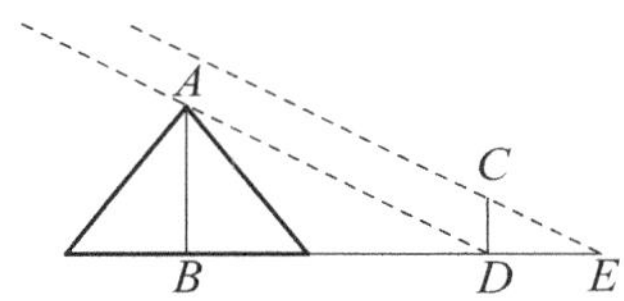

Since the right-angled triangles *ABD* and *CDE* were similar, their corresponding sides were in the same ratio, so $AB : CD = BD : DE$. From this, since *CD* was known and *BD*, *DE* could be measured, Thales calculated the height of the pyramid to be $AB = \frac{CD \cdot BD}{DE}$.

As another example, standing on the top of a tower with his eye at *A* and sighting a ship at *C*, Thales calculated "the distance of ships at sea" (Proclus, *In primum Euclidis Elementorum librum commentarii*, 352.16–17, ed. Friedlein).

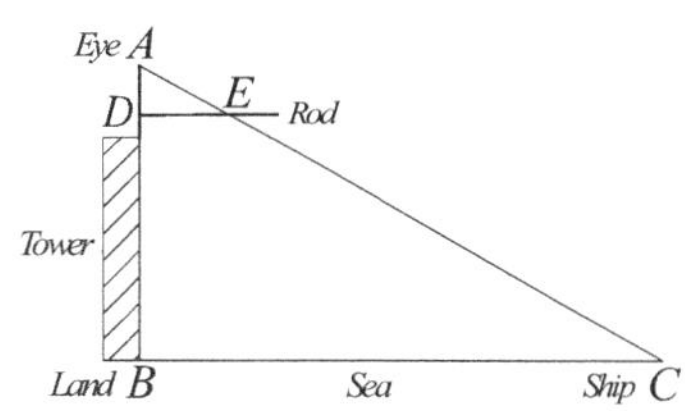

Indeed, since the right-angled triangles *ABC* and *ADE* were similar, their corresponding sides were in the same ratio, so $BC : DE = AB : AD$. From this, since *AB*, *AD* were known and *DE* could be measured, Thales calculated the distance of the ship at sea to be $BC = \frac{DE \cdot AB}{AD}$.

The mathematical problems that arise may belong to current parts of mathematics, but may also belong to none of the current parts. When successful, they and their solutions may even give rise to new parts of mathematics.

8.10 Mathematical Problem Posing

How are mathematical problems posed? Posing a mathematical problem requires skill because, with an inappropriate formulation, the search for a solution to a problem may have difficulty in starting, or may take the wrong course. Indeed, the formulation of the problem suggests which data to choose and which to reject, in which direction to look for a hypothesis and in which not to look for it. When one manages to formulate a mathematical problem, often one is already well ahead in research. For this reason, it is said that a well posed problem is already half solved. Cantor even says: "In re mathematica ars proponendi quaestionem pluris facienda est quam solvendi [In mathematics, the art of problem posing must be held more valuable than that of problem solving]" (Cantor 1932, 31).

The formulation of a mathematical problem transforms a puzzle, for which one wishes to obtain a solution, into an amenable task. This is an operation which is neither simple nor brief. To carry it out, two things are needed.

First, one must analyze the problem. The analysis will provide the terms of the problem, namely the conditions that have to be met when one tries to solve it. Generally, the terms of the problem are determined through a long and complex process.

Second, one must find an appropriate representation for the terms of the problem. Finding an appropriate representation is very important, because it can make relevant things explicit and can facilitate the finding of a solution to the problem.

However, the formulation of a mathematical problem does not guarantee by itself that one will be able to solve it. It may happen that, when one formulates the problem, the means to solve it are not yet available, or the formulation of the problem depends on assumptions that have not yet been made explicit. In these cases, finding a solution to the problem may be too difficult, and all efforts to find it may be in vain. On the other hand, it can be a waste to play safe, formulating a mathematical problem only because its solution is easy. A mediocre result may be an unsatisfactory reward for modest effort.

Thus, the important problems encountered in mathematical research are essentially different from formal puzzles (such as the missionaries and cannibals problem), from problems of formal deduction (which require you to deduce a given proposition from given axioms by given deduction rules), and from the problems given to students as an exercise in teaching. Of these problems, it is known from the beginning that they admit a solution by currently available means. On the contrary, this is not generally known of the problems encountered in mathematical research.

Nevertheless, even when a mathematical problem has not been solved yet, and it is not known whether it will be possible to solve it in a foreseeable future, its formulation, by itself, may yield a growth of knowledge.

For, the formulation of a mathematical problem can give rise to new problems. For example, the problem of whether every natural number can be written as the sum of four squares was solved by Lagrange a century and a half after Bachet formulated it in his Latin translation of Diophantus's *Arithmetica*. But, before being solved, the problem gave rise to Waring's problem which was solved only at the beginning of the twentieth century by Hilbert. Moreover, the formulation of a mathematical problem may give rise to new theories. For example, the formulation of the problem posed by Fermat's conjecture stimulated the development of Kummer's theory of ideal numbers, and Waring's and Goldbach's problems stimulated the development of analytic number theory.

The reason why the formulation of a mathematical problem, by itself, may yield a growth of knowledge, is that it suggests that there may be some relations between the data, and this is in itself a growth of knowledge. A solution to the problem establishes that such relations actually exist, and may also show why they exist.

Moreover, although the formulation of a mathematical problem does not guarantee that one will be able to solve it, the trust in the solvability of the problem may produce an emotional drive to get deeply involved with it. One can only hope to solve a problem whose solution one passionately desires. The search for a solution has a strong emotional element that leads one to engage intensively in a problem, even facing years of hard work and bitter disappointments. Eliminating emotions would deprive the development of mathematical knowledge of one of its most vital sustaining forces.

The emotional element has an important part also in the choice of mathematical problems. Only a small part of mathematical problems that can be posed is interesting to mathematicians, and the emotional element acts as a guide to distinguish what is interesting from what is not interesting, what is worth investigating from what is not worth investigating. Without a scale of interest of problems, one cannot discover anything valuable.

8.11 Mathematical Problem Solving

How are mathematical problems solved? They are solved starting from the problem, through the following steps.

(1) We examine the problem in order to understand it. Understanding the problem is a preliminary condition for thinking of ways and means to find a solution to it.

(2) To understand the problem, we choose a suitable notation to express it. This can be not only useful but even indispensable, to understanding.

(3) We consider the problem from different perspectives, highlighting different aspects of it. This can lead to seeing relations between the problem and other questions.

(4) We describe what general features a solution to the problem should have. This is useful because, by focusing on such conditions, we may greatly reduce the search space where a hypothesis can be located.

(5) We investigate whether there is any solved problem to which our problem may be related. For, a procedure which has been successful with the latter problem might be useful, with suitable changes, to solve our problem.

(6) We formulate a hypothesis on the basis of the general features that a solution to the problem should have. The hypothesis is obtained by some non-deductive rule. Which non-deductive rule is to be used depends on the problem, but different non-deductive rules may be used to find the hypothesis.

(7) If none of the hypotheses that we have obtained permits us to solve the problem, we examine whether they all depend on some common assumption which prevents the solution.

(8) If we arrive at a hypothesis that permits us to solve the problem, we examine whether the hypothesis is plausible, namely whether the arguments for it are stronger than the arguments against it, on the basis of experience. This is essential because the hypothesis might permit us to solve the problem for the wrong reason, in particular, because it contradicts experience, and from a contradiction, anything can be inferred.

(9) To examine whether the hypothesis is plausible, we draw consequences from it. The consequences will then be compared with experience.

(10) If the hypothesis is not plausible, we investigate why it is not plausible. This might provide indications as to how to modify the hypothesis or to formulate a new one.

(11) Even when the hypothesis is plausible, this is inconclusive for its acceptance. For, the latter also depends on other factors, such as fruitfulness, namely the ability to open new lines of research. When the hypothesis is seen to be not only plausible but also fruitful, it tends to consolidate and become stable.

(12) However, even when the hypothesis consolidates and becomes stable, this is not the end of the investigation. For, the hypothesis is in turn a problem that requires to be solved. To solve it, we return to step (1), and so on.

From this description, it is clear that solving a mathematical problem is a process in which one goes from the problem to others of increasing depth, and is an infinitely proceeding process.

8.12 The Analytic View of Theories and Big Data

In terms of the analytic view of theories, we can further discuss the argument of Big Data, already considered in Chap. 4, according to which the traditional scientific method, based on hypotheses or theories, is totally inadequate and must be replaced with the Big Data approach.

It is widely assumed that the Big Data approach requires a complete break from Cartesian logic, because the latter restricts logic to deductive reasoning, while the Big Data approach is based on inductive reasoning. According to this assumption, inductive reasoning is the non-scientific part of human thought, and is the exact opposite of deductive thinking. While deductive reasoning always comes to an end, inductive reasoning generally produces no finished status, and it is possible to continue the reasoning indefinitely.

Thus, Malle claims that the Big Data approach "requires a complete break from Cartesian logic" (Malle 2013). For, the latter is based on "deductive thinking" (ibid.). On the contrary, the Big Data approach is based on "inductive reasoning," which is "the non-scientific part of human thought" (ibid.). The non-scientific part, because in the traditional scientific way of thought "a consequence is logically deducted from a hypothesis. It is then checked and tested" (ibid.). On the contrary, the Big Data approach is based on inductive reasoning, because "the analysis of huge amounts of data mainly focuses on finding correlations" (ibid.). The correlations will be found by inductive algorithms, and "induction allows algorithms to reproduce observed phenomena by generalizing beyond their scope," but "without trying to make models out of them" (ibid.). This "is far from Cartesian principles," because inductive thinking is "the exact opposite of deductive thinking" (ibid.). Therefore, "either we reason in the context of a deductive approach, or else we firmly choose an inductive approach" (ibid.). The inductive approach "leads to several consequences. For example, while deductive reasoning always comes to an end, inductive reasoning generally produces no finished status," it "is possible to continue the reasoning indefinitely" (ibid.). Nevertheless, inductive reasoning "produces an imperfect but useful knowledge" (ibid.).

But, if by 'Cartesian logic' Malle means Descartes's conception of inference, his claims about Cartesian logic are invalid. For, Descartes does not restrict logic to deductive reasoning. Indeed, he says that, when we must consider very many cases, "often our intellectual capacity is insufficient to enable us to encompass all of them in a single intuition; in which event we must be content with the level of certainty which the operation" of induction "allows" (Descartes 1996, X, 389). So "induction" is necessary, because in those cases our knowledge "cannot be reduced" to deduction and, "having cast off all fetters of syllogisms," we are only "left with this one path," induction, "to which we must stick" (ibid.).

But, independently of what Malle means by 'Cartesian logic', his claims are invalid in terms of the analytic view of theories.

Thus, it is invalid to say that inductive reasoning is the non-scientific part of human thought. According to the analytic view of theories, inductive reasoning is a main part of the building of theories. In the latter, induction plays an important role, being one of the main non-deductive rules through which hypotheses are obtained, so inductive reasoning is an important tool for scientific discovery.

Also, it is invalid to say that inductive thinking is the exact opposite of deductive thinking. According to the analytic view of theories, induction and deduction work side by side in the building of theories. For, every step upward through which a hypothesis is obtained by non-deductive reasoning, is accompanied by a step downward through which a solution to the problem is deduced from the hypothesis.

Also, it is invalid to say that, while deductive reasoning always comes to an end, inductive reasoning generally produces no finished status, and it is possible to continue the reasoning indefinitely. According to the analytic view of theories, the non-deductive reasoning through which a hypothesis is obtained always comes to an end. What does not come to an end is the building of a theory as a whole. It does not come to an end because, beyond every hypothesis, one can always find a deeper hypothesis that provides a more comprehensive explanation, thus opening the way to an ever-growing deepening of the problem.

References

Bourbaki, Nicolas. 1949. Foundations of mathematics for the working mathematician. *The Journal of Symbolic Logic* 14: 1–8.

Bourbaki, Nicholas. 1950. The architecture of mathematics. *The American Mathematical Monthly* 57: 221–232.

Bourbaki, Nicolas. 2004. *Elements of mathematics: Theory of sets*. Berlin: Springer.

Cantor, Georg. 1932. *Gesammelte Abhandungen mathematischen und philosophischen Inhalts*. Berlin: Springer.

Descartes, René. 1996. *Oeuvres*. Paris: Vrin.

Goodman, Nicolas D. 1991. Modernizing the philosophy of mathematics. *Synthese* 88: 119–126.

Malle, Jean-Pierre. 2013. Big data: Farewell to Cartesian thinking? *Paris Innovation Review*, 15 March. http://parisinnovationreview.com/articles-en/big-data-farewell-to-cartesian-thinking

Part III
The Mathematical Process

Chapter 9
Objects

Abstract According to heuristic philosophy of mathematics, one of the tasks of the philosophy of mathematics is to give an answer to the question: What is the nature of mathematical objects? This question has received several answers. The chapter discusses them and argues that they are inadequate. Then, it offers an alternative answer: mathematical objects are hypotheses introduced to solve mathematical problems by the analytic method.

Keywords Mathematical objects as logical objects · Simplifications · Mental constructions · Independently existing objects · Abstractions · Structures · Fictions · Idealizations of physical bodies · Idealizations of operations · Heuristic view of mathematical objects

9.1 What Mathematics Is About

As we have seen in Chap. 3, according to heuristic philosophy of mathematics, one of the tasks of the philosophy of mathematics is to give an answer to the question: What is the nature of mathematical objects?

This question goes back to antiquity. Herodotus narrates that Egypt's king Sesostris "distributed the land among all the Egyptians, giving an equal square parcel of land to each man, and made this his source of revenue, imposing the payment of a yearly tax" (Herodotus, *Historiae*, II.109, ed. Hude). If the Nile, by overflowing, "robbed any man of a part of his land, the man would come to the king and declare what had happened" (ibid.). Then "the king would send men to look into it and measure how much less the land had become, so that thereafter the man should pay less, in proportion to the tax originally imposed" (ibid.). This "led to the discovery of geometry, which then passed into Greece" (ibid.). Thus, according to Herodotus, geometry had an empirical origin, because it was born from the land measuring in Egypt.

But, passing into Greece, geometry changed nature, turning into an abstract discipline. According to Proclus, "Pythagoras transformed the study of geometry

C. Cellucci, *The Making of Mathematics*, Synthese Library 448,
https://doi.org/10.1007/978-3-030-89731-4_9

into a liberal discipline, tracing its first principles and studying its problems from a purely abstract and theoretical viewpoint" (Proclus, *In primum Euclidis Elementorum librum commentarii*, 65.15–19, ed. Friedlein).

This change in geometry involved a change in mathematical objects. Because of its concrete nature, Egyptian geometry was about such things as parcels of land, hence mathematical objects were physical things, knowable by the senses. Conversely, because of its abstract nature, Greek geometry was about abstract things, so mathematical objects were non-physical things, not knowable by the senses.

Plato interpreted this change by assuming that mathematical objects were independently existing entities, non-physical and non-mental, knowable only by intuition.

Indeed, Plato says that, although mathematicians "use visible figures and make their claims about them," mathematicians "are not thinking of them but of those ideas of which they are likeness. They pursue their inquiry for the sake of the square itself and the diagonal itself, and not for the sake of the diagonal they draw, and similarly with the others" (Plato, *Respublica*, VI 510 d 5–e 1). They use the visible figures they draw "only as images, but what they really seek is to get sight of those ideas which can be seen only by the mind" (ibid., VI 510 e 3–511 a 1). Indeed, "there really exist these ideas in themselves, that we cannot perceive with the senses but only with intuition" (Plato, *Timaeus*, 51 d 4–5). They have a kind of reality that "is always the same, ungenerated and imperishable," it "is not visible nor perceptible by any sense," and "only intuition has been granted to contemplate it" (ibid., 52 a 1–4).

Since antiquity, the question 'What is the nature of mathematical objects?' has received several answers by the three big foundationalist schools and their direct or indirect descendants: mathematical objects are logical objects, simplifications, mental constructions, independently existing entities, abstractions, structures, fictions, idealizations of physical bodies, or idealizations of operations. The chapter discusses these answers and argues that they are inadequate. Then, it offers an alternative answer: mathematical objects are hypotheses introduced to solve mathematical problems by the analytic method.

9.2 Mathematical Objects as Logical Objects

An answer to the question 'What is the nature of mathematical objects?' is that the objects of arithmetic are logical objects, because they are definable in logical terms, and all properties of the objects of arithmetic are deducible from logical principles. This is the view of mathematical objects of logicism (see Chap. 2). Indeed:

(1) Frege says that "the objects of arithmetic" are definable in logical terms, so they are "logical objects" (Frege 2013, II, 149). Then "the prime problem of arithmetic" is the question: "How are we" to conceive "logical objects, in particular, the numbers?" (ibid., II, Afterword, 265).

(2) All properties of the objects of arithmetic are deducible from logical principles, so "there is no such thing as a peculiarly arithmetical mode of inference that cannot be reduced to the general inference modes of logic" (Frege 1984, 113).

(3) Therefore, "no sharp boundary can be drawn between logic and arithmetic. Considered from a scientific point of view, both together constitute a unified science" (ibid., 112).

The view that mathematical objects are logical objects, however, has the following shortcomings.

(1) Frege says that the prime problem of arithmetic is the question of how we are to conceive logical objects, in particular, numbers. To answer this question, Frege assumes that an assertion about number "is an assertion about a concept" (Frege 1960, 61). So, to define number, he defines the number of a concept. To this purpose, he assumes what is now known as Hume's principle: For any concepts *F* and *G*, the number of *F* is identical to the number of *G* if and only if *F* and *G* are equinumerous (namely, there is a one-to-one correspondence between *F* and *G*). Then he defines '*n* is a number' as '*n* is the number of *F*, for some concept *F*'. This, however, encounters the problem that, with this definition of number, "we can never – to take a crude example – decide" whether Julius Caesar "is a number or is not" (ibid., 68). To solve this problem, Frege defines 'the number of *F*' as the extension of the concept 'equinumerous with *F*'. And he assumes the Basic Law V (already described in Chap. 2). But this, in turn, encounters the problem that, by the Basic Law V, "we cannot decide yet whether an object that is not given to us as" an extension, is an extension, or under which concept it may fall, hence "the reference of a name such as" 'the extension of *F*' "is by no means completely determined" (Frege 2013, I, 16).

(2) Frege says that there is no arithmetic mode of inference that cannot be reduced to the general inference modes of logic. But his attempt to reduce all arithmetic modes of inference to the general inference modes of logic ended in failure. As we have already seen in Chapter 2, the Basic Law V, which was essential to this reduction, leads to Russell's paradox. After receiving the letter in which Russell made known to him the paradox, Frege himself admitted defeat: "Hardly anything more unwelcome can befall a scientific writer than to have one of the foundations of his edifice shaken after the work is finished. This is the position into which I was put by a letter from Mr Bertrand Russell" (Frege 2013, II, Afterword, 253). And "even now, I do not see" how "the numbers can be apprehended as logical objects" if it is not "permissible to pass from a concept to its extension" (ibid.). So "my efforts to become clear about what is meant by number have resulted in failure" (Frege 1979, 263). Therefore, "I have had to abandon the view that arithmetic is a branch of logic" (Frege 1969, 298).

(3) Frege says that no sharp boundary can be drawn between logic and arithmetic. Of course, that the Basic Law V leads to a paradox, does not exclude by itself that arithmetical laws could be reduced to some basic logical laws L different from those of Frege's logical system and not leading to a paradox. Such possibility, however, is

excluded by Gödel's first incompleteness theorem, by which there are arithmetic truths that cannot be deduced from basic logical laws L, so there are arithmetic truths that are not logical truths.

Hale and Wright argue that Gödel's first incompleteness theorem "might just as well" be "taken as revealing, not that there are arithmetic truths which are not" logical truths, "but that logical truth likewise defies complete deductive characterization" (Hale and Wright 2001, 4, footnote 12). For, by the strong incompleteness theorem for second-order logic, not all logically valid second-order propositions are deducible from L. So, the fact that there are arithmetic truths that are not deducible from L, does not mean that there are arithmetic truths that are not logical truths. Such arithmetic truths might as well be logically valid propositions that are not deducible from L, and hence might be logical truths. Therefore, if "the logicist is entitled to take logic as including second- and perhaps higher-order logic," then "Gödel's incompleteness result has no specific bearing on the logicist project" (ibid.).

This argument, however, is invalid because the arithmetic truths that cannot be deduced from L by Gödel's first incompleteness theorem are first-order propositions and, by the completeness theorem for first-order logic, all logically valid first-order propositions are deducible from L. So, if the arithmetic truths in question were logically valid, they would be deducible from L. Since they are not deducible from L, they cannot be logical truths.

In light of the above shortcomings, we may conclude that the view that mathematical objects are logical objects is untenable.

9.3 Mathematical Objects as Simplifications

Another answer to the question 'What is the nature of mathematical objects?' is that finitary mathematics is about certain concrete objects, while infinitary mathematics is also about certain abstract objects which are only ideal objects. The use of ideal objects is useful, because it permits to simplify the demonstration of properties of concrete spatio-temporal objects. But the use of ideal objects is also justified, because the consistency of the ideal propositions about them can be proved by a consistency proof based on the intuitive mode of thought. This is the view of mathematical objects of formalism (see Chap. 2). Indeed:

(1) Hilbert says that mathematics is about certain "concrete objects that exist intuitively as an immediate experience before all thought" (Hilbert 1996c, 1150). But mathematics also makes use of "the actual infinite" (Hilbert 1967, 373). This permits to simplify proving properties of concrete objects, which "results in the simplification and completion of the theory" (Hilbert 1996b, 1144). However, "the problem arises" of "the consistency of the axioms" (Hilbert 1996a, 1119). To solve the problem, the theory must be "rigorously formalized" (Hilbert 1996b, 1137). This is possible because a theory can always be formalized.

(2) Once the theory is formalized, one must prove the consistency of the axioms, and must prove it by a proof that "lies within the province of intuition" (Hilbert 1967, 383). This is possible because one can always prove the consistency of a theory by a proof of that kind.

(3) This will guarantee that "mathematical statements are in fact incontestable and ultimate truths" (Hilbert 1996a, 1121). For, in a consistent theory, no false statement is provable.

The view that mathematical objects are means of simplification, however, has the following shortcomings.

(1) Hilbert assumes that a theory can always be formalized. But this assumption is refuted by Gödel's first incompleteness theorem, by which there will always be propositions of the theory that are true but cannot be demonstrated in the formalized theory.

(2) Hilbert assumes that one can always prove the consistency of a theory by a proof that lies within the province of intuition. But this assumption is refuted by Gödel's second incompleteness theorem, by which no such proof is possible.

(3) Hilbert assumes that, in a consistent theory, no false statement is provable. But this assumption is refuted by the theorem on the false extensions, by which any consistent sufficiently strong formal system has a consistent extension in which some false proposition is provable (see Chap. 2).

In light of the above shortcomings, we may conclude that the view that mathematical objects are means of simplification is untenable.

9.4 Mathematical Objects as Mental Constructions

Another answer to the question 'What is the nature of mathematical objects?' is that they are mental constructions. This does not mean that mathematical objects are merely ideas conceived by the mind, but rather that they are entities constructed out of the basic intuition of mathematics, the intuition of time. This is the view of mathematical objects of intuitionism (see Chap. 2). Indeed:

(1) Brouwer criticizes infinitary mathematics because it "completely abandons any geometric or arithmetical intuition and confines its subject matter exclusively to the domain of mathematical language" (Brouwer 1998, 55). It assumes that mathematical objects "have mathematical existence" simply when "they have been represented in spoken or written language together with" the "laws upon which their development depends" (Brouwer 1975, 125).

(2) On the contrary, mathematical objects are entities constructed "out of the basic intuition of mathematics," namely "out of the intuition of time" (ibid., 108). They are "languageless constructions which arise from the self-unfolding of the basic intuition" (ibid., 443). Such constructions are the same in all human beings, because "it is an essential hypothesis for human understanding that the structure of" construction "is the same for all individuals" (ibid., 599).

(3) In principle, "no mathematics can exist which has not been intuitively built up" (ibid., 52). The "only possible foundation of mathematics must be sought in this construction under the obligation carefully to watch which constructions intuition allows and which not" (ibid.). This requires "a new practice of creative mathematical work" (ibid., 477). In fact, out of the basic intuition of mathematics, intuitionism "has already erected anew several of the theories of mathematics proper in unshakable certainty" (Brouwer 1967, 492).

The view that mathematical objects are mental constructions based on intuition, however, has the following shortcomings.

(1) Brouwer criticizes infinitary mathematics because it completely abandons any geometrical or arithmetical intuition. However, his criticism is not motivated by any necessity of mathematical practice, but only by the purely ideological assumption that we can obtain knowledge of mathematics only if mathematics is constructed by direct intuition. This assumption implies that important parts and results of modern mathematics must be rejected. As Kitcher and Aspray say, "the result was a mutilation of mathematics, in which long, ugly proofs of intuitionistic analogs replaced the simple and elegant classical proofs and theorems" (Kitcher and Aspray 1988, 7).

(2) Brouwer says that mathematical objects are entities constructed out of the intuition of time, and that such constructions are the same in all human beings. But, as Gillies points out, this contrasts, for example, with the fact that, "if we look at number systems historically, we find" that "different, and generally more primitive, systems existed for centuries in many parts of the world. Often there was a crucial innovation in some relatively small group, and this change gradually spread elsewhere," so "the system of natural numbers" was "constructed slowly and often painfully over long periods of time by human beings" (Gillies 2014, 26). Generally, "some mathematical concepts are counter-intuitive rather than natural, and are the result of centuries of trial and error rather than easy inferences from an a priori intuition" (ibid.).

(3) Brouwer says that, in principle, no mathematics can exist which has not been intuitively built up. But this implies that there cannot exist certain mathematical objects which are important to physics. In particular, as already mentioned in Chap. 2, there cannot exist everywhere defined discontinuous functions. This is not an oversight, but deliberate. For, Brouwer declares that he does not care a bit for the use of mathematics in physics, since mathematics is not aimed at the "expansion of human domination over nature," mathematics is aimed at beauty, and "things as such are not beautiful, nor is their domination by shrewdness" (Brouwer 1975, 483). Indeed, mathematics "is an art, its application to the world an evil parasite" (Borwein, Bailey, and Girgensohn 2004, 295).

In light of the above shortcomings, we may conclude that the view that mathematical objects are mental constructions based on intuition is untenable.

9.5 Mathematical Objects as Independently Existing Entities

Another answer to the question 'What is the nature of mathematical objects?' is that they are independently existing entities. As we have seen above, this view of mathematical objects was first formulated by Plato, and hence goes under the name 'mathematical platonism'.

Mathematical platonism has a very large following among mathematicians because, as Dieudonné says, "each mathematician" has "the feeling" that "he is working with something real. This sensation is probably an illusion, but is very convenient" (Dieudonné 1970, 145).

However, several mathematicians adhere to mathematical platonism not merely because they have the feeling that they are working with something real, but because they believe that they are actually working with something real.

Thus, Hermite says: "I believe that numbers and functions of analysis are not the arbitrary product of our minds" but "exist outside of ourselves, with the same character of necessity as the objects of objective reality; and that we encounter, or discover them, and that we study them as do the physicists, chemists and zoologists, etc." (Hermite 1905, 398).

Hardy says: "I believe that mathematical reality lies outside us, that our function is to discover or observe it, and that the theorems which we prove, and which we describe grandiloquently as our 'creations', are simply our notes of our observations" (Hardy 1992, 123–124). Indeed, a mathematician is "an observer, a man who gazes at a distant range of mountains and notes down his observations" (Hardy 1929, 18). When "he sees a peak he believes that it is there simply because he sees it," if he wishes a pupil "to see it, he points to it," and when the "pupil also sees it, the research, the argument, the proof is finished" (ibid.).

Thom says that "mathematicians should have the courage of their most profound convictions and thus affirm that mathematical forms indeed have an existence that is independent of the mind considering them" (Thom 1971, 696). In fact, "the hypothesis stating that Platonic ideas give shape to the universe is the most natural and, philosophically, the most economical," although "at any given moment, mathematicians have only an incomplete and fragmentary vision of this world of ideas" (ibid., 697).

9.6 Gödel on Mathematical Objects as Independently Existing Entities

Although the view of mathematical objects as independently existing entities was first formulated by Plato, Gödel has given its most articulated formulation. Indeed:

(1) Gödel says that mathematics "describes a non-sensual reality, which exists independently both of the acts and of the dispositions of the human mind and is only perceived, and probably perceived very incompletely, by the human mind" (Gödel 1986–2002, III, 323). Such reality consists of sets, "the objects of transfinite set theory" (ibid., II, 267). Sets do "not belong to the physical world," and "even their indirect connection with physical experience is very loose" (ibid.). They "exist objectively and independently of our mental acts and decisions" (ibid., III, 311).

A set can be thought of as "something obtainable from" the empty set $\varnothing$ "by iterated application of the operation 'set of'," where iterated application "is meant to include transfinite iteration" (ibid., II, 259 and footnote 13). If, by the operation 'set of', it is meant the power set operation $P(X)$ that collects all the subsets of a set X into one set, then a set is a member of the hierarchy defined by: (i) $V_0 = \varnothing$; (ii) $V_{\alpha+1} = P(V_\alpha)$, for any ordinal α; (iii) $V_\lambda = \bigcup_{\alpha<\lambda} V_\alpha$, for any limit ordinal λ.

The hierarchy of sets is commonly represented in a way that is reminiscent of the traditional representations of Dante's *Inferno*, such as the following, which occurs in a sixteenth century edition.

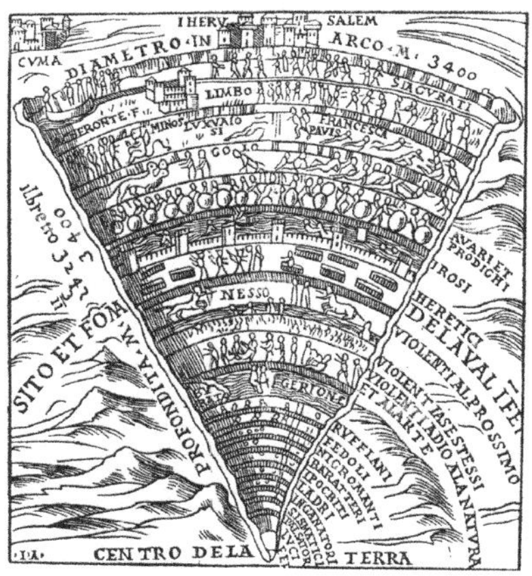

Indeed, the hierarchy of sets is commonly represented as follows.

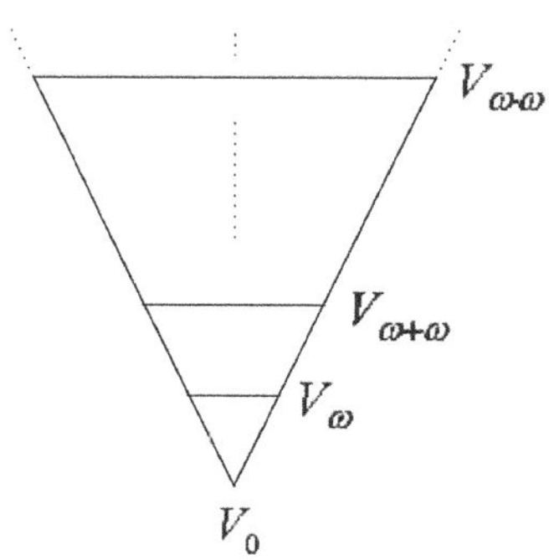

(2) But, "despite their remoteness from sense experience, we do have something like a perception also of the objects of set theory," namely "mathematical intuition" (ibid., II, 268). The latter is a superior faculty, not "derived from subconscious induction or Darwinian adaptation" (ibid., III, 354).

(3) Mathematical intuition "is sufficiently clear to produce the axioms of set theory and an open series of extensions of them" (ibid., II, 268). It gives insights both "for obtaining unambiguous answers to the questions of transfinite set theory," such as Cantor's continuum hypothesis, and "for the solution of the problems of finitary number theory (of the type of Goldbach's conjecture)" (ibid., II, 269).

(4) Through mathematical intuition, sets "are known precisely, and general laws" about them "can be recognized with certainty" (ibid., III, 312, footnote 18). For, through it, we can recognize "the truth of the axioms concerning them" (ibid., II, 258). Their truth can be recognized only by intuition, because for the "axioms there exists no other" rational "foundation except that they" can "directly be perceived to be true" by "an intuition of the objects falling under them" (ibid., III, 346–347).

(5) We can arrive at directly perceiving the axioms to be true by the procedure which "consists in focusing more sharply on the concepts concerned" (ibid., III, 383). Indeed, "by directing our attention in a certain way," we will get an intuition of those concepts, and hence a "comprehensive and secure insight" into "the axioms that hold for them" (ibid.). For, "cultivating (deepening) knowledge of the abstract concepts" will "produce in us a new state of consciousness in which we describe in detail the basic concepts we use in our thought" (ibid.).

(6) But, besides mathematical intuition, there is "another (though only probable) criterion of truth of mathematical axioms, namely their fruitfulness in mathematics and, one may add, possibly also in physics" (ibid., II, 269). Here fruitfulness means "fruitfulness in consequences, in particular in 'verifiable' consequences, i.e., consequences demonstrable without the new axiom" (ibid., II, 261). The "simplest case of an application of the criterion under discussion arises when some set-theoretical axiom has number-theoretical consequences verifiable by computation up to any given integer" (ibid.).

The view that mathematical objects are independently existing entities, however, has the following shortcomings.

(1) Gödel says that mathematics describes a non-sensual reality, which exists independently both of the acts and of the dispositions of the human mind, and is only perceived by the human mind. This raises the question: How, then, can mathematics be useful and applied in physics? Gödel's answer to this question is that mathematics "might help us considerably in knowing" the physical reality "if in some respect, or as to certain parts, it happened to be similar or isomorphic to it. In fact, this would correspond closely to the manner in which mathematics is applied in theoretical physics" (ibid., III, 353–354, footnote 44). Thus, according to Gödel, mathematics can be applied in theoretical physics because it is isomorphic to physical reality. But this conflicts with the fact that different mathematical theories can be used to deal

with the same physical phenomena. For example, on the one hand, Weyl says that group theory is exactly what is "necessary for an adequate description of the quantum mechanical relations" (Weyl 1950, xxi). On the other hand, von Neumann says that Hilbert space theory is exactly what is needed for an adequate description of the quantum mechanical relations, because it permits to present "quantum mechanics in a unified representation" which "is mathematically rigorous" (von Neumann 1955, vii).

(2) Gödel says that, despite their remoteness from sense experience, we do have something like a perception also of the objects of set theory, namely, mathematical intuition. This raises the question: How can we have this kind of perception? Gödel assumes that, in order "to make the handling of abstract impressions (as opposed to sense impressions) possible," we must have "some physical organ," and "such a physical organ must be closely related to the neural center for language" (Wang 1996, 233). But there is no evidence that such a physical organ exists, and indeed, its existence makes no sense in evolutionary terms, because the highly abstract objects of set theory were first considered only in the nineteenth century. As to the claim that the physical organ in question must be closely related to the neural center for language, experiments have shown that "mathematics and language are functionally and neuroanatomically independent" (Brannon 2005, 3178).

(3) Gödel says that mathematical intuition is sufficiently clear to produce the axioms of set theory and an open series of extensions of them. But this conflicts with the fact that, at least since Kant, it has been held that "an intuition is a singular representation" (Kant 1992, 589). How, then, can mathematical intuition produce the axioms of set theory and an open series of extensions of them, if generally the axioms are not singular propositions but universal propositions? Moreover, mathematical intuition has given mathematicians no insight for obtaining unambiguous answers to the questions of transfinite set theory, such as Cantor's continuum hypothesis, and for the solution of problems of finitary number theory, of the type of Goldbach's conjecture.

(4) Gödel says that, through mathematical intuition, sets are known precisely, and general laws about them can be recognized with certainty. But this is refuted by Gödel's second incompleteness theorem, by which the consistency of no axiom system for set theory can be recognized with certainty. Moreover, as argued in Chap. 2, intuition is unreliable, so it is unrealistic to appeal to it as a foundation for the axioms of sets theory and the whole of mathematics.

(5) Gödel says that we can arrive at directly perceiving the axioms concerning sets to be true by the procedure which consists in focusing more sharply on the concepts concerned. But this is invalid. For, suppose that, by focusing more sharply on the concept of set Σ, we get an intuition of that concept. Let S be a formal system for set theory whose axioms this intuition ensures us to be true of Σ. Thus, Σ is a model of S, hence S is consistent. Then, by Gödel's first incompleteness theorem, there is a proposition A of S that is true of Σ but is unprovable in S. Since A is unprovable in S, the formal system $S' = S \cup \{\neg A\}$ is consistent, and hence has a

model, say Σ'. Now, $\neg A$ is true of Σ' and hence A is false of Σ'. So, Σ and Σ' are both models of S, but A is true of Σ and false of Σ', hence Σ and Σ' are not equivalent. Suppose now that again, as Gödel says, by focusing more sharply on the concept of set Σ', we get an intuition of that concept. Then we have two different intuitions, one ensuring us that Σ is the concept of set, and the other one ensuring us that Σ' is the concept of set, where the proposition A is true of Σ and false of Σ'. Which of Σ and Σ' is the genuine concept of set? Gödel's procedure provides no answer, so through it we cannot arrive at directly perceiving which axioms for sets are true.

(6) Gödel says that, besides mathematical intuition, there is another (though only probable) criterion of truth of mathematical axioms, namely their fruitfulness in consequences, and the simplest case of application of this criterion of truth is when some set-theoretical axiom has number-theoretical consequences verifiable by computation up to any given integer. But this is invalid. As Gödel himself admits, this criterion of truth "cannot yet be applied to the specifically set-theoretic axioms (such as those referring to great cardinal numbers), because very little is known about their consequences in other fields" (Gödel 1986–2002, II, 269). What is more important, this criterion of truth fails, because from false axioms it is possible to deduce true consequences. To assert that the axioms are true, it would be necessary to demonstrate that all the consequences that can be deduced from them are true, because a single false consequence would be sufficient to invalidate the axioms. But, as it will be argued in Chap. 10, demonstrating that all the consequences that can be deduced from the axioms are true is generally impossible.

In light of the above shortcomings, we may conclude that the view that mathematical objects are independently existing entities is untenable.

9.7 Mathematical Objects as Abstractions

Another answer to the question 'What is the nature of mathematical objects?' is that they are objects obtained from physical bodies by abstraction. Abstraction is the process by which one considers only some attributes and leaves out the other ones. This view of mathematical objects has been put forward by Aristotle.

Indeed, Aristotle says that "the mathematician carries out his investigation about things reached by abstraction. For, he investigates things leaving out all sensible qualities, such as weight, lightness, hardness," and "considering only quantity and continuity, in one, two or three dimensions, and the properties of these as quantitative and continuous; and he does not consider them in any other respect" (Aristotle, *Metaphysica*, K 3, 1061 a 28–35).

Then, the mathematician investigates the same things as the physicist, since "the physicist concerns himself with all the properties, active and passive, of physical bodies" (Aristotle, *De Anima*, A 1, 403 b 11–12). And "physical bodies contain surfaces and volumes, lines and points, and these are the subject-matter of mathematics" (Aristotle, *Physica*, B 2, 193 b 24–26).

However "the mathematician, though he too treats of these things, nevertheless does not treat them as the limits of a physical body; nor does he consider the attributes indicated as the attributes of such bodies," but he "separates them" from physical bodies; for, "in thought, they are separable from physical motion" (ibid., B 2, 193 b 33–34). Namely, the mathematician "thinks of mathematical objects, which are not separate, as being separate, when he thinks of them" (Aristotle, *De Anima*, Γ 7, 431 b 15–16). Thus, the mathematician deals with attributes that "are not considered as properties of any particular physical body and are obtained by abstraction" (Aristotle, *De Anima*, A 1, 403 b 14–15).

On the other hand, "since, as it appears, there is nothing that exists as separate from the sensible magnitudes, it is in the sensible forms that one finds" both the things "that are spoken of in abstraction," namely the mathematical objects, and "the properties and affections of sensible objects. For this reason, one would not be able to learn or understand anything if he had not the faculty of perception" (ibid., Γ 8, 432 a 3–8). So, although "one thinks of mathematical objects as separate, when one thinks of them," in fact they "are not separate" (ibid., Γ 7, 431 b 15–17).

The view that mathematical objects are objects obtained from physical bodies by abstraction, however, has the following shortcomings.

(1) By abstraction from physical bodies one cannot obtain objects, such as actual infinite sets, having attributes that do not occur in physical bodies. As Hersh says, "the first few natural numbers are abstracted from physical sets. But the really big natural numbers are not found in nature. The set of all natural numbers, N, is an infinite set, not found in nature. We made it up" (Hersh 2001, 4). Moreover, we "study infinitely differentiable infinite-dimensional manifolds of infinite connectivity. These are not found in the physical world. Not to mention the 'big' sets of contemporary set-theorists" (ibid.). That these mathematical objects cannot be obtained by abstraction from physical bodies depends on the fact that, as Kant says, "abstraction is only the negative condition under which universal representations can be generated" (Kant 1992, 593). Through it, "nothing is produced, but rather left out" (ibid., 487). Therefore, "abstraction does not add anything," but "rather cuts off everything that does not belong to the concept" (ibid., 351).

(2) By abstraction from physical bodies one cannot obtain objects having that exactness that is proper to mathematical objects. For example, by abstraction from an edge of a table one cannot obtain a straight line, but only a line that is slightly zigzag and with some, however small, breadth.

(3) By abstraction from physical bodies one cannot obtain objects having that determinateness that is proper to mathematical objects. As Frege says, suppose that "there are a black and a white cat sitting side by side before us. We do not attend to their colour, and they become colourless – but they still sit side by side. We do not attend to their posture, and they cease to sit without, however, assuming a different posture – but each of them is still in its place. We no longer attend to the place, and they cease to occupy one – but they continue presumably to be separate" (Frege 1984, 197–198). By these successive abstractions, from the black and the white cat we have perhaps obtained "a general concept of cat" (ibid., 198). But, "by continued application of this procedure, each object is transformed into a more and more

bloodless phantom. As the end result, we obtain from each object a something emptied of all content; but the something obtained from one object differs nevertheless from the something obtained from another object, even though it is not easy to say how" (ibid.).

(3) By abstraction from physical bodies one cannot obtain mathematical objects when the attributes considered are subordinate to those which are left out. For example, suppose that there are a black and a white cat sitting side by side before us, and that we do not attend to what distinguishes them from all the other things. We thus obtain the concept of two 'something'. But, from the concept of two 'something', if we do not attend to 'something', we do not obtain the concept of two, since then there would not be left any more things of which one could say that there are two of them. Therefore, Frege says that if, "in considering a white cat and a black cat, I disregard the properties which serve to distinguish them," I do not obtain the concept 'two' but rather "the concept 'cat'" (Frege 1960, 45). The "concept 'cat', no doubt, which we have arrived at by abstraction, no longer includes the special characteristics of either, but of it, for just that reason, there is only one" (ibid., 45–46).

(5) By abstraction from physical bodies one can arrive at extravagant conclusions. As Frege ironically says, the supporters of abstraction consider directing attention away from something an operation that "is especially effective. We attend less to a property, and it disappears. By thus making one characteristic mark after another disappear, we obtain more and more abstract concepts," therefore "inattention is a most effective logical power; this is presumably why professors are absent-minded" (Frege 1984, 197). But directing attention away from something leads to absurdities. For example, "if, abstracting from the distinction between my house and that of my neighbour, I were to regard them as the same, and consequently act in the other house as in my own, then the erroneousness of my abstraction would soon be made clear to me" (Frege 2013, II, 107).

In light of the above shortcomings, we may conclude that the view that mathematical objects are objects obtained from physical bodies by abstraction is untenable.

9.8 Mathematical Objects as Structures

Another answer to the question 'What is the nature of mathematical objects?' is that mathematics is not about objects but about structures, where a structure is a collection of objects with certain relations among them, highlighting the interrelationships among the objects rather than the objects themselves. Mathematical objects are simply positions in structures. This view of mathematical objects has been put forward mainly by Shapiro, and Resnik.

(1) Thus, Shapiro says that "mathematics is the science of structure" (Shapiro 2000, 257). A structure is "a collection of objects with certain relations among

them," highlighting "the interrelationships among the objects, and ignoring any features of them that do not affect how they relate to other objects in the system" (ibid., 259). Thus, "the subject-matter of arithmetic is a single abstract structure," the structure of natural numbers, which is "the pattern common to any infinite collection of objects that has a successor relation, a unique initial object, and satisfies the induction principle" (ibid., 258). Mathematical objects are simply positions in a structure. Thus, "the number 2 is no more and no less than the second position in the natural number structure; and 6 is the sixth position. Neither of them has any independence from the structure in which they are positions, and as positions in this structure, neither number is independent of each other" (ibid.).

(2) Mathematical objects exist if they can be modelled in the set-theoretic hierarchy. For, "a structure can be characterized by an axiomatization, which is an implicit definition" (Shapiro 2004, 32). However, not every implicit definition "successfully characterizes a structure," to characterize it an implicit definition must fulfil two conditions, the "existence condition," that "at least one structure" satisfies "the axioms," and the "uniqueness condition," that "at most one structure is described" (Shapiro 2000, 285). The uniqueness condition "depends on the underlying logic of the axiomatization," if the latter is taken to be second-order logic, then "there are unique characterizations of rich mathematical structures" (ibid.). As to the existence condition, "in mathematics as practised, set theory" is "taken to be the ultimate court of appeal for existence questions," so "a certain type of mathematical objects" exist if "objects of this type can be found or modelled in the set-theoretic hierarchy" (ibid., 288).

(3) Ultimately, "a structure is an equivalence class in the set-theoretic hierarchy" (Shapiro 1997, 92). For, "isomorphism and structure-equivalence are equivalence relations," so "one can informally take a structure to be an isomorphism type or a structure-equivalence type" (ibid.).

(4) We can learn about the structure of natural numbers by studying an exemplification of it. Indeed, a characterization of a structure by an axiomatization "is successful if it is categorical," namely "if all its models are isomorphic" (Shapiro 2004, 32), Then, "since isomorphic models are equivalent, the relevant properties of any model of the axiomatization are the same, and so, in a sense, any model is as good as any other. We can learn about the structure by studying an exemplification of it" (ibid.). This holds, in particular, of the structure of natural numbers. For, the characterization of the structure of natural numbers by the axioms of second-order Peano arithmetic PA^2 is successful, because the axioms of PA^2 are categorical (see Chap. 2). Then, we can learn about the structure of natural numbers by studying an exemplification of it. Indeed, "all that matters about the natural numbers is their relationship to one another" (ibid.).

The view that mathematical objects are structures, however, has the following shortcomings.

(1) Shapiro says that mathematics is the science of structure, mathematical objects are simply positions in a structure. But so he identifies mathematics with the

mathematics of a particular school, the Göttingen's school, which, through van der Waerden's *Moderne Algebra*, influenced Bourbaki. (On the Göttingen's school, see Ferreirós 1999, Ferreirós and Gray 2006, Rowe 2018, Roselló 2019; on Bourbaki, see Aczel 2006, Mashaal 2006). Contrary to the claim that mathematics is the science of structure, there are many mathematical questions, such as the distribution of prime numbers or the transcendence of π, which are not questions about a structure.

(2) Shapiro says that mathematical objects exist if they can be modelled in the set-theoretic hierarchy. But, as he admits, "for the structuralist, the set-theoretic hierarchy is just another structure. What makes us think that set theory itself is coherent?" (Shapiro 2000, 288). To be sure, "we cannot justify the coherence of set theory itself by modelling it in the set-theoretic hierarchy" (ibid.). The circle would be too blatant. Despite this, Shapiro claims that, since "the coherence of set theory is presupposed by much of the foundational activity in contemporary mathematics," contemporary "mathematics presupposes that satisfiability (in the set-theoretic hierarchy) is sufficient for existence" (ibid., 288–289). We must "accept this presupposition and make use of it like everyone else, and are in no better (and no worse) of a position to justify it" (ibid., 289). Such "presupposition is not vicious, even if it lacks external justification" (Shapiro 1997, 136). Contrary to Shapiro's claim, such presupposition is indeed vicious. Shapiro's argument is a petitio principii, because it is similar to the argument that we must accept the existence of God because the existence of God is presupposed by the existence of the world.

(3) Shapiro says that, ultimately, a structure is an equivalence class in the set-theoretic hierarchy. But then the primitive concept of mathematics is the concept of set, not the concept of structure. To avoid this problem, Shapiro formulates an axiomatic theory of structures, but the latter is entirely similar to set theory, and neither clarifies nor solves any problem that set theory does not solve. Shapiro himself admits that "the set-theoretic hierarchy and the realm of structures are little more than notational variants of each other" (Shapiro 2004, 20). Then the concept of structure is just the concept of set in disguise.

(4) Shapiro says that we can learn about the structure of natural numbers by studying an exemplification of it. But, as we have seen in Chap. 2, Zermelo's and von Neumann's definitions of natural numbers provide two exemplifications of the structure of natural numbers and, according to Shapiro, all that matters about the natural numbers is their relationship to one another. Clearly, in both such exemplifications, the relationship of natural numbers to one another is the membership relation, but the membership relation has different properties in the exemplifications in question, even properties inconsistent with each other. Thus, on the one hand, in Zermelo's exemplification $1 \notin 3$, on the other hand, in von Neumann's exemplification, $1 \in 3$. What is, then, the right exemplification of the structure of natural numbers? Structuralism can give no answer.

In light of the above shortcomings, we may conclude that the view that mathematical objects are structures is untenable.

9.9 Mathematical Objects as Fictions

Another answer to the question 'What is the nature of mathematical objects?' is that they are fictions. This view of mathematical objects has been put forward mainly by Field, and Balaguer.

(1) Thus, Field says that all there is to mathematics is that "we have a good story about natural numbers, another good story about sets, and so forth" (Field 1989, 22). For, mathematical objects are fictions, like the characters of a story. A mathematical proposition is true only in the same sense in which the proposition "'Oliver Twist lived in London' is true: the latter is true only in the sense that it is true according to a certain well-known story, and the former is true only in the sense that it is true according to standard mathematics" (ibid., 3). But, although a mathematical proposition is true only in this sense, mathematics is useful because we can use it "as a device for drawing conclusions" about the physical world "much more easily than we could draw them by a direct proof" (Field 1980, 28). On the other hand, since mathematics is about fictional objects, "it would be extremely surprising if it were to be discovered that standard mathematics implied that there are at least 10^6 non-mathematical objects in the universe, or that the Paris Commune was defeated" (ibid., 13). To make sure that this will not occur, it is necessary to prove that mathematics is conservative over physics. Namely, it is necessary to prove that the conclusions we arrive at by adding a mathematical theory to a physical theory which makes no reference to mathematical entities, "are not genuinely new, they are already derivable in a more long-winded fashion from the premises" of the physical theory, "without recourse to the mathematical entities" (ibid., 10–11). This proof will guarantee that any deduction which "can be made with the help of mathematics could be made (usually more long-windedly) without it" (ibid., x). Then, one will be entitled to assert that "mathematics is not really indispensable after all" (Field 1989, 26). And one will be "free to use any mathematics he likes for deducing consequences, as long as" one can always dispense with "the mathematics he uses" (Field 1980, 14).

(2) Now, it can actually be proved that mathematics is conservative over physics. For, "there are some fairly general strategies that can be employed to purge" physical "theories of all reference to mathematical entities" (Field 1989, 18).

(3) Thus, there is "a rewritten version of Newtonian physics" in which individual variables range over concrete objects, which amounts to "formulating it without mathematical entities" (ibid., 18, footnote 11).

The view that mathematical objects are fictions, however, has the following shortcomings.

(1) Field says that all there is to mathematics is that we have a good story about natural numbers, another good story about sets, and so forth, and that a mathematical proposition is true only in the same sense in which the proposition 'Oliver Twist lived in London' is true. But Dickens did not first introduce the character Oliver Twist and then investigated him, discovering that Oliver was a small boy whose mother died the moment he was born, who lived in a poorhouse and ended up getting involved with a gang of thieves, but eventually was adopted by a kind-hearted and

rich man. Dickens made up the character Oliver Twist with all his features through his story. On the contrary, a mathematician first introduces certain mathematical objects and then investigates them discovering some of their properties, afterwards some other mathematician discovers other properties, and so on. Therefore, mathematical objects are essentially different from characters in a story.

(2) Field says that it can actually be proved that mathematics is conservative over physics. But this is invalid. For, let M be a mathematical theory that is added to a physical theory N which makes no reference to mathematical entities. To prove that mathematics is conservative over physics, one should prove that, if a proposition A of N is provable in $N + M$, then A is provable already in N. But, as Shapiro points out, since there is a one-to-one correspondence between space-time points and quadruples of real numbers, "it is possible to model the natural numbers in space-time and, in effect, to do arithmetic in N. To put it differently, the natural number structure is exemplified in the universe of space-time" (Shapiro 1983, 526). Then "the Gödel's incompleteness theorems apply" (ibid., 527). So, there is a proposition A of N which is provable in $N + M$ but not in N. Hence, the mathematical theory M is not conservative over the physical theory N. This refutes the fictionalist "argument for the conservativeness of mathematics over physics" (ibid., 528).

(3) Field says that there is a rewritten version of Newtonian physics in which individual variables range over concrete objects. But this is invalid. For, to make a rewritten version of Newtonian physics in which individual variables range over concrete objects, one has to introduce entities having properties of mathematical entities, rather than properties of concrete objects. Thus, the postulates of Field's theory of space-time imply that there is a one-to-one correspondence between space-time points, conceived of as physical entities, concrete and not abstract, and quadruples of real numbers. Since real numbers are uncountable, this involves postulating uncountably many physical entities. According to Field, this is not objectionable, nor "does it become any more objectionable when one postulates that these physical entities obey structural assumptions analogous to the ones that platonists postulate for the real numbers" (Field 1980, 31). But then what Field calls 'physical entities' are really mathematical entities, so the rewritten version of Newtonian physics does not actually purge Newtonian physics of all reference to mathematical entities.

In light of the above shortcomings, we may conclude that the view that mathematical objects are fictions is untenable.

9.10 Mathematical Objects as Idealizations of Physical Bodies

Another answer to the question 'What is the nature of mathematical objects?' is that they are objects obtained from physical bodies not, as Aristotle says, by abstraction, but by idealization. Idealization is the process by which one pushes certain attributes of physical bodies to the limit. This view of mathematical objects has been put forward by Husserl. Indeed:

(1) Husserl says that "the mathematical realm" is "a world of ideal objects" (Husserl 2008, 49). These ideal objects, the mathematical objects, are the result of "idealizing the world of bodies in respect to what has spatiotemporal shape in this world," so idealization "created ideal objects" (Husserl 1970, 32). Being the result of idealizing the world of bodies, mathematical objects are "something which cannot be 'seen'," they are "essences drawn immediately from intuition simpliciter" as "ideal essences" (Husserl 1983, 166). The ideal essences are "ideal 'limits' which it is essentially impossible to find in any sensuous intuition" (ibid., 167).

(2) Idealization "is fundamentally different in its essence from the seizing upon an essence by simple 'abstraction'" (ibid., 167). For, the ideal essences which are the result of an idealization, being ideal 'limits' which it is essentially impossible to find in any sensuous intuition, have attributes that do not occur in physical bodies. On the contrary, the objects obtained by abstraction from physical bodies, have attributes that occur in physical bodies.

The view that mathematical objects are idealizations of physical bodies, however, has the following shortcomings.

(1) Husserl says that the mathematical realm is a world of ideal objects. This raises the question: Why, then, is mathematics applicable to the real world? Husserl's answer is that mathematics is applicable to it because, "through Galileo's mathematization of nature, nature itself is idealized" and "becomes" a "mathematical manifold" (Husserl 1970, 23). Galileo substitutes a "mathematically substructed world of idealities for the only real world, the one that is actually given through perception, that is ever experienced and experienceable" (ibid., 49). And Galileo conceives the "idea of a rational infinite totality of being with a rational science systematically mastering it" (ibid., 22). Thus, however, Husserl does not account for the applicability of mathematics to the real world, but only for the applicability of mathematics to a mathematically substructed world of idealities. So the question is not solved but only transformed into the question: If mathematics is applicable to a mathematically substructed world of idealities, why is mathematics then applicable to the real world?

(2) Husserl says that idealization is fundamentally different in its essence from the seizing upon an essence by simple abstraction. Thus, being the result of idealizing the world of bodies, mathematical objects are ideal limits which it is essentially impossible to find in any sensuous intuition, so they have attributes that do not occur in physical bodies. Then, however, Husserl not only does not account for the applicability of mathematics to the real world, but even trivializes the question of the applicability. For, if mathematical objects have attributes that do not occur in physical bodies, then all mathematical propositions are vacuously true in the real world. For example, consider the proposition 'In any triangle, the sum of the internal angles is equal to two right angles'. This proposition is of the form $A(x) \to B(x)$, where $A(x)$ expresses 'x is a triangle' and $B(x)$ expresses 'The sum of the internal angles of x is equal to two right angles'. Now, since, according to Husserl, triangles have attributes that do not occur in physical bodies, $A(x)$ is false and hence

$A(x) \rightarrow B(x)$ is vacuously true in the real world. For the same reason, even the opposite proposition, 'In any triangle, the sum of the internal angles is not equal to two right angles', being of the form $A(x) \rightarrow \neg B(x)$, is vacuously true in the real world. This is absurd.

In light of the above shortcomings, we may conclude that the view that mathematical objects are idealizations of physical bodies is untenable.

9.11 Mathematical Objects as Idealizations of Operations of Collecting

Another answer to the question 'What is the nature of mathematical objects?' is that they are idealizations of operations of collecting. This view of mathematical objects has been put forward by Kitcher. Indeed:

(1) Kitcher says that mathematical objects are the result of an idealization of "operations of collecting" which "we are able to perform with respect to any objects" (Kitcher 1984, 12). Or, "we may personify the idealization, by thinking" of mathematical objects as the result of operations of collecting "of an ideal subject, whose status as an ideal subject resides in her freedom from certain accidental limitations imposed on us" (ibid., 109). So, "the ideal subject is an idealization of ourselves" (ibid., 111). The relation between the operations of the ideal subject and "the actual operations of human agents parallels that between the laws of ideal gases and the actual gases which exist in our world" (ibid.). The ideal gases do not exist in our world, and yet "actual gases approximately satisfy the condition set down in the Boyle-Charles Law" of ideal gases; moreover, "if the molecules of gases were of negligible size and if there were no intermolecular forces then gases would obey the Boyle-Charles Law" (ibid., 117). This "prompts us to abstract from certain features of the actual situation, introducing the notion of an ideal gas to describe how actual gases would behave if complicating factors were removed" (ibid.). Much in the same way, the ideal subject does not exist in our world, and yet the actual operations of collecting of human agents approximate the operations of the ideal subject, and moreover, if human agents were free from their biological limitations, their operations would coincide with those of the ideal subject. This prompts us to "specify the capacities of the ideal agent by abstracting from the incidental limitations on our own collective practice" (ibid.).

(2) In particular, the description of the hierarchy of sets "may be viewed as a literal account of the iterated constructive activity of the ideal mathematical subject," specifically, of "the iterated performance of one fundamental operation, that of collecting" (ibid., 133). Of course, since "the sequence of stages at which sets are formed is highly superdenumerable," this involves that, if "each of the stages corresponds to an instant in the life of the constructive subject," then "the subject's

activity is carried out in a medium analogous to time, but far richer than time. (Call it 'supertime')" (ibid., 146). This means "to idealize still further from our own thoroughly finite performances" (ibid., 147). But such idealization "is no different in principle from earlier idealizations in which we abstract from our own mortality or from our inability to survey infinite domains," so "the conception of the ideal subject as an idealization of ourselves does not lapse when we release the subject from the constraints of our time" (ibid.).

(3) To talk of an ideal subject "is not to suppose that there is a mysterious being with superhuman powers" (ibid., 110). It is only to suppose that "mathematical truths are true in virtue of" conditions "which actually are satisfied by nothing at all but are approximately satisfied by operations we perform (including physical operations)" (ibid.).

The view that mathematical objects are idealizations of operations of collecting, however, has the following shortcomings.

(1) Kitcher says that mathematical objects are the result of operations of collecting of an ideal subject that is an idealization of ourselves, where the operations of the ideal subject stand to our actual operations as the laws of ideal gases stand to the actual gases which exist in our world. But this is invalid, because the actual gases which exist in our world very closely approximate the behaviour of ideal gases at high temperatures and low pressures, and the degree of approximation can be calculated. On the contrary, the ideal subject is supposed to be able to perform certain operations, such as all possible collecting of the members of an uncountable totality, which are beyond the possibilities of any human agent.

(2) Kitcher says that the description of the hierarchy of sets may be viewed as a literal account of the iterated constructive activity of the ideal subject. Although this involves that the activity of the ideal subject is carried out not in our time but in a supertime, the conception of the ideal subject as an idealization of ourselves does not lapse when we release the subject from the constraints of our time. But this is invalid. For while, as already said, the actual gases which exist in our world very closely approximate the behaviour of ideal gases at high temperatures and low pressures, on the contrary, the structure of time by no means approximates the structure of the supertime, because the latter would be highly superdenumerable.

(3) Kitcher says that to talk of an ideal subject is not to suppose that there is a mysterious being with superhuman powers. But this is invalid, because an ideal subject capable of performing superoperations in a supertime should have powers not only superhuman but even superdivine. Indeed, as Parsons points out, the mind of such an ideal subject "would differ not only from finite minds but also from the divine mind as conceived in philosophical theology" (Parsons 2005, 273). For, in the latter, either the divine mind "is thought of as in time, and therefore as doing things in an order with the same structure as that in which finite beings operate, or its eternity is interpreted as complete liberation from succession" (ibid.).

In light of the above shortcomings, we may conclude that the view that mathematical objects are idealizations of operations of collecting is untenable.

9.12 The Basis for an Alternative View of Mathematical Objects

Since all views of mathematical objects discussed above are untenable, an alternative view is required. However odd it may sound, the basis for an alternative view of mathematical objects can be found in Plato.

As we have seen above, for Plato, mathematical objects are independently existing entities, non-physical and non-mental, knowable only by intuition. Therefore, "one very common view is that Plato was a mathematical Platonist in the contemporary sense of the phrase, holding that there is a realm of mathematical reality that is not constructed by but is discovered by mathematicians" (White 2006, 230).

But Plato was a mathematical platonist only as to the knowledge of mathematical objects a mathematician can have after death. He was not a mathematical platonist as to the knowledge of mathematical objects a mathematician can have while he is alive.

Indeed, as we have seen in Chap. 5, according to Plato, a mathematician can know mathematical objects as independently existing entities by intuition only after death. As long as he has a body, he "will not be able to go that far" (Plato, *Respublica*, VII 533 1–2). He can only introduce mathematical objects as hypotheses and investigate their properties by the analytic method.

Indeed, Plato says that "those who deal with geometry, arithmetic, and the like, hypothesize the odd and the even, geometrical figures, the three kinds of angles, and other things akin to these, depending on the subject of their investigation. They take these as known and assume them as hypotheses" (Plato, *Republic*, VI 510 c 2–6). And they investigate their properties by the analytic method. So, they solve mathematical problems not by intuition but by discursive thought, hence it is "discursive thought, not intuition, the state of geometry and the like, namely thought intermediate between opinion and intuition" (ibid., VI 511 d 2–5).

Therefore, according to Plato, the actual making of mathematics is not accounted for by mathematical platonism, but by saying that, to solve mathematical problems, mathematicians introduce mathematical objects as hypotheses, and investigate their properties by the analytic method. In this sense, the basis for an alternative view of mathematical objects can be found in Plato.

9.13 The Heuristic View of Mathematical Objects

The alternative view of mathematical objects, a basis for which can be found in Plato, is the heuristic view. According to it, mathematical objects are hypotheses human beings make to solve mathematical problems by the analytic method. Like all hypotheses in the analytic method, the hypotheses must be plausible.

Only mathematical objects introduced by hypotheses that are plausible can be said to 'exist'. This, however, does not mean that they exist literally, in the physical or in a superphysical world. For, mathematical objects as hypotheses are only ideas conceived by the mind. On the other hand, not all ideas conceived by the mind can be accepted as mathematical objects, but only those introduced by hypotheses which are plausible. Therefore, that mathematical objects exist, means that they are introduced by hypotheses that are plausible.

Since mathematical objects are hypotheses human beings make to solve mathematical problems by the analytic method, mathematical objects are means of investigation.

9.14 Some Remarks About the Heuristic View

Some remarks about the heuristic view of mathematical objects can be useful.

(1) That mathematical objects are hypotheses human beings make to solve mathematical problems by the analytic method, does not mean that mathematical knowledge is not objective. Hart claims that "the objectivity we expect of mathematical truths" seems "to require independent objects," because "mathematical truth seems to require the existence of objects which are neither physical nor mental" (Hart 1996, 3). Just "as truths of history require the existence of historical figures like Caesar, so truths of mathematics require the existence of prime numbers like 13" (ibid., 2). But these claims are unfounded. The objectivity of mathematical knowledge cannot require that mathematical objects be independently existing entities because, as argued above, the view that mathematical objects are independently existing entities is untenable. The objectivity of mathematical knowledge can only require that mathematical objects be hypotheses that are plausible, namely such that the arguments for them are stronger than the arguments against them, on the basis of the existing knowledge. This does not diminish the value of mathematical knowledge, because plausible knowledge is the most we can achieve.

(2) That mathematical objects are hypotheses human beings make to solve mathematical problems by the analytic method, does not mean that they are mental constructions in Brouwer's sense. For, hypotheses are ideas formulated by discursive thought, mental constructions in Brouwer's sense are constructions out of the intuition of time. Then, there is an essential difference between mathematical objects as hypotheses and mathematical objects as mental constructions. For example, actual infinite sets can be introduced as hypotheses to solve mathematical problems, but cannot be mental constructions in Brouwer's sense, and hence must be excluded from mathematics, because they are "certainly not built up mathematically" (Brouwer 1975, 80).

(3) That mathematical objects are hypotheses human beings make to solve mathematical problems by the analytic method, does not mean that they are fictions.

For, being hypotheses human beings make to solve mathematical problems by the analytic method, they must agree with experience, and are accepted only as long as no strong arguments against them emerge. Conversely, fictions are not introduced to solve problems, so they are not supposed to agree with experience, and are accepted despite our being aware that they have no relation to reality. Thus Vaihinger, the founder of philosophical fictionalism, says that, "whereas every hypothesis seeks to be an adequate expression of some reality still unknown and to mirror this objective reality correctly, the fiction is advanced with the consciousness that it is an inadequate" manner "of conception, whose coincidence with reality is, from the start, excluded" (Vaihinger 2008, 268).

(4) That mathematical objects are hypotheses human beings make to solve mathematical problems by the analytic method, implies that they are mental-cultural human products. To solve some mathematical problem, one or more mathematicians may conceive the idea of some new mathematical objects, and use that idea as hypothesis to solve the problem by the analytic method. Afterwards, other mathematicians may find the hypothesis useful to solve other problems by the analytic method. And so on. Thus mathematical objects, once they have been conceived by the mind of one or more mathematicians, become part of the cultural process as cultural products. As White says, the introduction of new mathematical objects is "an event that takes place" both "in one or more nervous systems" and "in the cultural tradition," where "the human nervous system" is "the catalyst which makes the cultural process possible" (White 1947, 303). Of course, like all other cultural products, newly introduced mathematical objects may encounter resistance. Thus, this has been the case of the number zero, negative numbers, infinitesimals, imaginary numbers, and actual infinite sets.

(5) The hypotheses by which mathematical objects are introduced characterize their identity, but do not entail that those mathematical objects exist – just as, saying that hippogriffs are animals having the foreparts of a griffin and the body of a horse, characterizes their identity, but does not entail that hippogriffs exist. Identity does not imply existence. To say that mathematical objects exist, one must show that the hypotheses by which they are introduced are plausible.

(6) That mathematical objects are hypotheses human beings make to solve mathematical problems by the analytic method, does not mean that the mathematician who introduces certain mathematical objects as hypotheses knows all their properties. Gödel says that "the creator necessarily knows all properties of his creatures, because they can't have any others except those he has given to them" (Gödel 1986–2002, III, 311). But it is not so. A painter does not necessarily know all properties of his creatures, otherwise there would be no need for art critics. It is one thing to introduce mathematical objects as hypotheses, and another thing to discover all their properties. Thus, as Penrose points out, "when Cardano introduced his complex numbers, he could have had no inkling of the many magical properties which were to follow – properties which go under various names, such as the Cauchy integral formula, the Riemann mapping theorem, the Lewy extension property"

(Penrose 1991, 96). Such properties, and many others, do not pertain to later modifications of the concept of complex number, but are "properties of the very numbers" that "Cardano had first encountered in about 1539" (ibid.). Cardano could have had no inkling of such properties because mathematical objects, entering into relation with other mathematical objects, acquire new properties, other than those the creator has given to them, and discovering them requires an investigation.

9.15 The Open-Ended Character of Mathematical Objects

It has been said above that the hypotheses by which mathematical objects are introduced characterize their identity. This, however, does not mean that the hypotheses characterize their identity completely and conclusively. Their identity may undergo changes, because mathematical objects are not closed in themselves, they are open to interactions with other mathematical objects, from which they may gain new determinations. So mathematical objects have an open-ended character.

Moreover, mathematical objects introduced to solve some problem may turn out to be useful also to solve other problems. Such objects, then, consolidate and acquire a stability that makes them amply independent of the problem for which they were originally introduced, and they become subjects of study themselves.

For example, Menaechmus introduced conic sections to solve the problem of finding two mean proportionals between two given lines (see Chapter 5). As Weil says, "Menaechmus solved the problem by conics (two parabolas or a parabola and a hyperbola)," so "there is no doubt that he invented them for this purpose. And the problem of doubling the cube comes down to that of finding two geometrical means between two known quantities" (Weil 2015, 114). In fact, "it is easy to imagine" Menaechmus's "process of discovery, because the cone consists of a circle with a variable diameter and the parabola gives the series of all the geometrical means between a fixed and a variable term" (ibid.).

After Menaechmus introduced conic sections to solve the problem of finding two mean proportionals between two given lines, conic sections turned out to be useful to solve other problems, and eventually became subjects of study themselves in Apollonius of Perga's treatise *Conica*, one of the greatest scientific works from the ancient world.

When mathematical objects become subjects of study themselves, they may appear so familiar to the mathematician that he may have the feeling that he is working with something real. But it is an illusion, which arises from the mathematician's great familiarity with them. Mathematical objects are not something real because, as argued above, the view that mathematical objects are independently existing entities is untenable.

Conversely, other mathematical objects, also introduced to solve some mathematical problem, do not consolidate and do not acquire stability.

For example, Newton and Leibniz introduced infinitesimals as hypotheses to solve problems involving continuous change. To that aim, Newton said it was useful to imagine "infinitely small additions" by which "quantities increase during each infinitely small interval of time" (Newton 1967–1981, III, 81). Leibniz said it was "useful to imagine infinitely small quantities, or, as Nicolaus Mercator called them, infinitesimals, which may not be taken to be zero when the assignable ratio between them is sought," but "are rejected whenever they are added to incomparably greater quantities" (Leibniz 1971, IV, 63). But infinitesimals did not consolidate and did not acquire stability, no precise definition of them was ever given, they were also assigned contradictory determinations. Therefore, they were the object of severe criticism from the very beginning.

Thus, Rolle said that, while "geometry had always been regarded as an exact science, and indeed as the source of the exactness that is spread throughout all other parts of mathematics," this "character of exactness no longer reigns in geometry ever since the new system of the infinitely small" quantities "has been mixed to it" (Rolle 1703, 312). Indeed, the new system of the infinitely small quantities "has produced nothing for the truth, and it would seem to me that it often conceals mistakes" (ibid.). Since, $x + dx = x$ implies that $dx = 0$, "all these infinite sequences of infinities, which the system provides, will only be nothings that are supposed to be infinitely contained within other nothings" (ibid., 324).

Berkeley said that "the mathematicians of this age embrace obscure notions and uncertain opinions, and are puzzled about them, contradicting each other and disputing like other men" (Berkeley 1953, 127, footnote 5). Thus, "Leibniz and his followers" make "no manner of scruple, first to suppose, and secondly to reject, quantities infinitely small: with what clearness in the apprehension and justness in the reasoning, any thinking man, who is not prejudiced in favour of those things, may easily discern" (Berkeley 1992, 179). Infinitesimals "are neither finite quantities, nor quantities infinitely small, nor yet nothing," so we may "call them the ghosts of departed quantities" (ibid., 199).

As a result, infinitesimals never crystallized and became subjects of study themselves, they hung fluttering between being and non-being, and in the nineteenth century disappeared from mathematics.

They were revived in the twentieth century by Robinson in a different form, through the development of non-standard analysis. However, their revival was not motivated by any necessity of mathematical practice, but only by the purely ideological aim to show that "Leibniz ideas can be fully vindicated" (Robinson 1966, 2). Indeed, non-standard analysis does not really add anything that cannot be easily achieved by standard methods of mathematical analysis. Moreover, as Connes points out, "as soon as you have a non-standard number, you get a non-measurable set," whereas "every set you can name is measurable," so it seems "utterly doomed to failure to try to use nonstandard analysis to do physics" (Goldstein and Skandalis 2007, 26).

References

Aczel, Amir D. 2006. *The artist and the mathematician: The story of Nicolas Bourbaki, the genius mathematician who never existed*. New York: Thunder's Mouth Press.

Berkeley, George. 1953. Siris. In *The works of George Berkeley Bishop of Cloyne*, ed. Arthur Aston Luce and Thomas Edmund Jessop, vol. 5, 25–164. London: Nelson.

———. 1992. *De motu* and *The analyst*. Dordrecht: Springer.

Borwein, Jonathan, David Bailey, and Roland Girgensohn. 2004. *Experimentation in mathematics: Computation paths to discovery*. Natick: A K Peters.

Brannon, Elizabeth. 2005. The independence of language and mathematical reasoning. *Proceedings of the National Academy of Science* 102: 3177–3178.

Brouwer, Luitzen Egbertus Jan. 1967. Intuitionistic reflections on formalism. In *From Frege to Gödel: A source book in mathematical logic, 1879–1931*, ed. Jean van Heijenoort, 490–492. Cambridge: Harvard University Press.

———. 1975. *Collected works 1: Philosophy and foundation of mathematics*. Amsterdam: North-Holland.

———. 1998. The structure of the continuum. In *From Brouwer to Hilbert: The debate on the foundations of mathematics in the 1920s*, ed. Paolo Mancosu, 54–63. Oxford: Oxford University Press.

Dieudonné, Jean. 1970. The work of Nicholas Bourbaki. *The American Mathematical Monthly* 77: 134–145.

Ferreirós, José. 1999. *Labyrinth of thought*. Cham: Birkhäuser.

Ferreirós, José, and Jeremy John Gray, eds. 2006. *The architecture of modern mathematics: Essays in history and philosophy*. Oxford: Oxford University Press.

Field, Hartree. 1980. *Science without numbers*. Oxford: Blackwell.

———. 1989. *Realism, mathematics and modality*. Oxford: Blackwell.

Frege, Gottlob. 1960. *The foundations of arithmetic*. New York: Harper.

———. 1969. *Nachgelassene Schriften*. Hamburg: Meiner.

———. 1979. *Posthumous writings*. Oxford: Blackwell.

———. 1984. *Collected papers on mathematics, logic, and philosophy*. Oxford: Blackwell.

———. 2013. *Basic laws of arithmetic*. Oxford: Oxford University Press.

Gillies, Donald. 2014. Should philosophers of mathematics make use of sociology? *Philosophia Mathematica* 22: 12–34.

Gödel, Kurt. 1986–2002. *Collected works*. Oxford: Oxford University Press.

Goldstein, Catherine, and Georges Skandalis. 2007. An interview with Alain Connes. *European Mathematical Society Newsletter* 63: 25–31.

Hale, Bob, and Crispin Wright. 2001. *The reason's proper study: Essays towards a neo-Fregean philosophy of mathematics*. Oxford: Oxford University Press.

Hardy, Godfrey Harold. 1929. Mathematical proof. *Mind* 38: 1–25.

———. 1992. *A mathematician's apology*. Cambridge: Cambridge University Press.

Hart, Wilbur Dyre. 1996. Introduction. In *The philosophy of mathematics*, ed. Wilbur Dyre Hart, 1–13. Oxford: Oxford University Press.

Hermite, Charles. 1905. Letter to Stieltjes, 13 May 1894. In Charles Hermite, and Thomas Jan Stieltjes, *Correspondance*, vol. 2, 398–400. Paris: Gauthier-Villars.

Hersh, Reuben. 2001. Reply to Martin Gardner. *The Mathematical Intelligencer* 23 (2): 3–5.

Hilbert, David. 1967. On the infinite. In *From Frege to Gödel: A source book in mathematical logic, 1879–1931*, ed. Jean van Heijenoort, 367–392. Cambridge: Harvard University Press.

———. 1996a. The new grounding of mathematics: First report. In *From Kant to Hilbert: A source book in the foundations of mathematics*, ed. William Ewald, vol. 2, 1117–1134. Oxford: Oxford University Press.

———. 1996b. The logical foundations of mathematics. In *From Kant to Hilbert: A source book in the foundations of mathematics*, ed. William Ewald, vol. 2, 1134–1148. Oxford: Oxford University Press.

———. 1996c. The grounding of elementary number theory. In *From Kant to Hilbert: A source book in the foundations of mathematics*, ed. William Ewald, vol. 2, 1149–1157. Oxford: Oxford University Press.

Husserl, Edmund. 1970. *The crisis of European sciences and transcendental phenomenology*. Evanston: Northwestern University Press.

———. 1983. *Ideas pertaining to a pure phenomenology and to a phenomenological philosophy I*. The Hague: Nijhoff.

———. 2008. *Introduction to logic and theory of knowledge: Lectures 1906/07*. Cham: Springer.

Kant, Immanuel. 1992. *Lectures on logic*. Cambridge: Cambridge University Press.

Kitcher, Philip. 1984. *The nature of mathematical knowledge*. Oxford: Oxford University Press.

Kitcher, Philip, and William Aspray. 1988. An opinionated introduction. In *History and philosophy of modern mathematics*, ed. William Aspray and Philip Kitcher, 3–57. Minneapolis: University of Minnesota Press.

Leibniz, Gottfried Wilhelm. 1971. *Mathematische Schriften*. Hildesheim: Olms.

Mashaal, Maurice. 2006. *Bourbaki: A secret society of mathematicians*. Providence: American Mathematical Society.

Newton, Isaac. 1967–1981. *Mathematical papers*. Cambridge: Cambridge University Press.

Parsons, Charles. 2005. *Mathematics in philosophy: Selected essays*. Ithaca: Cornell University Press.

Penrose, Roger. 1991. *The emperor's new mind: Concerning computers, minds, and the laws of physics*. London: Penguin.

Robinson, Abraham. 1966. *Non-standard analysis*. Amsterdam: North-Holland.

Rolle, Michel. 1703. Du nouveau système de l'infini. *Mémoires de mathématiques et de physique de l'Académie Royale des Sciences*: 312–336.

Roselló, Joan. 2019. *Hilbert, Göttingen and the development of modern mathematics*. Newcastle upon Tyne: Cambridge Scholars Publishing.

Rowe, David E. 2018. *A richer picture of mathematics: The Göttingen tradition and beyond*. Cham: Springer.

Shapiro, Stewart. 1983. Conservativeness and incompleteness. *The Journal of Philosophy* 80: 521–531.

———. 1997. *Philosophy of mathematics: Structure and ontology*. Oxford: Oxford University Press.

———. 2000. *Thinking about mathematics: The philosophy of mathematics*. Oxford: Oxford University Press.

———. 2004. Foundations of mathematics: Metaphysics, epistemology, structure. *The Philosophical Quarterly* 54: 16–37.

Thom, René. 1971. 'Modern' mathematics: An educational and philosophic error? *American Scientist* 59 (6): 695–699.

Vaihinger, Hans. 2008. *The philosophy of 'as if'*. London: Routledge.

von Neumann, John. 1955. *Mathematical foundations of quantum mechanics*. Princeton: Princeton University Press.

Wang, Hao. 1996. *A logical journey: From Gödel to philosophy*. Cambridge: The MIT Press.

Weil, Simone. 2015. Letters to André Weil, 1940. In Simone Weil, *Seventy letters: Personal and intellectual windows on a thinker*, 112–127. Eugene: Wipf and Stock.

Weyl, Hermann. 1950. *The theory of groups and quantum mechanics*. Mineola: Dover.

White, Leslie A. 1947. The locus of mathematical reality: An anthropological footnote. *Philosophy of Science* 14: 289–303.

White, Michael J. 2006. Plato and mathematics. In *A companion to Plato*, ed. Hugh H. Benson, 228–243. Malden: Blackwell.

Chapter 10
Demonstrations

Abstract According to heuristic philosophy of mathematics, one of the tasks of the philosophy of mathematics is to give an answer to the question: What is the nature of mathematical demonstrations? This question has received several answers. The chapter discusses them and argues that they are inadequate. Then, it offers an alternative answer: mathematical demonstration is analytic demonstration.

Keywords Axiomatic demonstration · Formal demonstration · Purity of method · Deductive demonstration · Analytic demonstration · Proof without cuts · Depth of demonstrations · Mathematical style

10.1 Different Views of Mathematical Demonstration

As we have seen in Chap. 3, according to heuristic philosophy of mathematics, one of the tasks of the philosophy of mathematics is to give an answer to the question: What is the nature of mathematical demonstrations?

This question has received several answers. The most significant ones are: mathematical demonstration is axiomatic demonstration, formal demonstration, or deductive demonstration.

The chapter discusses these answers and argues that they are inadequate. Then it offers an alternative answer: mathematical demonstration is analytic demonstration.

10.2 Axiomatic Demonstration

An axiomatic demonstration is a deduction of a proposition from given principles which are true, in some sense of 'true'.

Mainstream philosophy of mathematics holds that mathematical demonstration is axiomatic demonstration. This follows from the fact that, according to mainstream philosophy of mathematics, the method of mathematics is the axiomatic method.

The main characters of axiomatic demonstration can be described as follows.

C. Cellucci, *The Making of Mathematics*, Synthese Library 448,
https://doi.org/10.1007/978-3-030-89731-4_10

(1) Axiomatic demonstration involves a downward path from the given principles to a proposition.

(2) Axiomatic demonstration establishes the truth of a proposition, and this is its function.

(3) Axiomatic demonstration can only make use of the given principles and propositions already deduced from them.

(4) Axiomatic demonstration cannot make any essential use of mathematical diagrams. This means that mathematical diagrams can be used, but only provided that in principle one could always do without them.

(5) Axiomatic demonstration is ultimately based on intuition. For, the ultimate justification for the principles consists in some kind of intuition.

(6) Axiomatic demonstration is rigorous demonstration. For, a demonstration is rigorous if and only if it starts from principles which are justified and deduces consequences from them.

10.3 Axiomatic Demonstration and Mathematicians

Many mathematicians maintain that mathematical demonstration is axiomatic demonstration.

Thus, Cupillari says that "a proof is a logical argument that establishes the truth of a statement beyond any doubt. A proof consists of a finite chain of steps, each one of them a logical consequence of the previous one" starting from "certain assumptions" (Cupillari 2005, 3).

Nathanson says that "theorems are deductions from axioms. Each line in a proof is a simple consequence of the previous lines of the proof, or of previously proved theorems. Our conclusions are true, unconditionally and eternally" (Nathanson 2008, 773).

Nicholson says that "a proof is a process by which a mathematical statement is shown to be true. Each step of the process follows logically from axioms, stated assumptions or previously established results until the desired mathematical result is reached" (Nicholson 2019, 105).

10.4 Formal Demonstration

A variant of the concept of axiomatic demonstration is that of formal demonstration.

A formal demonstration is a deduction of a proposition from given axioms which are true in some sense of 'true', by means of given formal deduction rules.

Several mathematicians maintain that rigorous mathematical demonstration is formal demonstration.

Thus, Hanna says that "the term rigorous proof or formal proof" is "understood here to mean a proof" such that "every definition, assumption, and rule of inference appealed in the proof has been, or could be, explicitly stated," and "every step in the chain of deductions which constitutes the proof is set out explicitly" (Hanna 1983, 3).

Mac Lane says that "mathematics consists of formal rules, formal systems, and formal definitions of concepts" (Mac Lane 1986, 441). A "mathematical proof is rigorous when it is (or could be) written" as "a sequence of inferences from" axioms, "each inference made according to one of the stated rules" (ibid., 377).

Hales says that "the ultimate standard of proof is a formal proof, which is nothing other than an unbroken chain of logical inferences from an explicit set of axioms" (Hales 2012, x). A "rigorous argument" is "a formal proof," because in a formal proof "all the intermediate logical steps are supplied, without exception" (Hales 2008, 1371). In a formal proof "every logical inference has been checked all the way back to the fundamental axioms of mathematics" (ibid.).

Moreover, several mathematicians maintain that every axiomatic demonstration can be translated into a formal demonstration.

Thus, Bourbaki says that, "as every one agrees today, the distinctive character of a mathematical text is that it can be formalized, i.e., translated into a certain kind of sign-language" (Bourbaki 1949, 3). Indeed, "the axiomatic method is, strictly speaking, nothing but this art of drawing up texts whose formalization is straightforward in principle" (Bourbaki 2004, 8).

Macintyre says that there is "an algorithm for translating the majority of conventional mathematical statements" into "one of the formalisms" of mathematical logic, and "moreover one could go onto translate (again mechanically in principle) classical informal proofs into formal proofs of some accepted formal system" (Macintyre 2005, 2420).

Gersting says that, although "proofs are usually not written a step at a time with formal justifications for each step," they "can be translated into a formal proof if required. In fact, the value of a formal proof is that it serves as a sort of insurance – if" a "proof cannot be translated into a formal proof, it should be viewed with great suspicion" (Gersting 2014, 99).

Sometimes, the view that every axiomatic demonstration can be translated into a formal demonstration, is even called the common view.

Thus, Detlefsen says that "the prevailing view of proof sees rigor as a necessary feature of proof and formalizability as a necessary condition of rigor" (Detlefsen 2008, 16). The "reasoning behind this view is straightforward: (i) proper proofs are proofs that either are or can readily be made rigorous; (ii) proofs that are or can readily be made rigorous are formalizable; therefore (iii) all proper proofs are formalizable. Call this" view the "common view" (ibid., 17).

Since the concept of formal demonstration is a variant of that of axiomatic demonstration, in what follows 'axiomatic demonstration' will be used to mean either axiomatic demonstration in general, or specifically formal demonstration. Which is meant will be clear from context.

10.5 Axiomatic Demonstration and Intuition

Several supporters of the view that mathematical demonstration is axiomatic demonstration, argue that the axioms are justified, directly or indirectly, by some kind of intuition.

Thus, as we have seen in Chap. 9, Gödel argues that we can arrive at directly perceiving the axioms to be true by the procedure that consists in focusing more sharply on the concepts concerned. So we will get an intuition of the concepts and a secure insight into the axioms that hold for them.

This, however, is invalid because, as already pointed out there, by that procedure we arrive at two different concepts of set, and the procedure is unable to tell us which of them is the genuine one, so through it we cannot arrive at directly perceiving which axioms for sets are true.

On the other hand, as we have seen in Chap. 2, Hilbert argues that the truth of the axioms is indirectly guaranteed by intuition, since it can be established by a demonstration of their consistency based on intuition.

This, however, is invalid because, as already pointed out there, by Gödel's second incompleteness theorem, it is generally impossible to give a demonstration of the consistency of the axioms based on intuition.

10.6 Axiomatic Demonstration and Justification from Consequences

Several supporters of the view that mathematical demonstration is axiomatic demonstration, maintain that the axioms are justified if many true consequences can be deduced from them. As we have seen in Chap. 9, even Gödel supports this justification of the axioms, in addition to the justification in terms of intuition. But, before Gödel, already Zermelo and Russell had supported it.

Thus, Zermelo says that "principles must be judged from the point of view of science", namely their consequences, "and not science from the point of view of principles fixed once and for all" (Zermelo 2010–2013, I, 135). For example, the principle of choice is justified because there is "a number of elementary and fundamental theorems and problems" that "could not be dealt with at all without the principle of choice" (ibid., I, 131). So long as "the principle of choice cannot be definitely refuted, no one has the right to prevent the representatives of productive science from continuing to use this 'hypothesis'," and "developing its consequences to the greatest extent" (ibid., I, 135).

Russell says that "the reason for accepting an axiom" is "always largely inductive, namely that many propositions which are nearly indubitable can be deduced from it, and that no equally plausible way is known by which these propositions could be true if the axiom were false, and nothing which is probably false can be deduced from it" (Whitehead and Russell 1925–1927, I, 59). Now, "in the case of the axiom of reducibility, the inductive evidence in its favour is very strong, since the reasonings which it permits and the results to which it leads are all such as appear valid" (ibid.). So, there seems "to be but the slenderest ground for fearing that the use of the axiom of reducibility may lead us into error" (ibid., I, 60).

This, however, is invalid because, to maintain that the axioms are justified if many true consequences can be deduced from them, is incompatible with the assumption of mainstream philosophy of mathematics, that the method of mathematics is the axiomatic method. For, in the axiomatic method, it is the axioms that justify the consequences deduced from them, not the other way round.

Moreover, the axioms must be true, in some sense of 'true'. But, to assert that the axioms are true, it would not be enough that many true consequences can be deduced from them. For, as Kant says, "if even only a single false consequence can be derived from a proposition, then this proposition is false" (Kant 1998, A 791/B 819). Therefore, to assert that the axioms are true, it would be necessary to demonstrate that all the consequences that can be deduced from the axioms are true.

But, demonstrating that all the consequences that can be deduced from the axioms are true, is generally impossible. As Kant says, "inferring the truth of a cognition from the truth of its consequences, would be allowed only if all of the possible consequences are true," but "this procedure is unusable, because to have insight into all possible consequences of any proposition that is assumed exceeds our powers" (Kant 1998, A790/B818).

Evidence for this is provided by the fact that the set of all consequences of the axioms of second-order Peano arithmetic PA^2 is not algorithmically enumerable. For, the axioms of PA^2 are categorical (see Chap. 2). Therefore, a proposition is a consequence of the axioms of PA^2 if and only if it is true in the intended interpretation of PA^2. But, by Tarski's undefinability theorem, the set of all propositions which are true in the intended interpretation of PA^2 is not definable in the latter, hence a fortiori it is not algorithmically enumerable. From this it follows that the set of all consequences of the axioms of PA^2 is not algorithmically enumerable.

This means that there is no algorithm, a fortiori no feasible procedure, for enumerating all consequences of the axioms of PA^2. Therefore, as Kant says, inferring the truth of the axioms of PA^2 from the truth of all their possible consequences exceeds our powers.

10.7 The Demand for Purity of Method

The underlying reason of axiomatic demonstration is what Hilbert calls "the demand for 'purity' of method" (Hilbert 1987, 125). Namely, the demand that the demonstration of a proposition of a given part of mathematics must not use propositions of other parts of mathematics.

The demand for purity of method goes back to Aristotle. Indeed, as we have seen in Chap. 4, Aristotle says that a proposition of one science cannot be demonstrated by another science, for example, it is not possible to demonstrate a geometrical proposition by arithmetic.

The demand for purity of method is reaffirmed by Hilbert. He says that it is necessary "everywhere to guarantee the purity of method," because "our understanding is not satisfied when, in a demonstration of a proposition of arithmetic, we draw on geometry, or in a demonstration of a geometrical truth we draw on function theory" (Hilbert 2004, 236).

The demand for purity of method is intrinsic in the concept of axiomatic demonstration. For, according to the latter, a demonstration of a proposition of a given part of mathematics is a deduction of the proposition from the axioms for that part of mathematics.

This depends on the fact that, as Hintikka says, in the axiomatic method you compress "all the truths about" a given "subject matter into" a "set of axioms" which "are supposed to tell you everything there is to be told about this subject matter" (Hintikka 1998, 1). Then "the rest of your work will consist in merely teasing out the logical consequences of the axioms. It suffices to study the axioms" (ibid.). For, "all substantive assumptions are codified in the axioms, whereas all the theorems are derived from the axioms by purely logical means" (ibid., 2). This is the underlying reason of axiomatic demonstration.

10.8 Failure of the Demand for Purity of Method

The demand for purity of method, however, fails because, by Gödel's first incompleteness theorem, it cannot be generally satisfied.

As Gödel himself points out, "a special case of" the first incompleteness theorem "is that there are arithmetic propositions which can be proved only by analytical methods, and, further, that there are arithmetic propositions which cannot be proved even by analysis but only by methods involving extremely large infinite cardinals and similar things" (Gödel 1986–2002, III, 48). Thus, in principle, such arithmetic propositions cannot be demonstrated without using propositions of other parts of mathematics.

This impossibility in principle has a practical correlate: the demand for purity of method is largely ignored in practice. An example of this is the solution to the problem posed by Fermat's conjecture which, as already pointed out in Chap. 2, uses hypotheses from various parts of mathematics. In fact, as Krajewski says, "it is accepted and common to look for a proof by taking advantage of other branches of mathematics than the one in which the problem is formulated" (Krajewski 2020, 160). Also, "merging methods and concepts of various branches is seen as valuable," and "new branches were created when similarities of constructions in different parts of mathematics were noticed and properly defined" (ibid.).

The reason why the demand for purity of method is largely ignored in practice, is that it can so much increase the complexity of a demonstration as to make a demonstration physically unfeasible.

For example, Boolos (1987) considers an axiomatic theory T based on the following axioms:

(1) $\forall n(f(n, 1) = s(1))$,

(2) $\forall x(f(1, s(x)) = s(s(f(1, x))))$,

(3) $\forall n \forall x(f(s(n), s(x)) = f(n, f(s(n), x)))$,

(4) $D(1)$,

(5) $\forall x(D(x) \rightarrow D(s(x)))$.

In the intended interpretation of T, the variables x and n range over the positive integers, 1 denotes the number one, s denotes the successor function, there is no particular interpretation intended for D, and f denotes the following Ackermann-style function $f(n, x)$ defined on the positive integers: $f(1, x) = 2x$, $f(n, 1) = 2$, and $f(n + 1, x + 1) = f(n, f(n + 1, x))$.

Boolos then considers the proposition:

(6) $D(f(s(s(s(s(1)))), s(s(s(s(1))))))$.

Now, there is a formal demonstration of (6) from the axioms (1)–(5) of T plus the axioms of set theory, in which "every symbol can easily be written down" (Boolos 1987, 1). But this formal demonstration, using the axioms of set theory in addition to the axioms (1)–(5) of T, does not satisfy the demand for purity of method.

On the other hand, there is a formal demonstration of (6) from the axioms (1)–(5) of T alone, which therefore satisfies the demand for purity of method. But "it is well beyond the bounds of physical possibility that any actual or conceivable creature or device could ever write down all the symbols" of this formal demonstration, because "there are far too many symbols in" it "for this to be possible" (ibid.). Indeed, no formal demonstration of (6) from the axioms (1)–(5) of T alone "could possibly be written down in full detail, in this universe" (ibid., 2). For, writing it down in full detail would require at least

$$\underbrace{2^{2^{\cdot^{\cdot^{\cdot^{2}}}}}}_{65536\text{ times}}$$

symbols. This is a gigantic number, much larger than 2^{80}, the estimated number of atoms in the visible universe. Thus a formal demonstration of (6) from the axioms (1)–(5) of T alone would be physically unfeasible.

This shows that, contrary to the demand for purity of method, to be physically feasible, the demonstration of a proposition of a given part of mathematics may essentially require the use of propositions of other parts of mathematics.

10.9 Axiomatic Demonstration and Gödel's Incompleteness Theorems

The view that mathematical demonstration is axiomatic demonstration is refuted by Gödel's incompleteness theorems.

For, an axiomatic demonstration of a proposition of a given part of mathematics can only make use of axioms of that part. But, by Gödel's first incompleteness theorem, an axiomatic demonstration of a proposition of a given part of mathematics may essentially require, in addition to axioms of that part, also axioms of other parts of mathematics.

Moreover, an axiomatic demonstration of a proposition of a given part of mathematics is supposed to establish the truth of that proposition, in some sense of 'true'. But, by Gödel's second incompleteness theorem, an axiomatic demonstration of a proposition of a given part of mathematics cannot establish the truth of that proposition, because the axioms from which the proposition is deduced cannot even be demonstrated to be consistent by absolutely reliable means.

10.10 Other Shortcomings of Axiomatic Demonstration

In addition to being refuted by Gödel's incompleteness theorems, the view that mathematical demonstration is axiomatic demonstration has the following other shortcomings.

(1) It cannot account for how mathematicians discover results. For, according to it, results are discovered starting from the principles and deducing consequences from them. But, as a matter of fact, most mathematical results are not discovered that way.

(2) It cannot account for the fact that a demonstration of a proposition of a given part of mathematics may require the use of propositions of other parts of mathematics. For, an axiomatic demonstration of a proposition of a given part of mathematics is a deduction from the principles of that part only.

(3) It cannot account for the fact that a demonstration yields something new. For, an axiomatic demonstration is a deduction from given principles, and deductive rules are non-ampliative.

(4) It cannot account for the fact that multiple demonstrations are sought for the same proposition. For, an axiomatic demonstration establishes the truth of a proposition, and this is its function. So, once the truth of a proposition has been established, there is no point in establishing it once again by another axiomatic demonstration, let alone by hundreds of them.

(5) It cannot account for the fact that different demonstrations of the same proposition may have different degrees of reliability. For, they are all deductions from principles that are true, so they are all equally reliable.

10.11 Axiomatic Demonstration and Euclid's *Elementa*

One of the reasons why mainstream philosophy of mathematics holds that mathematical demonstration is axiomatic demonstration, is that the latter is the basis of textbooks such as Euclid's *Elementa* or Bourbaki's *Éléments de mathématique*, and mainstream philosophy of mathematics assumes that such textbooks are a paradigm for mathematics.

But this assumption is invalid, if by 'a paradigm for mathematics' it is meant 'a paradigm for mathematical research'. For, axiomatic demonstration does not serve to find solutions to problems, hence to advance knowledge, but only to present, justify, and teach solutions already found. Therefore, textbooks such as Euclid's *Elementa* or Bourbaki's *Éléments de mathématique* are not a paradigm for mathematical research.

That axiomatic demonstration does not serve to find solutions to problems, hence to advance knowledge, but only to present, justify, and teach solutions already found, is already clear from Euclid.

As Proclus tells us, Euclid "put together the *Elementa* collecting many of the theorems of Eudoxus, perfecting many others by Theaetetus, and bringing to incontrovertible demonstration things which had only been somewhat loosely established by his predecessors" (Proclus, *In primum Euclidis Elementorum librum commentarii*, 68.7–10, ed. Friedlein). However, in the *Elementa*, Euclid "did not bring in everything he could have collected, but only what could serve as an introduction" (ibid., 610.8–9). He only included "theorems and problems that are worked out for the instruction of beginners" (ibid., 69.6–7). He omitted matters that "are unsuitable for a selection of elements, because they are difficult and capable of unlimited complexity" (ibid., 74.21–22). Thus, he did not include "the material concerning unordered irrationals that Apollonius has copiously elaborated" (ibid., 74.22–24).

From this, it is clear that the *Elementa* were intended to be an introductory textbook, and axiomatic demonstration was not meant to serve to find solutions to problems, hence to advance knowledge, but only to present, justify, and teach solutions already found. In fact, in his own mathematical research work, Euclid did not use the axiomatic method but only the analytic-synthetic method. In particular, this is the method Euclid used in his lost works *Porismata* and *Loci ad superficiem*, in which he seems to have attempted to start a stereometric study of the conics.

Indeed, Pappus tells us that these Euclid's works were part of "the so-called analytic corpus," a body of works of the golden age of Greek geometry that is "a special resource prepared for the use of those who, after going through the ordinary elements, want to acquire a power in geometry of solving problems set to them" (Pappus, *Collectio*, VII, 634.3–7, ed. Hultsch). The analytic corpus "is the work of three men, Euclid, the author of the *Elementa*, Apollonius of Perga, and Aristaeus the Elder, and proceeds by the method of analysis and synthesis" (ibid., VII, 634.8–11). Thus, Euclid's works included in the analytic corpus did not proceed by the axiomatic method, but by the method of analysis and synthesis.

10.12 Axiomatic Demonstration and Bourbaki's *Éléments*

That axiomatic demonstration does not serve to find solutions to problems, hence to advance knowledge, but only to present, justify, and teach solutions already found, is also clear from Bourbaki.

As Dieudonné tells us, Bourbaki's *Éléments de mathématique* were intended to "encompass the principal ideas of modern mathematics" (Dieudonné 1970, 136). Specifically, Bourbaki wanted "to give an account of modern mathematics which satisfies the idea of establishing a center from which all the rest unfolds" (ibid, 141). This made it "necessary to eliminate many things," such as group theory, because "in group theory" one "cannot say that we have a general method of attack. We have several of them, and one always has the impression that one is working like a craftsman, by accumulating a series of stratagems" (ibid.). Therefore, Bourbaki proposed to set forth only those parts of mathematics that "can be shaped into a coherent theory, logically arranged, easily set forth and easily used" (ibid.).

This shows that the *Éléments de mathématique* were intended to be a textbook, and confirms that axiomatic demonstration is intended to be a didactic argument, not a means to advance mathematical research. Indeed, Dieudonné also tells us that "Bourbaki has no pretension of being a work stimulating to research," he "can allow himself to write on dead theories, things which have been definitely settled and which only need to be gleaned" (ibid., 144). Bourbaki does not aim "to give stimulating speeches on the open problems of the new mathematics, because these open problems are in general much farther than Bourbaki can go. This is living mathematics and Bourbaki does not touch living mathematics" (ibid., 145).

10.13 Axiomatic Demonstration as Paradigm for Mathematical Teaching

It might be thought that, even though textbooks such as Euclid's *Elementa* or Bourbaki's *Éléments de mathématique* are not a paradigm for mathematical research, nevertheless they are a paradigm for mathematical teaching. In fact, since antiquity, several people have claimed that only the axiomatic method is adequate as a method of mathematical teaching.

Thus, Proclus says that, while the analytic method is concerned with problem solving, the axiomatic method is concerned with theorem proving, and "theorems are superior to problems in dignity of nature" (Proclus, *In primum Euclidis Elementorum librum commentarii*, 243.15–16, ed. Friedlein). Admittedly, all of geometry "operates by way of problems; but, so far as it borders on the highest science," geometry "rises by way of theorems from problems to theorems, from secondary to primary things, from things which more regard arts to those endued with a greater power of producing science" (ibid., 243.17–21). As Euclid's *Elementa* show, "starting from the elements we shall be able to know the other parts of" geometry, "while without them we will be unable to understand its complexity, and the learning of the rest will be beyond us" (ibid., 71.9–13).

Leibniz says that only the axiomatic method is adequate as a "method of teaching" (Leibniz 1923–, VI.4, 579). The analytic method is inadequate to that purpose, because "the maximally perfect scientific method does not start," as the analytic method does, "from things which are posterior by nature" (ibid., VI.4, 580). It starts, as the synthetic method does, "from maximally simple and general notions and truths," and "descends to special and composite ones," following "the laws of synthesis or combinatory art, which shows" how "theorems arise by joining definitions with each other and with axioms" (ibid.). Therefore, "it is better to produce a synthesis, since that work is of permanent value, whereas when we begin an analysis on account of particular problems we often do what has been done before" (ibid., VI.4, 544).

Pólya says that there are two kinds of demonstration, namely "properly so-called logical" proof, namely axiomatic demonstration, and "heuristic proof" (Pólya 1941, 450). Namely, analytic demonstration. While logical proof "establishes the truth" of a theorem, heuristic proof "cannot demonstrate the truth" of "a theorem; it can only augment or diminish our confidence in a theorem which is still only a conjecture neither proved nor disproved" (ibid.). Logical proof "is definitive, it establishes irrefutably the truth of the theorem – once for all," on the contrary "heuristic proof is provisional" (ibid.). The heuristic proof "I find today increases my confidence" in the theorem, but such confidence "may be shaken tomorrow by another heuristic proof and definitely shattered the following day by the rigorous refutation of the theorem under consideration" (ibid.). Therefore, the logical proof "appears generally by itself on the pages of mathematical treatises; the heuristic" proof "which in general guided the invention of the logical" proof "is omitted" (ibid.). Then, the logical proof should be taught "in the high school," for, logical proofs "are the hallmark of mathematics" (Pólya 1981, II, 126).

But, already in the sixteenth century, Aconcio disputed the claim that the axiomatic method is adequate as a method of mathematical teaching.

Indeed, Aconcio says that discovery is obtained by "the resolutive method," namely analysis, which proceeds "from the end or effect to the proximate causes" (Aconcio 1558, 87). Thus, it proceeds "from the things that are more known to those that are less known" (ibid., 121). Teaching must proceed in the same order as discovery, because "it would be much worse if the thing were explained in terms of those things which are more removed from our knowledge than the thing itself" (ibid., 46). So the order "must be the same in teaching and in" research, "namely, it must always proceed from the things that are more known to those that are less known" (ibid., 114). It "is not enough that the things that have been discovered in the order of resolution might be more easily taught in the inverse order" (ibid., 120). Namely, proceeding from the causes to the effect, by synthesis. For, the things in question must not be taught in the order that is easier for the teacher, but in the order that makes them "more easily understood" by the student," and this is accomplished if they "are explained in the same order as the one in which they were discovered" (ibid.).

10.14 Limitations of Axiomatic Demonstration for Mathematical Teaching

Even several contemporary mathematicians have disputed the claim that the axiomatic method is adequate as a method for mathematical teaching.

Thus, Kline says that axiomatic demonstration "deceives the student" by leading him "to believe that mathematics is created by geniuses who start with axioms and reason directly from the axioms to the theorems. The student feels humbled and baffled" (Kline 1970, 273–274). Axiomatic demonstration "omits the real activity," it "does dress up this activity but conceals the flesh and blood," therefore it "is often so artificial that is meaningless" (ibid., 272).

Campbell says that "it is a great disservice to students to emphasize how the Greeks used deductive reasoning as a procedure to obtain truth" (Campbell 1976, 342). Thus, "students are misled into believing that" Euclid's presentations of the results of geometry "are the actual reasoning rather than the end product of reasoning," so "Euclid becomes a curse to reasoning, not a blessing" (ibid.).

Rota says that "the student needs to develop an understanding, however partial and imperfect" (Rota 1997, 242). But "what an axiomatic presentation of a piece of mathematics conceals is at least as relevant to the understanding of mathematics as what an axiomatic presentation pretends to state" (ibid., 148).

Then it is invalid to say, as Laugwitz does, that even if "relevant growth of knowledge normally goes with" analytic demonstration "rather than with some rigid framework supplied in the guise of" axiomatic demonstration, axiomatic demonstration "is vital for the consolidation and transmission of knowledge in the form of textbooks" (Laugwitz 2000, 178).

This is invalid because, by omitting the real mathematical process, textbooks based on axiomatic demonstration misrepresent the nature of the subject. To represent the latter authentically, the real mathematical process should be included.

Thus, Lakatos says that "some working mathematicians" claim "that the introduction of" analytic demonstration "would require the rewriting of textbooks, and would make them so long that one could never read them to the end. Papers would become much longer too. The answer to this pedestrian argument is: let us try" (Lakatos 1976, 144). Even if, with analytic demonstration, papers would become much longer, on the positive side, papers "would be much fewer too, as the statement of the problem-situation would too obviously display the pointlessness of quite a few of them" (ibid., 144, footnote 1).

Hersh says that, "if you have made a publishable discovery, the process by which you found out your result probably won't be included in your article" to avoid receiving "a rejection note from the editor: 'We don't have space for all good papers we are receiving, even without irrelevant heuristics'" (Hersh 2014, 82). This "is very unfortunate," because "the final polished deductive proof may conceal a key insight, that could have made the mathematical result more meaningful and accessible," therefore "it's important to advocate publication of the heuristic side of mathematical discovery, along with the deductive proof" (ibid.).

10.15 Deductive Demonstration

Another concept of demonstration is deductive demonstration. A deductive demonstration is a deduction of a proposition from established mathematics – the body of mathematics that is currently accepted as the basis for demonstrations of mathematicians. Like axiomatic demonstration, deductive demonstration serves to present, justify, and teach already acquired propositions.

Hersh and Prawitz maintain that mathematical demonstration is deductive demonstration.

Thus, Hersh says that "axiomatic proof" cannot "establish anything, because axioms are not established, they are simply postulated" (Hersh 2014, 78). So, proof cannot "start from a pre-ordained set of axioms. It starts from relevant pieces of established mathematics" (ibid., 73). Established mathematics "is the body of mathematics that is accepted as the basis for mathematicians' proofs. It includes proved statements 'in the literature'" (ibid., 75). Starting from relevant pieces of established mathematics, proof deduces "new mathematical results," which are then "incorporated into established mathematics" (ibid., 73). Therefore, proof is "deductive proof" (ibid., 82). Deductive proof, namely deduction from established mathematics, is "proof as it is understood by mathematicians" (ibid., 74). For, deductive proof "is the method by which established mathematics becomes established," and only deductive proof "legitimates a result as 'established'" (ibid., 82). So "deductive proof is the standard for acceptance of one's findings into the body of established mathematics" (ibid., 82–83).

Prawitz says that, "after Gödel's incompleteness theorem, one cannot think that mathematics is rightly characterized by saying that it aims at proving theorems in given axiomatic systems" (Prawitz 2014, 74). But "this is not an argument against the general view that in mathematics one solves problems by deductive proofs from premisses held to be true" (ibid., 75). Mathematics, "after its deductive turn in ancient Greek, is essentially a deductive science" (ibid., 78). Indeed, "since the Greeks, most mathematicians understand themselves as not making mathematical assertions unless they believe it has been established deductively" (ibid., 88). Of course, this requires that proofs "ultimately rest on initial premisses for which we have conclusive grounds" (ibid., 87). Now, we have conclusive grounds for the initial premisses only if the initial premisses "express obvious truths," so "ultimately all assertions are consequently seen as resting on deductive proofs whose initial premisses are assertions of obviously true sentences" (ibid., 88). Therefore, "mathematical knowledge is obtained by deductive inferences from truths that are considered to be obvious"(ibid., 84).

10.16 Shortcomings of Deductive Demonstration

The view that mathematical demonstration is deductive demonstration, however, has the following shortcomings.

(1) It does not permit to demonstrate propositions that cannot be deduced from established mathematics.

For, if mathematical demonstration is deductive demonstration, then only deductions from established mathematics are allowed in mathematics, therefore it is impossible to demonstrate propositions that cannot be deduced from established mathematics. This conflicts with the fact that major advances in mathematics essentially require introducing hypotheses that cannot be deduced from established mathematics. Feferman says: "Experience shows that significant progress" in mathematics "depends to an enormous extent on creative ingenuity in the exploitation of accepted principles rather than essentially new principles" (Feferman 2006, 439). But this is invalid. For example, by their creation of the calculus of infinitesimals, Leibniz and Newton were able to demonstrate results that could not be deduced from the then established mathematics. This essentially required introducing hypotheses that could not be deduced from the then established mathematics, and were even fiercely opposed by established mathematicians, such as Rolle (see Chap. 9). As another example, by his creation of set theory, Cantor was able to demonstrate results that could not be deduced from the then established mathematics. This essentially required introducing hypotheses that could not be deduced from the then established mathematics, and were even fiercely opposed by established mathematicians. Thus, Kronecker said that Cantor was "a corrupter of youth" (Schoenflies 1927, 2). Schwarz said that Cantor's work was "a sickly confusion," and hoped that "one could succeed to occupy the unhappy young man with concrete problems, otherwise he will certainly come to a bad end!" (Schwarz 1967, 255–256). Hermite said that "the impression that Cantor's memoirs produce on us is desolating;" even "recognizing that he has opened a new field of research, none of us is tempted to follow him. It is impossible for us to find, among the results that are susceptible of comprehension, just one that possesses a real interest" (Hermite 1984, 209).

(2) It implies that mathematicians can be replaced by computers completely, so there is no need for mathematicians at all.

For, as already mentioned in Chap. 3, there is an algorithm for enumerating all deductions from given premisses. So, if mathematical demonstration is deductive demonstration, then the algorithm will enumerate all deductive demonstrations from established mathematics. Of course, enumerating all deductive demonstrations from established mathematics is a time consuming and inefficient method for discovering demonstrations. Therefore, as Turing says, in practice "we do not really want to make" demonstrations "by hunting through enumerations for them, but by hitting on one and then checking up to see that it is right. However, this method is always" in principle "replaceable by the longer method if one has got a method of checking up" (Turing 2004, 212). Given enough time and space, the algorithm will eventually discover the desired deductive demonstration. This means that, if mathematical demonstration is deductive demonstration, then in principle mathematicians can be replaced by computers completely. Gowers even says that, in fact, "over the next hundred years or so, computers will" be "eventually supplanting" mathematicians "completely" (Gowers 2002, 134). But this conflicts with Gödel's first incompleteness theorem, by which, as already Gödel pointed out, "it is not possible to mechanise mathematical reasoning, i.e., it will never be possible to replace the mathematician by a machine" (Gödel 1986–2002, III, 164).

(3) It implies that all mathematical propositions can ultimately be deduced from obvious truth, such as $1 + 1 = 2$ or $2 + 1 = 3$ or $2 - 1 = 1$.

For, if mathematical demonstration is deductive demonstration, then, tracing all the deductive chains all the way back, we will ultimately arrive at some propositions that cannot have been established by deductive proof themselves. They cannot have been postulated either, because deductive proof is not axiomatic proof. It only remains that they are obvious truth, such as $1 + 1 = 2$ or $2 + 1 = 3$ or $2 - 1 = 1$, which are a product of the ability that newborn babies innately have as a result of natural selection (see Chap. 17). In fact, Hersh says that this ability provides "a bodily (neurological or biological) foundation for arithmetic" (Hersh 2014, 141). Then, all mathematical propositions can ultimately be deduced from obvious truth, such as $1 + 1 = 2$ or $2 + 1 = 3$ or $2 - 1 = 1$. But, since deduction is non-ampliative, this implies that all of mathematics is ultimately contained in such obvious truth. This is absurd.

10.17 Axiomatic or Deductive Demonstration, and Rhetoric

Both supporters of the view that mathematical demonstration is axiomatic demonstration, and supporters of the view that mathematical demonstration is deductive demonstration, say that demonstration has a rhetorical function. For, if demonstration serves to present, justify, and teach already acquired propositions, then the function of demonstration is to convince the audience that the propositions are true.

Thus, Hardy says: "Proofs are what Littlewood and I call 'gas', rhetorical flourishes designed to affect psychology, pictures on the board in the lecture, devices to stimulate the imagination of pupils" (Hardy 1929, 18).

Davis and Hersh say: "Mathematical proof has its rhetorical moments and its rhetorical elements" (Davis and Hersh 2014, 182). Indeed, mathematical proof is aimed "to convince the intended audience," consequently, "our confidence in the correctness of our results is not absolute, nor is it fundamentally different in kind from our confidence in our judgments of the realities of ordinary daily life" (ibid., 188).

Kitcher says: "The presentation of the proof" serves "a rhetorical function" (Kitcher 1991, 7). Indeed, "given two presentations of the same proof, one may be effective in prompting members of a particular audience to undergo" the "process of belief formation, while the other may not" (ibid.).

10.18 Demonstration as Subsidiary

Clearly, if demonstration has a rhetorical function, then it cannot be the driving force of mathematical research, it can play only a subsidiary role.

Some mathematicians take this position to the extremes. They claim that demonstration has no role at all in mathematical research, the latter is exclusively a matter of intuition.

Thus, Enriques says: "Intuition is the aristocratic way of discovery, rigour the plebeian way" (Hodge 1948–1949, 14). Indeed, "we aristocrats do not need proofs. Proofs are for you commoners" (Parikh 2009, 16 footnote).

In fact, his former student Conforto tells us that Enriques did not "feel the need for a logical proof of some property because he 'saw'; and this made him sure of the truth of the proposition in question and completely satisfied him" (Conforto 1947, 116). During a walk with Enriques, when Conforto declared that he "did not see the truth of a statement" that Enriques "considered evident," Enriques "stopped dead," he "twirled his walking stick and, pointing it at a puppy on a windowsill, said to me: Don't you see it? For me, it is as if you told me that you don't see that puppy!" (ibid,).

But to claim that demonstration has no role at all in mathematical research, the latter is exclusively a matter of intuition, is to downgrade demonstration to the status of a mere pedagogical tool.

10.19 Analytic Demonstration

Because of the shortcomings of the concepts of axiomatic demonstration and deductive demonstration, an alternative concept of mathematical demonstration is needed. Such is the concept of analytic demonstration.

An analytic demonstration consists, first, in a non-deductive derivation of a hypothesis from a problem and possibly other data, where the hypothesis is a sufficient condition for the solution of the problem, namely, such that a solution to the problem can be deduced from it, and the hypothesis is plausible, namely such that the arguments for it are stronger than the arguments against it, on the basis of experience. Then, an analytic demonstration consists in a non-deductive derivation of a new hypothesis from the previous hypothesis, considered in turn as a problem, and possibly from other data, where the new hypothesis is a sufficient condition for the solution of the problem posed by the previous hypothesis and is plausible. And so on.

Heuristic philosophy of mathematics holds that mathematical demonstration is analytic demonstration. This follows from the fact that, according to heuristic philosophy of mathematics, the method of mathematics is the analytic method.

10.20 Characters of Analytic Demonstration

The main characters of analytic demonstration can be described as follows.

(1) Analytic demonstration involves both an upward path, from the problem to plausible hypotheses that are sufficient conditions for its solution, and a downward path, from plausible hypotheses to the problem.

(2) Analytic demonstration discovers hypotheses that are sufficient conditions for a solution to the problem and are plausible. So, analytic demonstration is both a means of discovery and a means of justification, a means of discovery because it discovers hypotheses, a means of justification because it shows that the hypotheses are plausible.

(3) Analytic demonstration can make use of hypotheses that cannot be deduced from established mathematics. Indeed, the hypotheses for the solution to the problem are obtained by non-deductive rules, and non-deductive rules are ampliative, so they can produce hypotheses that cannot be deduced from established mathematics.

(4) Analytic demonstration can make essential use of mathematical diagrams. For, in an analytic demonstration, the hypotheses for the solution to the problem are obtained from the problem, and possibly other data including data acquired from mathematical diagrams, by some non-deductive rule. Thus analytic demonstration can be words only, or words and diagrams, or even diagrams only.

(5) Analytic demonstration is not based on intuition, because in the analytic method intuition has no role in solving problems. It has no role in the formulation of hypotheses, because the latter are obtained from the problem, and possibly other data already available, by some non-deductive rule, so not by intuition but by inference. It has no role in the justification of hypotheses, because their plausibility is established by means of the plausibility test procedure described in Chap. 5, so not by intuition but by inference.

(6) Analytic demonstration is as rigorous as axiomatic demonstration. Of course, analytic demonstration deduces a solution to a problem from hypotheses that cannot be said to be true but only plausible. But axiomatic demonstration too deduces a result from principles that, by Gödel's second incompleteness theorem, cannot be demonstrated to be true by absolutely reliable means, and hence can only be plausible.

10.21 Analytic Demonstration and Gödel's Incompleteness Theorems

Gödel's incompleteness theorems not only do not refute, but even confirm the view that mathematical demonstration is analytic demonstration.

For, by Gödel's first incompleteness theorem, the hypotheses to solve a problem of a given part of mathematics may have to be sought in another part of mathematics. This not only does not refute, but even confirms the claim that mathematical demonstration is analytic demonstration because, in an analytic demonstration, the hypotheses need not belong to the field of the problem but may belong to other fields.

Moreover, by Gödel's second incompleteness theorem, the mathematical knowledge resulting from the deduction of propositions from given hypotheses cannot be demonstrated to be firmly grounded. This not only does not refute, but even confirms the claim that mathematical demonstration is analytic demonstration, because an analytic demonstration only establishes the plausibility of the solution to a problem. By Gödel's second incompleteness theorem, plausibility is all we can achieve.

10.22 The Point of Analytic Demonstration

The point of analytic demonstration is apparent from the fact that, as we have seen above, Gödel's incompleteness theorems, on the one hand, refute the view that mathematical demonstration is axiomatic demonstration, and, on the other hand, not only do not refute, but even confirm the view that mathematical demonstration is analytic demonstration.

The point of analytic demonstration is also apparent from the fact that analytic demonstration plays a role in mathematics much more fundamental than axiomatic demonstration. As Friend says, "all proofs can be viewed as analytic" (Friend 2014, 208). And "it is better to view proofs as analytic because under this view we look for much more in a proof than truth and rigour" (ibid., 214). Analytic proofs "lead us to much more fundamental types of exploration than we would have engaged in had we viewed proofs as axiomatic" (ibid., 205). For, while axiomatic demonstration does not serve to find solutions to problems, hence to advance knowledge, but only to present, justify, and teach solutions already found, analytic demonstration serves to find solutions to problems, hence to advance knowledge.

10.23 Analytic Demonstration vs. Axiomatic Demonstration

In addition to not being refuted by Gödel's incompleteness theorems, the view that mathematical demonstration is analytic demonstration does not have the other shortcomings of the view that mathematical demonstration is axiomatic demonstration.

(1) It can account for how mathematicians discover results. An example of this is the demonstration of Fermat's conjecture.

(2) It can account for the fact that a demonstration of a proposition of a given part of mathematics may require the use of propositions of other parts of mathematics. For, an analytic demonstration may use hypotheses from any part of mathematics.

(3) It can account for the fact that a demonstration yields something new. For, an analytic demonstration involves finding hypotheses by means of non-deductive rules, which are ampliative

(4) It can account for the fact that multiple demonstrations are sought for the same proposition. For, each analytic demonstration establishes new connections between the proposition and the existing knowledge, so each analytic demonstration may lead to progress even when other demonstrations are already known.

(5) It can account for the fact that different demonstrations of the same proposition may have different degrees of reliability. For, they may be based on hypotheses with a different degree of plausibility, so they need not be all equally reliable.

10.24 Analytic Demonstration vs. Deductive Demonstration

The view that mathematical demonstration is analytic demonstration does not either have the shortcomings of the view that mathematical demonstration is deductive demonstration.

(1) It permits to demonstrate propositions that cannot be deduced from established mathematics.

For, the hypotheses for demonstrating a proposition are obtained by non-deductive rules, and non-deductive rules are ampliative, so they can produce hypotheses that cannot be deduced from established mathematics. Therefore, the hypotheses can permit to demonstrate propositions that cannot be deduced from established mathematics.

(2) It does not imply that mathematicians can be replaced by computers completely.

For, there is no algorithm capable of producing hypotheses that are sufficient conditions for the solution to a problem and are plausible. Producing them may involve creating an entirely new framework, which goes essentially beyond established mathematics.

(3) It does not imply that all mathematical propositions can ultimately be deduced from obvious truths, such as $1 + 1 = 2$ or $2 + 1 = 3$ or $2 - 1 = 1$.

For, the mathematical problems that are being solved take usually the form of universal propositions, and no universal proposition can be deduced from particular propositions such as $1 + 1 = 2$ or $2 + 1 = 3$ or $2 - 1 = 1$.

10.25 Analytic Demonstration and Revolutions in Mathematics

The view that mathematical demonstration is analytic demonstration does not have a further negative implication of the view that mathematical demonstration is deductive demonstration, namely that it cannot account for the fact that that there are revolutions in mathematics. On the contrary, such view can account for it.

Several mathematicians and historians of mathematics deny that there are revolutions in mathematics. According to them, the development of mathematics is a linear process, in which new knowledge simply is juxtaposed to the existing knowledge.

Thus, Duhem says that, while in physics "experiment constantly causes new disagreements to break out between laws and facts," and "physicists constantly touch up and modify laws in order that they may more faithfully represent facts," mathematics merely "adds new final and indisputable propositions to the final and indisputable propositions it already possessed" (Duhem 1954, 177).

Crowe says that "revolutions never occur in mathematics," for example "Euclid was not deposed by, but reigns along with, the various non-Euclidean geometries" (Crowe 1975, 165).

Dieudonné says that while, in the natural sciences, when a theory becomes inadequate, it is "modified, if necessary by completely overturning its foundations," there is "nothing like this in mathematics" (Dieudonné 1998, 161). In it, the change "which takes place between one generation and another is" not a revolution, but only "a reorganization" (ibid., 162).

But the claim that there are no revolutions in mathematics is unjustified. The development of mathematics is not merely a linear process, in which new knowledge simply is juxtaposed to the existing knowledge. As Gillies says, there are revolutions in it, in the sense of innovations which "change mathematics (or a branch of mathematics) in a profound and far-reaching way" (Gillies 1992, 6).

This raises the question: Is there an objective criterion to decide whether a development in mathematics is a revolution, in the sense of an innovation which changes mathematics (or a branch of mathematics) in a profound and far-reaching way?

An answer to this question can be given on the basis of what has been said above about major advances in mathematics, namely that they essentially require introducing hypotheses that cannot be deduced from established mathematics. Then, a development in mathematics is a revolution if it essentially requires introducing hypotheses that cannot be deduced from established mathematics.

In terms of this, two kinds of mathematics can be distinguished: normal mathematics and revolutionary mathematics. Normal mathematics is mathematics that does not require introducing hypotheses that cannot be deduced from established mathematics. Revolutionary mathematics is mathematics that essentially requires introducing hypotheses that cannot be deduced from established mathematics. Normal mathematics is very extensive, being the kind of research most mathematicians do all the time. Revolutionary mathematics is much more limited, involving an essential change in the fundamental hypotheses, which is comparatively rare.

The view that mathematical demonstration is deductive demonstration cannot account for the fact that there are revolutions in mathematics, because a deductive demonstration consists in a deduction of a proposition from premisses belonging to established mathematics, and deduction is non-ampliative. Therefore, this view can account only for normal mathematics, not for revolutionary mathematics. Indeed, if mathematical demonstration were deductive demonstration, there would be no revolutionary mathematics at all, all of mathematics would be normal.

On the contrary, the view that mathematical demonstration is analytic demonstration can account for the fact that there are revolutions in mathematics, because an analytic demonstration can make use of hypotheses that cannot be deduced from established mathematics. Therefore, this view can account for both normal mathematics and revolutionary mathematics.

10.26 Objection to Analytic Demonstration as Means of Discovery

Against the view that mathematical demonstration is analytic demonstration, Netz objects that, "just as an analysis ends up with the added construction of the solution, so a synthesis ends up with the solution" (Netz 2000, 143). Analysis is not "a method of discovery in any privileged sense" (ibid.). With respect to synthesis, analysis creates "the illusion that the solution is necessary and emerges naturally out of the problem," but this is only an illusion because, "given a synthesis," we always "are capable of producing, post factum, an analysis which makes the synthesis appear necessary" (ibid., 156).

This objection, however, is invalid. Indeed, it is invalid to say that, just as an analysis ends up with the solution to the problem, so a synthesis ends up with the solution to the problem. For, a synthesis starts from principles that must be discovered, but cannot be discovered by synthesis because this would involve an infinite regress, they must be discovered by analysis.

Moreover, there is no algorithm for discovering hypotheses, and hence for obtaining a solution by analysis, while there is an algorithm for enumerating all deductions from given principles, and hence for obtaining a solution by synthesis.

Also, it is invalid to say that, given a synthesis, we can always produce, post factum, an analysis that makes the synthesis appear necessary. As Newton points out, when propositions discovered by analysis are demonstrated by synthesis, it is difficult to see the analysis by which those propositions were found out, because synthesis hides analysis (see Chap. 3).

10.27 Objection to Analytic Demonstration as Means of Justification

To the view that mathematical demonstration is analytic demonstration, Hersh objects that an analytic demonstration, "relying on examples, analogy and induction" is "not demonstrative" (Hersh 2014, 58). For, "analogy and induction" only "grant plausibility, not 'established' status" (ibid., 81). Analytic demonstration "does not establish the result, even if" the result "has been verified in millions of individual cases" (ibid., 82). Only deductive demonstration "connects some proposed 'result' to the body of established mathematics" (ibid., 81). The result of an analytic demonstration is not a proved theorem but an unproved conjecture, and "the distinction between a proved theorem and an unproved conjecture is the central, characteristic feature of mathematics, as practiced from Euclid to this day" (ibid., 83).

This objection, however, is invalid because the result of an analytic demonstration is not an unproved conjecture. The validity of the result does not depend on analogy or induction, the latter are only means to find hypotheses. In an analytic demonstration, while the hypotheses are arrived at by non-deductive rules, they are not justified by such rules, but only by the fact that they are plausible. Establishing that they are plausible requires an independent argument, such as the plausibility test procedure (see Chap. 5). Therefore, the hypotheses used in an analytic demonstration, and the propositions deduced from them, are not unproved conjectures but plausible propositions. And, by Gödel's second incompleteness theorem, plausibility is all we can achieve.

10.28 Analytic Demonstration and Subformula Property

Sometimes the concept of analytic demonstration is confused with Gentzen's concept of proof with the subformula property, namely proof such that all formulas occurring in it are subformulas of formulas that occur in the conclusion.

Thus, Poggiolesi says that the concept of proof with the subformula property "can be thought of as a formalisation of" the concept of "analytic proof" of Greek geometers "in the framework of the sequent calculus" (Poggiolesi 2012, 444). For, an analytic proof of Greek geometers is a proof that is "self-contained: every element which occurs in the proof will also occur in the conclusion" (ibid., 445). And a proof with the subformula property is such that "every formula which occurs in it is a subformula of the formulas which occur in the conclusion" (ibid.). Since, by Gentzen's cut elimination theorem, every proof in the sequent calculus can be transformed into a proof with the subformula property, the cut elimination theorem can be said to "support the analytic method" (ibid., 446).

But this is invalid. An analytic demonstration is not a demonstration such that every element which occurs in it will also occur in the conclusion. For, in an analytic demonstration, the hypotheses to solve a problem need not belong to the same part of mathematics as the problem, they may belong to other parts of mathematics, so elements may occur in the demonstration that do not occur in the conclusion. Therefore, the concept of analytic demonstration is different from the concept of proof without cuts.

Poggiolesi also says that "a good proof system" is one which satisfies "the analyticity property" (Poggiolesi 2012, 444). Now, "a methodological requirement strictly connected with" the analyticity property consists "in the so-called 'purity of method'," namely the demand that "mathematicians should not use in their proofs concepts belonging to a theoretical domain different from that employed initially" (Poggiolesi 2011, 13). Therefore, a good proof system is one which satisfies the demand for purity of method.

But this is invalid, because a proof system which satisfies the demand for purity of method can be very inefficient. As it appears from Boolos's example of formal demonstration in the axiomatic theory *T* considered above, the demand for purity of method can so much increase the complexity of a demonstration as to make it physically unfeasible. Moreover, as already pointed out above, the demand for purity of method cannot be satisfied by Gödel's first incompleteness theorem. Therefore, a proof system which satisfies the demand for purity of method cannot be a good proof system.

10.29 Analytic Demonstration and Depth of Demonstrations

In terms of the concept of analytic demonstration, it is possible to deal with the question of depth of demonstrations.

A demonstration can be said to be deep if it connects different parts of mathematics or opens the way to new parts of mathematics. Then axiomatic demonstration cannot be said to be deep, because it remains constrained within the confines of a closed system, so it cannot connect different parts or opens the way to new parts of mathematics. Conversely, an analytic demonstration can be said to be deep, because it can reveal unexpected connections between different parts of mathematics which may suggest new perspectives and new problems, and even opens the way to new parts of mathematics.

Byers says that the depth of a solution to a problem "is a measure of the creativity that accompanies its birth" (Byers 2007, 374). And a solution to a problem is creative if it "is not mechanical – if it does not involve juggling a number of predetermined elements according to predetermined rules. It involves the emergence of a novel way of looking at the original situation" (ibid., 388).

Then, axiomatic demonstration is non-creative, because a solution to a problem by an axiomatic demonstration is obtained from predetermined principles by predetermined rules, the rules, being deductive, introduce elements which are predetermined because they are contained in the principles, and moreover, there is an algorithm for enumerating all deductions from the principles. On the contrary, analytic demonstration is creative, because a solution to a problem by an analytic demonstration is obtained from the problem by non-predetermined rules, the rules, being non-deductive, introduce elements which are not predetermined because they are not contained in the problem, and moreover, there is no algorithm for enumerating all the hypotheses for solving a problem.

10.30 Analytic Demonstration and Mathematical Style

In terms of the concept of analytic demonstration, it is also possible to deal with the question of mathematical style.

Mathematical style is closely tied to mathematical method. As Chasles says, "style, in fact, is so intimately welded to the spirit of the methods, that it must advance with them; just as, if style has anticipated the methods, it must also powerfully influence them and the general progress of science" (Chasles 1875, 207).

That mathematical style is closely tied to mathematical method is clear from Lakatos's distinction between two mathematical styles: the deductivist style and the heuristic style.

The deductivist style "starts with a painstakingly stated list of axioms" and "definitions," the "list of axioms and definitions is followed by carefully worded theorems," and "the theorem is followed by the proof" (Lakatos 1976, 142).

Supporters of the deductivist style can be distinguished into two groups. The members of the first group hold that "deduction is the heuristic pattern in mathematics," so "the logic of discovery is deduction" (ibid., 143).

The members of the second group hold that only the "presentation of mathematical discoveries" proceeds deductively, "mathematical discovery does not proceed deductively," it is "a completely non-rational affair" (ibid.).

A member of the first group is Musgrave who, as we have seen in Chap. 3, claims that there is a logic of discovery, and it is deductive logic. But, as argued there, this claim is invalid.

A member of the second group is Dieudonné who, as we have seen in Chap. 2, claims that the presentation of mathematical discoveries proceeds deductively, mathematical discovery itself does not proceed deductively, it is a completely non-rational affair, being based on intuition. But these claims are invalid. The claim that the presentation of mathematical discoveries proceeds deductively conflicts with the strong incompleteness theorem for second-order logic, by which there is no consistent formal system for second-order logic capable of deducing all second-order logical consequences of any given set of propositions. So, there can be mathematical discoveries whose presentation cannot proceed deductively. The claim that mathematical discovery is a completely non-rational affair, being based on intuition, conflicts with the fact that unrestrained intuition can generate a multitude of different conjectures, and choosing the right one involves a rational process.

Lakatos says that the deductivist style presents definitions, theorems, and proofs "out of the blue, in an artificial and authoritarian way. It hides" the factors "which led to their discovery" (ibid., 144). To it, Lakatos opposes the heuristic style, which "highlights these factors. It emphasises the problem situation; it emphasises the 'logic' which gave birth to the new" definition, theorem, or proof, thus showing "the advantage of introducing heuristic elements into mathematical style" (ibid.).

Lakatos's formulation of the heuristic style, however, is inadequate. For, Lakatos analyzes the factors that lead to discovery in terms of his method of proof and refutations, whose limitations have been highlighted in Chap. 3. An adequate formulation of the heuristic style is given only by analytic demonstration, because analytic demonstration tells you how to arrive at a hypothesis and how to show that it is plausible.

References

Aconcio, Iacopo. 1558. *De methodo, hoc est, de recta investigandarum tradend-arumque scientiarum ratione*. Basel: Perna.

Boolos, George. 1987. A curious inference. *Journal of Philosophical Logic* 16: 1–12.

Bourbaki, Nicholas. 1949. Foundations of mathematics for the working mathematician. *The Journal of Symbolic Logic* 14: 1–8.

———. 2004. *Elements of mathematics: Theory of sets*. Berlin: Springer.

Byers, William. 2007. *How mathematicians think: Using ambiguity, contradiction, and paradox to create mathematics*. Princeton: Princeton University Press.

Campbell, Douglas. 1976. *The whole craft of number*. Boston: Prindle, Weber & Schmidt.

Chasles, Michel. 1875. *Aperçu historique sur l'origine et le développement des méthodes en géométrie*. Paris: Gauthier-Villars.

Conforto, Fabio. 1947. Intuizione visiva degli enti algebrici. *Periodico di Matematiche* 25: 115–116.

Crowe, Michael J. 1975. Ten 'laws' concerning patterns of change in the history of mathematics. *Historia Mathematica* 2: 161–166.

Cupillari, Antonella. 2005. *The nuts and bolts of proofs*. Burlington: Elsevier Academic Press.

Davis, Philip J., and Reuben Hersh. 2014. Rhetoric and mathematics. In Reuben Hersh, *Experiencing mathematics: What do we do, when we do mathematics?* 177–189. Providence: American Mathematical Society.

Detlefsen, M. 2008. Proof: Its nature and significance. In *Proof and other dilemmas: Mathematics and philosophy*, ed. Bonnie Gold and Roger A. Simon, 3–32. Washington: The Mathematical Association of America.

Dieudonné, Jean. 1970. The work of Nicholas Bourbaki. *The American Mathematical Monthly* 77: 134–145.

———. 1998. *Mathematics: The music of reason*. Berlin: Springer.

Duhem, Pierre. 1954. *The aim and structure of physical theory*. Princeton: Princeton University Press.

Feferman, Solomon. 2006. The impact of the incompleteness theorems on mathematics. *Notices of the American Mathematical Society* 53: 434–439.

Friend, Michèle. 2014. *Pluralism in mathematics: A new position in philosophy of mathematics*. Dordrecht: Springer.

Gersting, Judith K. 2014. *Mathematical structures for computer science*. New York: Freeman.

Gillies, Donald. 1992. Introduction. In *Revolutions in mathematics*, ed. Donald Gillies, 1–14. Oxford: Oxford University Press.

Gödel, Kurt. 1986–2002. *Collected works*. Oxford: Oxford University Press.

Gowers, Timothy. 2002. *Mathematics: A very short introduction*. Oxford: Oxford University Press.

Hales, Thomas C. 2008. Formal proof. *Notices of the American Mathematical Society* 55: 1370–1380.

———. 2012. *Dense sphere packings: A blueprint for formal proofs*. Cambridge: Cambridge University Press.
Hanna, Gila. 1983. *Rigorous proof in mathematics education*. Toronto: OISE Press.
Hardy, Godfrey Harold. 1929. Mathematical proof. *Mind* 38: 1–25.
Hermite, Charles. 1984. Lettres à Gösta Mittag-Leffler (1874–1883). *Cahiers du séminaire d'histoire des mathématiques* 5: 49–285.
Hersh, Reuben. 2014. *Experiencing mathematics: What do we do, when we do mathematics?* Providence: American Mathematical Society.
Hilbert, David. 1987. *Grundlagen der Geometrie*. Stuttgart: Teubner.
———. 2004. Grundlagen der Euklidischen Geometrie. In *David Hilbert's lectures on the foundations of geometry 1891–1902*, ed. Michael Hallett and Ulrich Majer, 221–286. Berlin: Springer.
Hintikka, Jaakko. 1998. *The principles of mathematics revisited*. Cambridge: Cambridge University Press.
Hodge, William. 1948–1949. Federigo Enriques obituary. In *Royal Society of Edinburgh Year Book 1948–1949*, 13–16.
Kant, Immanuel. 1998. *Critique of pure reason*. Cambridge: Cambridge University Press.
Kitcher, Philip. 1991. Persuasion. In *Persuading science: The art of scientific rhetoric*, ed. Marcello Pera and William Shea, 3–27. Sagamore Beach: Science History Publications.
Kline, Morris. 1970. Logic versus pedagogy. *The American Mathematical Monthly* 77: 264–282.
Krajewski, Stanislaw. 2020. Anti-foundationalist philosophy of mathematics and mathematical proofs. *Studia Humana* 9: 154–164.
Lakatos, Imre. 1976. *Proofs and refutations: The logic of mathematical discovery*. Cambridge: Cambridge University Press.
Laugwitz, Detlef. 2000. Controversies about numbers and functions. In *The growth of mathematical knowledge*, ed. Emily Grosholz and Herbert Breger, 177–198. Dordrecht: Springer.
Leibniz, Gottfried Wilhelm. 1923–. *Sämtliche Schriften und Briefe*. Berlin: Academie Verlag.
Macintyre, Angus. 2005. The mathematical significance of proof theory. *Philosophical Transactions of the Royal Society* 363: 2419–2435.
Mac Lane, Saunders. 1986. *Mathematics: Form and function*. Berlin: Springer.
Nathanson, Melvyn B. 2008. Desperately seeking mathematical truth. *Notices of the American Mathematical Society* 55: 773.
Netz, Reviel. 2000. Why did Greek mathematicians publish their analyses? In *Ancient & medieval traditions in the exact sciences*, ed. Patrick Suppes, Julis M. Moravcsik, and Henry Mendell, 139–157. Stanford: CSLI Publications.
Nicholson, Neil R. 2019. *A transition to proof: An introduction to advanced mathematics*. Boca Raton: CRC Press.
Parikh, Carol. 2009. *The unreal life of Oscar Zariski*. New York: Springer.
Poggiolesi, Francesca. 2011. *Gentzen calculi for modal propositional logic*. Dordrecht: Springer.
———. 2012. On the importance of being analytic: The paradigmatic case of the logic of proofs. *Logique et Analyse* 219: 443–461.
Pólya, George. 1941. Heuristic reasoning and the theory of probability. *The American Mathematical Monthly* 48: 450–465.
———. 1981. *Mathematical discovery: On understanding, learning, and teaching problem solving*. New York: Wiley.
Prawitz, Dag. 2014. The status of mathematical knowledge. In *From a heuristic point of view*, ed. Cesare Cozzo and Emiliano Ippoliti, 73–90. Newcastle upon Tyne: Cambridge Scholars Publishing.
Rota, Gian-Carlo. 1997. *Indiscrete thoughts*. Cham: Birkhäuser.
Schoenflies, Arthur. 1927. Die Krisis in Cantor's mathematischem Schaffen. *Acta Mathematica* 50: 1–23.

Schwarz, Hermann Amandus. 1967. Extract from a letter to Weierstrass, 17 October 1887. In Herbert Meschkowski, *Probleme des Unendlichen: Werk und Leben Georg Cantors*, 255–256. Berlin: Springer.
Turing, Alan Mathison. 2004. Systems of logic based on ordinals. In *The essential Turing*, ed. Brian Jack Copeland, 146–204. Oxford: Oxford University Press.
Whitehead, Alfred North, and Bertrand Russell. 1925–1927. *Principia Mathematica*. Cambridge: Cambridge University Press.
Zermelo, Ernest. 2010–2013. *Collected works – Gesammelte Werke*. Berlin: Springer.

Chapter 11
Definitions

Abstract According to heuristic philosophy of mathematics, one of the tasks of the philosophy of mathematics is to give an answer to the question: What is the nature of mathematical definitions? In the past century, the prevailing answer to this question has been the stipulative view of mathematical definition. The latter has been so absolutely prevailing and considered not problematic that the nature of mathematical definitions has not been much discussed. Nevertheless, the stipulative view is inadequate. The chapter highlights its shortcomings and puts forward an alternative view, the heuristic view of mathematical definition.

Keywords Stipulative view of mathematical definition · Heuristic view of mathematical definition · Heuristic differences between definitions · Role of definitions in proofs · Justification of definitions

11.1 The Stipulative View of Mathematical Definition

As we have seen in Chap. 3, according to heuristic philosophy of mathematics, one of the tasks of the philosophy of mathematics is to give an answer to the question: What is the nature of mathematical definitions?

In the past century, the prevailing answer to this question has been the stipulative view of mathematical definition, according to which:

(1) A definition merely stipulates the meaning of a term;
(2) A definition is an abbreviation;
(3) A definition is always correct;
(4) A definition can always be eliminated;
(5) A definition says nothing about the existence of the thing defined.

The stipulative view has been so absolutely prevailing and considered not problematic that the nature of mathematical definitions has not been much discussed. The main exception is Lakatos 1976. Recent works on the subject are Brown 2008, Chap. 7; Tappenden 2008, 2011; Werndl 2009.

C. Cellucci, *The Making of Mathematics*, Synthese Library 448,
https://doi.org/10.1007/978-3-030-89731-4_11

The nature of mathematical definitions has not been much discussed because, if mathematical definition is what the stipulative view says it is, then mathematical definition is a marginal topic in the philosophy of mathematics.

However, despite its being absolutely prevailing and considered not problematic, the stipulative view is inadequate. Mathematical definition is not a marginal topic in the philosophy of mathematics. As Harris says, "the progress" in a mathematical theory "is reflected as much in its definitions as in its theorems" (Harris 1992, 99). Rota even says that "paradoxically, we could state that the most valuable part of mathematics consists of the definitions" (Rota 1996, 290).

The chapter highlights the shortcomings of the stipulative view and offers an alternative view, the heuristic view of mathematical definition.

Already Lakatos 1976 pointed out some of the shortcomings of the stipulative view, and formulated a heuristic view. Lakatos's heuristic view, however, is inadequate because it is part of his method of proof and refutations, whose limitations have been already pointed out in Chap. 3. The chapter puts forward a heuristic view that hopefully is not subject to them.

11.2 Pascal's Formulation of the Stipulative View

A fully articulated formulation of the stipulative view of mathematical definitions was first given by Pascal. His formulation is paradigmatic.

According to Pascal, the method of mathematics is the axiomatic method, and the principles are about certain terms that are left undefined.

Indeed, Pascal says that, in each part of mathematics, "in pushing our researches further and further, we arrive necessarily at certain primitive terms" so clear that they "can no longer be defined, and at certain principles so clear that we can find no others that can serve to demonstrate them" (Pascal 1904–1914, IX, 246). These primitive terms and principles are the basis of the method of mathematics, which consists "in not defining things clear and understood by all mankind," namely the primitive terms, "and in defining all the rest; and in not demonstrating all the things known to mankind," namely the principles, "and in demonstrating all the rest" (ibid., IX, 247).

The terms other than the primitive ones are defined by means of definitions in accordance with the stipulative view, which Pascal characterizes as follows.

(1) A definition merely stipulates the meaning of a term. In mathematics "only those definitions are recognized which logicians call 'nominal', namely only impositions of names upon things which have been clearly designated by terms perfectly known" (ibid., IX, 242–243).

(2) A definition is an abbreviation. Mathematicians "only impose names upon things in order to abbreviate discourse" (ibid., IX, 244). The only "usefulness and use" of definitions "is to clarify and abbreviate discourse, expressing, by the name we impose, what could not be otherwise expressed but by several terms" (ibid., IX, 243).

(3) A definition is always correct. Indeed, "definitions are very free, and are never subject to be contradicted; for, nothing is more permissible than to give whatever name we please to a thing which we have clearly designated" (ibid., IX, 243–244).

(4) A definition can always be eliminated. For, one can always "substitute mentally the definition in place of the thing defined" (ibid., IX, 244).

(5) A definition says nothing about the existence of the thing defined. For, "one must always define things before investigating whether they are possible or not" (ibid., II, 184). So, the definition of a thing comes before the investigation of whether the thing is possible or not. Hence, "it is evident that there is no necessary connection between the definition of a thing and the assurance of its existence; and that we can just as well define an impossible thing as a true thing. Thus we can call rectilinear and rectangular a triangle we should imagine as having two right angles, and show afterwards that such a triangle is impossible" (ibid., II, 185).

11.3 Mathematical Logic and the Stipulative View

In the past century, most formulations of the view of mathematical definitions have been mere linguistic variants of Pascal's formulation. This is clear, in particular, from the formulations of the founding fathers of mathematical logic: Frege, Peano, Russell, and Hilbert. For example, Frege characterizes mathematical definitions as follows.

(1) A definition merely stipulates the meaning of a term. Indeed, "a definition is an arbitrary stipulation by which a new sign" is "introduced to take the place of a complex expression whose sense we know" (Frege 1979, 211).

(2) A definition is an abbreviation. Definitions "merely introduce abbreviative notations (names)" (Frege 2013, I, vi). Their use "is to bring about an extrinsic simplification by stipulating an abbreviation" (Frege 1967, 55). This "and this alone is the use of definitions in mathematics" (Frege 1984, 274).

(3) A definition is always correct. For, "a definition does not assert anything," so a definition is not "something that is in need of proof or of some other confirmation of its truth" (Frege 1980, 36).

(4) A definition can always be eliminated. A definition is "something wholly inessential and dispensable" because, "if the definiens occurs in a sentence and we replace it by the definiendum, this does not affect the thought at all" (Frege 1979, 208). Therefore, "nothing follows from" a definition "that could not also be inferred without it" (Frege 1967, 55).

(5) A definition says nothing about the existence of the thing defined. The requirement on the definition of a concept, that there be objects falling under the concept, "encounters great difficulties," because "the only apparent way to show" that "a concept has this property, is to cite an object falling under the concept. But to do that, you already need the concept" (Frege 1979, 179). Therefore, "the admissibility of a concept is entirely independent of the question whether objects fall under it, and if so which, or in other words, whether there be objects, and if so which, of which it can be truly asserted" (ibid.).

Following the lead of Frege, Peano, Russell, and Hilbert, the stipulative view has become the standard view of mathematical definitions in mathematical logic textbooks (see, for example, Tarski 1994, 30–32; Mates 1972, 197–203; Suppes 1957, Chap. 8).

11.4 The Rise and Establishment of the Stipulative View

Even though a fully articulated formulation of the stipulative view was first given by Pascal, the core of this view goes back to antiquity. Indeed, Aristotle says: "One type of definition will be the statement of what a name, or some other name-like expression, means, for example, what triangle means" (Aristotle, *Analytica Posteriora*, B 10, 93 b 30–32).

However, Aristotle considers this view of definition only to criticize it. Indeed, he says that, if a definition were merely a statement of what a name means, "then a definition would be an expression which means the same as a name. But this is absurd. For, first, there would be definitions of non-substances and of things which do not exist, because it is possible to mean also things which do not exist" (Aristotle, *Analytica Posteriora*, B 7, 92 b 27–30). Second, "all expressions would be definitions. For, it is possible to assign a name to any expression whatever, so all our discourses would be definitions," for example, the entire poem "*Iliad* would be a definition" (ibid., B 7, 92 b 30–32). Third, "no demonstration could demonstrate that this name makes known this thing; therefore, definitions could not make known this thing either" (ibid., B 7, 92 b 32–34).

To this view of definition, Aristotle opposes his own view of definition as statement of the essence. As we have seen in Chap. 4, for Aristotle, the object of science is to know the essence of things. Then, in particular, "a definition is said to be a statement of the essence of a thing" (Aristotle, *Analytica Posteriora*, B 10, 93 b 29).

The stipulative view of definition is revived in the early modern period as a result of Galileo's change of the object of science with respect to Aristotle. As we have seen in Chap. 4, for Galileo, the object of science is no longer to know the essence of things, but only to know certain phenomenal properties of things, mathematical in kind. Then, a definition is not a statement of the essence of a thing, "the definitions of mathematicians" are only "an imposition of names, or we might say abbreviations of speech" (Galilei 1968, VIII, 74). They are arbitrary stipulations and, "being arbitrary, can never be bad" (ibid., IV, 700).

Pascal's fully articulated formulation of the stipulative view is part of the process of revival of the stipulative view of definition in the early modern period as a result of Galileo's change of the object of science with respect to Aristotle. Indeed, like Galileo, Pascal says that the definitions of mathematicians are only impositions of names upon things in order to abbreviate discourse, they are very free, so they are arbitrary stipulations, and are never subjected to be contradicted, so they can never be bad.

As already said, in the past century the stipulative view of definition has been prevailing. A main factor in this has been the passage from the material axiomatic method to the formal axiomatic method.

Thus, Hilbert says that a "theory has nothing to do with the real objects and with the intuitive content of knowledge; it is a pure thought construct, of which one cannot say that is true or false" (Hilbert 2013, 435). Therefore, "the axioms can be taken quite arbitrarily" (Hilbert 2004, 563). They are only subject to the condition that they must "never lead to contradictions" (Hilbert 1996, 1093). And the definitions are arbitrary stipulations, because a definition is "a mere explanation of signs" and is only "the introduction of an abbreviating symbol for a complex expression" (Hilbert and Bernays 1968–1970, I, 292).

11.5 Shortcomings of the Stipulative View

Despite its prevailing in the past century, the stipulative view of mathematical definition has serious shortcomings.

(1) By claiming that a definition merely stipulates the meaning of a term, the stipulative view cannot account for the fact that finding a suitable definition can make the difference in discovering a solution to a problem.

For example, let us consider the definition of a sphere. There are two historically important kinds of definition of a sphere, one to be found in Plato, Aristotle, and Theodosios, and the other to be found in Euclid.

Plato, Aristotle, and Theodosios define a sphere as a solid figure with every point on its surface equidistant from its centre.

Indeed, Plato says that a sphere is a solid figure "which is equally distant in all directions from the centre to the extremities" (Plato, *Timaeus*, 33 b 4–5).

Aristotle says that a sphere is a solid figure such that "all its extremities are equally distant from the centre" (Aristotle, *De coelo*, II 14, 297 a 24).

Theodosios says that a sphere is "a solid figure contained by one surface, such that all the straight lines falling upon it from one point among those lying inside the figure are equal to each other" (Theodosios, *Sphaerica*, I, Def. 1, ed. Heiberg).

On the contrary, Euclid defines a sphere as the solid figure generated when a semicircle is being rotated about its diameter.

Indeed, Euclid says that a sphere is "the figure comprehended when, the diameter of a semicircle remaining fixed, the semicircle is carried around and restored again to the same position from which it began to be moved" (Euclid, *Elementa*, XI, Definition 14).

On the other hand, however, Euclid defines a circle in a way similar to the way in which Plato, Aristotle, and Theodosios define a sphere.

Indeed, Euclid says that a circle is "a plane figure contained by one line, such that all the straight lines falling upon it from one point among those lying inside the figure are equal to each other" (Euclid, *Elementa*, I, Definition 15).

Why does Euclid define a sphere in a way different from the way in which he defines a circle? This is because Euclid's definition of a sphere plays an important heuristic role in Euclid's solution to the problem about the five Platonic figures, it guides him to find a solution to the problem (see below).

This does not mean that Euclid could not have found a solution to the problem if he had defined a sphere in a way similar to the way in which he defines a circle, it only means that, then, finding a solution would have been more difficult. Therefore, Euclid's definition of a sphere made the difference in discovering a solution.

This cannot be accounted for if, as the stipulative view claims, a definition merely stipulates the meaning of a term.

(2) By claiming that a definition is an abbreviation, the stipulative view cannot account for the fact that mathematicians often use concepts for a long time, even centuries, before they can find a suitable definition for them.

For example, as Grabiner says, "Fermat implicitly used" the derivative; "Newton and Leibniz discovered it; Taylor, Euler, Maclaurin developed it; Lagrange named and characterized it; and only at the end of this long period of development did Cauchy and Weierstrass define it" (Grabiner 2010, 159–160). So "the historical order of development of the derivative is the reverse of the usual order of textbook exposition," in which "one starts with a definition, then explores some results;" indeed a "definition is often the end, rather than the beginning, of a subject" (ibid., 160).

This cannot be accounted for if, as the stipulative view claims, a definition is only an abbreviation. For, an abbreviation is only "the effect of our will to want to represent a group of signs by a simpler expression" (Peano 1957–1959, II, 173). So, it cannot require centuries to find it.

(3) By claiming that a definition is always correct, the stipulative view cannot account for the fact that mathematicians often give definitions which afterwards turn out to be incorrect.

For example, Jordan defines 'curve' as "the succession of points represented by the equations $x = f(t)$, $y = \varphi(t)$, where f, φ are functions of an independent variable t. If these functions are continuous, the curve will be called 'continuous'" (Jordan 1887, 587). Jordan's definition is based on the idea that a curve is the place of successive positions of a moving point, in particular, a continuous curve is the place of successive positions of a continuously moving point. The successive positions of the point in the plane will be completely specified by the two coordinates x and y as functions of time t, so a curve will be represented by a system of two equations $x = f(t)$, $y = \varphi(t)$, where f, φ are functions – a continuous curve if those functions are continuous.

Jordan's definition was widely accepted. However, in 1890, Peano defined something that is a continuous curve according to Jordan's definition, but "goes through every point of a square" (Peano 1973, 144).

Peano's definition of the curve is an arithmetical one, but afterwards a geometrical representation has been given. The latter can be described as follows (see Moore 1900, 77). Divide a square into nine equal subsquares, and draw a continuous curve through the diagonals of the subsquares in the manner shown in the following figure, where the corners of each right-angular twist in the curve have been rounded for illustrative purposes.

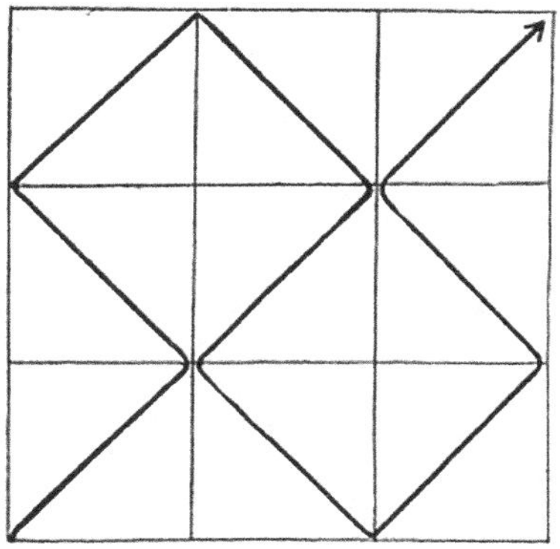

Then divide each of the nine subsquares into nine equal sub-subsquares, and draw a continuous curve through the diagonals of the sub-subsquares in the manner shown in the following figure, where once again the corners of each right-angular twist in the curve have been rounded for illustrative purposes.

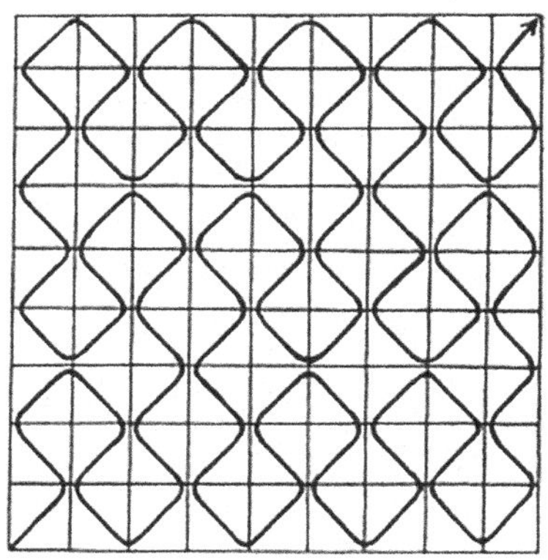

If this process is continued indefinitely, the curve will run through all the points of the square at least once.

Then, Jordan's definition of curve or continuous curve is incorrect. For, a curve is a one-dimensional object, a square is a two-dimensional object, and Peano's curve, while being a curve, is also a square.

This cannot be accounted for if, as the stipulative view claims, a definition is always correct. For, then a definition is not something that is in need of demonstration or of some other confirmation of its truth, since "it is not true or false, being the expression of a volition, not of a proposition" (Whitehead and Russell 1925–1927, I, 11).

(4) By claiming that a definition can always be eliminated, the stipulative view cannot account for the fact that it is not generally the case that a definition can be eliminated.

For example, in standard textbooks on number theory, division is defined as follows: "Let $a \neq 0$; let b be arbitrary. Then b is said to be divisible by a if there is a number q such that $b = qa$. This q, namely $q = \frac{b}{a}$, is then uniquely determined" (Landau 1958, 11). So, the definition of division is:

(A) If $a \neq 0$, then $\frac{b}{a} = q$ is an abbreviation for $b = qa$.

But, if (A) is the definition of division, then it cannot be eliminated. For example, it cannot be eliminated from contexts such as $\frac{1}{0} \neq 2$. Alternatively, we can take $\frac{b}{0} = 0$ for all b, replacing (A) with the definition:

(B) $\frac{b}{a} = q$ is an abbreviation for: If $a \neq 0$ then $b = qa$, and if $a = 0$ then $q = 0$.

If (B) is the definition of division, then it can always be eliminated. But, since $\frac{b}{0} = 0$ for all b, this leads to the violation of some basic algebraic laws, such as $\frac{b}{a} + \frac{d}{c} = \frac{bc+da}{ac}$. For, then we have $\frac{1}{1} + \frac{1}{0} = 1 + 0 = 1 \neq 0 = \frac{1}{0} = \frac{1 \cdot 0 + 1 \cdot 1}{1 \cdot 0}$.

This cannot be accounted for if, as the stipulative view claims, a definition can always be eliminated. For, then, "if difficulties are encountered in the elimination, this proves that the definition was not well devised. Indeed, this method of substitution" of the definiens for the definiendum "is one of the best for recognizing the exactness of a definition" (Peano 1973, 245). But this claim contrasts with the fact that definition (A), which cannot be eliminated, is well devised, while definition (B), which can be eliminated, is not well devised.

(5) By claiming that a definition says nothing about the existence of the thing defined, the stipulative view cannot account for the fact that in mathematics it is assumed that the thing defined exists.

For example, in standard calculus textbooks, the derivative $f'(x)$ of a function f is introduced as follows: "The derivative $f'(x)$ is defined by the equation $f'(x) = \lim_{h \to 0} \frac{f(x+h)-f(x)}{h}$ provided the limit exists" (Apostol 1967–1969, I, 160). Thus the definition introduces the derivative of a function as an existing thing.

The reason why in mathematics it is assumed that the thing defined exists is that, as Leibniz says, "if we give some definition, and it is not clear from it that the idea which we ascribe to the thing is possible, we cannot trust the demonstrations which we have deduced from the definition" (Leibniz 1923–, VI.4, 158). For, "if that idea happens to involve a contradiction, it can be the case that contradictories are true even of it at the same time, and thus our demonstrations will be useless. From this, it is clear that definitions are not arbitrary" (ibid.).

This cannot be accounted for if, as the stipulative view claims, a definition says nothing about the existence of the thing defined. For then "one need not prove the existence of that which is defined" (Peano 1957–1959, II, 173).

11.6 Some Remarks About the Stipulative View

That the stipulative view has the shortcomings described above does not mean that no definition can merely stipulate the meaning of a term. For example, when one says that x^n means $\underbrace{x \times x \ldots \times x}_{n \text{ times}}$, this merely stipulates the meaning of the term x^n, and yet is a perfectly legitimate and useful definition because, for great n, the expression $\underbrace{x \times x \ldots \times x}_{n \text{ times}}$ would become so lengthy as to be typographically unmanageable.

Rather, that the stipulative view has the shortcomings described above means that it is invalid to assume, as Pascal, Frege, Peano, Russell, and Hilbert do, that all definitions merely stipulate the meaning of a term.

This assumption is commonly made by mainstream philosophers of mathematics.

Thus, Quine says that any definition "is a convention of notational abbreviation" which "introduces some specific expression," called "the 'definiendum', as arbitrary shorthand for some complex expression," called "the 'definiens'" (Quine 1966, 71). The "signs thus introduced are wholly arbitrary; all that is required of a definition is that it be theoretically immaterial, i.e., that the shorthand which it introduces admit in every case of unambiguous elimination in favor of the antecedent longhand" (ibid.).

Curry says that any definition is "a convention" by which "we introduce a new symbol or symbol combination called a 'definiendum', with the stipulation that it is to stand for some other symbol combination, called the definiens, whose meaning is already known" (Curry 1977, 106). Thus definitions are "abbreviative linguistic devices, which are theoretically eliminable" by "successive replacement of definienda by their respective definientia," although they are "practically necessary in order to cut our discussions down to manageable size" (ibid.).

The assumption in question is also commonly made by authors of mathematical textbooks.

Thus, Preparata and Yeh say that a "definition has the function of stipulating what meaning is to be attributed to a concept which has not occurred previously" (Preparata and Yeh 1973, 3).

Hunter says that "a definition is a statement that stipulates the meaning of a new term, symbol, or object" (Hunter 2015, 42).

But the assumption that all definitions merely stipulate the meaning of a term is invalid. Definitions that merely stipulate the meaning of a term are legitimate, but play a minor role in mathematics. For, every such "definition is an equality" since it "has the form: the defined = the thing defining, where the defined is a new sign or word or sentence or proposition, and the defining is an expression composed of known things. The definition expresses the convention of using the defined in place of the longer defining phrase" (Peano 1973, 235). This contrasts, for example, with the fact that, as Netz points out, "most definitions" in Euclid's *Elementa* and *Data* and in the works of Autolycus, Archimedes and Apollonius, "do not prescribe equivalences between expressions (which can then serve to abbreviate, no more).

They specify the situations under which properties are considered to belong to objects" (Netz 1999, 95). Namely, they specify "conditions where a property (independently understood) is assumed to belong to an object (independently understood)" (ibid., 93). Thus they are neither equalities nor express conventions.

Every definition which is an equality and expresses a convention does not extend our knowledge because it is "a nominal definition" (Peano 1973, 243). Namely, it is a definition of a name by which a thing is indicated, not a definition of the thing itself, and a name is inessential. As Shakespeare says: "That which we call a rose | By any other name would smell as sweet" (Shakespeare, *Romeo and Juliet*, Act 2, Scene 2, vv. 43–44).

This contrasts with the fact that many definitions extend our knowledge. Such definitions are neither equalities nor conventions, but the result of extensive investigations, which may involve a long series of demonstrations about some subject. Only of such definitions it can be said that "definitions are – and are recognized to be – among the most important features of mathematics" (Thomas 2014a, 251).

Somewhat inconsistently, even some supporters of the stipulative view admit this. Thus, Russell says that, "in spite of the fact that definitions are theoretically superfluous, it is nevertheless true that they often convey more important information than is contained in the propositions in which they are used" (Whitehead and Russell 1925–1927, I, 11). For, a definition may give "definiteness to an idea which had previously been more or less vague" (ibid., I, 12). Therefore, "definitions are what is most important, and what most deserves" prolonged "attention" (ibid.).

This cannot be accounted for if it is assumed that all definitions merely stipulate the meaning of a term.

11.7 The Heuristic View of Mathematical Definition

Since the stipulative view of mathematical definition has the shortcomings described above, a different approach to mathematical definition is necessary.

An alternative view of mathematical definition is the heuristic view. The heuristic view of mathematical definition derives from the heuristic view of mathematical objects (see Chap. 9). If mathematical objects are hypotheses human beings make to solve mathematical problems by the analytic method, then so are the mathematical definitions which introduce them. Namely, mathematical definitions are hypotheses which are made to solve mathematical problems by the analytic method. Like all hypotheses in the analytic method, such hypotheses must be plausible.

Specifically, a mathematical definition is the hypothesis that there exists something which satisfies the condition stated in the definiens, and the hypothesis must be plausible.

For example, Euclid's definition of a circle considered above is the hypothesis that there exists a plane figure contained by one line, such that all the straight lines falling upon it from one point among those lying inside the figure are equal to each other. This hypothesis is assumed to be plausible, because Euclid does not merely say that a circle is a plane figure with these properties, but also assumes that such a figure exists, since he says: "Let it have been postulated" that "a circle can be drawn with any centre and radius" (Euclid, *Elementa*, I, Postulate 3).

This agrees with Mill's view that Euclid's definition of a circle, "being analysed, consists of two propositions; the one an assumption with respect to a matter of fact," namely, "a figure may exist, having all the points in the line which bounds it equally distant from a single point within it;" the other one the fixation of "the meaning of a name," namely, "any figure possessing this property is called a circle" (Mill 1963–1986, VII, 145). The first proposition, namely the assumption, is that to which every Euclid's demonstration based on this definition "really appeals," since without this assumption "the demonstration could not stand" (ibid.). Conversely, the second proposition, namely the fixation of the meaning of a name, is inessential, because "whether that figure be called a circle or not is quite immaterial" (ibid.).

The same holds of all other Euclid's definitions. For each of them, Euclid does not merely say that the thing defined is an object with such-and-such properties, he also assumes that such a thing exists.

11.8 Evidence for the Heuristic View

The heuristic view does not have the shortcomings of the stipulative view. This can be seen as follows.

(1) The heuristic view can account for the fact that finding a suitable mathematical definition can make the difference in discovering a solution to a problem.

For, a mathematical definition is a hypothesis, and finding a suitable hypothesis is a crucial step towards finding a solution to a problem.

(2) The heuristic view can account for the fact that mathematicians often use concepts for a long time, even centuries, before they can find a suitable definition for them.

For, a mathematical definition is not merely the expression of a volition, finding it may require as much effort and time as finding any hypothesis capable of yielding a solution to a problem.

(3) The heuristic view can account for the fact that mathematicians often give definitions which afterwards turn out to be incorrect.

For, a mathematical definition is a hypothesis that is plausible, and a hypothesis that is plausible at one stage may become implausible at a later stage, when the arguments against the hypothesis become stronger than those for it.

(4) The heuristic view can account for the fact that it is not generally the case that a mathematical definition can be eliminated.

For, a mathematical definition is a hypothesis which plays an essential role in the solution to a problem, therefore, it cannot be eliminated. Of course, different mathematical definitions may yield different solutions to the same problem, but each of them plays an essential role in the respective solution to the problem, therefore, it cannot be eliminated from the respective solution.

(5) The heuristic view can account for the fact that in mathematics it is assumed that the thing defined exists.

For, a mathematical definition is the hypothesis that there exists something which satisfies the condition stated in the definiens, and the hypothesis must be plausible.

11.9 Heuristic Differences Between Definitions

Further evidence for the heuristic view of mathematical definition is provided by the following important facts, that the stipulative view of mathematical definition cannot account for:

(a) Two extensionally equivalent definitions of the same concept can have different heuristic values.

(b) Definitions in mathematics are not starting points but arrival points in the solution to problems.

(c) There is no circularity of definitions and theorems.

(d) Many definitions in mathematics are demonstration-generated.

(e) Definitions in mathematics can be justified.

The remainder of this chapter will be devoted to arguing that, contrary to the stipulative view of mathematical definition, the heuristic view of mathematical definition can account for these facts.

11.10 Heuristic Values of Extensionally Equivalent Definitions

The heuristic view of mathematical definition can account for the fact that two extensionally equivalent definitions of the same concept can have different heuristic values.

An example of this is provided by the two definitions of a sphere already considered above. They are extensionally equivalent but have different heuristic values, because they represent different perspectives on a problem, and hence may not be equally effective as heuristic guides to find a solution to the problem.

Indeed, as already mentioned, Euclid's definition of a sphere plays an important heuristic role in Euclid's solution to the problem about the five Platonic figures: (1) To construct each of the five so-called Platonic figures or regular solids – the tetrahedron, the cube, the octahedron, the dodecahedron, and the icosahedron. (2) To comprehend it in a given sphere. (3) To characterize the relation between the length of an edge of the figure and the length of the diameter of the given sphere.

Euclid deals with this problem in Euclid, *Elementa*, XIII, Propositions 13–17. Then, in the final proposition of the *Elementa*, he sets out the edges of the five Platonic figures, he compares them, and concludes with an argument that "no other figures, besides the said five figures, can be constructed" (Euclid, *Elementa*, XIII, Proposition 18).

As an example of Euclid's approach, let us consider Euclid's solution to the problem: "To construct a cube and to comprehend it in a sphere," and "to prove that the square on the diameter of the sphere is triple the square on the side of the cube" (Euclid, *Elementa*, XIII, Proposition 15).

Let AB be the diameter of the given sphere. Cut AB at C so that $AC = 2BC$. Let the semicircle ADB be described on AB, draw CD from C at right angles to AB, and join BD.

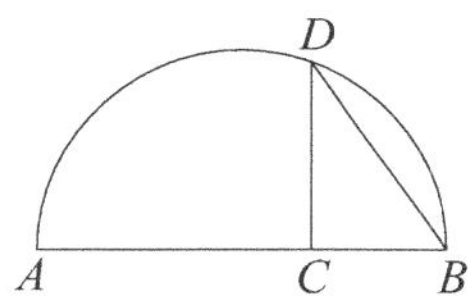

Let the cube having side $KE = BD$ be laid out. Join KG and EG.

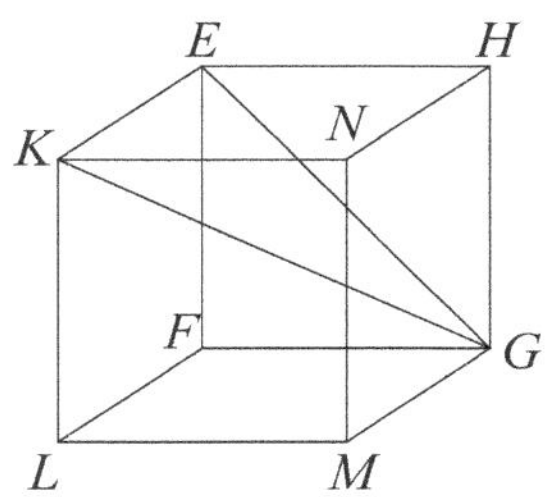

Since KE is at right angles to EH and EF, KE is also at right angles to the plane EG and of course to the straight line EG, so the angle KEG is a right angle. Therefore, the semicircle described on KG will pass through vertex E.

Again, since GF is at right angles to FL and FE, GF is also at right angles to the plane FK, and of course, if we join FK, to the straight line FK, so the angle KFG is a right angle. Therefore, the semicircle described on KG will pass through the vertex F.

Similarly, it can be seen that the semicircle described on KG will pass through the remaining vertices of the cube.

If, then, KG remaining fixed, the semicircle described on KG is carried around and restored again to the same position from which it began to be moved, the cube will be comprehended in a sphere with a diameter KG.

To prove that the cube is comprehended in the given sphere, we must show that $KG = AB$.

Now, by two applications of the Pythagorean theorem, we have $KG^2 = KE^2 + EG^2 = KE^2 + 2EF^2 = 3KE^2$. So, $KG^2 = 3KE^2$.

Clearly, $AB : BC = AB^2 : AB \cdot BC$. On the other hand, $AB : BD = BD : BC$, hence $AB \cdot BC = BD^2$. Therefore, $AB : BC = AB^2 : BD^2$. But, since $AC = 2BC$, we have $AB = 3BC$. So, $3BC : BC = AB^2 : BD^2$, hence $AB^2 = 3BD^2$.

From $KG^2 = 3KE^2$, $KE = BD$, and $AB^2 = 3BD^2$, it follows that $KG^2 = AB^2$, so $KG = AB$. Thus the cube is comprehended in the given sphere.

Finally, to prove that the square on the diameter of the given sphere is triple the square on the side of the cube, we need only observe that, from $AB^2 = 3BD^2$and $KE = BD$, it follows $AB^2 = 3KE^2$.

From the proof of *Elementa*, XIII, Proposition 15, we can see why Euclid gives the definition of a sphere in that form. The crux of the proof is to show that the cube can be comprehended in a sphere, and Euclid's definition immediately suggests how to do that. Simply, the cube will be comprehended in the sphere obtained when, a diagonal of the cube remaining fixed, the semicircle described on the diagonal is carried around and restored again to the same position from which it began to be moved. By Euclid's definition, this will generate a sphere comprehending the cube.

That Euclid's definition of a sphere immediately suggests how to show that the cube can be comprehended in a sphere, also explains another peculiarity of Euclid's proof. Although Euclid's construction is equivalent to showing that the cube can be comprehended in the given sphere, Euclid does not show this directly. He first shows that the cube can be comprehended in the sphere obtained when, a diagonal of the cube remaining fixed, the semicircle described on the diagonal is carried around and restored again to the same position from which it began to be moved. He does so because Euclid's definition of a sphere immediately suggests how to show this. Then he shows that the sphere so generated is equal to the given sphere.

11.11 The Resistance to Recognizing the Heuristic Value of Definitions

Euclid's solution to the problem about the five Platonic figures has been criticized from the Middle Age until today, because it is based on Euclid's definition of a sphere, which refers to motion.

Thus, Omar Khayyam says that an obvious definition of a sphere is that "it is a solid figure which is contained by a single surface, inside of which is a point such that all the straight lines drawn from it to the containing surface are equal" (Rashed 2017, 7, footnote 18). On the contrary, with his definition of a sphere, "Euclid deflected from this definition, towards what he said by a lack of discernment and by negligence" (ibid.). If Euclid's definition of a sphere "had a meaning, one would have defined the circle by saying, that the circle is a plane figure generated by the rotation of a straight line in a plane surface, in such a manner that one of its two extremities remains fixed in its place while the other gets ultimately to the starting point of the motion" (ibid.). But, quite rightly, "people have deflected from that kind of definition because of the existence of motion," and "we must follow their tracks and be consistent with the demonstrative rules and the universal laws mentioned in the works on logic" (ibid.).

Mueller says: "Euclid's treatment of the regular solids in terms of a construction and comprehension within the sphere generated by a revolving semicircle obscures the intuitive basis of what he is doing" (Mueller 2006, 254).

The opposition to Euclid's definition of sphere derives from two factors. The first factor is the influence of Aristotle's view of definition as statement of the essence. Since the essence of a thing is supposed to be stable and unchanging, a definition cannot involve movement. Indeed, Aristotle says that "the objects of mathematics are without movement" (Aristotle, *Metaphysica*, A 8, 989 b 32).

The second factor is the non-recognition of the heuristic value of definitions. As we have seen, Euclid's definition of a sphere plays an important heuristic role in Euclid's solution to the problem about the five Platonic figures, because it guides him to find a solution to the problem. This is all the more important as Euclid "believed the goal of the whole *Elementa* to be the construction of the so-called Platonic figures" (Proclus, *In primum Euclidis Elementorum librum commentarii*, 68.21–23, ed. Friedlein). Indeed, Euclid's solution to the problem about the five Platonic figures "appears to be the culmination of the work" of the *Elementa* "as a whole," or, more precisely, "one might say that geometrically that is where the work is headed" (Thomas 2014b, 234). Euclid's belief in the centrality of his solution to the problem about the five Platonic figures is justified because, as Byers says, that solution is "one of the most beautiful and remarkable results in geometry" (Byers 2007, 221).

The fact that Euclid realized that a dynamic definition of sphere, rather than a static one, would have more easily allowed him to achieve the goal of the whole *Elementa*, shows that he was perfectly aware that two extensionally equivalent definitions of the same concept may have different heuristic values.

The heuristic view can account for this fact because, according to it, definitions are hypotheses, two hypotheses may be extensionally equivalent without being intensionally equivalent, so they can represent different perspectives on a problem, and hence may not be equally effective in solving it.

11.12 The Place of Definitions in Mathematical Research

The heuristic view of mathematical definition can account for the fact that definitions in mathematics are not starting points but arrival points in the solution to problems.

Definitions in mathematics are often opposed to definitions in philosophy by saying that, while definitions are starting points in mathematics, they are arrival points in philosophy.

Thus, Rota says that, "whereas mathematics starts with a definition, philosophy ends with a definition" (Rota 1997, 97).

Dummett says that mathematics starts with a definition, because mathematicians first "formulate a precise concept under which it may be reasonably claimed that every case determinately either falls or does not fall. Having done so, their argumentation will proceed within the boundaries of the definitions they have adopted" (Dummett 2010, 11). Conversely, philosophy ends with a definition, because the philosopher's reasoning "is not carried out, as the mathematician's is, within a framework of concepts already made precise" (ibid.). For, "the philosopher's only resource is the analysis of concepts we already possess, but about which we are confused; he seeks to remove that confusion," and eventually arrives at definitions that "clarify the concepts" (ibid.).

But this is invalid. Definitions are not starting points in mathematics, otherwise it would be incomprehensible why, as pointed out above, mathematicians often use concepts for a long time, even centuries, before they can find a suitable definition for them. The heuristic view of mathematical definition can account for this fact because, according to it, definitions are hypotheses that are made to solve problems by the analytic method and, in the analytic method, the hypotheses are not starting points but arrival points in the solution to a problem.

In fact, contrary to the claim that definitions are starting points in mathematics, definitions are not starting points, not only for the heuristic view, but also for the stipulative view. For, according to the latter, a definition merely states what a name means, so, as Bernardi says, "a definition cannot be a starting point for proving something" (Bernardi 2014, 47). Indeed, according to mainstream philosophy of mathematics, of which the stipulative view of mathematical definition is part, "any proof is supported by axioms (and already known theorems), and not by any definitions," so a theorem is "proved starting only from axioms," and not "from both axioms and definitions" (ibid.). Thus definitions are not starting points. On the other hand, according to mainstream philosophy of mathematics, definitions are not arrival points either, because they are not the result of an investigation. Rather, they are intermediate points between axioms and theorems, which are "useful only for recalling the meaning of a term, namely for substituting a term with the proprieties used in defining that term" (ibid.).

11.13 The Alleged Circularity of Definitions and Theorems

The heuristic view of mathematical definition can also account for the fact that there is no circularity of definitions and theorems.

Rota says that "the theorems of mathematics motivate the definitions as much as the definitions motivate the theorems. A good definition is 'justified' by the theorems that can be proved with it, just as the proof of the theorem is 'justified' by appealing to a previously given definition" (Rota 1997, 97). So, there is "a hidden circularity in formal mathematical exposition. The theorems are proved starting with definitions; but the definitions themselves are motivated by the theorems" (ibid.).

The only way for the stipulative view to avoid this circularity would be to give an absolute justification of the theorems. This is what Hilbert attempted to do with his foundationalist program, but he failed because, by Gödel's incompleteness theorems, no absolute justification of the theorems is possible. Therefore, with the stipulative view, the circularity of definitions and theorems is unavoidable.

Conversely, with the heuristic view, no circularity of definitions and theorems arises.

Indeed, as Descartes says, "it is not circular to prove a cause by several effects that are known otherwise" to agree with experience, "and then conversely to prove certain other effects by this cause" (Descartes 1996, II, 198). For, the effects in question are known independently to agree with experience, and "we are free to make any assumption" with "the sole proviso that all the consequences of our assumption must agree with our experience" (ibid., VIII-1, 101).

Then, if definitions are hypotheses that are made to solve problems by the analytic method, there is no circularity of definitions and theorems. For, a definition is not justified by the theorems that are proved with it. Like all other hypotheses, it is justified by the fact that it is plausible, namely that the arguments for it are stronger than the arguments against it. Thus a definition is justified by things other than the theorems that are proved with it.

11.14 Demonstration-Generated Definitions

The heuristic view of mathematical definition can also account for the fact that many definitions in mathematics are demonstration-generated.

This was pointed out already by Saccheri. He says that a mathematical definition that is not merely stipulative "is generally the result of a long series of demonstrations about some subject," since it is obtained "after a long investigation of the properties, or predicates, belonging to some subject" (Saccheri 1701, 120).

Lakatos tries to account for the fact that many definitions in mathematics are demonstration-generated, in terms of his method of proof and refutation. He says that mathematical definitions are "proof-generated" because are obtained from naive conjectures "through proofs and refutations" (Lakatos 1976, 89). But, as argued in Chap. 3, the method of proof and refutation does not provide a basis for methodology qua logic of discovery. Therefore, Lakatos does not account for the fact that many definitions in mathematics are demonstration-generated.

On the contrary, the heuristic view of mathematical definitions accounts for it. For, according to it, a definition is a hypothesis that is plausible. If the inquiry brings up new data incompatible with the hypothesis, then, like all hypotheses in the analytic method, the definition is modified or replaced with a new one, obtained from the problem, and possibly other data already available, by some non-deductive rule, and plausible. And so on. So definitions are demonstration-generated.

11.15 The Justification of Definitions

The heuristic view of mathematical definition can also account for the fact that definitions in mathematics can be justified.

Werndl asks: "In what ways are definitions in mathematics justified, and are these kinds of justification reasonable? By a justification of a definition I mean a reason provided for the definition" (Werndl 2009, 313).

The stipulative view cannot provide a significant answer to these questions because, according to it, definitions are arbitrary stipulations. On the contrary, the heuristic view provides a satisfactory answer. For, according to it, definitions in mathematics are justified because they are hypotheses that are plausible. Then, definitions are justified because the arguments for them are stronger than those against them.

This kind of justification is reasonable because it satisfies Aristotle's criterion of reasonableness for justification already considered in Chap. 5, namely that, for each thesis concerning some given subject, we must examine the arguments for it and the arguments against it.

11.16 Adequacy of the Heuristic View of Mathematical Definition

Since the heuristic view of mathematical definition can account for several important facts that the stipulative view cannot, it seems fair to say that the heuristic view is adequate and the stipulative view is inadequate.

This is only a special case of the general fact that, as argued in Chap. 3, heuristic philosophy of mathematics is adequate and mainstream philosophy of mathematics is inadequate. Definition makes this particularly clear. Indeed while, for mainstream philosophy of mathematics, definition is a marginal topic, for heuristic philosophy of mathematics, definition is a central topic in the philosophy of mathematics.

References

Apostol, Tom M. 1967–1969. *Calculus*. New York: Wiley.

Bernardi, Claudio. 2014. What mathematical logic says about the foundations of mathematics? In *From a heuristic point of view*, ed. Emiliano Ippoliti and Cesare Cozzo, 41–53. Newcastle upon Tyne: Cambridge Scholars Publishing.

Brown, James Robert. 2008. *Philosophy of mathematics: A contemporary introduction to the world of proofs and pictures*. London: Routledge.

Byers, William. 2007. *How mathematicians think: Using ambiguity, contradiction, and paradox to create mathematics*. Princeton: Princeton University Press.

Curry, Haskell B. 1977. *Foundations of mathematical logic*. Mineola: Dover.
Descartes, René. 1996. *Œuvres*. Paris: Vrin.
Dummett, Michael. 2010. *The nature and future of philosophy*. New York: Columbia University Press.
Frege, Gottlob. 1967. *Begriffsschrift*, a formula language, modeled upon that of arithmetic, for pure thought. In *From Frege to Gödel: A source book in mathematical logic, 1879–1931*, ed. Jean van Heijenoort, 5–82. Cambridge: Harvard University Press.
———. 1979. *Posthumous writings*. Oxford: Blackwell.
———. 1980. *Philosophical and mathematical correspondence*. Oxford: Blackwell.
———. 1984. *Collected papers on mathematics, logic, and philosophy*. Oxford: Blackwell.
———. 2013. *Basic laws of arithmetic*. Oxford: Oxford University Press.
Galilei, Galileo. 1968. *Opere*. Barbera: Florence.
Grabiner, Judith V. 2010. *A historian looks back: The calculus as algebra and selected writings*. Washington: The Mathematical Association of America.
Harris, Joe. 1992. Developments in algebraic geometry. In *Mathematics in the twenty-first century: 1988 centennial symposium August 8–12*, ed. Felix E. Browder, 89–100. Providence: American Mathematical Society.
Hilbert, David. 1996. On the concept of number. In *From Kant to Hilbert: A source book in the foundations of mathematics*, ed. William Ewald, 1092–1095. Oxford: Oxford University Press.
———. 2004. Grundlagen der Geometrie. In David Hilbert, *Lectures on the foundations of geometry 1891–1902*, 540–602. Berlin: Springer.
———. 2013. *Lectures on the foundations of arithmetic and logic 1917–1933*. Heidelberg: Springer.
Hilbert, David, and Paul Bernays. 1968–1970. *Grundlagen der Mathematik*. Berlin: Springer.
Hunter, David James. 2015. *Essentials of discrete mathematics*. Burlington: Jones & Bartlett.
Jordan, Camille. 1887. *Cours d'analyse*, vol. 3. Paris: Gauthier-Villars.
Lakatos, Imre. 1976. *Proofs and refutations: The logic of mathematical discovery*. Cambridge: Cambridge University Press.
Landau, Ernst. 1958. *Elementary number theory*. New York: Chelsea.
Leibniz, Gottfried Wilhelm. 1923–. *Sämtliche Schriften und Briefe*. Berlin: Academie Verlag.
Mates, Benson. 1972. *Elementary logic*. Oxford: Oxford University Press.
Mill, John Stuart. 1963–1986. *Collected works*. Toronto: University of Toronto Press.
Moore, Eliakim Hastings. 1900. On certain crinkly curves. *Transactions of the American Mathematical Society* 1: 72–90.
Mueller, Ian. 2006. *Philosophy of mathematics and deductive structure in Euclid's Elements*. Mineola: Dover.
Netz, Reviel. 1999. *The shaping of deduction in Greek mathematics: A study in cognitive history*. Cambridge: Cambridge University Press.
Pascal, Blaise. 1904–1914. *Oeuvres*. Paris: Hachette.
Peano, Giuseppe. 1957–1959. *Opere scelte*. Rome: Cremonese.
———. 1973. *Selected works*. London: Allen & Unwin.
Preparata, Franco P., and Raymond Tzuu-Yau Yeh. 1973. *Introduction to discrete stuctures for computer science and engineering*. Boston: Addison-Wesley.
Quine, Willard Orman. 1966. *The ways of paradox, and other essays*. New York: Random House.
Rashed, Roshdi. *2017. Ibn al-Haytham's geometrical methods and the philosophy of mathematics*. London: Routledge.
Rota, Gian-Carlo. 1996. Report on the present state of combinatorics. *Discrete Mathematics* 153: 289–303.
———. 1997. *Indiscrete thoughts*. Cham: Birkhäuser.
Saccheri, Giovanni Girolamo. 1701. *Logica demonstrativa*. Pavia: Magri.
Suppes, Patrick. 1957. *Introduction to logic*. New York: Van Nostrand Reinhold.

Tappenden, James. 2008. Mathematical concepts and definitions. In *The philosophy of mathematical practice*, ed. Paolo Mancosu, 256–275. Oxford: Oxford University Press.

———. 2011. Définitions mathématiques pour philosophes. *Les Études philosophiques* 97: 179–191.

Tarski, Alfred. 1994. *Introduction to logic and to the methodology of the deductive sciences*. Oxford: Oxford University Press.

Thomas, Robert. 2014a. Reflections on the objectivity of mathematics. In *From a heuristic point of view*, ed. Emiliano Ippoliti and Cesare Cozzo, 241–256. Newcastle upon Tyne: Cambridge Scholars Publishing.

———. 2014b. Acts of geometrical construction in the *Spherics* of Theodosios. In *From Alexandria, through Baghdad*, ed. Nathan Sidoli and Glen Van Brummelen, 227–237. Berlin: Springer.

Werndl, Charlotte. 2009. Justifying definitions in mathematics – Going beyond Lakatos. *Philosophia Mathematica* 17: 313–340.

Whitehead, Alfred North, and Bertrand Russell. 1925–1927. *Principia Mathematica*. Cambridge: Cambridge University Press.

Chapter 12
Diagrams

Abstract According to heuristic philosophy of mathematics, one of the tasks of the philosophy of mathematics is to give an answer to the question: What is the nature of mathematical diagrams? In the last few decades, the prevailing answer to this question has been the axiomatic view of mathematical diagrams. The chapter discusses this answer and argues that it is inadequate. Then, it puts forward an alternative answer, the heuristic view of mathematical diagrams.

Keywords Arguments against mathematical diagrams · Views about mathematical diagrams · Axiomatic view of mathematical diagrams · Mathematical diagrams and intuition · Heuristic view of mathematical diagrams · Heuristic view and objections to mathematical diagrams · Adequacy of the heuristic view

12.1 The Status of Mathematical Diagrams

As we have seen in Chap. 3, according to heuristic philosophy of mathematics, one of the tasks of the philosophy of mathematics is to give an answer to the question: What is the nature of mathematical diagrams?

In Greek mathematics, mathematical diagrams were considered a legitimate component of demonstrations (see Netzt 1999, Chapter 1; Manders 2008; Saito and Sidoli 2012; Catton and Montelle 2012; Macbeth 2014, Chapter 2; Carter 2019).

On the contrary, in the early modern period, doubts have been raised about their role in demonstrations. The question has been variously discussed. Then, from the end of the nineteenth century, many mathematicians have declared that demonstrations must be independent of mathematical diagrams.

Thus, Pasch says that, "if geometry is to be really deductive, the deduction process must" everywhere "be independent of diagrams; only the relations between the geometrical terms laid down in the propositions and definitions employed may be taken into account" (Pasch 1882, 98).

Nevertheless, mathematical diagrams have continued to be used in demonstrations in several areas of mathematics, though without a precise justification.

C. Cellucci, *The Making of Mathematics*, Synthese Library 448,
https://doi.org/10.1007/978-3-030-89731-4_12

In the last few decades, however, there has been an attempt to give a justification for the use of mathematical diagrams in demonstration, by fitting it into the axiomatic method. This has given rise to the axiomatic view of mathematical diagrams. At present, the latter is the prevailing answer to the question of the nature of mathematical diagrams.

However, the chapter discusses this answer to the question of the nature of mathematical diagrams and argues that it is inadequate. Then it offers an alternative answer, the heuristic view of mathematical diagrams.

12.2 The Tradition of Mathematical Diagrams

As said above, in the early modern period, doubts have been raised about the role of mathematical diagrams in demonstrations. The question has been variously discussed.

On the one hand, some people, such as Newton and Locke, affirm that mathematical diagrams are a legitimate component of demonstrations.

Thus, Newton says that "geometric constructions" are "ones which are achieved by means of straight lines and circles and any figures in the plane" (Newton 1967–1981, VII, 303). By introducing algebraic terms in geometry, Descartes and his followers "have lost the simplicity in which all elegance in geometry consists" (Newton 1967–1981, V, 429). For, algebraic calculations, "progressing by means of" algebraic "operations alone, very often express in an intolerably roundabout way quantities which in geometry are designated by the drawing of a single line" (ibid., IV, 421). Indeed, "it is not its equation but its description which produces a geometrical curve" (ibid., V, 425). Anyone "who examines the constructions of problems by the straight line and circle devised by the first geometers will readily perceive that geometry was contrived as a means of escaping the tediousness of calculation by the ready drawing of lines" (ibid., V, 429).

Locke says that "diagrams drawn on paper" are "not liable to the uncertainty that words carry in their signification. An angle, circle, or square, drawn in lines, lies open to the view, and cannot be mistaken: it remains unchangeable, and may at leisure be considered and examined, and the demonstration" may "be revised, and all the parts of it may be gone over more than once, without any danger of the least change" (Locke 1975, 550). In particular, in the case of a long demonstration, "the memory would often have great difficulty" to "record and retain the several ideas about which the demonstration is made," and "the great help against this, which mathematicians find in diagrams and figures, which remain unalterable in their draughts, is very apparent" (ibid., 551).

On the other hand, other people, such as Leibniz and Wolff, deny that mathematical diagrams are a legitimate component of demonstrations.

Thus, Leibniz says: "I do not agree at all that in mathematics the particular demonstrations on the diagram which one draws" give "certainty," in fact, "it is not the diagrams that provide the demonstration to the geometers," the "force of the demonstration is independent of the diagram drawn, whose only role is to make it easier to understand what one wants to say and to fix one's attention" (Leibniz 1965, V, 342). It is "the universal propositions, namely, the definitions, the axioms, and the theorems already demonstrated, that make up the reasoning, and would sustain it even if there were no diagram" (ibid.). Indeed, "the force of the demonstrations depends upon intelligible notions and truths, the only ones capable of letting us judge what is necessary" (ibid., VI, 504).

Wolff says that "a demonstration is a proof which can be analyzed into formal syllogisms as its only constituents such that the conclusion of the previous one forms the premise of the subsequent one, and no syllogism assumes a premise that is not already previously proved as correct" (Wolff 1716, 501). Indeed, "in geometrical demonstrations one thinks in syllogisms that are properly formulated," and "in mathematics" generally "nothing is found but by such syllogisms properly formulated" (Wolff 1736, 97). Even "in other disciplines," if "one wants to demonstrate and expose something in the mathematical way, syllogisms properly formulated must lead to the goal," because, "by the help of these syllogisms, one can resist the most subtle errors" (ibid.).

12.3 Kant on Mathematical Diagrams

Unlike Leibniz and Wolff, Kant reaffirms that mathematical diagrams are a legitimate component of demonstrations.

According to Kant, "mathematical cognition" is cognition "from the construction of concepts," where "to construct a concept means to exhibit a priori the intuition corresponding to it" (Kant 1998, A713/B741). In particular, to construct a geometrical concept means to exhibit a figure corresponding to it, where "the individual drawn figure is empirical, and nevertheless serves to express the concept without damage to its universality, for in the case of this empirical intuition we have taken account only of the action of constructing the concept" (ibid., A714/B742). For example, we construct the concept of a triangle by exhibiting a figure corresponding to it, where in the figure we take account only of the action of constructing the concept, "to which many determinations, e.g., those of the magnitude of the sides and the angles, are entirely indifferent, and thus we have abstracted from these differences, which do not alter the concept of the triangle" (ibid.).

The construction of concepts is essential to mathematics. Thus, "give a philosopher the concept of a triangle, and let him try to find out in his way how the sum of its angles might be related to a right angle" (ibid., A716/B744). To carry out this task, "he has nothing but the concept of a figure enclosed by three straight lines, and in it the concept of equally many angles. Now he may reflect on this concept as long as he wants, yet he will never produce anything new" (ibid.).

Conversely, "let the geometer take up this question" (ibid.). He will proceed like in Euclid's demonstration that, in any triangle, the sum of the internal angles is equal to two right angles (Euclid, *Elementa*, I, Proposition 32). Thus, the geometer "begins at once to construct a triangle" (Kant 1998, A716/B744). Namely, he draws a triangle *ABC*. Then, "since he knows that two right angles together are exactly equal to all of the adjacent angles that can be drawn at one point on a straight line, he extends one side of his triangle," *BC*, to *D*, "and obtains two adjacent angles," *ACB* and *ACD*, "that together are equal to two right ones" (ibid.). Now, "he divides the external one of these angles," *ACD*, "by drawing a line," *CE*, "parallel to" the opposite side of the triangle" (ibid.). Namely, *AB*.

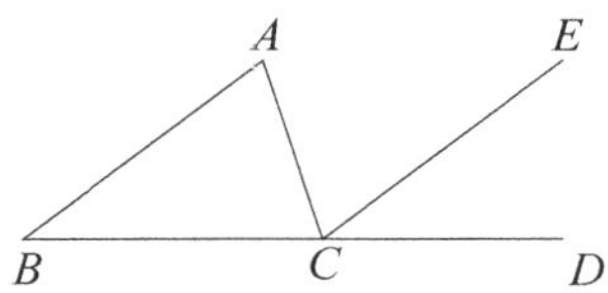

Since *AB* is parallel to *CE* and *AC* falls upon them, "there arises an external adjacent angle," *ACE*, "which is equal to an internal one" (ibid.). Namely, to the angle *BAC*. For, *ACE* and *BAC* are alternate angles, hence they are equal to one another (Euclid, *Elementa*, I, Proposition 29). Again, since *AB* is parallel to *CE* and *BD* falls upon them, there arises an external angle, *ECD*, which is equal to an internal one, namely, to angle *ABC*. For, *ECD* and *ABC* are alternate angles, hence they are equal to one another (Euclid, *Elementa*, I, Proposition 29). Then the whole angle *ACD*, which is the sum of angles *ACE*, *ECD*, is equal to the sum of the angles *BAC*, *ABC*.

Add the angle *ACB* to each. Then the sum of the angles *ACD*, *ACB* is equal to the sum of the angles *BAC*, *ABC*, *ACB*. But the sum of the angles *ACD*, *ACB* is equal to two right angles (Euclid, *Elementa*, I, Proposition 13). Therefore, the sum of the angles *BAC*, *ABC*, *ACB* is equal to two right angles. Namely, in *ABC*, the sum of the internal angles is equal to two right angles.

Thus, "through a chain of inferences that is always guided by intuition," the geometer "arrives at a fully illuminating and at the same time general solution of the question" (ibid., A717/B745). The chain of inferences is a demonstration, because the word "demonstration derives from 'monstrare', to display, to lay before the eyes," and this word "can actually only be used" of demonstrations such as that of Euclid, "where the object is exhibited in intuition" (Kant 1992, 340).

According to Kant a demonstration, thus understood, is essential to establish any mathematical proposition. The latter cannot be established by a logical proof, namely a logical deduction from axioms, because logic "serves of course merely for passing judgment and for correcting our cognition, but not for expanding it" (ibid., 529). A demonstration is something different from a logical proof, namely a logical deduction from axioms, because it "proceeds through the intuition of the object" (Kant 1998, A735/B763). Therefore, the use of the figure is essential to mathematics.

12.4 The Intuition Argument

After Kant, however, several arguments have been advanced against the view that mathematical diagrams are a legitimate component of demonstrations.

A first argument is the intuition argument, according to which mathematical diagrams involve an appeal to intuition, and a demonstration involving an appeal to intuition lacks rigour, therefore it is defective.

Thus, Ayer says that "the use made of diagrams in geometrical treatises shows that geometrical reasoning is not purely abstract and logical, but depends on our intuition of the properties of figures" (Ayer 1952, 83). However, "the appeal to intuition, though generally of psychological value, is also a source of danger to the geometer" and must be avoided in "a truly rigorous geometrical proof" (ibid.). In particular, "Euclid himself was guilty" of appealing to intuition, and consequently "the presence of the figure is essential to some of his proofs. This shows that his system is not, as he presents it, completely rigorous" (ibid.)

12.5 The Quantification Logic Argument

Another argument against the view that mathematical diagrams are a legitimate component of demonstrations is the quantification logic argument, according to which the use of diagrams seemed necessary until the nineteenth century, because the only logic available was syllogistic logic, which was inadequate to mathematics. But, with the development of quantification logic, such use is no longer necessary, because quantification logic is adequate to mathematics.

Thus, Hintikka says that Kant's assumption that the use of the figure is essential to mathematics was due to the fact that "Kant's notion of logic was extraordinarily narrow, comprised essentially only of syllogistic reasoning" (Hintikka 2020, 86). And syllogistic reasoning is inadequate to mathematics, as it is clear from the fact that "the history of the philosophy of mathematics is dotted by unsuccessful attempts to turn Euclid into a syllogistic form" (Hintikka 2020, 87). These attempts have been unsuccessful because "geometrical reasoning is quantificational and relational," while syllogistic reasoning "suffices only for monadic reasoning. It cannot cope with dependent quantifiers," therefore "Euclid (or Kant)" had "to reduce somehow their reasoning with dependent quantifiers to reasoning with single quantifiers" (ibid.). To this end, they resorted to the use of the figure. Indeed, their "argumentation refers to" a "figure" since otherwise they "would have had to argue in terms of dependent quantifiers" (ibid., 88).

However, referring to a figure amounts to an operation of "instantiation, introduction of objects to exemplify certain geometrical concepts" (ibid., 89). This operation is "codified in what we call first-order logic (quantification theory)," and specifically in "the rules of instantiation that are the backbone of our contemporary first-order logic, that is, the rules of existential and universal instantiation" (ibid., 86). Therefore, there is no need to refer to a figure, what is needed is "logical instantiation and nothing more" (ibid.).

12.6 The Incorrectly Drawn Diagram Argument

Another argument against the view that mathematical diagrams are a legitimate component of demonstrations is the incorrectly drawn diagram argument, according to which incorrectly drawn figures can lead to absurd propositions.

Thus, Klein says that, since Euclid "placed so little importance upon correct geometric drawing, there is real danger that a student of Euclid may, because of a falsely drawn figure, come to a false conclusion. It is in this way that" numerous "formally correct proofs of false theorems" arise, "which rest on figures which are wrongly drawn" (Klein 2016, II, 228).

For example, let us consider the following 'demonstration' of the proposition: From a point not on a line, two distinct perpendiculars can be drawn to the line. For, draw any two circles, intersecting at points A and B, and draw the diameters AC and AD. Then draw CD intersecting the circles at points E and F, respectively, and draw AE and AF.

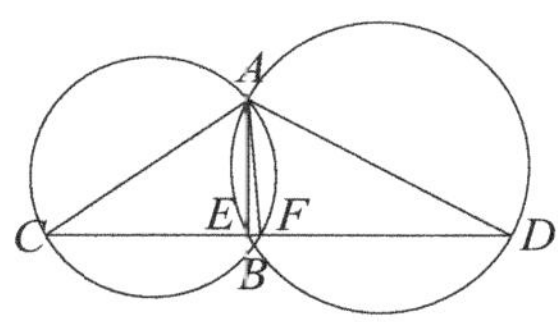

Since AFC and AED are triangles in a semicircle, they must be right-angled. Therefore AE and AF must be perpendiculars to the line CD from a point A not on the line CD. Thus from a point not on a line two distinct perpendiculars can be drawn to the line.

12.7 The Limit Diagram Argument

Another argument against the view that mathematical diagrams are a legitimate component of demonstrations is the limit diagram argument, according to which even correctly drawn diagram, in particular, diagrams about limits, can lead to absurd propositions.

Thus, Giaquinto says that "visual thinking is unreliable for situations in which limits are involved, and so it is not a means of discovery in those situations, let alone a means of proof" (Giaquinto 2008, 35).

For example, let us consider the following 'demonstration' of the proposition: The sum of the two legs of a right-angled triangle is equal to the hypotenuse. For, let ABC be a right-angled triangle, divide its hypotenuse BC into n equal parts, where n is a natural number, and through every point of division draw a pair of rectilinear segments, one parallel to the leg AB, the other parallel to the leg AC. Thus we obtain a steplike line from B to C.

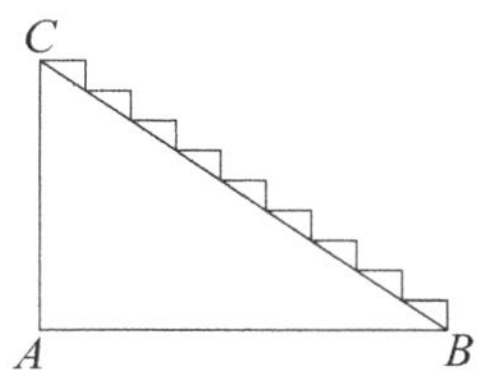

Let S_n be the sum of all the links of the steplike line from B to C. Since the sum of all the segments drawn parallel to one of the legs is equal to that leg, for any n, $S_n = AB + AC$, so:

$$\lim_{n\to\infty} S_n = AB + AC. \tag{12.1}$$

On the other hand, as n increases, the steplike line from B to C will get closer and closer to the hypotenuse BC, and in the limit will coincide with it, so:

$$\lim_{n\to\infty} S_n = BC. \tag{12.2}$$

From (12.1) and (12.2) it follows that $AB + AC = BC$. Thus the sum of the legs of a right-angled triangle is equal to the hypotenuse.

12.8 The Particularity Argument

Another argument against the view that mathematical diagrams are a legitimate component of demonstrations is the particularity argument, according to which, since mathematical diagrams are particular figures, inferring general conclusions from them can lead to absurd propositions.

Thus, Tennant says that "one is cautioned, and corrected" about "the mistake of assuming as given information that is true only of the triangle that one has happened to draw, but which could well be false of other triangles that one equally well might have drawn in its stead" (Tennant 1986, 304). A mathematical diagram, for example a drawn triangle, "is only an heuristic to prompt certain trains of inference," it "is dispensable," and "has no proper place in the proof as such. For the proof is a syntactic object consisting only of sentenced arranged in a finite and inspectable array" (ibid.).

For example, let us consider the following 'demonstration' of the proposition: The three perpendicular bisectors (that is, the perpendicular passing through the midpoints) of the sides of a triangle intersect at a single point inside the triangle. For, let *ABC* be any triangle, and let *O* be the point of intersection of the perpendicular bisectors of the sides *AB* and *AC*. From *O* draw *OE* to the midpoint *E* of *BC*. Join the vertices *A*, *B*, and *C* to *O*.

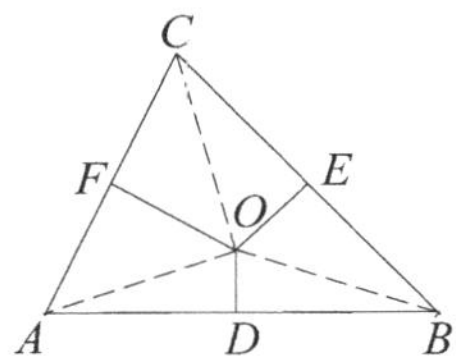

The triangles *AOF* and *COF* are equal, because two corresponding sides and the included angles are equal, hence $OA = OC$.

The triangles *AOD* and *BOD* are equal, because two corresponding sides and the included angles are equal, hence $OA = OB$.

From $OA = OC$ and $OA = OB$ it follows $OB = OC$. Thus *O* lies equidistant from *B* and *C*, and *OE* is the perpendicular to *BC*. Since *E* is the midpoint of *BC*, *OE* is the perpendicular bisector of the side *BC*. Thus *O* is the point of intersection of the perpendicular bisectors of the sides *AB*, *AC*, and *BC*, and is inside the triangle *ABC*.

12.9 The View of the Logical Dispensability of Mathematical Diagrams

The above arguments against the view that mathematical diagrams are a legitimate component of demonstrations, have given rise to two different, though related, views about mathematical diagrams.

The first view about mathematical diagrams is that of the logical dispensability of mathematical diagrams. This view does not exclude the use of mathematical diagrams in demonstrations, but requires to make sure that any such use is logically dispensable, namely, that a demonstration using mathematical diagrams might always be replaced by a deduction from the axioms.

A full expression of this view is Hilbert's *Grundlagen der Geometrie*, which is an "attempt to choose for geometry a simple and complete set of independent axioms and to deduce from these the most important geometrical theorems" (Hilbert 1987, 1). The use of figures "facilitates the understanding, and is a fruitful means of discovering new propositions" (Hilbert 2004a, 75). So, although "one could also abstain from using figures," we "will not do this. Rather, we will use figures often" (Hilbert 2004b, 541). But figures "can easily be misleading" (Hilbert 2004a, 75). For, they involve an appeal to intuition, since "geometrical figures are signs or mnemonic symbols of space intuition" (Hilbert 1996a, 1100). Therefore, "we will never rely on them" (Hilbert 2004b, 541). Indeed, "a theorem is only proved when the proof is completely independent of the figure. The proof must call step by step on the preceding axioms" (Hilbert 2004a, 75). Thus, one must make sure that every proof using figures can be replaced by another one that is figure-free, and hence that the use of figures "is by no means necessary" (ibid.).

Despite these declarations of intent, however, Hilbert does not make sure that every demonstration in the *Grundlagen der Geometrie* using figures can be replaced by another one that is figure-free. This is the case even of the very first demonstration in the *Grundlagen der Geometrie*, which purports to show that "for any two points *A* and *C* there always exists at least one point *D* on the line *AC* that lies between *A* and *C*" (Hilbert 1987, 5).

In the demonstration, Hilbert uses the following four axioms (see Hilbert 1987, 3–5):

I 3: There exist at least two points on a line. There exist at least three points that do not lie on a line.

II 2: For two points *A* and *C*, there always exists at least one point *B* on the line *AC* such that *C* lies between *A* and *B*.

II 3: Of any three points on a line, there exists no more than one that lies between the other two.

II 4: Let *A*, *B*, *C* be three points which do not lie on a line, and let *a* be a line in the plane *ABC* which does not pass through any of the points *A*, *B*, *C*: then, if the line *a* passes through a point of the segment *AB*, it also passes through a point of the segment *AC*, or through a point of the segment *BC*. (In other words, if a line *a* enters a triangle, is also comes out again).

Hilbert's demonstration is as follows. By axiom I 3, there exists a point *E* not on the line *AC*. By axiom II 2, on the line *AE* there exists a point *F* such that *E* is a point of the segment *AF*. By axioms II 2 and II 3, on the line *FC* there exists a point *G* which does not lie on the segment *FC*. By axiom II 4, the line *EG* must intersect the segment *AC* at a point *D*. Therefore, there exists at least one point *D* on the line *AC* that lies between *A* and *C*.

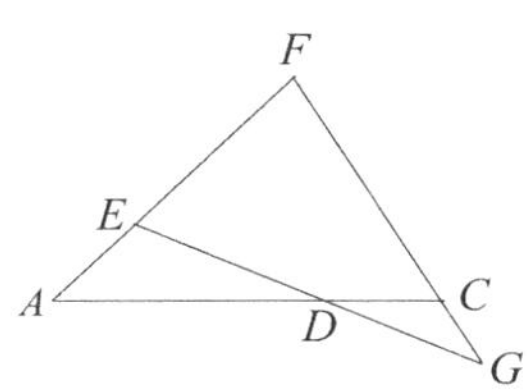

Hilbert demonstration, however, is not a deduction from the axioms I 3, II 2, II 3 and II 4, because it uses various facts which are not deduced from them but are read off from the figure. For example, Hilbert uses axiom II 4 to conclude, from the fact that the three points *A*, *F*, *C* do not lie on a line, and *EG* is a line in the plane *AFC*

which does not pass through any of the points *A*, *F*, *C* but passes through a point of the segment *AF*, that the line *EG* passes through a point of the segment *AC*. But axiom II 4 does not authorize Hilbert to conclude this. It authorizes him only to conclude that *EG* passes either through a point of the segment *AC* or through a point of the segment *FC*. To conclude that *EG* passes through a point of the segment *AC*, Hilbert should demonstrate that *EG* cannot pass through any point of the segment *FC*. But Hilbert does not demonstrate this, he tacitly reads it off from the figure.

Thus Hilbert does not make sure that even the very first demonstration of the *Grundlagen der Geometrie* can be replaced by another one that is figure-free, and hence that the use of the figure is by no means necessary.

Admittedly, Hilbert's demonstration could be replaced by a deduction from axioms, but this would require introducing additional axioms (see Meikle and Fleuriot 2003). Moreover, the resulting demonstration would be much more complex and essentially different from Hilbert's demonstration, and would not be perspicuous. The same holds of other demonstrations in the *Grundlagen der Geometrie* which make use of figures.

This has several implications. First, it shows that the view of the logical dispensability of mathematical diagrams is problematic.

Second, it shows that there is no foundation for the widespread opinion that, in the *Grundlagen der Geometrie*, "Hilbert gave a complete axiomatization of geometry and used only the axioms cited in the further development of the subject," so he gave "the first purely deductive treatment of geometry" (Menzler-Trott 2007, 296).

Third, it shows that there is no foundation for the claim that quantification logic is adequate to mathematics. This claim assumes that a demonstration involving figures can be replaced by a logical proof. But Hilbert's demonstration can be replaced only with a deduction from axioms that is vastly less perspicuous than Hilbert's demonstration. So Kant's appeal to figures can be disposed of only at the cost of an essential loss of perspicuity in the demonstration. What is more important, the claim that quantification logic is adequate to mathematics is invalid by the strong incompleteness theorem for second-order logic (see below).

12.10 The View of the Formal Dispensability of Mathematical Diagrams

The second view about mathematical diagrams is the view of the formal dispensability of mathematical diagrams. This view does not exclude the use of mathematical diagrams in demonstrations, but requires making sure that it is formally dispensable, namely, that a demonstration using mathematical diagrams might always be replaced by a formal derivation in some formal system.

A full expression of this view is Feferman's Thesis: "Every proof of a mathematical theorem that may involve diagrams and other possible devices can be replaced by" a "statements-only proof," and such proof "can in principle be formalized, i.e., has a counterpart formal derivation in some formal system" (Feferman 2012, 380).

The view expressed by Feferman's Thesis is widespread among mathematicians. Wiedijk even says that "all mathematicians will start using formalization for their proofs," and "referees will insist on getting a formalized version before they want to look at a paper" (Wiedijk 2008, 1414). And "Voevodsky predicted" that, in a few years, "journals will only accept articles accompanied by their machine-verifiable equivalents" (Harris 2015, 60).

That the view expressed by Feferman's Thesis is widespread among mathematicians, depends on the tacit assumption that the correctness of formal derivations in a formal system can be mechanically verified, without "need to fall back upon intuition or meaning" (Hilbert 1967, 475). This guarantees against the possibility of error, because formalizations "bring mathematical concept-formations and inferences into such a form that they are irrefutable" (Hilbert 1996b, 1152).

This tacit assumption, however, is invalid because it is confronted with two major obstacles.

First, the process of transforming a demonstration into a formal derivation is non-mechanical. This particularly holds of a demonstration based on diagrams. To formally dispense with mathematical diagrams, one needs to transform the reasoning based on them into a formal derivation, but such transformation is patently non-mechanical, because it involves the analysis of a figure. Thus, one cannot exclude the possibility that the transformation leads to a pseudo validation of an alleged theorem. This contradicts the claim that formal derivations bring mathematical concept-formations and inferences into such a form that they are irrefutable.

Second, mechanically verifying the correctness of a formal derivation requires mechanically verifying the correctness of the computer program for mechanically verifying the correctness of the formal derivation. But, by Rice's theorem, there is no algorithm for deciding non-trivial questions on computer programs, such as their correctness (see, for example, Robič 2015, Theorem 9.4).

Thus, Hersh says that "it just is not the case that a doubtful proof would become certain by being formalized. On the contrary, the doubtfulness of the proof would then be replaced by the doubtfulness of the coding and programming" (Hersh 1979, 43).

Hacking says that "Voevodsky's speculation" that "it would soon be possible to design proof checkers" which "could effectively verify correctness of proofs written in the appropriate machine-readable language," is "an extreme version of leibnizian" computer-generated and computer-checked "proof" (Hacking 2014, 25). According to it, "an author submits a paper with a proof or proof sketch, together with a programme for checking the proof, and a confirmation that, when run, the computer says, 'OK'" (ibid., 25–26). But this leads to a regress, because "who checks that the programme is sound? (ibid., 26).

Even if the previous two major obstacles could be removed, two other major ones would remain.

First, to assert that every demonstration of a given part of mathematics can be represented as a formal derivation, there should be a formal system in which formal demonstrations for all true propositions of that part of mathematics could be given. But, by Gödel's first incompleteness theorem, there is no such formal system.

Second, to assert that appealing to formalization guarantees against the possibility of error, it would be necessary to demonstrate the consistency of the formal system in which formal demonstrations for all true propositions of the field in question could be given, and to demonstrate it by absolutely reliable means. But, by Gödel's second incompleteness theorem, this is impossible.

Therefore, the view of the formal dispensability of mathematical diagrams is invalid, even more than the view of the logical dispensability of mathematical diagrams.

12.11 Mathematical Diagrams and Axiomatic Method

While disagreeing about the role of mathematical diagrams in mathematics, both Kant and the supporters of the views of the logical or formal dispensability of mathematical diagrams believe that the method of mathematics is the axiomatic method, that mathematics is true and certain, and that the basis of its truth and certainty is intuition.

The only difference between Kant and the supporters of the views in question is that they refer to two different versions of the axiomatic method.

Kant refers to the material axiomatic method. Indeed, he says that the method of mathematics is "the synthetic," namely axiomatic "method, which proceeds from the first grounds of a cognition and stops at the last consequences" (Kant 1992, 515). The first grounds of mathematical cognition are the axioms, which must be true of a fixed system of things, specified in advance, because axioms are directed at things that are "determinately given in pure intuition a priori" (Kant 1998, A724/B752). Since axioms are directed at things that are determinately given in pure intuition a priori, mathematics is true and certain, so "as soon as a proposition is maintained by mathematics it is infallible" (Kant 1992, 342).

Conversely, the supporters of the views of the logical or formal dispensability of mathematical diagrams refer to the formal axiomatic method. Thus, as we have seen in Chap. 6, Hilbert says that the method of mathematics is the axiomatic method, which proceeds from axioms that need not be true of a fixed system of things, specified in advance, but can be chosen arbitrarily. They are only subject to the condition that they must be demonstrated to be consistent by a demonstration based on the intuitive mode of thought. On this basis, mathematics is true and certain, and indeed is the paragon of reliability and truth.

But the view that the method of mathematics is the axiomatic method, that mathematics is true and certain, and that the basis of its truth and certainty is intuition, is invalid.

On the one hand, the view that the method of mathematics is the axiomatic method is invalid by Gödel's first incompleteness theorem. For, by the latter, for any consistent, sufficiently strong, formal system, there are propositions of the system that are true but cannot be deduced from the principles of the system.

On the other hand, the view that mathematics is true and certain, and that the basis of its absolute truth and certainty is intuition, is invalid by Gödel's second incompleteness theorem. For, by the latter, for any consistent sufficiently strong formal system, it is impossible to demonstrate, by absolutely reliable means, in particular by means based on intuition, that the principles of the system are consistent, let alone that they are true.

12.12 The Axiomatic View of Mathematical Diagrams

Despite Gödel's incompleteness theorems, however, as we have seen in Chap. 2, several mathematicians still believe that the method of mathematics is the axiomatic method, that mathematics is true and certain, and that the basis of its truth and certainty is intuition.

In particular, the disregard of the reasons that make it impossible to hold that the method of mathematics is the axiomatic method has led to the attempt, mentioned at the beginning of this chapter, to make the use of mathematical diagrams in demonstrations rigorous by fitting it into the axiomatic method. This has given rise to the axiomatic view of mathematical diagrams.

According to the axiomatic view of mathematical diagrams, all thought is linguistic in nature and cannot occur apart from language. Mathematical diagrams can be used in mathematics insofar as their relevant properties can be translated into formal languages and formal systems.

The basis for the axiomatic view of mathematical diagrams has been laid down by Hilbert himself. He says that geometrical figures can play the role of geometrical symbols, but "in order that these geometrical figures may be incorporated in the general treasure of mathematical symbols, a rigorous axiomatic investigation of their conceptual content is necessary" (Hilbert 1996a, 1101). For, "the use of geometrical signs as a means of strict proof presupposes the exact knowledge and complete mastery of the axioms which underlie those figures" (ibid., 1100–1101).

According to the supporters of the axiomatic view of mathematical diagrams, one can actually have formal languages and formal systems into which the relevant properties of mathematical diagrams can be translated.

Thus, Barwise and Etchemendy say that "one can have rigorous, logically sound (and complete) formal systems based on diagrams," so "there is no principled distinction between inference formalisms that use text and those that use diagrams" (Barwise and Etchemendy 1996, 181).

Avigad, Dean and Mumma say that one can have formal systems which spell out "precisely what inferences can be 'read off' from the diagram" (Avigad et al. 2009, 701).

The axiomatic view of mathematical diagrams, however, is inadequate. For, it does not deal with mathematical diagrams qua diagrams, and is unable to account for the role of mathematical diagrams as means of discovery because formal systems are not means of discovery. As Hanna and Sidoli say, the axiomatic view of mathematical diagrams is unable to account for the fact that mathematical diagrams qua diagrams "can often inspire the theorem to be proved and point out approaches to the construction of the proof itself" (Hanna and Sidoli 2007, 74).

The axiomatic view of mathematical diagrams could at most deal with mathematical diagrams as means of justification. But, in this capacity, such view is subject to the limitations of the axiomatic method, starting with Gödel's incompleteness theorems. In particular, the necessity to demonstrate by absolutely reliable means that the formal systems resulting from translating particular mathematical diagrams are consistent, is not a mere abstract necessity but a real one. For example, Miller 2012 demonstrated that Mumma's formal system for Euclid's diagrams is inconsistent.

12.13 Axiomatic View, Romanticism, and Thought as Linguistic

If the axiomatic view of mathematical diagrams does not deal with mathematical diagrams qua diagrams and is unable to account for the role of diagrams as means of discovery, it is because, as we have seen, such view assumes that all thought is linguistic in nature and cannot occur apart from language.

This assumption is a result of the impact of Romanticism on mathematics. In Chap. 6 it has been said that the impact of Romanticism on mathematics is especially clear from the formal axiomatic method. The axiomatic view of mathematical diagrams is an aspect of this impact.

The German proto-Romantic Hamann says that "our entire ability to think rests on language" (Hamann 1949–1957, III, 286). For, "reason is language, logos" (Hamann 1955–1975, V, 177). Therefore, "without language we would have no reason" (Hamann 1949–1957, III, 231).

Much in the same way, the axiomatic view of mathematical diagrams assumes that all thought is linguistic in nature and cannot occur apart from language. On this basis, it says that mathematical diagrams can be used in mathematics insofar as their relevant properties can be translated into formulas and inference rules of a formal system.

But the assumption that all thought is linguistic in nature and cannot occur apart from language, is unfounded. It conflicts with the experience of several mathematicians, who say that language plays no role in their creative thought.

Thus, Hadamard says: "Words are totally absent from my mind when I really think," and "do not reappear in my consciousness before I have accomplished or given up the research" (Hadamard 1954, 75).

Ulam says: "When I think about mathematical ideas, I see the abstract notions in symbolic pictures," consisting of "visual assemblages," so I "think pictorially, not verbally" (Ulam 1991, 183). Such thinking is "not in terms of words or syllogisms or signs," but in terms of "symbolic pictures" with "linkages between them forming a sort of meta- or super-logic with its own rules" (ibid.).

12.14 The Heuristic View of Mathematical Diagrams

Unlike the axiomatic view of mathematical diagrams, Carter 2019 and Starikova 2010 deal with the role of mathematical diagrams as means of discovery in certain parts of contemporary mathematics. They, however, do not deal with the general question of what is the method of mathematics and how mathematical diagrams fit into it. To deal with this question, an alternative view of mathematical diagrams is necessary. Such alternative view is provided by the heuristic view of mathematical diagrams.

According to the heuristic view of mathematical diagrams, mathematical diagrams are figures such that, on the basis of data acquired from them, one formulates hypotheses which are sufficient conditions for solving problems by the analytic method.

Examples of the use of mathematical diagrams in solving problems by the analytic method are implicit in several solutions to problems considered in previous chapters. Here is another simple example.

Let A be an arbitrary point on a diameter of the circle. Draw a perpendicular AB to the diameter, connecting A to the circumference, point B. Then draw a perpendicular BC at B intersecting, at point C, the diameter perpendicular to the original diameter.

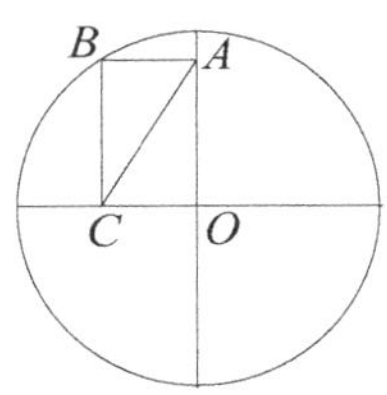

We consider the following problem.

Length of the line AC: Given the length d of the diameter of the circle, find the length of the line AC.

Solving the problem may seem difficult, because the only numerical information available is the length d of the diameter of the circle. However, draw radius OB.

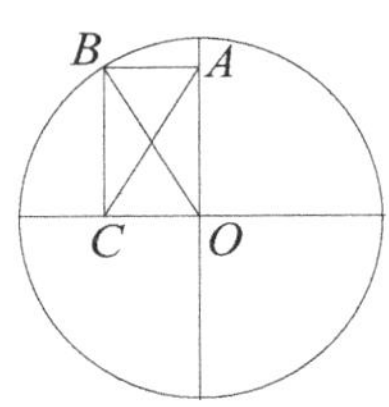

By analyzing the diagram and the conditions under which the problem would be solved, we non-deductively arrive at the following hypothesis:

(I) The diagonals of a rectangle are equal.

Hypothesis (I) is a sufficient condition for solving the problem. For, all angles of the quadrilateral *BAOC* are right angles (being included between perpendicular lines), so *BAOC* is a rectangle, hence *AC* and *OB* are the diagonals of a rectangle. Then, by hypothesis (I), length of $AC =$ length of $OB = d/2$ (*OB* being a radius). This solves the problem.

Hypothesis (I) is plausible, because the arguments for it are stronger than those against it. But hypothesis (I) is itself a problem that must be solved. To solve it, by analyzing the diagram and the conditions under which the problem would be solved, we arrive at the following hypothesis:

(II) Two right-angled triangles with the corresponding legs equal are equal.

Hypothesis (II) is a sufficient condition for solving the problem. For, in the right-angled triangles *BOC* and *ACO*, $BC = AO$ (being opposite sides of a rectangle), and $CO = CO$ (being the same side). Then, by hypothesis (II), the right-angled triangles *BOC* and *ACO* are equal, so *AC* and *OB*, the diagonals of the rectangle *BAOC*, are equal. Namely hypothesis (I). This solves the problem. The solution increases the plausibility of hypothesis (I).

Hypothesis (II) is plausible, because the arguments for it are stronger than those against it. But hypothesis (II) is itself a problem that must be solved. To solve it, it is necessary to formulate another hypothesis. And so on.

12.15 Heuristic View and Intuition Argument

On the heuristic view, mathematical diagrams are not subject to the above arguments against the view that mathematical diagrams are essential to mathematics.

First, on the heuristic view, mathematical diagrams are not subject to the intuition argument, because intuition plays no role in the analytic method. Indeed, as already said in Chap. 5, intuition plays no role in the discovery of hypotheses, because they are obtained from the problem, and possibly other data already available, including data acquired from a mathematical diagram, by some non-deductive rules, so not by intuition but by inference. Intuition plays no role in the justification of hypotheses, because their plausibility is established by comparing the arguments for them and the arguments against them on the basis of experience, so not by intuition but by inference.

In particular, intuition plays no role in the use of mathematical diagrams. This is clear from the fact that, in the philosophical tradition, intuition is conceived as the immediate apprehension of an object that is given to us.

Thus, Kant says that, "in whatever way and through whatever means a cognition may relate to objects, that through which it relates immediately to them" is "intuition. This, however, takes place only insofar as the object is given to us" (Kant 1998, A19/B33).

Now, one could hardly say that the use of mathematical diagrams in the analytic method is based on the immediate apprehension of an object that is given to us. A mathematical diagram is not an object that is given to us, it is made by the mathematician by a number of operations, through which he gives a representation of concepts and makes constructions on them. From such representations and constructions, the mathematician gathers data from which he obtains hypotheses by some non-deductive rule.

One might think that, rather than intuition, perception plays an important role in the use of mathematical diagrams. But, although undoubtedly perception plays some role in the use of mathematical diagrams, such role is a limited one. For, a mathematical diagram can be merely imagined. Moreover, Greek mathematicians were able to develop an extensive mathematics using very roughly drawn mathematical diagrams. This appears, for example, from one of the oldest surviving diagrams of Euclid's *Elementa*, illustrating Book II, Proposition 5.

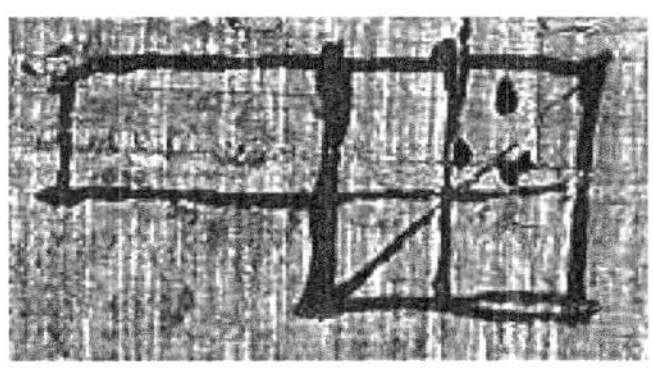

As Bueno says, a mathematical diagram need not "convey faithfully every aspect of the relations among the relevant mathematical objects. What is crucial is that the central, relevant relations" are "properly displayed in the diagram" (Bueno 2016, 11). On the other hand, "the non-relevant relations need not be properly represented at all: diagrams often misrepresent many features of the objects under consideration" (ibid.).

12.16 Heuristic View and Quantification Logic Argument

On the heuristic view, mathematical diagrams are not subject to the quantification logic argument, because the assumption that quantification logic is adequate to mathematics is invalid.

Indeed, by the strong incompleteness theorem for second-order logic, there is no consistent formal system for second-order logic capable of deducing all second-order logical consequences of any given set of propositions. Therefore, quantification logic is not an adequate theory of deductive reasoning, and hence is inadequate to mathematics, because much mathematics requires second-order logic.

Moreover since, by the strong incompleteness theorem for second-order logic, there is no consistent formal system for second-order logic capable of deducing all second-order logical consequences of any given set of propositions, to deduce them it is necessary to discover new rules. Therefore, the problem of discovery is essential to logic.

But quantification logic cannot say anything about the problem of discovery because it is deductive logic, and deductive logic is non-ampliative, namely the conclusion of a deduction is implicitly contained in the premisses.

Only the analytic method can say something about the problem of discovery, because the analytic method is both a means of discovery and a means of justification, and intuition plays no role in it, only inference.

12.17 Heuristic View and Incorrectly Drawn Diagram Argument

On the heuristic view, mathematical diagrams are not subject to the incorrectly drawn diagram argument, because the assumption that incorrectly drawn diagrams can lead to absurd propositions is invalid. For, such absurd propositions arise from a misuse of mathematical diagrams.

Thus, in the case of the 'demonstration' of the proposition that from a point not on a line two distinct perpendiculars can be drawn to the line, this can be seen as follows. Draw *CB*, *DB*, and *AB*. Since *CBA* and *DBA* are triangles in a semicircle, they are both right-angled. By Euclid's fifth postulate, from a point not on a line a single perpendicular can be drawn to the line. Therefore, the perpendiculars to *AB* at point *B*, *CB* and *DB*, are simply segments of the same line, *CBD*. This shows that, when *CD* was first drawn, it should have intersected the circles not in two points, *E* and *F*, but rather at one point, *B*, the point of intersection of the circles. Without the existence of points *E* and *F*, the fallacious demonstration could not have been produced. Thus the absurd conclusion is eliminated.

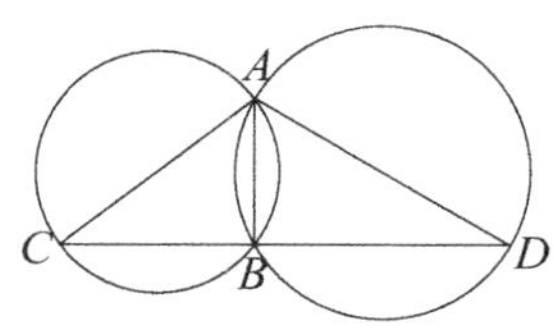

This shows that the absurd proposition that from a point not on a line two distinct perpendiculars can be drawn to the line, arises from a misuse of a mathematical diagram.

The situation is similar to that of the absurd propositions of the calculus of infinitesimals of Leibniz and Newton, or of set theory, which arise from a misuse of concepts.

Errors arising from misuses of concepts are ubiquitous in mathematics. So if, following Hilbert, one says that mathematical diagrams can easily be misleading and therefore we will never rely on them, then, much in the same way, one should say that concepts can easily be misleading and therefore we will never rely on them. But this is absurd. One can only say that we must take care that mathematical diagrams are properly drawn and properly analyzed, just as we must take care that concepts are carefully formulated and properly analyzed.

12.18 Heuristic View and Limit Diagram Argument

On the heuristic view, mathematical diagrams are not subject to the limit diagram argument, because the assumption that even correctly drawn diagrams, specifically mathematical diagrams about limits, can lead to absurd propositions, is invalid. For, such absurd propositions arise from reading into the diagrams something which is not contained therein.

Thus, in the case of the 'demonstration' of the proposition that the sum of the two legs of a right-angled triangle is equal to the hypotenuse, this can be seen as follows. By the definition of limit, $\lim_{n\to\infty} S_n = BC$ if and only if, for every $\varepsilon > 0$, there is an integer N such that $| S_n - BC | < \varepsilon$ whenever $n \geq N$. On the contrary, for any n, $S_n = AB + AC$, so S_n is a constant, and hence it is impossible that $\lim_{n\to\infty} S_n = BC$. Therefore $\lim_{n\to\infty} S_n = AB + AC$, but $\lim_{n\to\infty} S_n = BC$ is incorrect. Thus the absurd conclusion is eliminated.

This shows that the absurd proposition that the sum of the two legs of a right-angled triangle is equal to the hypotenuse, arises from reading into the diagram something which is not contained therein, namely that, as n increases, the steplike line from B to C in the limit will coincide with BC. The diagram only authorizes us to read in it that, as n increases, the steplike line from B to C will look ever less and less different from BC. It does not authorize us to say that in the limit it will coincide with BC.

Of course, recognizing this requires experience. As Bråting and Pejlare say, "our mathematical experience, as well as the context, is important while interpreting the visualization and 'seeing' the relation," indeed only "with experience we can learn to interpret the visualization in different ways, depending on what is asked for" (Bråting and Pejlare 2008, 354).

12.19 Heuristic View and Particularity Argument

On the heuristic view, mathematical diagrams are not subject to the particularity argument, because the assumption that, since mathematical diagrams are particular figures, inferring general conclusions from them can lead to absurd propositions, is invalid. For, such absurd propositions arise from overlooking the non-particularity condition, that a demonstration must make no use of the particularity of the diagram.

Thus, in the case of the 'demonstration' of the proposition that the perpendicular bisectors of the sides of a triangle intersect at a single point inside the triangle, this can be seen as follows. In the demonstration it is assumed, without argument, that the point of intersection O of the perpendicular bisectors of the sides AB and AC is internal to ABC. This, however, only holds of acute triangles. Indeed, it can be shown that the perpendicular bisectors of an obtuse triangle intersect at a single point outside the triangle. This is apparent from the following figure:

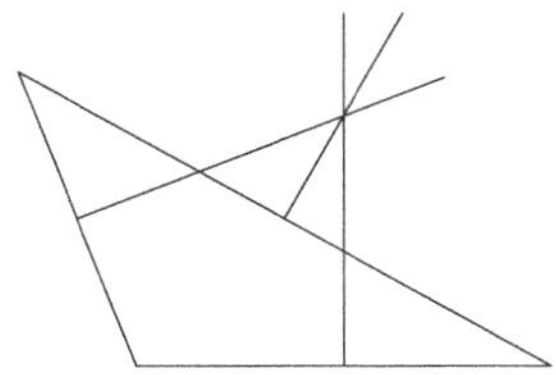

Therefore, what the demonstration really shows is not that the perpendicular bisectors of the sides of a triangle intersect at a single point inside the triangle, but only that the perpendicular bisectors of the sides of a triangle intersect at a single point. Thus the absurd conclusion is eliminated.

The non-particularity condition, that a demonstration must make no use of the particularity of the diagram, is already stated by Aristotle, who says that "the geometer calls this a foot-long line, this a straight line, and says that they are breadthless, though they are not: yet he does not use these things as something from which to make deductions" (Aristotle, *Analytica Priora*, A 41, 49 b 35–37).

The non-particularity condition can also be the basis for an alternative formulation of the universal generalization rule (see Cellucci 2009).

12.20 Adequacy of the Heuristic View of Mathematical Diagrams

It has been argued above that, if the axiomatic view of mathematical diagrams does not deal with mathematical diagrams qua diagrams and is unable to account for the role of mathematical diagrams as means of discovery, it is because such view assumes that all thought is linguistic in nature and cannot occur apart from language.

The heuristic view of mathematical diagrams does not make this assumption. For, it is based on the analytic method and, as we have seen in Chap. 5, in the analytic method the hypotheses for the solution to a problem are obtained from the problem and possibly other data, including data acquired from mathematical diagrams, by some non-deductive rule. Therefore, the heuristic view of mathematical diagrams deals with mathematical diagrams qua diagrams, and is able to account for the role of mathematical diagrams as means of discovery, because the analytic method is a method of discovery.

This has a significant implication for the notion of demonstration. On the axiomatic view of mathematical diagrams, a demonstration consists only of words. Conversely, on the heuristic view of mathematical diagrams, as said in Chap. 10, a demonstration can be words only, or words and diagrams, or even diagrams only.

On the basis of what has been said above, it seems fair to conclude that, unlike the axiomatic view of mathematical diagrams, the heuristic view provides an adequate account of diagrams in mathematics.

References

Avigad, Jeremy, Edward Dean, and John Mumma. 2009. A formal system for Euclid's *Elements*. *The Review of Symbolic Logic* 2: 700–768.

Ayer, Alfred Jules. 1952. *Language, truth and logic*. New York: Dover.

Barwise, John, and John Etchemendy. 1996. Heterogeneous logic. In *Logical reasoning with diagrams*, ed. Gerard Allwein and John Barwise, 179–200. Oxford: Oxford University Press.

Bråting, Kajsa, and Johanna Pejlare. 2008. Visualization in mathematics. *Erkenntnis* 68: 345–358.

Bueno, Otávio. 2016. Visual reasoning in science and mathematics. In *Model-based reasoning in science and technology*, ed. Lorenzo Magnani and Claudia Casadio, 3–19. Cham: Springer.

Carter, Jessica. 2019. Exploring the fruitfulness of diagrams in mathematics. *Synthese* 196: 4011–4032.

Catton, Philip, and Clemency Montelle. 2012. To diagram, to demonstrate: To do, to see, and to judge in Greek geometry. *Philosophia Mathematica* 20: 25–57.

Cellucci, Carlo. 2009. The universal generalization problem. *Logique et Analyse* 205: 3–20.

Feferman, Solomon. 2012. And so on ... : Reasoning with infinite diagrams. *Synthese* 186: 371–386.

Giaquinto, Marcus. 2008. Visualizing in mathematics. In *The philosophy of mathematical practice*, ed. Paolo Mancosu, 22–42. Oxford: Oxford University Press.

Hacking, Ian. 2014. *Why is there philosophy of mathematics at all?* Cambridge: Cambridge University Press.

Hadamard, Jacques. 1954. *The psychology of invention in the mathematical field*. Mineola: Dover.

Hamann, Johann Georg. 1949–1957. *Sämtliche Werke*. Wien: Herder.

———. 1955–1975. *Briefwechsel*. Frankfurt: Insel.

Hanna, Gila, and Nathan Sidoli. 2007. Visualization and proof: A brief survey of philosophical perspectives. *ZDM Mathematics Education* 39: 73–78.

Harris, Michael. 2015. *Mathematics without apologies: Portrait of a problematic vocation*. Princeton: Princeton University Press.

Hersh, Reuben. 1979. Some proposals for reviving the philosophy of mathematics. *Advances in Mathematics* 31: 31–50.

Hilbert, David. 1967. The foundations of mathematics. In *From Frege to Gödel: A source book in mathematical logic 1879–1931*, ed. Jean van Heijenoort, 464–479. Cambridge: Harvard University Press.

———. 1987. *Grundlagen der Geometrie*. Stuttgart: Teubner.

———. 1996a. From 'mathematical problems'. In *From Kant to Hilbert: A source book in the foundations of mathematics*, ed. William Ewald, vol. 2, 1096–1105. Oxford: Oxford University Press.

———. 1996b. The grounding of elementary number theory. In *From Kant to Hilbert: A source book in the foundations of mathematics*, ed. William Ewald, vol. 2, 1149–1157. Oxford: Oxford University Press.

———. 2004a. Die Grundlagen der Geometrie. In *David Hilbert's lectures on the foundations of geometry 1891–1902*, ed. Michael Hallett and Ulrich Majer, 72–123. Berlin: Springer.

———. 2004b. Grundlagen der Geometrie. In *David Hilbert's lectures on the foundations of geometry 1891–1902*, ed. Michael Hallett and Ulrich Majer, 540–602. Berlin: Springer.

Hintikka, Jaakko. 2020. Kant's theory of mathematics: What theory of what mathematics? In *Kant's philosophy of mathematics*, ed. Carl Posy, and Ofra Rechter, vol. 1: *The critical philosophy and its roots*, 85–102. Cambridge: Cambridge University Press.

Kant, Immanuel. 1992. *Lectures on logic*. Cambridge: Cambridge University Press.

———. 1998. *Critique of pure reason*. Cambridge: Cambridge University Press.

Klein, Felix. 2016. *Elementary mathematics from a higher standpoint*. Berlin: Springer.

Leibniz, Gottfried Wilhelm. 1965. *Die Philosophischen Schriften*. Hildesheim: Olms.

Locke, John. 1975. *An essay concerning human understanding*. Oxford: Oxford University Press.

Macbeth, Danielle. 2014. *Realizing reason: A narrative of truth and knowing*. Oxford: Oxford University Press.
Manders, Ken. 2008. The Euclidean diagram. In *The philosophy of mathematical practice*, ed. Paolo Mancosu, 80–133. Oxford: Oxford University Press.
Meikle, Laura I., and Jacques D. Fleuriot. 2003. Formalizing Hilbert's Grundlagen in Isabelle/Isar. In *Theorem proving in higher order logics*, ed. David Basin and Burkhart Wolff, 319–334. Berlin: Springer.
Menzler-Trott, Eckart. 2007. *Logic's lost genius: The life of Gerhard Gentzen*. Providence: American Mathematical Society.
Miller, Nathaniel. 2012. On the inconsistency of Mumma's Eu. *Notre Dame Journal of Formal Logic* 53: 27–54.
Netz, Reviel. 1999. *The shaping of deduction in Greek mathematics: A study in cognitive history*. Cambridge: Cambridge University Press.
Newton, Isaac. 1967–1981. *The mathematical papers*. Cambridge: Cambridge University Press.
Pasch, Moritz. 1882. *Vorlesungen über neuere Geometrie*. Leipzig: Teubner.
Robič, Borut. 2015. *The foundations of computability theory*. Berlin: Springer.
Saito, Ken, and Nathan Sidoli. 2012. Diagrams and arguments in ancient Greek mathematics: Lessons drawn from comparison of the manuscript diagrams with those in modern critical editions. In *The history of mathematical proof in ancient traditions*, ed. Karine Chemla, 135–162. Cambridge: Cambridge University Press.
Starikova, Irina. 2010. Why do mathematicians need different ways of presenting mathematical objects? The case of Cayley graphs. *Topoi* 29: 41–51.
Tennant, Neil. 1986. The withering away of formal semantics? *Mind & Language* 1: 302–318.
Ulam, Stanislaw M. 1991. *Adventures of a mathematician*. Berkeley and Los Angeles: University of California Press.
Wiedijk, Freek. 2008. Formal proof: Getting started. *Notices of the American Mathematical Society* 55: 1408–1414.
Wolff, Christian. 1716. *Mathematisches lexicon*. Leipzig: Bleditschens.
———. 1736. *Vernünftige Gedanken von den Kräften des menschlichen Verstandes*. Leipzig: Kanger.

Chapter 13
Notations

Abstract According to heuristic philosophy of mathematics, one of the tasks of the philosophy of mathematics is to give an answer to the question: What is the nature of mathematical notations? The prevailing answer to this question is the precision-conciseness view of mathematical notations. The chapter discusses this answer and argues that it is inadequate. Then it offers an alternative answer, the heuristic view of mathematical notations.

Keywords Precision-conciseness view of mathematical notations · Heuristic view of mathematical notations · Heuristic power of mathematical notations · Symbolic notations · Diagrammatic notations

13.1 The Precision-Conciseness View of Mathematical Notations

As we have seen in Chap. 3, according to heuristic philosophy of mathematics, one of the tasks of the philosophy of mathematics is to give an answer to the question: What is the nature of mathematical notations?

The formulation of a mathematical problem has an important part in finding a solution to the problem, because it transforms a puzzle for which we wish to obtain a resolution into an approachable task. This requires, first, an analysis of the terms of the problem and, second, a representation of the terms of the problem.

Often, a representation of the terms of the problem is given relying heavily upon ordinary language. But ordinary language is inconvenient in this respect, because it is ambiguous and verbose. To remedy this, mathematical notations are used. This raises the question of the nature of mathematical notations.

In the past century, the prevailing answer to this question has been the precision-conciseness view of mathematical notations, according to which mathematical notations are just representations that are precise and concise.

C. Cellucci, *The Making of Mathematics*, Synthese Library 448,
https://doi.org/10.1007/978-3-030-89731-4_13

Thus, Pólya says that a mathematical notation is a representation that is "precise" and "concise" (Pólya 2004, 135). Precise, because "a good notation should be unambiguous" (ibid., 136). Concise, because a good notation should achieve economy of expression to be "easy to remember and easy to recognize" (ibid., 137).

Enderton says that a mathematical notation is a representation that "has the advantages of" both "preciseness (so that expressions are less ambiguous)" and "conciseness (so that expressions are shorter)" (Enderton 1977, 13).

Biletch, Yu, and Kay say that a mathematical notation is a representation that satisfies the criteria of "consistency or unambiguity," and "conciseness" (Biletch et al. 2015, 18). Consistency or unambiguity involves that "each object or concept should only be represented by one symbol or one class of symbols" and "each symbol should only correspond to one object" (ibid.). Conciseness involves "choosing the shortest form of expression without compromising consistency" (ibid.).

The chapter discusses this answer to the question of the nature of mathematical notations and argues that it is inadequate. Then, it offers an alternative answer, the heuristic view of mathematical notations.

13.2 Precision-Conciseness View and Inessentiality

According to the precision-conciseness view of mathematical notations, since mathematical notations are just representations that are precise and concise, the role of mathematical notations in mathematics is only to achieve precision and conciseness, they have no other function. In particular mathematical notations are not essential to mathematics.

Thus, van der Waerden says that someone who considers mathematical notations to be not only precise and concise but essential to mathematics, "like many non-mathematicians, grossly overestimates the importance of symbolism in mathematics" (van der Waerden 1976, 205). These "people see our papers full of formulae, and they think that these formulae are an essential part of mathematical thinking," but "we, working mathematicians, know that in many cases the formulae are not at all essential, only convenient" (ibid.).

Devlin says that today "mathematics is presented to us by way of symbolic expression. But just how essential are those symbols?" (Devlin 2013, 88). Now, doing mathematics "is primarily a thinking process – something that takes place mostly in" the mathematicians's "head. Even when" mathematicians "are asked to 'show all their work', the collection of symbolic expressions that they write down is not necessarily the same as the process that goes on in their minds when they do math" (ibid., 89). Therefore, "the symbols are not essential" (ibid.).

Tagliasco and Vincenzi say that while, "for the communication of mathematical discourse, a good notation is not only useful, but also essential," conversely, "for the development of mathematics a good notation is useful, but not essential" (Tagliasco and Vincenzi 1998, 159).

13.3 The Philosophical Neglect of Mathematical Notations

In the past century, since the precision-conciseness view has been the prevailing answer to the question of the nature of mathematical notations, and according to it mathematical notations are not essential to mathematics, mathematical notations have not been the object of much philosophical reflection.

There are general histories of mathematical notations, such as Cajori 1993, and Mazur 2014, histories of special kinds of mathematical notations, such as Aczel 2015, Barrow 2001, Cajori 1923, Chrisomalis 2010, Glaser 1971, Ifrah 2000, Kaplan 2000, Massa Esteve 2012, Menninger 1992, Seife 2000, Sen and Agarwal 2016, and Serfati 2005. But, in the philosophy of mathematics, there has been little attention to mathematical notations, except for special aspects of them.

Significant philosophical works on mathematical notations are the book chapters Brown 2008, Chapter 6, and Colyvan 2012, Chapter 8; the articles on notations in formal languages Macbeth 2013, Dutilh Novaes 2012, and Schlimm 2018a; the articles on the role of mathematical notations in cognition Muntersbjorn 1999, De Cruz and de Smedt 2013, Dutilh Novaes 2013, and Schlimm 2018b. Nevertheless, the fact remains that in the philosophy of mathematics the question of the nature of mathematical notations has not been much discussed.

This is somewhat surprising because the prevailing answer to the question of the nature of mathematical notations, the precision-conciseness view, is inadequate.

13.4 Shortcomings of the Precision-Conciseness View

Indeed, the assumption of the precision-conciseness view that mathematical notations are not essential to mathematics, contrasts with the actual status of notations in mathematics.

As Colyvan says, "good notation does serious work in mathematics" (Colyvan 2012, 145). Mathematics would not "be served equally well by alternative notations," so "getting the notation right features prominently in mathematical practice," hence "especially important here is the development of an account of how good notation can advance mathematics" (ibid.).

But the precision-conciseness view of notations is unable to give such an account.

An example of this is provided by Newton's derivative notation $\dot{y}$ (Newton 1967–1981, VIII, 130), and Leibniz's derivative notation $\frac{dy}{dx}$ (Leibniz 1971, II, 196).

These notations were both precise and concise but, while for some questions either of them could be used, for many others Newton's notation was ill-suited. Indeed, as Montelle says, Newton's notation was "cumbersome because of its orientation around the notion of time. Any expression containing a derivative not taken with respect to time was quite involved. For example, to express the derivative $\frac{dz}{dx}$, Newton had to write $\dot{z} : \dot{x}$ which literally meant $\frac{dz}{dt} : \frac{dx}{dt}$" (Montelle 2011, 153).

What is more important, Newton's notation was unable "to allow for a richer development in the field of calculus, and to incorporate and represent for the equivalents in multi-variate calculus and calculus of variations" (ibid.).

Newton himself did not seem to realize the importance of notation. Indeed, concerning the difference between his derivative notation and Leibniz's derivative notation, Newton says: "But these are only ways of notation and signify nothing to the method itself, which may be without them" (Newton 1850, 175).

In spite of its shortcomings, "English mathematicians of the eighteenth century remained loyal to Newton's notation" (Montelle 2011, 153). As a result, since Leibniz's notation was free from those shortcomings, most developments of the calculus in the eighteenth century took place on continental Europe. Therefore, eventually, Newton's "tradition lost favor" (ibid.).

The derivative notation is an example of how good notation can advance mathematics. But the precision-conciseness view of notations is unable to account for this, because Newton's notation and Leibniz's notation were both precise and concise, and, according to the precision-conciseness view, to be precise and concise is all that is required of a representation to be a notation. Nevertheless, such notations were not equally suited for some questions.

To account for how good notation can advance mathematics, an alternative view of mathematical notations is needed.

13.5 The Heuristic View of Mathematical Notations

An alternative view of mathematical notations is the heuristic view of mathematical notations. According to it, mathematical notations are representations that are not only precise and concise, but can also be indispensable to discover solutions to problems, so they are essential to mathematics. The role of notations in mathematics is not only to achieve precision and conciseness, but to discover solutions to problems, and the choice of a suitable notation may be crucial to this purpose.

Dutilh Novaes asks whether "mathematical notations" may be "required for mathematical discovery" (Dutilh Novaes 2013, 48). The heuristic view of mathematical notations gives an affirmative answer to this question.

This contrasts with the precision-conciseness view of mathematical notations, according to which, as Colyvan says, "it's mathematical objects that matter, not the notations we use for them," so "a mathematical object by any other name would be just as useful" (Colyvan 2012, 144). But this conflicts with the fact that "some names are more revealing about that which they name" (ibid., 145).

Indeed, notations are not merely names, and not any notation will do. The choice of a suitable notation may be crucial to discover solutions to problems. As Châtelet says, "there exist 'objects'" whose "presentation, by a new notation, shakes up a whole mathematical continent" (Châtelet 2000, 8). Therefore, notations have heuristic power.

13.6 Zero Notation

An example of the heuristic power of notations is the notation for zero. Such notation eventually led to the formation of the concept of zero, adding it to the other natural numbers. (On the history of zero and the notation for zero, see Aczel 2015, Barrow 2001, Kaplan 2000, Seife 2000, and Sen and Agarwal 2016).

The number zero has been the object of extreme judgments. On the one hand, zero has been belittled as a mere nothing, a sheer nonentity. Thus, when in Shakespeare's *King Lear*, the Fool wants to signify that, by giving away his kingdom to the two daughters who flatter him, and banishing the third one who loves him, Lear has become a mere nothing, the Fool tells him: "Now thou art an O without a figure. I am better than thou art, now. I am a fool; thou art nothing" (Shakespeare, *King Lear*, Act 1, Scene 4, vv. 197–199). By telling Lear 'Now thou art an O without a figure', the Fool means to say: You are a zero without a number in front of it to give it value, so you are nothing.

On the other hand, zero has been exalted as the pivot on which all parts of reality swing. Thus, Kaplan says: "Look through" zero "and you will see the world. For," the "shining tools of mathematics let us follow the tackling course everything takes through everything else – and all of their parts swing on the smallest of pivots, zero" (Kaplan 2000, 1).

Clearly, these judgments are far-fetched. Zero is neither a mere nothing, a sheer nonentity, nor the pivot on which all parts of reality swing. This does not mean that zero is unimportant. Indeed, the notation for zero, 0, which stands for the number zero, has rendered at least three substantial services to mathematics.

As Whitehead says, the first service the notation for zero has rendered has been to make the decimal "notation possible – no slight service" (Whitehead 2017, 36). Indeed, by serving as a placeholder in the decimal notation, it has made possible to write all numbers, no matter how big or small, using ten symbols alone. But, although initially 0 served as a mere placeholder, the desire to assimilate the meaning of the sign 0 to that of the symbols 1, 2, ..., 9, which represented numbers, eventually led to regard 0 as a symbol representing a number, the number zero. However, the adoption of zero as a number was neither quick nor straightforward, and even today the number "that creates the most difficulties" to children "is zero" (Lengnink and Schlimm 2010, 256).

The second service the notation for zero has rendered has been to permit to express all equations in the same form, writing all symbols representing variables, definite numbers other than zero, and operations, on the left-hand side of the symbol =, and the symbol 0 on the right-hand side. Thus, Descartes says that equations are "sums composed of several terms, partly known, partly unknown, some of which are equal to the rest; or rather, all of which taken together are equal to nothing," namely to 0, "for it is often best to consider them in this form" (Descartes 1996, VI, 444). Indeed, to consider them in this form "made possible the growth of the modern conception of 'algebraic form'," for example, it made possible to say that $x^2 - 3x + 2 = 0$ and $x^2 - 4 = 0$ are "equations of the same form" since they are both "quadratic equations" (Whitehead 2017, 37).

The third service the notation for zero has rendered has been to permit to ignore minor differences of form. For example, among $x^2 - 3x + 2 = 0$ and $x^2 - 4 = 0$ there is the difference that x, as distinct from x^2, appears in the former but not in the latter. This difference is minor in comparison with the fact that they are both quadratic equations, and the difference can be obliterated using the properties that, for all x, $0 \times x = 0$ and $x + 0 = x$. Then $x^2 - 4 = 0$ can be written in the form $x^2 + (0 \times x) - 4 = 0$, so it "belongs to the same general form as does $x^2 - 3x + 2 = 0$" (ibid., 38).

This shows that the notation for zero has made "possible types of investigation which would have been impossible without it" (ibid.).

13.7 Decimal Notation

Another example of the heuristic power of notations is the decimal notation, of which the notation for zero is part. Brown even says that the decimal notation is "perhaps the greatest notational invention of all time" (Brown 2008, 85).

As already mentioned above, the decimal notation has made it possible to write all numbers, no matter how big or small, using ten symbols alone. In his *Liber Abbaci* of 1202, Leonardo Pisano (also known as Fibonacci, short for 'filius Bonacci'), noted this by saying: "The nine Indian figures are: 9 8 7 6 5 4 3 2 1. With these nine figures, and with the sign 0 which the Arabs call zephir, any number whatsoever is written, as is demonstrated below" (Leonardo Pisano 2002, 17).

At the time of Leonardo Pisano, the number notation most in use in Europe was the Roman notation. Leonardo Pisano was the most influential of all medieval writers in introducing the decimal notation in Europe. But, despite its advantages, the decimal notation met with resistance which was not easily overcome. Thus, in 1299, article CII of the statute of the Arte del Cambio [Guild of Money Changers] in Florence stated: "Be it established and ordained that nobody from this Guild shall dare or allow that he or another write or let write in his book or notebook or in any part of it, in which he writes debits and credits, anything which is understood by means of or in the letters of the abacus," namely in decimal figures," but rather "he shall write it openly and in full by way of" Roman "letters" (Arte del Cambio 1955, 72–73).

Eventually, however, the decimal notation became established, primarily because of its advantages for business purposes. On account of their close association with the transactions of merchants, numerals in decimal notation were even popularly called "figure mercantesche [mercantile figures]" (Università dei Mercanti 1554, folio 57 recto).

The decimal notation is based on the fact that it can be shown that any natural number a can be written uniquely in the form

$$a = a_n 10^n + a_{n-1} 10^{n-1} + \ldots + a_1 10 + a_0$$

where the a_i's are the $n + 1$ digits of a in base 10 and $a_n \neq 0$. Such form is usually abbreviated as $a = a_n a_{n-1} \ldots a_1 a_0$.

The decimal notation allows us to easily discover algorithms for the basic arithmetic operations, the algorithms we currently use.

For example, the decimal notation allows us to easily discover an algorithm for the addition of two numbers a and b in decimal notation, a with $n + 1$ digits, and b with $m + 1$ digits. Without loss of generality we may assume that $a \geq b$ and that, by possibly inserting $n - m$ zeros in b,

$$a = a_n 10^n + a_{n-1} 10^{n-1} + \ldots + a_1 10 + a_0$$

$$b = b_n 10^n + b_{n-1} 10^{n-1} + \ldots + b_1 10 + b_0.$$

Then,

$$a + b = (a_n + b_n)10^n + (a_{n-1} + b_{n-1})10^{n-1} + \ldots + (a_1 + b_1)10 + (a_0 + b_0).$$

So, the problem of computing $a + b$ reduces to that computing $a_i + b_i$, for $0 \leq i \leq n$, taking account of the possible carryover. The solution to the latter problem is given by the addition table:

+	0	1	2	3	4	5	6	7	8	9
0	0	1	2	3	4	5	6	7	8	9
1	1	2	3	4	5	6	7	8	9	10
2	2	3	4	5	6	7	8	9	10	11
3	3	4	5	6	7	8	9	10	11	12
4	4	5	6	7	8	9	10	11	12	13
5	5	6	7	8	9	10	11	12	13	14
6	6	7	8	9	10	11	12	13	14	15
7	7	8	9	10	11	12	13	14	15	16
8	8	9	10	11	12	13	14	15	16	17
9	9	10	11	12	13	14	15	16	17	18

Similarly, the decimal notation allows us to easily discover algorithms for subtraction, multiplication, and division. Such algorithms strictly depend on the decimal notation.

This does not mean that it is not possible to formulate algorithms for addition, subtraction, multiplication, and division in other notations. In particular, it is possible to formulate algorithms for them in Roman notation, although it is unclear to what extent the Romans actually knew and used such algorithms (see Detlefsen, Erlandson, Heston, and Young 1976; Kennedy 1981; Maher and Makowski 2001; Macbeth 2014, Chapter 2; Schlimm and Neth 2008; Turner 1951). However, calculations in Roman notation are more cumbersome than calculations in decimal notation.

13.8 Algebraic Notation

Another example of the heuristic power of notations is Viète's algebraic notation, which is at the origin of algebra. Grabiner even says that "general symbolic notation of the type we now take for granted was introduced in 1591 by the French mathematician François Viète," and "proved to be the greatest instrument of discovery in the history of mathematics" (Grabiner 1974, 357).

Viète says that, unlike numerical logistic, namely Diophantus's arithmetic, algebra "no longer employs its logic on numbers" (Viète 1646, 1). It uses its logic through a new symbolic logistic involving reasoning with variables, which "is far more fruitful and powerful than numerical logistic for comparing magnitudes with one another" (ibid.). While numerical logistic "operates with numbers," symbolic logistic, namely algebra, operates with "symbols or signs for things as, say, the letters of the alphabet" (ibid., 4). In particular, it lets "the given magnitudes be distinguished from the undetermined unknowns by a constant, permanent and very clear symbol, as, for instance, by designating the unknown magnitudes by the letter *A* and the other vowels *E*, *I*, *O*, *U*, *Y* and the given magnitudes by the letters *B*, *G*, *D* and the other consonants" (ibid., 8).

As an example of the heuristic power of Viète's algebraic notation, let us compare Diophantus's and Viète's solutions to the first problem of Diophantus's *Arithmetica*.

On the one hand, Diophantus formulates the problem as follows: "To divide a given number into two numbers having a given difference" (Diophantus, *Opera omnia*, I, 16, ed. Tannery). Diophantus's solution, in modernized notation, is: "Let the given number be 100, the given difference be 40" (ibid.). Moreover, let "the lesser number $= x$, so the greater number will be $x + 40$. Hence the sum of both will be $2x + 40$. But the given number is 100. Therefore $100 = 2x + 40$" (ibid.). By subtracting 40 from 100, "$2x = 60$ remains, whence $x = 30$. Therefore, the lesser number $= 30$, the greater $= 70$" (ibid.).

On the other hand, Viète formulates the problem as follows: "Given the difference between two roots and their sum, to find the roots" (Viète 1646, 42). Viète's solution is: "Let *B* be the difference between the two roots, and let *D* be their sum" (ibid.). Moreover, "let the smaller root be *A*. The greater will then be $A + B$. Therefore, the sum of the roots is $2A + B$. But this has been given as *D*. Hence $2A + B$ is equal to *D*. And, by transposition, $2A$ is equal to $D - B$, hence, dividing both members by 2, *A* is equal to $\frac{1}{2}D - \frac{1}{2}B$" (ibid.). Or, "let the greater root be *E*. The smaller will then be $E - B$. Therefore, the sum of the roots is $2E - B$. But this has been given as *D*. Hence $2E - B$ is equal to *D*. And, by transposition, $2E$ is equal to $D + B$, hence, dividing both members by 2, *E* is equal to $\frac{1}{2}D + \frac{1}{2}B$" (ibid.). Thus it has been shown that "half the sum of the roots minus half their difference is equal to the smaller root," and half the sum of the roots plus half their difference is equal "to the greater" (ibid.). For example, "let *B* be 40 and *D* be 100. Then *A* is 30 and *E* is 70" (ibid.).

While Diophantus's solution only concerns particular numbers, Viète's solution holds for all numbers. As Meskens says, "although Diophantus promises a general solution, he solves the problem using specific values" (Meskens 2010, 159). The "proposed numbers and parameters do indeed lead to a solution, but the reasoning behind their selection is never explained. The reader can only guess as to whether or not Diophantus had a general algorithm at his disposal" (ibid.). Conversely, "in Viète's algebraic application," Diophantus's "problems are resolved in a general way. His structure of proof is such that there are no specific problems to resolve on the way, just ordinary equations" (ibid.). The introduction of the algebraic notation helps Viète to discover a general solution to the problem, and indeed is indispensable to that aim.

Viète's algebraic notation was significantly improved by Descartes, who employed the first letters of the alphabet *a*, *b*, *c*, etc. to designate the known quantities, and the last letters *z*, *y*, *x*, etc. to designate the unknown ones. But "the decisive step was taken by Viète" (Kleiner 2007, 8). With his algebraic notation, he not only "transformed algebra from a study of the specific to the general," but "created a symbolic science that would apply widely, assisting in both the discovery and the demonstration of results" (ibid., 9).

13.9 Exponential Notation

Another example of the heuristic power of notations is Descartes's exponential notation. As Cajori says, "the modern symbolism for powers of numbers was introduced by René Descartes in his *La géométrie*, Paris, 1637" and "his notation spread rapidly" (Cajori 1913, 13).

Although Viète, with his algebraic notation, created a symbolic science, his algebraic notation required substantial improvements. One of them concerns the exponential notation.

Viète designates the powers of a magnitude by saying that "the first of the ladder of magnitudes is side [latus] or root [radix]," the second is "square [quadratum]," the third is "cube [cubus]," the fourth is "squared-square [quadrato-quadratum]," the fifth is "squared-cube [quadrato-cubus]," the sixth is "cubed-cube [cubo-cubus]," the seventh is "squared-squared-cube [quadrato-quadrato-cubus]," the eight is "squared-cubed-cube [quadrato-cubo-cubus]," the ninth is "cubed-cubed-cube [cubo-cubo-cubus]. And the remaining ones are to be denominated from these by this series and method" (Viète 1646, 3).

In terms of this designation of the powers of a magnitude, Viète establishes that "a side" multiplied "by itself produces a square. A side" multiplied "by a square produces a cube. A side" multiplied "by a cube produces a squared-square. A side" multiplied "by a squared-square produces a squared-cube. A side" multiplied "by a squared-cube produces a cubed-cube. And likewise the other way around" (ibid., 6). Moreover, "a square" multiplied "by itself produces a squared-square. A square" multiplied "by a cube produces a squared-cube. A square" multiplied "by a squared-square produces a cubed-cube. And likewise the other way around" (ibid.).

In addition, "a cube" multiplied "by itself produces a cubed-cube. A cube" multiplied "by a squared-square produces a squared-squared-cube. A cube" multiplied "by a squared-cube produces a squared-cubed-cube. A cube" multiplied "by a cubed-cube produces a cubed-cubed-cube. And likewise the other way around, and so on in the same order" (ibid.).

One could hardly say, however, that Viète's designation of the powers of a magnitude has a heuristic power in finding these properties of multiplying a magnitude by a magnitude.

Different is the case of Descartes's exponential notation. He introduces it as follows: "I write" the expression "aa or a^2 to multiply a by itself; and a^3 to multiply it once again by a, and so on to infinity" (Descartes 1996, VI, 371).

Now, Descartes's exponential notation almost immediately suggests the laws $a^n \cdot a^m = a^{n+m}$, $\frac{a^n}{a^m} = a^{n-m}$ for $a \neq 0$, and $(a^n)^m = a^{n \cdot m}$, where n and m are integers. With just one step further, the law $\frac{a^n}{a^m} = a^{n-m}$ with $n = 0$ and $m = 1$ suggests that $a^{-1} = \frac{1}{a}$. And the law $a^n \cdot a^m = a^{n+m}$ extended to fractions suggests that $a^{\frac{1}{2}} \cdot a^{\frac{1}{2}} = a^{\frac{1}{2}+\frac{1}{2}} = a^1$ and hence $a^{\frac{1}{2}} = \sqrt{a}$. This shows the heuristic power of Descartes's exponential notation.

13.10 Derivative Notation

Another example of the heuristic power of notations is the already mentioned Leibniz's derivative notation $\frac{dy}{dx}$.

Such notation suggests an analogy between derivatives and fractions. This contrasts with the fact that, according to the current formulation of the calculus, the derivative is a limit, not a fraction. Nevertheless, Leibniz's derivative notation "remains the most widely used and best known notation" because, in addition to being "extremely flexible," it "clearly expresses the fact that the derivative is very like a fraction" (Pender et al. 2012, 250). This is the basis of the heuristic power of Leibniz's derivative notation.

For example, through the analogy between derivatives and fractions, Leibniz's derivative notation helps to discover the chain rule: $\frac{dy}{dx} = \frac{dy}{du} \cdot \frac{du}{dx}$.

Indeed, it is through this analogy that Leibniz himself discovered the chain rule, as it is apparent from the fact that he called the latter the rule of "compound fractions [compositae fractiones]" (Leibniz 1923–, VII.5, 614). Specifically, Leibniz uses the rule in question to calculate the derivative of $\sqrt{a + bz + cz^2}$ as the composite of the square root function and the function $a + bz + cz^2$. Indeed, he says that, to calculate such derivative, one lets $a + bz + cz^2 = x$, one calculates the derivative of $\sqrt{x}$, and one multiplies by $\frac{dx}{dz}$. This is just the chain rule.

As Velleman says, "it is not surprising that derivatives act in some ways like fractions, because they are close to fractions. If we let Δx stand for a small change in x," then the derivatives in equation $\frac{dy}{dx} = \frac{dy}{du} \cdot \frac{du}{dx}$ "are close to the fractions $\Delta y/\Delta x$, $\Delta y/\Delta u$ and $\Delta u/\Delta x$, and since these really are fractions, we are justified in saying that $\frac{\Delta y}{\Delta x} = \frac{\Delta y}{\Delta u} \cdot \frac{\Delta u}{\Delta x}$" (Velleman 2017, 177).

Similarly, we are justified in saying that $\frac{\Delta y}{\Delta x} \cdot \frac{\Delta x}{\Delta y} = 1$, hence $\frac{\Delta y}{\Delta x} = \frac{1}{\frac{\Delta x}{\Delta y}}$. Thus, through the analogy between derivatives and fractions, Leibniz's derivative notation helps to discover the inverse function theorem: $\frac{dy}{dx} = \frac{1}{\frac{dx}{dy}}$.

Leibniz's derivative notation also extends to the second-order partial derivative notation $\frac{\partial^2 u}{\partial x \partial y}$, the derivative taken first with respect to x and then with respect to y. Leibniz himself foreshadowed this notation (see Leibniz 1971, II, 261). Such notation suggests an analogy between second-order partial derivatives and fractions. Through this analogy, the second-order partial derivative notation helps to discover the theorem of symmetry of second-order partial derivatives: $\frac{\partial^2 u}{\partial x \partial y} = \frac{\partial^2 u}{\partial y \partial x}$.

That, by suggesting an analogy between derivatives and fractions, Leibniz's derivative notation has heuristic power, explains why Leibniz's notation is still the most widely used and best known notation for the derivative, despite the fact that, according to the current formulation of the calculus, the derivative is a limit, not a fraction.

13.11 Kinds of Notations

Notations have been considered above without distinguishing between kinds of notations. However, it is possible to distinguish between at least three kinds of notations: symbolic notations, diagrammatic notations, and a mix of symbolic and diagrammatic notations. The character of symbolic notations and diagrammatic notations can be described as follows.

Symbolic notations express the terms of a problem by means of symbols, namely words, letters, or special characters.

Diagrammatic notations express the terms of a problem by means of mathematical diagrams, namely graphical arrangements.

Often it is assumed that notations consist exclusively of symbolic notations, since figures are considered to be something different from notations.

For example, Pólya says that, "in all sorts of problems, but especially in mathematical problems which are not too simple, suitable notation and geometrical figures are a great and often indispensable help" (Pólya 2004, 132). For, "when we have found" a "notation that we can easily retain," or "a figure that we can easily imagine," we "can reasonably believe that we have made some progress" (ibid., 183). Thus, Pólya assumes that geometrical figures are something different from notations.

This assumption, however, is invalid. For example, as already said in Chap. 7, when the Pythagoreans say 'Let *ABC* be a triangle', what they mean to say is 'Let *ABC* stand for a triangle'. So, what they mean to say is that the drawn triangle *ABC* is a diagrammatic notation for a triangle.

This had already been made quite clear by both Plato and Aristotle.

Thus, Plato says that mathematicians “make use of the visible forms and reason about them, though they are not thinking of these, but of those things which they resemble; not of the figures they draw, but of the square as such and the diagonal as such, and so on” (Plato, *Respublica*, VI 510 d 5–8).

Aristotle says that “the geometer does not draw any conclusion from the fact that the lines which he has himself described are thus and so; rather, he relies on the things indicated by these lines” (Aristotle, *Analytica Posteriora*, A 10, 77 a 1–3).

That, when saying ‘Let *ABC* be a triangle’, the Pythagoreans mean to say that the drawn triangle *ABC* is a diagrammatic notation for a triangle, shows that the Pythagoreans consider geometrical figures to be notations.

Therefore, it is invalid to assume that notations consist exclusively of symbolic notations because figures are something different from notations.

Châtelet says that, in several cases, we can make “the notation concrete by identifying it with a diagram” (Châtelet 2006, 36). Thus Châtelet underlines that “notations and diagrams are not so distinct, each potentially playing a significant role in mathematical ontogenesis” (de Freitas and Sinclair 2014, 213).

13.12 The Role of Symbolic and Diagrammatic Notations

Clearly, diagrammatic notations are most useful when the terms of the problem involve spatial concepts, while symbolic notations are most useful when the terms of the problem do not involve spatial concepts.

However, symbolic notations and diagrammatic notations are not opposed. Aristotle even uses the term ‘diagramma’ as a name for ‘geometrical proposition’, for example when he says that “even geometrical propositions [diagrammata] are discovered by an actualization” (Aristotle, *Metaphysica*, Θ 9, 1051 a 21–22).

In fact, mathematicians frequently use a mix of symbolic and diagrammatic notations. Even when the terms of the problem involve spatial concepts, a mix of symbolic and diagrammatic notations is often necessary.

An example of this is Leonardo da Vinci’s diagrammatic representation of the demonstration of Euclid, *Elementa*, Book I, Proposition 7. Leonardo’s drawing (in *Codex Atlanticus*, folio 483 verso, https://www.codex-atlanticus.it/#/Detail?detail=483) consists of 13 figures, displayed right to left because Leonardo was left-handed and wrote backward. For perspicuity, below the 13 figures have been rearranged left to right, somewhat edited, and polished.

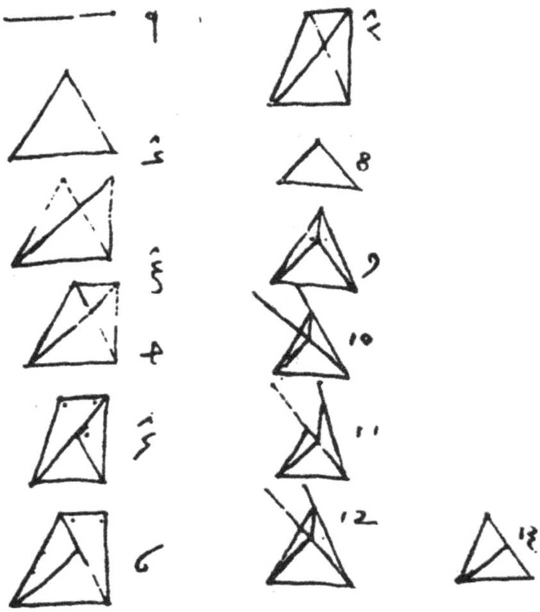

Despite rearrangement, editing, and polishing, from the sequence of the 13 figures alone it would be impossible to understand the demonstration they represent. The demonstration can be understood only if symbols are attached to the figures and a suitable text is added.

This example explains why mathematics frequently uses a mix of symbolic and diagrammatic notations.

13.13 Symbolic Notations and Lettered Diagrams

There are many ways in which mathematics uses a mix of symbolic and diagrammatic notations.

First, the mathematical diagrams which are used to express the terms of a problem are often lettered diagrams, namely diagrams with symbols attached to them.

Lettered diagrams have been extensively used from the very beginning of mathematics as discipline. As Netz says, "Greek mathematical exchanges, as a rule, were accompanied by something like lettered diagrams" (Netz 1999, 14). In particular, Euclid constantly uses lettered diagrams to express the terms of a problem.

However, already Aristotle uses lettered diagrams, for example, to express the terms of the problem to demonstrate that the angles at the base of an isosceles triangle are equal. Aristotle's text contains no diagram, but there is no doubt that he refers to a diagram, because he speaks of "geometrical diagrams such as the one that is used to demonstrate that the angles at the base of an isosceles triangle are equal" (Aristotle, *Analytica Priora*, A 24, 41 b 14–15). There is also no doubt that the diagram to which Aristotle refers is a lettered diagram, because in his text he uses letters *A*, *B*, *C*, *D*, *E*, *F* to indicate points, lines, or angles of the diagram.

More disputed is the question to which lettered diagram Aristotle actually refers. The most likely interpretation is that of Aristotle's ancient commentators such as Philoponus (see Philoponus, *In Aristotelis Analytica Priora commentaria*, 253.28–254.23, ed. Wallies). On the basis of it, the lettered diagram to which Aristotle refers is the following.

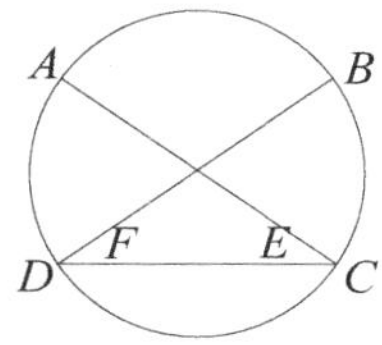

This accounts for Aristotle's text: "Let the lines *A*, *B* be drawn towards the centre" (Aristotle, *Analytica Priora*, A 24, 41 b 15). Now, "the angles *AC* and *BD* are equal" because "in general the angles in a semicircle are equal," and "the angles *C* and *D* are equal" because "this holds for every angle of a segment" (ibid., A 24, 41 b 16–18). Moreover, "the remaining angles *E* and *F* are equal, because equal parts," *C* and *D*, "have been subtracted from the whole angles" *AC* and *BD* "that are themselves equal," and "when equals are subtracted from equals, the remainders are equal" (ibid., A 24, 41 b 20–22).

Thus, according to this interpretation, the isosceles triangle is constructed by drawing two straight lines *A* and *B* towards the centre of a circle, and then connecting the points, *C* and *D*, at which the lines intersect with the circumference of the circle. The angles *AC* and *BD* are the angles formed by the lines *AC* and *BD* with the circumference of the circle. The angles *C* and *D* are the angles formed by the base of the triangle with the circumference of the circle. The angles *E* and *F* are the angles at the base of the triangle. Now, the angles *AC* and *BD* are equal, because they are angles in a semicircle. The angles *C* and *D* are equal, because they are angles of a segment. Then, the angles *E* and *F* are equal, because they are the result of subtracting equal angles, *C* and *D*, from equal angles, *AC* and *BD*. And, when equals are subtracted from equals, the remainders are equal.

Note that this demonstration is different from Euclid's demonstration of *Elementa*, Book I, Proposition 5. Perhaps it might be the demonstration that occurred in some of the elements of geometry composed, before Aristotle's time, by Hippocrates of Chios, or, at Aristotle's time, by Leon or Theudius.

13.14 Diagrammatic Notations, Spatial, and Non-spatial Concepts

Diagrammatic notations can be useful even when the problem does not involve spatial concepts.

For example, consider the problem of demonstrating that $\sqrt{2}$ is an irrational number. This requires demonstrating that there are no natural numbers m and n such that $\sqrt{2} = m/n$, or equivalently $m^2 = 2n^2$, so the problem does not involve spatial concepts. Nevertheless, as shown by Apostol (2000), using a diagram can be useful to solve the problem.

Suppose that $\sqrt{2}$ is a rational number. Let m and n be the smallest natural numbers such that $m^2 = 2n^2$. For such m and n, by the Pythagorean theorem, there is an isosceles right-angled triangle ABC with hypotenuse $AC = m$ and legs $AB = BC = n$.

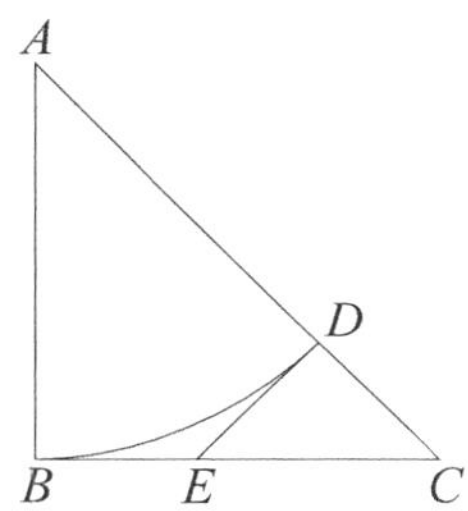

Draw a circular arc BD centred at A. Then $AD = AB = n$.

Draw a line DE perpendicular to AC at point D. Then the angle EDC is a right angle and, since the triangle ABC is an isosceles right-angled triangle, the angle DCE is half a right angle. Therefore, the triangle CDE is an isosceles right-angled triangle.

Now $DC = AC - AD = m - n$, hence also $DE = m - n$. Since BE and DE are tangents to the same arc from the same point E, $BE = DE$, so $BE = m - n$, hence $EC = BC - BE = n - (m - n) = 2n - m$.

Since CDE is an isosceles right-angled triangle with hypotenuse $EC = 2n - m$ and legs $DC = DE = m - n$, by the Pythagorean theorem, $(2n - m)^2 = 2(m - n)^2$. But $2n - m < m$ and $m - n < n$. This contradicts the assumption that m and n are the smallest natural numbers such that $m^2 = 2n^2$. Therefore, $\sqrt{2}$ is an irrational number.

Other solutions to the problem of demonstrating that $\sqrt{2}$ is an irrational number, not using diagrams, will be considered in Chap. 14.

13.15 Diagrammatic Use of Symbolic Notations

Third, symbolic notations themselves can be used to form mathematical diagrams for solving problems.

An example of this is provided by the following solution to the problem 'What is the sum of the first x odd numbers x^2, for arbitrary given x?' by the method of the 10 year old Gauss (see Chap. 7). Imagine the first x odd numbers, $1, 3, 5, \ldots, 2x - 1$, to be displayed in two rows in reverse order:

1	3	5	...	$2x - 5$	$2x - 3$	$2x - 1$
$2x - 1$	$2x - 3$	$2x - 5$	...	5	3	1

The sum of each column is $2x$ and there are x columns, so the total sum is $2x \cdot x$, therefore the sum of the first x odd numbers, for arbitrary given x, is $x \cdot x$, namely x^2.

In this solution to the problem, the numbers 1, 3, 5, . . ., $2x - 1$, displayed twice in two rows in reverse order, are used to form a mathematical diagram to solve the problem.

As already pointed out in Chap. 7 with respect to Gauss's solution, this solution does not use mathematical induction. It is a uniform argument that works for arbitrary given x, the kind of argument which Herbrand calls a prototype.

Another example of the fact that symbolic notations themselves can be used to form mathematical diagrams for solving problems is provided by the following solution to the problem of finding the sum of the absolutely convergent series $S = 1/2 + 1/4 + 1/8 + 1/16 + \ldots$. Now, $2S = 1 + 1/2 + 1/4 + 1/8 + 1/16 + \ldots$. Suppose the terms of $2S$ and the terms of S to be displayed in two rows, with the ith term of S under the $(i + 1)$th term of $2S$:

1	+	1/2	+	1/4	+	1/8	+	1/16	+	...
0	+	1/2	+	1/4	+	1/8	+	1/16	+	...

Subtracting the first column yields 1, and subtracting each of the other columns yields 0, therefore $S = 2S - S = 1$.

In this solution to the problem, the terms of $2S$ and the terms of S, displayed in two rows with the ith term in S under the $(i + 1)$th term in $2S$, are used to form a mathematical diagram to solve the problem.

These examples show that symbolic notations themselves can be used to form mathematical diagrams for solving problems.

13.16 Adequacy of the Heuristic View of Mathematical Notations

It has been argued above that mathematical notations are not merely names and not any notation will do, because the choice of a suitable notation may be crucial to discover solutions to problems.

Only the heuristic view of mathematical notations can account for this. According to it, the role of notations in mathematics is not only to achieve precision and conciseness but to discover solutions to problems, and the choice of a suitable notation may be crucial to this purpose.

In particular, since notations may consist of symbolic notations, diagrammatic notations, or a mix of symbolic and diagrammatic notations, notations may be crucial to discover solutions to problems in different ways.

Symbolic notations may be crucial to discover solutions to problems if they consist of symbols which, by their special character, may suggest hypotheses that are sufficient conditions for solving the problems.

Diagrammatic notations may be crucial to discover solutions to problems if they consist of figures which, by their special character, may suggest hypotheses that are sufficient conditions for solving the problems.

Notations which are a mix of symbolic and diagrammatic notations may be crucial to discover solutions to problems if they consist of a mix of symbols and figures which, by their special character, may suggest hypotheses that are sufficient conditions for solving the problems.

Since notations may be crucial to discover solutions to problems and only the heuristic view is able to account for this, the heuristic view is adequate.

References

Aczel, Amir Dan. 2015. *Finding zero*. New York: Palgrave Macmillan.

Apostol, Tom M. 2000. Irrationality of the square root of two – A geometric proof. *The American Mathematical Monthly* 107 (9): 841–842.

Arte del Cambio. 1955. *Statuti dell'Arte del Cambio di Firenze (1299–1316): Con aggiunte e correzioni fino al 1320*. Florence: Olschki.

Barrow, John D. 2001. *The book of nothing*. New York: Vintage Books.

Biletch, Barry David, Yu Hongji, and Kathleen Rose Kay. 2015. *An analysis of mathematical notation: For better or for worse*. Worchester: Worchester Polytechnic Institute.

Brown, James Robert. 2008. *Philosophy of mathematics: A contemporary introduction to the world of proofs and pictures*. London: Routledge.

Cajori, Florian. 1913. History of the exponential and logarithmic concepts I. *The American Mathematical Monthly* 20: 5–14.

———. 1923. The history of notations of the calculus. *Annals of Mathematics* 25 (1): 1–46.

———. 1993. *A history of mathematical notation*. Mineola: Dover.

Châtelet, Gilles. 2000. *Figuring space: Philosophy, mathematics and physics*. Cham: Springer.

———. 2006. Interlacing the singularity, the diagram and the metaphor. In *Virtual mathematics: The logic of difference*, ed. Simon B. Duffy, 31–45. Manchester: Clinamen.

Chrisomalis, Stephen. 2010. *Numerical notation: A comparative history*. Cambridge: Cambridge University Press.

Colyvan, Mark. 2012. *An introduction to the philosophy of mathematics*. Cambridge: Cambridge University Press.

De Cruz, Helen, and Johan de Smedt. 2013. Mathematical symbols as epistemic actions. *Synthese* 190: 3–19.

de Freitas, Elizabeth, and Nathalie Sinclair. 2014. *Mathematics and the body: Material entanglements in the classroom*. Cambridge: Cambridge University Press.

Descartes, René. 1996. *Œuvres*. Paris: Vrin.

Detlefsen, Michael, Douglas K. Erlandson, J. Clark Heston, and Charles M. Young. 1976. Computation with Roman numerals. *Archive for History of Exact Sciences* 15: 141–148.

Devlin, Keith. 2013. The music of math games. *American Scientist* 101: 87–91.

Dutilh Novaes, Catarina. 2012. *Formal languages in logic: A philosophical and cognitive analysis*. Cambridge: Cambridge University Press.

———. 2013. Mathematical reasoning and external symbolic systems. *Logique et Analyse* 221: 45–65.
Enderton, Herbert B. 1977. *Elements of set theory*. New York: Academic.
Glaser, Anton. 1971. *History of binary and other nondecimal numeration*. Los Angeles: Tomash Publishers.
Grabiner, Judith V. 1974. Is mathematical truth time-dependent? *The American Mathematical Monthly* 81: 354–365.
Ifrah, Georges. 2000. *The universal history of numbers: From prehistory to the invention of the computer*. New York: Wiley.
Kaplan, Robert. 2000. *The nothing that is: A natural history of zero*. Oxford: Oxford University Press.
Kennedy, James G. 1981. Arithmetic with Roman numerals. *The American Mathematical Monthly* 88 (1): 29–32.
Kleiner, Israel. 2007. *A history of abstract algebra*. Cham: Birkhäuser.
Leibniz, Gottfried Wilhelm. 1923–. *Sämtliche Schriften und Briefe*. Berlin: Academie Verlag.
———. 1971. *Mathematische Schriften*. Hildesheim: Olms.
Lengnink, Katja, and Dirk Schlimm. 2010. Learning and understanding numeral system: Semantic aspects of number representations from an educational perspective. In *PhiMSAMP. Philosophy of mathematics: Sociological aspects and mathematical practice*, ed. Benedikt Löwe and Thomas Müller, 235–264. London: College Publications.
Leonardo Pisano. 2002. *Liber abaci*. Cham: Springer.
Macbeth, Danielle. 2013. Writing reason. *Logique et Analyse* 221: 25–44.
———. 2014. *Realizing reason: A narrative of truth and knowing*. Oxford: Oxford University Press.
Maher, David W., and John F. Makowski. 2001. Literacy evidence for Roman arithmetic with fractions. *Classical Philology* 96: 376–399.
Massa Esteve, Maria Rosa. 2012. The role of symbolic language on the transformation of mathematics. *Philosophica* 87: 153–193.
Mazur, Joseph. 2014. *Enlightening symbols: A short history of mathematical notation and its hidden powers*. Princeton: Princeton University Press.
Menninger, Karl. 1992. *Number words and number symbols: A cultural history of numbers*. Mineola: Dover.
Meskens, Ad. 2010. *Travelling mathematics – The fate of Diophantos' arithmetic*. Cham: Springer.
Montelle, Clemency. 2011. A 'symbolic' history of the derivative. In *Mathematical time capsules*, ed. Dick Jardine and Amy Shell-Gellasch, 151–158. Washington, DC: The Mathematical Association of America.
Muntersbjorn, Madeline M. 1999. Naturalism, notation, and the metaphysics of mathematics. *Philosophia Mathematica* 7: 178–199.
Netz, Reviel. 1999. *The shaping of deduction in Greek mathematics: A study in cognitive history*. Cambridge: Cambridge University Press.
Newton, Isaac. 1850. Letter to John Keill, 11 May 1714. In *Correspondence of Sir Isaac Newton and Professor Cotes: Including letters of other eminent men*, ed. Joseph Edleston, 174–175. London/Cambridge: Parker/Deighton.
———. 1967–1981. *The mathematical papers*. Cambridge: Cambridge University Press.
Pender, William, David Saddler, Julia Shea, and Derek Ward. 2012. *Cambridge 3 unit mathematics year 11 enhanced version*. Cambridge: Cambridge University Press.
Pólya, George. 2004. *How to solve it: A new aspect of mathematical method*. Princeton: Princeton University Press.
Schlimm, Dirk. 2018a. On Frege's *Begriffsschrift* notation for propositional logic: Design principles and trade-offs. *History and Philosophy of Logic* 39 (1): 53–79.
———. 2018b. Numbers through numerals: The constitutive role of eternal representations. In *Naturalizing logico-mathematical knowledge: Approaches from philosophy, psychology and cognitive science*, ed. Sorin Bangu, 195–217. New York: Routledge.

Schlimm, Dirk, and Hansjörg Neth. 2008. Modeling ancient and modern arithmetic practices: Addition and multiplication with Arabic and Roman numerals. In *Proceedings of the 30th annual meeting of the Cognitive Science Society*, ed. Vladimir Sloutsky, Bradley Love, and Kateri McRae, 2097–2102. Austin: Cognitive Science Society.

Seife, Charles. 2000. *Zero: The biography of a dangerous idea*. New York: Viking.

Sen, Syamal K., and Ravi P. Agarwal. 2016. *A landmark discovery, the dreadful void, and the ultimate mind*. London: Academic.

Serfati, Michel. 2005. *La révolution symbolique: La constitution de l'écriture symbolique mathématique*. Paris: Petra.

Tagliasco, Vincenzo, and Antonio Vincenzi. 1998. *Dietro le formule...: I discorsi della logica e della matematica*. Turin: Bollati Boringhieri.

Turner, J. Hilton. 1951. Roman elementary mathematics: The operations. *The Classical Journal* 47 (63–74): 106–108.

Università dei Mercanti. 1554. *Statuti della honoranda Università de' mercanti della inclita città di Bologna, riformati l'anno MDL*. Bologna: Anselmo Giaccarello.

van der Waerden, Bartel Leendert. 1976. Defence of a 'shocking' point of view. *Archive for History of Exact Sciences* 15: 199–210.

Velleman, Daniel J. 2017. *Calculus: A rigorous first course*. Mineola: Dover.

Viète, François. 1646. *Opera mathematica*. Leiden: Ex Officina Bonaventurae & Abrahami Elzeviriorum.

Whitehead, Alfred North. 2017. *An introduction to mathematics*. Mineola: Dover.

Part IV
The Functionality of Mathematics

Chapter 14
Explanations

Abstract According to heuristic philosophy of mathematics, one of the tasks of the philosophy of mathematics is to give an answer to the question: What is the nature of mathematical explanations? Two different kinds of mathematical explanations can be distinguished: mathematical explanations of mathematical facts, and mathematical explanations of empirical facts. The chapter argues that there are mathematical explanations of mathematical facts, but there are no mathematical explanations of empirical facts.

Keywords Intra-mathematical explanations · Extra-mathematical explanations · Explanatory axiomatic demonstration · Explanatory analytic demonstration · Top-down explanatory demonstration · Bottom-up explanatory demonstration · Explanatory demonstration and understanding · Illusoriness of extra-mathematical explanations

14.1 Intra-Mathematical and Extra-Mathematical Explanations

As we have seen in Chap. 3, according to heuristic philosophy of mathematics, one of the tasks of the philosophy of mathematics is to give an answer to the question: What is the nature of mathematical explanations? Two different kinds of mathematical explanations can be distinguished: mathematical explanations of mathematical facts, or intra-mathematical explanations, and mathematical explanations of empirical facts, or extra-mathematical explanations.

The chapter argues that there are intra-mathematical explanations, and that they consist primarily of explanatory demonstrations, there being an objective difference between explanatory demonstrations and non-explanatory demonstrations. Two kinds of explanatory demonstrations are distinguished: top-down explanatory demonstrations, which are based on the axiomatic method, and bottom-up explanatory demonstrations, which are based on the analytic method. These two kinds of explanatory demonstrations are relevant to mathematics in different ways, both produce understanding but different kinds of understanding.

C. Cellucci, *The Making of Mathematics*, Synthese Library 448,
https://doi.org/10.1007/978-3-030-89731-4_14

While the chapter argues that there are intra-mathematical explanations, on the other hand it argues that there are no extra-mathematical explanations.

14.2 Intra-Mathematical Explanations and Mathematicians

First, we consider intra-mathematical explanations. Many mathematicians hold that there are intra-mathematical explanations and that they consist primarily of explanatory demonstrations, there being an objective difference between explanatory demonstrations and non-explanatory demonstrations.

Thus, Rota says that "not all proofs give satisfying reasons why a conjecture should be true. Verification is proof, but verification may not give the reason" (Rota 1997, 138–139). The reason is given only by a proof that "gives away the secret of the theorem," namely one that "leads us to perceive the inevitability of the statement being proved" (ibid., 132).

Niss says that "we should distinguish between proofs that justify" and "proofs that explain" (Niss 2006, 57). Often "one is led to accept the truth of a proposition as a result of a correct deduction without really having obtained an insight into why the steps of the proof reveal the truth of the proposition," on the other hand, "some proofs not only justify but also explain why a proposition in true" (ibid.).

Byers says that "certain proofs" are such that "one is forced to accept the validity of the argument and therefore that the theorem is true, while nevertheless remaining in the dark as to why it is true. Verifying a proof is one thing and understanding it is quite another" (Byers 2007, 336). So "not all proofs are created equal," and "a 'good' proof" is "one that brings out clearly the reason why" (ibid., 337).

Auslander says that a "role of proof is explanation. This is what concerns most mathematicians" because, "almost by definition, a proof is supposed to explain the result. Now, it must be admitted that not all proofs meet this standard," and "such considerations have often led to the development of new, more understandable, proofs" (Auslander 2008, 66).

Gowers says that, "when one looks at the actual practice of mathematics, it becomes clear that proofs are far more than mere certificates of truth" (Gowers 2007, 37). They "are valued for their explanatory power, and a new proof of a theorem can provide crucial insights" (Gowers and Nielsen 2009, 879).

14.3 Objections to Intra-Mathematical Explanations

Contrary to mathematicians, several philosophers object to the view that there are intra-mathematical explanations and that they consist primarily of explanatory demonstrations. Their arguments, however, are invalid. Here are the main ones.

(1) Mathematicians rarely describe themselves as explaining, in fact they do not pursue explanations of mathematical facts.

Thus, Resnik and Kushner say that "the production of explanations" is "not an acknowledged goal of mathematical research. Mathematicians rarely describe themselves as explaining," in fact "the practice of explaining mathematical phenomena has been barely acknowledged" (Resnik and Kushner 1987, 151).

Zelcer says that "there are relatively few references to explanation in mainstream mathematical literature" (Zelcer 2013, 179). In fact, "the desire to pursue explanations of mathematical facts did not motivate new mathematics" (ibid., 180).

But this argument is contradicted by the assertions of the mathematicians cited above. They show that mathematicians do not content themselves with demonstrations which only guarantee that the theorem is true, they look for demonstrations that show why it is true and hence have explanatory power, so they pursue explanations of mathematical facts. The argument is also contradicted by empirical studies which show that "mathematical explanations do occur in scholarly mathematical practice, as indicated by the occurrence of explanation indicators in research articles published in mathematics journals" (Mizrahi 2020).

(2) The notion of explanatory demonstration is not viable, because there is no objective difference between explanatory demonstrations and non-explanatory demonstrations.

Thus, Resnik and Kushner say that "the notion of explanatory proof is not viable" (Resnik and Kushner 1987, 156). There is no objective difference between explanatory and non-explanatory proofs because, for any result, one might "count all its proofs as explanatory" (ibid., 153). For, any proof answers at least one why-question, namely, the question "why a result is true (rather than false)" (ibid.).

Zelcer says that there are no "explanatory proofs which qualitatively differ from non-explanatory proofs" (Zelcer 2013, 176). In fact, "mathematical proofs are different" from "explanations" (ibid., 190). For, "explanations account for how and/or why," while "proofs in mathematics, however interesting or sophisticated, are epistemic devices that show how theorems fit together and function within a mathematical system," but by no means "account for the necessary fact of their existence or truth" (ibid.).

But this argument is contradicted by the assertions of the mathematicians cited above. They show that there is an objective difference between explanatory demonstrations and non-explanatory demonstrations. In particular, it is simply factually wrong that one might count all proofs of a result as explanatory on the ground that any proof answers the question why a result is true. As matter of fact, many proofs merely show that the result is true, they do not explain why it is true.

(3) Intra-mathematical explanations need not consist only of explanatory demonstrations.

Thus, Colyvan says that, "while most of the discussion of intra-mathematical explanation has focussed on proofs, these are not the only loci of explanation," we "should also expect to see explanation in reductions of one theory to another and in various generalisations of a theory" (Colyvan 2012a, 85).

Lange says that "not all mathematical explanations consist of proofs of the facts they explain," so "any adequate theory of mathematical explanation must account for mathematical explanations that are not proofs" (Lange 2018, 1287).

But this argument is not really an objection because, saying that intra-mathematical explanations consist primarily of explanatory demonstrations, does not mean that there are no other kinds of intra-mathematical explanation. It only means that the search for explanatory demonstrations plays a primary role in mathematics.

(4) A kind of intra-mathematical explanation that does not consist of explanatory demonstrations is the reduction of arithmetic to set theory. For example, the reduction explains why the multiplication of natural numbers is commutative.

Thus, Kitcher says that "reducing arithmetic to set theory has explanatory, as well as ontological, value. For, in the light of the reduction, our understanding is advanced through exhibition of the kinship between theorems of arithmetic and theorems in other developments of set theory" (Kitcher 1978, 123).

Maddy says that an example of "the explanatory power of set theory" is that, if you "ask why multiplication is commutative" and "take a set theoretic perspective," then "an answer is forthcoming" (Maddy 1981, 498-499). Multiplication is commutative because, "if A and B are sets, then there is a one-to-one correspondence between the cartesian products $A \times B$ and $B \times A$. The central idea in the proof of this fact is the old observation that a rectangle of n rows of m dots contains $n \cdot m$ dots, but turned on its side it contains $m \cdot n$ dots" (ibid., 499). This "explains why multiplication is commutative" (ibid.).

But this argument is invalid because the reduction of arithmetic to set theory does not explain what natural numbers are, nor why the multiplication of natural numbers is commutative.

First, as we have seen in Chap. 2, Zermelo identifies natural numbers 0, 1, 2, 3, ...with the sets $\varnothing$, $\{\varnothing\}$, $\{\{\varnothing\}\}$, $\{\{\{\varnothing\}\}\}$, ... , while von Neumann's identifies them with the sets $\varnothing$, $\{\varnothing\}$, $\{\varnothing,\{\varnothing\}\}$, $\{\varnothing,\{\varnothing\},\{\varnothing,\{\varnothing\}\}\}$, ... And, as we have seen in Chap. 2, Maddy herself admits that there is no reason deep enough to say that one identification rather than the other one uncovers the true identity of the natural numbers. This amounts to admitting that the reduction of arithmetic to set theory does not explain what natural numbers are.

Second, Maddy says that, by taking a set theoretic perspective, one can explain why multiplication is commutative. For, one can give a demonstration that, for any sets A and B, there is a one-to-one correspondence between the cartesian products $A \times B$ and $B \times A$, where the central idea of the demonstration is that a rectangle of n rows of m dots contains $n \cdot m$ dots, but turned on its side it contains $m \cdot n$ dots. This means that the central idea of the demonstration is the following visual demonstration:

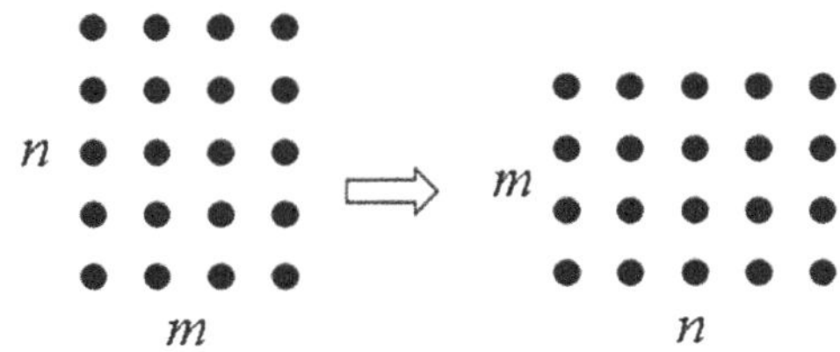

Then, however, the explanation of why the multiplication of natural numbers is commutative is reached not by taking a set theoretic perspective, but by appealing to a visual demonstration. What is explanatory is the visual demonstration, not the set theoretic perspective. So, it is not the reduction of arithmetic to set theory that explains why the multiplication of natural numbers is commutative.

14.4 Demonstrations and Explanatoriness

The view that intra-mathematical explanations consist primarily of explanatory demonstrations needs specification. For, as argued in Chap. 10, there are two basic notions of demonstration, axiomatic demonstration and analytic demonstration, which have different purposes.

The purpose of axiomatic demonstration is to present, justify, and teach already acquired propositions.

The purpose of analytic demonstration is to discover hypotheses that are sufficient conditions for the solution of problems, and are plausible.

That there are two basic notions of demonstration, axiomatic demonstration and analytic demonstration, raises the questions: Which axiomatic demonstration is explanatory? Which analytic demonstration is explanatory?

14.5 Aristotle on Explanatory Axiomatic Demonstration

An answer to the question 'Which axiomatic demonstration is explanatory?' has first been given by Aristotle.

Aristotle says that a demonstration is a deduction which "proceeds from premisses that are true and prime" (Aristotle, *Analytica Posteriora*, A 2, 71 b 20–21). Namely, a deduction which proceeds from principles, because "to proceed from prime premisses is to proceed from principles" (ibid., A 2, 72 a 5–6). All propositions are demonstrated from principles, because "one cannot demonstrate anything except from its own principles" (ibid., A 9, 75 b 37–38). Thus, according to Aristotle, demonstration is axiomatic demonstration.

The purpose of axiomatic demonstration is to present, justify, and teach already acquired propositions. Indeed, axiomatic demonstrations are didactic arguments, because "didactic arguments are those that deduce" propositions "from the principles proper to each subject matter" (Aristotle, *De Sophisticis Elenchis*, 2, 165 b 1–2). Didactic arguments do not yield new knowledge, because they "produce their teaching through what we already know" (Aristotle, *Analytica Posteriora*, A 2, 71 a 6–7).

Not only, according to Aristotle, demonstration is axiomatic demonstration, but two different kinds of axiomatic demonstration can be distinguished, 'demonstration that', and 'demonstration why'. For, there are "differences between a 'demonstration that' [oti] and a 'demonstration why' [dioti]" (Aristotle, *Analytica Posteriora*, A 13, 78 b 33–34). In a 'demonstration that' "the cause is not stated" (ibid., A 13, 78 b 14–15). In a 'demonstration why' the cause is stated, so "a demonstration through the cause is a 'demonstration why'" (ibid., B 16, 98 b 19–20).

What does Aristotle mean by 'cause'? According to Aristotle, "things are called causes in many ways" (Aristotle, *Physica*, B 3, 195 a 4). Now, when saying that in a 'demonstration that' the cause is not stated, while in a 'demonstration why' the cause is stated, by 'cause' Aristotle means "the reason why" (ibid., B 3, 194 b 19). In what follows, the term 'cause' will be meant in this sense.

Since in a 'demonstration that' the cause is not stated, a demonstration which does not show why something is the case is non-explanatory. Conversely, since in a 'demonstration why' the cause is stated, a demonstration which shows why something is the case is explanatory. Since a 'demonstration that' does not show why something is the case, it follows that 'demonstration why', and only it, is explanatory.

Thus Aristotle's answer to the question 'Which axiomatic demonstration is explanatory?' is: 'Demonstration why', and only it, is explanatory.

14.6 Bolzano on Explanatory Axiomatic Demonstration

That 'demonstration why', and only it, is explanatory, has been reaffirmed several times since Aristotle.

For example, Bolzano says that the "proof of a truth" is "the derivation of it from those truths which must be considered as the ground for it," namely from the "axioms" (Bolzano 2004, 110). So, according to Bolzano, proof is axiomatic proof.

Not only, according to Bolzano, proof is axiomatic proof, but two different kinds of axiomatic proof can be distinguished. For, "the objective ground of a truth" is different "from the subjective means through which we come to know it" (Bolzano 2014, § 525). So, axiomatic proofs that are derivations from subjective grounds can be distinguished from axiomatic proofs that are derivations from objective grounds. Axiomatic proofs that are derivations from subjective grounds "only aim at certainty" and may "be called 'certifications'," while axiomatic proofs that are derivations from objective grounds may be called "'groundings'" (ibid.).

Already "Aristotle and, after him, for many centuries the Scholastics, distinguished between two kinds of proofs, those which show only the 'oti', i.e., which show that something is the case, and those which show the 'dioti', i.e., which show why something is the case" (ibid., § 198). Certifications correspond to Aristotle's 'demonstrations that', namely, demonstrations in which the cause is not stated, while groundings correspond to Aristotle's 'demonstrations why', namely, demonstrations in which the cause is stated. For, "the concepts of ground and consequence are intimately related to those of 'cause' and 'effect'" (ibid., § 168).

According to Bolzano, the distinction between certifications and groundings is an objective one.

For example, "the proof of the first proposition of Euclid's *Elements* (the possibility of an equilateral triangle)" is only a certification, because it does not "really indicate the objective ground of this truth" (ibid., § 525). For, it is not true that "an equilateral triangle" is "only possible because those two circles intersect," on the contrary, "those circles intersect because there is an equilateral triangle" (ibid.).

Different is the character of Bolzano's own proof of the intermediate value theorem, according to which "every continuous function of x which is positive for one value of x, and negative for another, must be zero for some intermediate value of x" (Bolzano 2004, 255). Bolzano's proof is "not a mere" certification "but the objective grounding of the truth to be proved" (ibid., 260).

14.7 Plato on Explanatory Analytic Demonstration

An answer to the question 'Which analytic demonstration is explanatory?' has first been given by Plato.

As we have seen in Chap. 5, Plato says that, to solve a problem, you should posit the hypothesis which you judge to be the strongest, then you should investigate its consequences, to see whether they are in accord, or are not in accord, with each other. Even if the consequences of the hypothesis are in accord with each other, you should give an account of the hypothesis itself, and you should give such an account in the same way, positing another hypothesis, whichever should seem best of the higher ones, and so on. For, solving a problem is an infinite task. Thus, according to Plato, demonstration is analytic demonstration.

The purpose of analytic demonstration is to discover hypotheses that are sufficient conditions for the solution of a problem and are plausible. This follows from the fact that the consequences of the hypotheses should be in accord with each other.

Analytic demonstration, and only it, is explanatory, because only it proceeds from the problem to be solved to hypotheses that are the cause of the solution, so only it proceeds "by working out the cause" (Plato, *Meno*, 98 a 3–4).

Thus Plato's answer to the question 'Which analytic demonstration is explanatory?' is: Analytic demonstration as such, and only it, is explanatory.

14.8 Descartes on Explanatory Analytic Demonstration

That analytic demonstration as such, and only it, is explanatory has been reaffirmed several times since Plato.

For example, Descartes says that, although axiomatic demonstration "demonstrates the conclusion clearly" by the "use of a long series of definitions, postulates, axioms, theorems, and problems," nevertheless axiomatic demonstration is not satisfying "nor appeases the minds of those who are eager to learn, since it does not show how the thing in question was discovered" (Descartes 1996, VII, 156). Indeed, axiomatic demonstration appears "discovered more through chance than through method," so by using it "we get out of the habit of using our reason" (ibid., X, 375). Thus, although axiomatic demonstration demonstrates the conclusion clearly, nevertheless it may not explain, and "there is a big difference between demonstrating and explaining" (ibid., II, 198)

Only analytic demonstration "shows the true way by means of which the thing was discovered methodically" (ibid., VII, 155). For, "the way and order of discovery is one thing, that of teaching another" (ibid., V, 153). The way and order of discovery is analytic demonstration, the way of teaching is axiomatic demonstration. By showing the true way by means of which the thing was discovered methodically, analytic demonstration explains. For, through analytic demonstration, "it is the causes which are demonstrated by the effects," and "the causes from which I deduce" the effects "serve not so much to demonstrate them as to explain them" (ibid., VI, 76).

14.9 Main Difference Between Axiomatic and Analytic Demonstration

In a sense, Aristotle's answer to the question 'Which axiomatic demonstration is explanatory?' and Plato's answer to the question 'Which analytic demonstration is explanatory?', are both valid, because they make it clear that an explanatory demonstration must indicate the reason why the result holds.

However, between Aristotle's answer and Plato's answer there is an important difference. On the one hand, by Aristotle's answer that 'demonstrations why', and only it, is explanatory, not all axiomatic demonstrations are explanatory. On the other hand, by Plato's answer that analytic demonstration as such, and only it, is explanatory, all analytic demonstrations are explanatory.

The difference is due to the fact that, in an axiomatic demonstration, the proposition demonstrated is deduced from given principles, and a proposition can be deduced from principles that are not the reason of the conclusion. Therefore, Aristotle distinguishes between 'demonstration that' and 'demonstration why'.

Conversely, in an analytic demonstration, the hypotheses by means of which the problem is solved are obtained from the problem, and possibly other data already available, by some non-deductive rule, and are the reason of the solution of the problem.

14.10 Top-Down and Bottom-Up Explanatory Demonstration

In terms of Aristotle's answer to the question 'Which axiomatic demonstration is explanatory?' and of Plato's answer to the question 'Which analytic demonstration is explanatory?', two kinds of explanatory demonstration can be distinguished: top-down explanatory demonstration and bottom-up explanatory demonstration. A top-down explanatory demonstration is an axiomatic demonstration that is a 'demonstration why', a bottom-up explanatory demonstration is an analytic demonstration.

Between top-down explanatory demonstration and bottom-up explanatory demonstration there is an important difference.

Since, in a top-down explanatory demonstration, the proposition demonstrated is deduced from given principles, and in a deductive inference the conclusion is contained in the premisses, in such a demonstration the explanans is implicitly given from the beginning. The aim of the research is to make it explicit.

Conversely since, in a bottom-up explanatory demonstration, the hypotheses by means of which the problem is solved are obtained from the problem, and possibly other data already available, by some non-deductive rule, and in a non-deductive inference the conclusion is not contained in the premisses, in such a demonstration the explanans is not implicitly given from the beginning. The aim of the research is to discover it.

That two kinds of explanatory demonstration can be distinguished shows that there is a serious lacuna in the present literature on intra-mathematical explanations. For, such literature totally ignores bottom-up explanatory demonstration, it exclusively considers top-down explanatory demonstration.

14.11 Examples of Top-Down Explanatory Demonstration

We consider some examples of top-down explanatory demonstration.

(A) Let us examine two demonstrations of the Pythagorean theorem: In a right-angled triangle, the square on the hypotenuse is equal to the sum of the squares on the legs.

Demonstration 1 (Euclid, *Elementa*, I, Proposition 47). Let ABC be a right-angled triangle, having the angle BAC right. We must show that the square on BC is equal to the sum of the squares on BA and AC.

Describe the square $BCED$ on BC, the square $GFBA$ on BA, and the square $HACK$ on AC (Euclid, *Elementa*, I, Proposition 46). Draw AL through point A parallel to BD or CE (Euclid, *Elementa*, I, Proposition 31). Join AD and FC (Euclid, *Elementa*, I, Postulate 1).

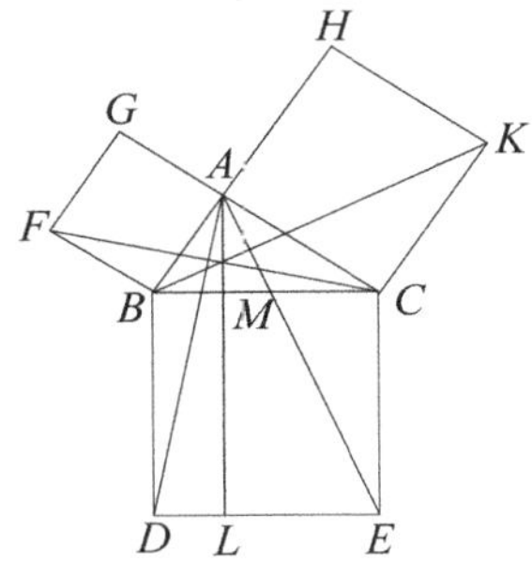

Since angles *BAC* and *BAG* are right angles, they sum to two right angles, therefore *CA* is in a straight line with *AG* (Euclid, *Elementa*, I, 14).

For the same reasons, *BA* is in a straight line with *AH*.

Since the angles *DBC* and *FBA* are right angles, they are equal (Euclid, *Elementa*, I, Postulate 4). Add the angle *ABC* to both. Then the whole angle *DBA* is equal to the whole angle *FBC* (Euclid, *Elementa*, I, Common Notion 2).

Since *DB* is equal to *BC*, and *FB* is equal to *BA*, the two sides *DB* and *BA* are equal to the two sides *CB* and *BF*, respectively. On the other hand, as we have seen, angle *DBA* is equal to angle *FBC*. Thus, the base *AD* is equal to the base *FC*, and the triangle *ABD* is equal to the triangle *FBC* (Euclid, *Elementa*, I, Proposition 4).

Now, the parallelogram *BDLM* is double the triangle *ABD*, because they have the same base *BD*, and are between the same parallels *BD* and *AL* (Euclid, *Elementa*, I, Proposition 41). And the square *GFBA* is double of triangle *FBC*, because they have the same base *FB* and are between the same parallels *FB* and *GC* (Euclid, *Elementa*, I, Proposition 41). But the doubles of equal things are equal to one another. Therefore, the parallelogram *BDLM* is also equal to the square *GFBA*.

Similarly, if *AE* and *BK* are joined, the parallelogram *MLEC* can be shown to be equal to the square *HACK*.

Therefore, the whole square *BDEC* is equal to the sum of the squares *GFBA* and *HACK* (Euclid, *Elementa*, I, Common Notion 2). Namely, the square on *BC* is equal to the sum of the squares on *BA* and *AC*.

Demonstration 2 (Derived from Euclid, *Elementa*, VI, Proposition 31). Let *ABC* be a right-angled triangle, having the angle *BAC* right. We must show that $BC^2 = AB^2 + AC^2$.

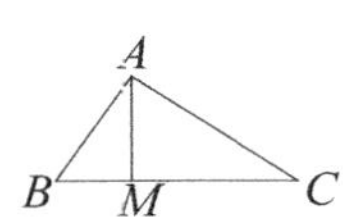

Draw a perpendicular *AM* from the right angle *BAC* to the base *BC*. The resulting triangles *ABM* and *AMC* are both similar to triangle *ABC*. Triangles *ABM* and *ABC* are similar, because angle *MBA* is common, and angles *BMA* and *BAC* are equal being both right, so the triangles have two of their angles equal. Triangles *AMC* and *ABC* are similar, because angle *MCA* is common, and angles *CMA* and *BAC* are equal being both right, so the triangles have two of their angles equal.

Since triangles *ABM* and *ABC* are similar, $AB : BM = BC : AB$, hence $AB^2 = BC \cdot BM$. Since triangles *AMC* and *ABC* are similar, $AC : MC = BC : AC$, hence $AC^2 = BC \cdot MC$. Then $AB^2 + AC^2 = BC \cdot BM + BC \cdot MC = BC \cdot (BM+MC) = BC^2$.

The relation between Demonstration 2 and Demonstration 1 can be seen as follows. In Demonstration 1, by construction, *BC* is equal to *DB*. So, in Demonstration 2, from $AB^2 = BC \cdot BM$, it follows $AB^2 = DB \cdot BM$, hence the square *GFBA* is equal to the parallelogram *BDLM*. Also, in Demonstration 1, by construction, *BC* is equal to *CE*. So, in Demonstration 2, from $AC^2 = BC \cdot MC$, it follows $AC^2 = CE \cdot MC$, hence the square *HACK* is equal to the parallelogram *MLEC*. Then the sum of the squares *GFBA* and *HACK* is equal to the sum of the parallelograms *BDLM* and *MLEC*, which is equal to the square *BCED*. Therefore the square *BCED* is equal to the sum of the squares *GFBA* and *HACK*.

Demonstration 1 is non-explanatory. As Schopenhauer says, "Euclid's stilted, indeed underhand, proof leaves us without an explanation of why" (Schopenhauer 2010, 98). For, "lines are often drawn without any indication of why," and the reader "must admit in astonishment what remains completely incomprehensible in its inner workings" (ibid., 96). On the contrary, Demonstration 2 is explanatory. For, the main fact on which it depends is that a right-angled triangle can be divided into two triangles that are similar to the original one, and this is the reason why, in a right-angled triangle, the square on the hypotenuse is equal to the sum of the squares on the legs. Indeed, right-angled triangles are the only kind of triangles that can be divided into two triangles that are similar to the original one.

(B) Let us examine two demonstrations that $\sqrt{2}$ is an irrational number.

Demonstration 1 (Euclid, *Opera Omnia*, III, Appendix to Book X, Proposition 27, ed. Heiberg). Suppose that $\sqrt{2}$ is a rational number. Then $\sqrt{2} = m/n$ for some natural numbers m and n which have no common factors. Thus m and n are not both even numbers. From $\sqrt{2} = m/n$ it follows $m^2 = 2n^2$. So, m^2 is an even number, hence m is an even number (since, if the square of a number is even, then the number itself is even). Thus, $m = 2p$ for some natural number p. Then from $m^2 = 2n^2$ it follows $4p^2 = 2n^2$, so $n^2 = 2p^2$. Hence n^2 is an even number, therefore n is an even number (again since, if the square of a number is even, then the number itself is even). Thus m and n are both even numbers. Contradiction. Therefore $\sqrt{2}$ is an irrational number.

Demonstration 2. Suppose that $\sqrt{2}$ is a rational number. Then $\sqrt{2} = m/n$ for some natural numbers m and n, so $m^2 = 2n^2$. Since $m^2 = m \cdot m$, in the prime factorization of m^2 each prime factor occurs an even number of times. Similarly for n^2. Then, in the prime factorization of $2n^2$ the prime factor 2 occurs an odd number of times. Hence, $m^2 \neq 2n^2$. Contradiction. Therefore $\sqrt{2}$ is an irrational number.

Demonstration 1 is non-explanatory. For, the main fact on which it depends is that, if the square of a number is even, then the number itself is even, and this is not the reason why it cannot be $m^2 = 2n^2$. On the contrary, Demonstration 2 is explanatory. For, the main fact on which it depends is that, in the prime factorization of m^2, the prime factor 2 appears an even number of times, while in the prime factorization of $2n^2$ it appears an odd number of times, and this is the reason why it cannot be $m^2 = 2n^2$. Thus, as Davis and Hersh say, Demonstration 2 "exposes the 'real' reason" (Davis and Hersh 1981, 301).

14.12 Examples of Bottom-Up Explanatory Demonstration

Examples of bottom-up explanatory demonstration are the analytic demonstrations described in Chap. 5. We reexamine some of them.

(i) Hippocrates of Chios gives a demonstration that the lunule obtained by circumscribing about a right-angled isosceles triangle a semicircle, and about the base a segment of a circle similar to those cut off by the sides, is equal to the triangle. To this aim, by analyzing the conditions under which the problem would be solved, Hippocrates of Chios non-deductively arrives at the hypothesis (I) 'Similar segments of circles are to each other as the squares on their bases' and shows that (I) is a sufficient condition to establish the result. Now, (I) is the reason why the lunule is equal to the triangle. Then, by analyzing the conditions under which the problem posed by (I) would be solved, Hippocrates of Chios non-deductively arrives at the hypothesis (II) 'Circles are to each other as the squares on their diameters' and shows that (II) is a sufficient condition to establish (I). Now, (II) is the reason why similar segments of circles are to each other as the squares on their bases. And so on.

(ii) Plato gives a demonstration that, for any given square, the side of the figure double in size is the diagonal of the given square. To this aim, by analyzing the conditions under which the problem would be solved, Plato non-deductively arrives at the hypothesis (I) 'A diagonal cuts each square in half', and shows that (I) is a sufficient condition to establish the result. Now, (I) is the reason why, for any given square, the side of the figure double in size is the diagonal of the given square. Then, by analyzing the conditions under which the problem posed by (I) would be solved, Plato non-deductively arrives at the hypothesis (II) 'If two triangles have their corresponding sides equal, then they are equal', and shows that (II) is a sufficient condition to establish (I). Now, (II) is the reason why a diagonal cuts each square in half. And so on.

14.13 Explanatory Demonstrations and Generality

It is widely thought that there is a strict relation between explanatory demonstrations and generality: the general demonstration is more explanatory.

This view goes back to Aristotle, who claims that the general demonstration is more explanatory because it "shows more the cause and the reason why" (Aristotle, *Analytica Posteriora*, A 24, 85 b 27). Therefore, the "general demonstration is better" (ibid., A 24, 85 b 26–27).

The view has been repeatedly reaffirmed since Aristotle. For example, concerning the two demonstrations that $\sqrt{2}$ is an irrational number considered above, Wang says that "we favour" Demonstration 2 over Demonstration 1 because "it serves to demonstrate the irrationality of square roots of all prime numbers, which gives the real reason for the irrationality of $\sqrt{2}$ (that is, a particular case)," on the contrary, Demonstration 1 "applies exclusively to a particular case" (Wang 2019, 234). Therefore, Demonstration 2 is "more useful or better, as it justifies the unification of more particular numbers under a more general concept or theorem" (ibid., 234–235).

Wang refers to the fact that Demonstration 2 remains valid if, instead of 2, we consider an arbitrary prime number p, so Demonstration 2 extends to a demonstration that $\sqrt{p}$ is an irrational number, for any prime number p.

But the view that the general demonstration is more explanatory is invalid. For, let us consider the following alternative demonstration that $\sqrt{p}$ is an irrational number, for any prime number p.

Demonstration 3. Suppose that there is a prime number p such that $\sqrt{p}$ is a rational number. Then $\sqrt{p} = m/n$, for some natural numbers m and n which have no common factors. From $\sqrt{p} = m/n$ it follows $m^2 = pn^2$, so n^2 is a factor of m^2. But, since m and n have no common factors, m^2 and n^2 also have no common factors. Hence, from the fact that n^2 is a factor of m^2, it follows that $n^2 = 1$. Then, from $m^2 = pn^2$ we obtain $m^2 = p$, so p is a square number. But no prime number is a square number, so p is not a square number. Contradiction. Therefore, $\sqrt{p}$ is an irrational number.

Like the extension of Demonstration 2 to an arbitrary prime number p, Demonstration 3 is general because it applies to all prime numbers, while Demonstration 1 applies only to a particular prime number, namely 2. But, while the extension of Demonstration 2 to an arbitrary prime number p is explanatory, Demonstration 3 is non-explanatory. For, the reason why $\sqrt{p}$ is an irrational number, for any prime number p, is not that, if m and n have no common factors, then m^2 and n^2 also have no common factors. So although Demonstration 3 is more general than Demonstration 1, it is non-explanatory.

This example shows that the belief that the general demonstration, qua general, is more explanatory, is invalid.

14.14 Explanatory Demonstration and Visual Demonstration

A special kind of demonstration is visual demonstration, namely demonstration consisting of a figure which shows that a proposition holds without need of a detailed argument in words.

Some visual demonstrations are explanatory. For example, such is the visual demonstration that the multiplication of natural numbers is commutative considered earlier in this chapter.

On the other hand, not all visual demonstrations are explanatory. For example, let us consider the following visual demonstration of the Pythagorean theorem, whose basic idea goes back to Thābit ibn Qurra (see Chemla 2005, 160–161; or Djebbar 2009, 7–8).

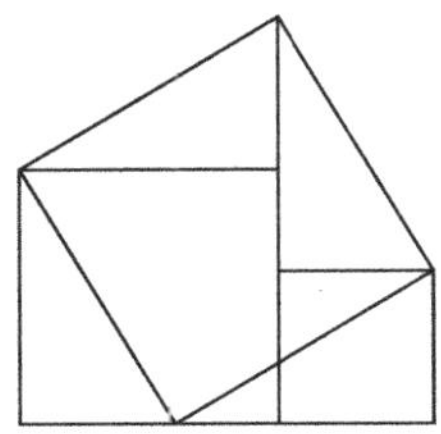

The figure contains four equal copies of the given right-angled triangle. Removing the two lower copies of the triangle yields the square on the hypotenuse, while removing the two upper copies of the triangle yields the squares on the legs. Therefore, the demonstration shows that the square on the hypotenuse is equal to the sum of the squares on the legs.

However, while demonstrating the Pythagorean theorem, the demonstration does not explain why the theorem holds, so it is not explanatory.

14.15 The Relevance of Explanatory Demonstration to Mathematics

Since there are two different kinds of explanatory demonstration, top-down and bottom-up, the question arises: What is the relevance of these two kinds of explanatory demonstrations to mathematics?

To answer this question, it must be considered that, as already said, axiomatic demonstration is meant to present, justify, and teach already acquired propositions, while analytic demonstration is meant to discover solutions to problems, thus acquiring new knowledge. Therefore, top-down explanatory demonstration and bottom-up explanatory demonstration are relevant to mathematics in different respects.

Top-down explanatory demonstration is relevant to finished mathematics, namely mathematics presented in finished form. For, it shows why a proposition is true, so it is effective in showing why the proposition must be accepted.

Bottom-up explanatory demonstration is relevant to the making of mathematics, in particular discovery. For, it shows how to obtain hypotheses for solving a problem, so it is effective in discovering a solution to the problem.

14.16 The Disregard of Bottom-Up Explanatory Demonstration

Despite the fact that bottom-up explanatory demonstration is relevant to the making of mathematics, current work on mathematical explanation disregards it, only allowing for top-down explanatory demonstration.

An example of this is Lehet's claim that, unlike explanatory definitions, explanatory proofs do not contribute to the production of mathematical knowledge.

Indeed, Lehet says that "a mathematical definition is explanatory if it makes the mathematical concept being defined more accessible – i.e., it explains some feature or property of the relevant concept" (Lehet 2021, 1173). By 'accessibility to the mathematical concept' Lehet means "familiarity with the mathematical object" (ibid., 1165–1166). For, she says that "in advanced mathematics, it is common that the relevant mathematical objects will be unfamiliar, and initially inaccessible, to us, but explanatory definitions enable us to become familiar with such objects" (ibid., 1175). So, an explanatory definition is one which makes the objects it defines "more familiar to us" (ibid., 1168).

With this notion of explanatory definition, Lehet claims that "explanatory definitions contribute to the production of" mathematical "knowledge because they give mathematicians the means to grasp" abstract "mathematical objects" (ibid., 1170). For, "in order to develop new conjectures and to progress within mathematics, it is necessary to develop a familiarity" with "advanced mathematical concepts, which is precisely what explanatory definitions do" (ibid., 1175). On the contrary, explanatory proofs do not contribute to the production of mathematical knowledge, but only "to the justification of" already acquired "mathematical knowledge. They explain 'why' a theorem is true and so explain why we are justified in claiming knowledge of that theorem," but they do not yield new theorems, because they "will not result in an increased accessibility of mathematical concepts" (ibid., 1170). So, "whereas explanatory definitions enable us to better grasp mathematical concepts themselves, explanatory proofs only enable us to grasp the justification of mathematical facts" (ibid.).

From Lehet's claim that explanatory proofs do not contribute to the production of mathematical knowledge, it is clear that by 'explanatory proof' she means 'top-down explanatory demonstration', not 'bottom-up explanatory demonstration'. For, while top-down explanatory demonstration only shows why a proposition is true, bottom-up explanatory demonstration shows how to obtain hypotheses for solving a problem, and hence contributes to the production of mathematical knowledge. So, Lehet's claim applies only to top-down explanatory demonstration. Hence, Lehet disregards bottom-up explanatory demonstration, and only allows for top-down explanatory demonstration.

Moreover, Lehet's notion of explanatory definition is inadequate because, according to it, a mathematical definition is explanatory if it 'explains' some feature or property of the relevant concept. Now, by 'explains' Lehet cannot merely mean 'states', because any definition, not only an explanatory one, states some feature or property of the relevant concept. But, if by 'explains' Lehet does not merely mean 'states' and properly means 'explains', then Lehet's notion of explanatory definition presupposes the concept of explanation, and hence is begging the question. Also, according to Lehet's notion, an explanatory definition is one which makes the objects it defines more familiar to us. But familiarity is a subjective and psychological notion, hence so is Lehet's notion of explanatory definition.

On the contrary, the notion of explanatory definition which is implicit in the heuristic view of mathematical definition described in Chap. 11, does not have these shortcomings. According to it, definitions are hypotheses which are made to solve mathematical problems by the analytic method, therefore, they contribute to the production of mathematical knowledge. But, contrary to Lehet's notion, this notion of definition is not begging the question, because it does not presuppose the concept of explanation, and is not subjective or psychological, because the notion of solving a mathematical problem by the analytic method is not subjective or psychological (see Chap. 5).

14.17 Explanation and Understanding

A concept that has traditionally been closely associated with explanation is understanding. Indeed, from antiquity it has been held that explanation provides understanding.

Thus, Plato says: "Don't you call a person a dialectician if he is able to give the explanation of the essence of each thing? And, as for the person who is unable to give an explanation, will not you say" that "he has no understanding of it?" (Plato, *Respublica*, VII 534 b 3–6).

That explanation provides understanding has been repeatedly reaffirmed since antiquity.

Thus, von Wright says that "ordinary usage does not make a sharp distinction between the words 'explain' and 'understand'. Practically every explanation, be it causal or teleological or of some other kind, can be said to further our understanding of things" (von Wright 1971, 6).

Kitcher says that "a theory of explanation should show us how scientific explanation advances our understanding" (Kitcher 1981a, 329). In particular, in mathematics, "the goal of explanations is to provide understanding," and we should "approach the topic of mathematical explanation via the concept of understanding – or, more exactly, via the concept of a failure in understanding" (Kitcher 1981b, 473).

Bangu says that "the very point of an explanation is to convey understanding," not subjective understanding but "objective understanding," the "kind of understanding" that "can only be achieved when such a correct, publicly available answer is available" (Bangu 2017a, 104–105).

Admittedly, some people have expressed scepticism as to the fact that in mathematics explanation provides understanding. Thus, von Neumann says that "in mathematics you don't understand things, you just get used to them" (Zukav 1980, 208 footnote).

But this sounds more like a joke than a serious assertion. For, if in mathematics one did not understand things, then the development of mathematics would be just a matter of chance, which would not account for the fast development of mathematics in certain periods.

14.18 What It Is to Understand

That explanation provides understanding raises the question of what one means by 'understanding'.

Thus, Poincaré asks: "What is it, to understand?" (Poincaré 2015, 430). For example, what is it to understand a theorem? and to understand a demonstration?

Poincaré's answer is that to understand a theorem is not merely "to recognize that one already knows the meaning of all the terms employed" (ibid., 431). To understand a theorem is to recognize the fitness of the terms employed in the theorem to each other.

On the other hand, to understand a demonstration is not merely "to examine successively each of the syllogisms composing it and to ascertain its correctness, its conformity to the rules of the game" (ibid., 430–431). To understand a demonstration is to recognize "this I know not what which makes the unity of the demonstration" (ibid., 436). This means that to understand a demonstration is to recognize the fitness of the parts of the demonstration to each other.

Thus, the understanding of a theorem is the recognition of the fitness of the terms employed in the theorem to each other.

The understanding of a demonstration is the recognition of the fitness of the parts of the demonstration to each other.

14.19 Top-Down and Bottom-Up Understanding

Since explanation provides understanding and there are two different kinds of explanatory demonstration, top-down and bottom-up, one can distiguish between two kinds of understanding of a demonstration: top-down understanding, and bottom-up understanding.

The top-down understanding of a demonstration is the recognition of the fitness of the parts of the demonstration to each other, which shows why a proposition is true. This is the understanding provided by top-down explanatory demonstrations. For example, Demonstration 2 of the Pythagorean theorem considered above provides this kind of understanding.

The top-down understanding of a demonstration, showing why a proposition is true, is relevant to finished mathematics.

The bottom-up understanding of a demonstration is the recognition of the fitness of the parts of the demonstration to each other, which shows how to obtain hypotheses for solving a problem. This is the understanding provided by bottom-up explanatory demonstrations. For example, Plato's demonstration that the diagonal of the given square will be the side of the square double in size (see Chap. 5), provides this kind of understanding.

The bottom-up understanding of a demonstration, showing how to obtain hypotheses for solving a problem, is relevant to the making of mathematics.

14.20 Extra-Mathematical Explanations and Applicability

After considering intra-mathematical explanations, we consider extra-mathematical explanations.

The claim that there are extra-mathematical explanations, namely mathematical explanations of empirical facts, means that there are explanations of facts external to mathematics, primarily physical phenomena, in which mathematics is not merely a useful tool, but is that in virtue of which the explanation is achieved.

The claim that there are extra-mathematical explanations must not be confused with the claim that mathematics is applicable to empirical facts.

Of course, the claim that there are extra-mathematical explanations presupposes that mathematics is applicable to empirical facts. Now, from the seventeenth century there have been many successful applications of mathematics to empirical facts, so the claim that mathematics is applicable to empirical facts is not problematic.

However, that mathematics is applicable to empirical facts, does not mean that the applications of mathematics to empirical facts give an explanation of them, and hence that there are extra-mathematical explanations. Indeed, in what follows it will be argued that there are no extra-mathematical explanations.

14.21 Two Claims About Extra-Mathematical Explanations

The claim that there are extra-mathematical explanations was first put forward by the Pythagoreans. Actually, the Pythagoreans put forward a very strong form of the claim: There are mathematical explanations of all empirical facts because all empirical facts are inherently mathematical.

Indeed, according to the Pythagoreans, the universe was formed "in accordance with number by the forethought of Him who created all things" (Nicomachus of Gerasa, *Introductio arithmetica*, A vi 1, 12.3–5, ed. Hoche). So, "number is the principle both as the material of things and as constituting their properties and states"

(Aristotle, *Metaphysica*, 986 a 16–17). Hence, "numbers are what is primary in all nature" and "all other things are assimilable to numbers in their nature" (ibid., A 5, 985 a 32–986 a 1). Therefore, everything can be explained on the basis of number, and "it would be impossible to understand or know anything without number" (Philolaus 44 B 4, ed. Diels-Kranz).

But already Aristotle pointed out that the Pythagoreans's very strong form of the claim was untenable. For example, the Pythagoreans believed the number ten "to be perfect and to embrace the whole nature of numbers," therefore they "said that the bodies in the heaven had to be ten" (Aristotle, *Metaphysica*, A 5, 986 a 8–10). But, "as the observed bodies are only nine, to meet this they invented a tenth body, the Counter-Earth" (ibid, A 5, 986 a 11–12). This conflicts with the fact that there is no Counter-Earth, the Pythagoreans invented it only to square with their own assumption that the number ten is perfect.

As an alternative to the Pythagoreans's very strong form of the claim, in the last few decades some people have put forward a more moderate form of the claim: There are mathematical explanations of some empirical facts because those empirical facts are inherently mathematical.

Certain problems concerning empirical facts are commonly cited as evidence for the more moderate form of the claim. The main ones are the honeycomb problem, the Magicicada problem, the strawberry problem, the Königsberg bridges problem, and the Kirkwood gaps problem.

14.22 The Honeycomb Problem

The honeycomb problem is: Why do bees make honeycombs with hexagonal cells rather than with cells of another figure?

From antiquity, the following mathematical explanation has been proposed for this problem. The reason why bees make honeycombs with hexagonal cells is that a hexagonal grid is the optimal way to divide a surface into regions of equal area with the least total perimeter. This minimizes the use of wax while maximizing the area for honey storage.

Thus, Varro says that bees make honeycombs with hexagonal cells, because "the geometers show that a hexagon inscribed in a circular figure encloses the greatest amount of space" (Varro, *De agricultura*, III, xvi.5).

Pappus says that bees make honeycombs with hexagonal cells "by virtue of a certain geometrical forethought" (Pappus, *Collectio*, V, 304.26, ed. Hultsch). For, "bees know" that "the hexagon is greater than the square and the triangle and will hold more honey for the same expenditure of material in constructing each" (ibid., V, 306.29–32).

The mathematical explanation of the honeycomb problem has been repeatedly reaffirmed since antiquity.

Thus, Kirby says that bees are "heaven-instructed mathematicians, who before any geometer could calculate under what form a cell would occupy the least space without diminishing its capacity," built "their hexagonal cells" and "were enabled, without study," to "construct the opposite story of combs" (Kirby 1835, II, 337).

The mathematical explanation of the honeycomb problem has been reaffirmed also recently, with the argument that such explanation is confirmed by Hales's honeycomb theorem: "Any partition of the plane into regions of equal area has perimeter at least that of the regular hexagonal honeycomb tiling" (Hales 2001, 1). So a hexagonal grid represents the optimal way to divide a surface into regions of equal area with the least total perimeter.

Thus, Devlin says that, "of all the different architectures" honeybees "could have used, Hales's theorem shows that the structure they use," namely a hexagonal cross section, "is the most efficient" (Devlin 2005, 77). For, it is the architecture that uses "the least amount of wax to build walls" (ibid., 73). So "the evolution of the honeybees incorporated a natural proof of" Hales's "result" (ibid., 77). Hales's theorem was proved only in 2001, but "the honeybees, in their own way," had "known the theorem all along" (ibid., 76). The "incredible precision with which the bees construct their honeycombs means they are natural geometers and engineers of the highest order" (ibid., 77).

The mathematical explanation of the honeycomb problem, however, is inadequate. The claim that this explanation is confirmed by Hales's honeycomb theorem is invalid, because the latter applies only to two-dimensional figures, not to three-dimensional ones. But the honeycomb actually "consists of prismatic cells having regular hexagonal openings while the bottom is closed by three equal rhombuses," where "the cells are situated between two parallel planes filling the space bounded by them without interstices" (Bleicher and Fejes Tóth 1965, 969).

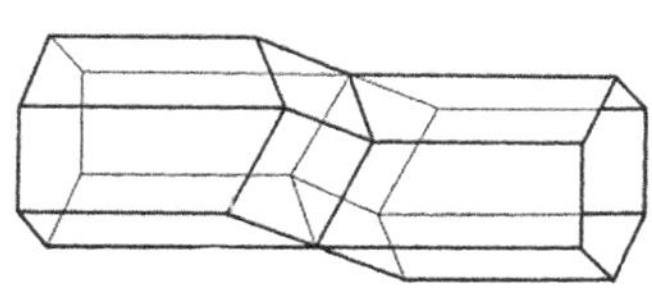

Therefore, the optimality problem of bee cells is not the optimality problem of their openings. As Fejes Tóth points out, if we call 'width' of the honeycomb the distance between the two parallel planes, then the optimality problem of bee cells is: "Among the open cells of" a given volume "generating a honeycomb (of any width), find that one of least surface-area" (Fejes Tóth 1964, 469).

Now, "is the solution of this problem a bee-cell? The answer is: No" (ibid., 470). Indeed, "for all values" of volume and width, "we know better cells than those of the bees" (Bleicher and Fejes Tóth 1965, 970). Thus, bee cells are not those using the least amount of wax to build walls.

Therefore, the mathematical explanation of the honeycomb problem does not account for why bees do make honeycombs with hexagonal cells rather than with cells of another figure. As Nazzi says, the "formation of the hexagons is still a matter of debate and a definitive conclusion has not yet been reached" (Nazzi 2016, 1).

14.23 The Magicicada Problem

The Magicicada problem is: Why is the life cycle of the cicada species 'Magicicada tredecim' and 'Magicicada septendecim' a prime number, 13 and 17 years respectively?

Magicicada tredecim and Magicicada septendecim are two species of cicadas widely distributed over eastern North-America, Magicicada tredecim generally southern, Magicicada septendecim generally northern. They spend most of their lives in the soil from depths of two to twenty-four inches, feeding on xylem fluids from the roots of deciduous forest trees. After 13 or 17 years, mature Magicicada tredecim nymphs and Magicicada septendecim nymphs, respectively, burrow upward in the soil, and in the springtime emerge synchronously in very high number, leaving their nymphal skins behind. Soon after emerging, males begin their constant singing to attract females, about ten days after the emergence female mate and begin depositing eggs. Both males and females live for approximately four to six weeks above ground, and most of them are usually dead by the beginning of July. About six to ten weeks after egg laying, eggs hatch and the newborn nymphs drop to the ground where they enter the soil and begin another life cycle of 13 and 17 years, respectively.

The following mathematical explanation has been proposed for this problem. The reason why the life cycle of Magicicada tredecim and Magicicada septendecim is a prime number, 13 and 17 years respectively, is that a life cycle of 13 or 17 years exceeds the life cycle of any predator, which is of 2–5 years. The fact that 13 and 17 are prime numbers minimizes the number of intersections with the life cycle of predators, because prime numbers maximize their least common multiple relative to all lower numbers. Therefore, a prime life cycle maximizes the number of years between successive intersections with the life cycle of predators with lower period lengths.

Thus, Gould says that the reason why Magicicada tredecim and Magicicada septendecim have a 13 and 17 years life cycle, respectively, is that 13 and 17 "are large enough to exceed the life cycle of any predator," and 13 and 17 are "also prime numbers" (Gould 1977, 102). Then, "consider a predator with a cycle of five years: if cicadas emerged every 15 years, each bloom would be hit by the predator. By cycling at a large prime number, cicadas minimize the number of coincidences (every 5×17, or 85 years, in this case). Thirteen- and 17-year cycles cannot be tracked by any smaller number" (ibid.).

The mathematical explanation of the Magicicada problem, however, is inadequate, because the prime numbers 13 and 17 appear in the description of the life cycle only because we measure the latter in years, which is a mere convention. If, instead of measuring it in years, we measured it, say, in months, then the life cycle would be 156 and 204 months, respectively, and neither 156 nor 204 is a prime number.

Baker and Colyvan argue that "biologists do not in fact use" months as "units in describing or discussing cicada life-cycles, which suggests that years are the most salient unit in this context" (Baker and Colyvan 2011, 329). But this argument is invalid because, if biologists do not use months as units in describing or discussing cicada life-cycles, it is not because years are the most salient unit in this context, but only because the finer measurement in seasons or months is not necessary. The rougher measurement in years is enough, and is more manageable because it involves smaller numbers. This has nothing to do with prime numbers.

Therefore, the mathematical explanation of the magicicada problem does not account for why the life cycle of Magicicada tredecim and Magicicada septendecim is a prime number.

14.24 The Strawberry Problem

The strawberry problem is: Why does Mother fail every time she tries to distribute exactly 23 strawberries evenly among her 3 children without cutting any?

The following mathematical explanation has been proposed for this problem. Mother fails because 23 cannot be divided evenly by 3.

Thus, Lange says that "the fact that 23 cannot be divided evenly by 3 explains why Mother fails every time she tries to distribute exactly 23 strawberries evenly among her 3 children without cutting any," and this explanation of an empirical fact is "distinctively mathematical" (Lange 2017, 6).

The mathematical explanation of the strawberry problem, however, is inadequate. As Bangu says, although the strawberry problem "does not look like a typical mathematical proposition," what is "crucial here is what the proposition says, that is, what the why-question actually asks" (Bangu 2017b). Now, with the why-question "we are not asking about" Mother, or about children, or "about strawberries either" (ibid.). The "reference to mothers and strawberries (and children) seems merely a façade or superfluous" (ibid.). The strawberry problem is actually a question "about numbers," it asks why "a certain number (of uncut strawberries) cannot be divided exactly (by Mother) by a certain number," so in fact it is a disguised "mathematical proposition" (ibid.).

Therefore, the mathematical explanation of the strawberry problem is not an extra-mathematical explanation, but really an intra-mathematical one. Instead of an extra-mathematical explanation, we have here simply an application of a mathematical result about natural numbers to strawberries.

14.25 The Königsberg Bridges Problem

At Kant's time, in Königsberg there were seven bridges, *a*, *b*, *c*, *d*, *e*, *f*, *g*, over the river Pregel, joining four areas *A*, *B*, *C*, *D*.

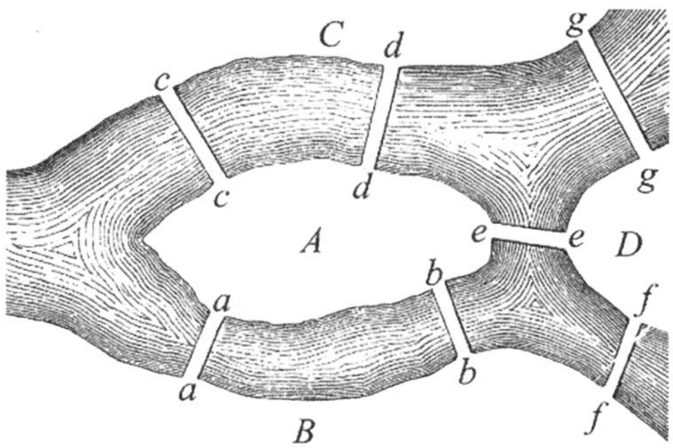

On Sundays afternoon, the citizens of Königsberg used to take walks over the bridges and, while crossing and re-crossing them many times, they asked themselves the question that is now called the Königsberg bridges problem: Why would it have been impossible to walk across all the seven bridges, crossing each bridge just once?

Euler proposed a mathematical explanation for this. The Königsberg bridges problem can be represented by the following graph.

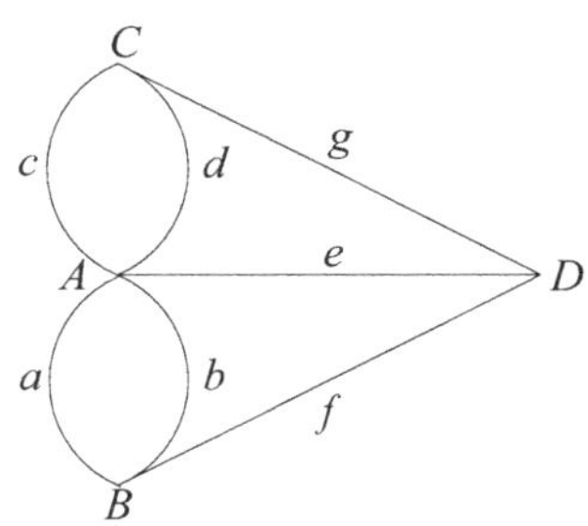

Let us call 'node' a point to or from which lines are drawn; 'edge' a line connecting two consecutive nodes; and 'route' a number of edges taken in consecutive order such that each edge is traversed just once. Euler demonstrated a theorem which, in terms of the graph, states: If there are more than two nodes at which an odd number of edges meet, the graph cannot be traversed completely in one route (Euler 1741, 139).

Now, in the graph, at node A five edges meet, and at each of the nodes B, C, and D three edges meet, so there are four nodes at which an odd number of edges meet. Therefore, by Euler's theorem, it would have been impossible to walk across all the seven bridges, crossing each bridge just once.

The mathematical explanation of the Königsberg bridges problem has been reaffirmed several times since Euler. Thus, Pincock says that, "knowing Euler's theorem, I could now explain why" there will be "no path that crosses every bridge exactly once and that returns to the starting point" (Pincock 2007, 259). Mathematics permits "to describe the bridges of Königsberg without knowing" the "details of the bridges' physical construction" (ibid., 263).

The mathematical explanation of the Königsberg bridges problem, however, is inadequate because it abstracts from all empirical conditions.

For example, suppose that someone had started walking on one of the bridges while a flood of the river Pregel destroyed some of the bridges, picking up debris and pushing it forcefully against the bridge, so as to cause its foundations to wash away and structural elements to break apart.

Or, as Lange says, suppose that "someone was poised to shoot anyone who tried to cross a given bridge" (Lange 2017, 8).

Or, more to the historical point, suppose that someone had started walking on one of the bridges on the day of August 1944 when a RAF bombing raid destroyed another of the bridges.

If one of these empirical facts had occurred, that fact, rather than Euler's theorem, would have been the explanation of the impossibility to walk across all the seven bridges, crossing each bridge just once.

Thus, like in the case of the strawberry problem, although the Königsberg bridges problem does not look like a typical mathematical proposition, what is crucial here is what the proposition says, namely what the why-question actually asks. Now, with the why-question, we are not asking about the citizens of Königsberg, or about the Königsberg bridges, the reference to the citizens of Königsberg and the Königsberg bridges seems merely a façade or superfluous. The Königsberg bridges problem is really a question about a graph, so in fact it is a disguised mathematical proposition.

Therefore, the mathematical explanation of the Königsberg bridges problem is not an extra-mathematical explanation, but actually an intra-mathematical one. Instead of an extra-mathematical explanation, we have here simply an application of a mathematical result about graphs to a graph.

14.26 The Kirkwood Gaps Problem

The Kirkwood gaps are regions in the main asteroid belt between Mars and Jupiter where there are relatively few asteroids. The Kirkwood gaps problem is: Why do Kirkwood gaps exist?

The following mathematical explanation has been proposed for this problem. The Kirkwood gaps exist because the corresponding orbital periods are eigenvalues of a certain vector operator.

Thus, Colyvan says that "the explanation of this important astronomical fact is provided by the mathematics of eigenvalues (that is, basic functional analysis)" (Colyvan 2010, 302). So we have "statements involving mathematical entities (the eigenvalues of the system) explaining physical phenomena (the relative absence of asteroids in the Kirkwood gaps)" (ibid., 303).

The mathematical explanation, however, is inadequate. As Bueno argues, "mathematics alone does not provide an explanation of the regularities involved in the Kirkwood gaps. The explanation is offered by the identification of the gravitational interaction between the asteroids, planets, and the Sun, subject to chaos" (Bueno 2012, 973). Indeed, "gravitational forces are clearly the relevant kind of physical interaction to explain the behaviour of the motions of these asteroids," the "eigenvalues of the system are not what explains that behaviour" (ibid.). Mathematics simply "provides a mathematical description of the relevant interactions" (ibid., 972). Moreover, the equations must be "interpreted in a suitable way" because, "without a suitable interpretation, the mathematics does not state anything about the physical world. Uninterpreted, the equations describe relations among various functions and, in certain cases, numbers. They do not specify relations among physical objects" (ibid., 973). The "explanation is ultimately achieved by the identification of suitable causal structures: the gravitational forces among the planets and asteroids" (ibid., 975).

Colyvan's reply to this argument is that, while it is true that "the gravitational fields of the Sun and Jupiter" are "what supplies the mechanism," nevertheless "the mechanism is one thing but the full explanation (arguably) involves more than just the mechanical details" (Colyvan 2012b, 1035). Of course, "the mathematics needs an interpretation," but, "far from this being an obstacle to mathematics explaining, this is exactly why mathematics is able to deliver the very general explanations I have in mind" (ibid.).

This reply, however, is inadequate, because the fact that the equations, by themselves, merely describe relations among various functions and, in certain cases, numbers, and do not specify relations among physical objects, means that the equations say nothing specific about the Kirkwood gaps.

14.27 No Empirical Facts Are Inherently Mathematical

From what has been said above it appears that even the more moderate form of the claim, that there are mathematical explanations of some empirical facts because those empirical facts are inherently mathematical, is untenable. Simply, there are no extra-mathematical explanations.

As Bueno says, "mathematics alone fails to specify what is going on at the physical level" because it "does not determine, it does not even constrain, the physical world. This is not surprising given that mathematics, that is, pure mathematics, does not state anything about the physical world" (Bueno 2012, 980).

Even theories that seem to be mathematical, such as Newtonian theory of the world, are not purely mathematical. As Guicciardini says, "some of Newton's demonstrative geometrical techniques are based on physical insights rather than mathematical procedures" (Guicciardini 1999, 96).

Newton himself says that natural sciences "depend as well on physicall principles as on mathematicall demonstrations" and the foundation of their propositions is "from experiments, and so but physicall: whence the propositions themselves can be esteemed no more than physicall principles of a science" (Newton 1959, 187).

Frege claims that geometry is applicable to "all that is spatially intuitable, whether actual or product of our fancy. The wildest visions of delirium, the boldest inventions of legend and poetry" remain, "so long as they remain intuitable, still subject to the axioms of geometry" (Frege 1960, 20).

But, that geometry is applicable to all these facts, does not mean that it provides an explanation for them. As already said above, the claim that there are extra-mathematical explanations must not be confused with the claim that mathematics is applicable to empirical facts. Mathematics is applicable to empirical facts, but there are no extra-mathematical explanations.

The claim that there are extra-mathematical explanations mistakes mathematics for something that is inherent in the world. But mathematics is not something which is inherent in the world, it is only one of the tools by which human beings make the world understandable to themselves.

14.28 Extra-Mathematical Explanations and Mathematical Platonism

Some supporters of the claim that there are extra-mathematical explanations endorse mathematical platonism.

To this purpose, they use the so-called "enhanced indispensability argument," which "argues for platonism on the grounds that mathematics play an indispensable explanatory role in science" (Baker 2017, 194).

The enhanced indispensability argument is: "(1) We ought rationally to believe in the existence of any entity that plays an indispensable explanatory role in our best scientific theories. (2) Mathematical objects play an indispensable explanatory role in science. (3) Hence, we ought rationally to believe in the existence of mathematical objects" (Baker 2009, 613).

The enhanced indispensability argument, however, is invalid, because the assumptions, (1) and (2), on which it is based, are unjustified.

Assumption (1) is unjustified, because the history of science shows that our best scientific theories at one time, are supplanted by other scientific theories at another time. Then, assumption (1) would imply that we ought rationally to believe in the existence of the mathematical objects that play an indispensable explanatory role in our best scientific theories at one time, but we ought not longer to believe in the existence of these mathematical objects when the best scientific theories at one time are supplanted by other scientific theories at another time. Therefore, our rational justification in believing in the existence of certain mathematical objects would depend on time. But this is incompatible with mathematical platonism, according to which mathematical objects exist independently of time.

Assumption (2) is unjustified, because it takes it for granted that there are extra-mathematical explanations. For, one can assert that mathematical objects play an indispensable explanatory role in science only if there are extra-mathematical explanations. But, as it has been argued above, there are no extra-mathematical explanations.

References

Auslander, Joseph. 2008. On the roles of proof in mathematics. In *Proof and other dilemmas: Mathematics and philosophy*, ed. Bonnie Gold and Roger A. Simon, 62–77. Washington: The Mathematical Association of America.

Baker, Alan. 2009. Mathematical explanation in science. *The British Journal for the Philosophy of Science* 60: 611–633.

———. 2017. Mathematics and explanatory generality. *Philosophia Mathematica* 25: 194–209.

Baker, Alan, and Mark Colyvan. 2011. Indexing and mathematical explanation. *Philosophia Mathematica* 19: 323–334.

Bangu, Sorin. 2017a. Scientific explanation and understanding: Unificationism reconsidered. *European Journal for Philosophy of Science* 7: 103–126.

———. 2017b. Review of Marc Lange, Because without cause: Non-causal explanations in science and mathematics. *The British Journal for the Philosophy of Science Review of Books* http://www.thebsps.org/bjps-rob/marc-lange-because-without-cause/

Bleicher, Michael Nathaniel, and László Fejes Tóth. 1965. Two-dimensional honeycombs. *The American Mathematical Monthly* 72: 969–973.

Bolzano, Bernard. 2004. *Mathematical works*. Oxford: Oxford University Press.

———. 2014. *Theory of science*. Oxford: Oxford University Press.

Bueno, Otávio. 2012. An easy road to nominalism. *Mind* 121: 967–982.

Byers, William. 2007. *How mathematicians think: Using ambiguity, contradiction, and paradox to create mathematics*. Princeton: Princeton University Press.

Chemla, Karine. 2005. Geometrical figures and generality in ancient China and beyond: Liu Hui and Zhao Shuang, Plato and Thabit ibn Qurra. *Science in Context* 18: 123–166.

Colyvan, Mark. 2010. There is no easy road to nominalism. *Mind* 119: 285–306.

———. 2012a. *An introduction to the philosophy of mathematics*. Cambridge: Cambridge University Press.

———. 2012b. Road work ahead: Heavy machinery on the easy road. *Mind* 121: 1031–1046.

Davis, Philip J., and Reuben Hersh. 1981. *The mathematical experience*. Cham: Birkhäuser.

Descartes, René. 1996. *Oeuvres*. Paris: Vrin.

Devlin, Keith. 2005. *The math instinct*. New York: Thunder's Mouth Press.

Djebbar, Ahmed. 2009. *Textes geometriques arabes (IXe – XVe siècles)*. Dijon: IREM.

Euler, Leonhard. 1741. Solutio problematis ad geometriam situs pertinentis. *Commentarii Academiae Scientiarum Imperialis Petropolitanae* 8: 128–140.

Fejes Tóth, László. 1964. What the bees know and what they do not know. *Bulletin of the American Mathematical Society* 70: 468–481.

Frege, Gottlob. 1960. *The foundations of arithmetic: A logico-mathematical enquiry into the concept of number*. New York: Harper.

Gould, Stephen Jay. 1977. *Ever since Darwin: Reflections on natural history*. New York: Norton.

Gowers, Timothy. 2007. Mathematics, memory and mental arithmetic. In *Mathematical knowledge*, ed. Mary Leng, Alexander Paseau, and Michael Potter, 33–58. Oxford: Oxford University Press.

Gowers, Timothy, and Michael Nielsen. 2009. Massively collaborative mathematics. *Nature* 461: 879–881.

Guicciardini, Niccolò. 1999. *Reading the 'Principia': The debate on Newton's mathematical methods for natural philosophy from 1687 to 1736*. Cambridge: Cambridge University Press.

Hales, Thomas Callister. 2001. The honeycomb conjecture. *Discrete & Computational Geometry* 25: 1–22.

Kirby, William. 1835. *On the power wisdom and goodness of God as manifested in the creation of animals and in their history habits and instincts*. London: Pickering.

Kitcher, Philip. 1978. The plight of the Platonist. *Noûs* 12: 119–136.

———. 1981a. Explanatory unification. *Philosophy of Science* 48: 507–531.

———. 1981b. Mathematical rigor – Who needs it? *Noûs* 15 (4): 469–473.

Lange, Marc. 2017. *Because without cause: Non-causal explanations in science and mathematics*. Oxford: Oxford University Press.

———. 2018. Mathematical explanations that are not proofs. *Erkenntnis* 83: 1285–1302.

Lehet, Ellen. 2021. Induction and explanatory definitions in mathematics. *Synthese* 198: 1161–1175.

Maddy, Penelope. 1981. Sets and numbers. *Noûs* 15: 495–511.

Mizrahi, Moti. 2020. Proof, explanation, and justification in mathematical practice. *Journal for General Philosophy of Science*. https://doi.org/10.1007/s10838-020-09521-7.

Nazzi, Francesco. 2016. The hexagonal shape of the honeycomb cells depends on the construction behavior of bees. *Scientific Reports* 6: 28341.

Newton, Isaac. 1959. Letter to Oldenburg, 11 June 1672. In Isaac Newton, *Correspondence*, I, 171–193. Cambridge: Cambridge University Press.

Niss, Mogens. 2006. The structure of mathematics and its influence on the learning process. In *New Mathematics Education Research and Practice*, ed. Jürgen Maasz and Wolfgang Schlöglmann, 51–62. Rotterdam: Sense Publishers.

Pincock, Christopher. 2007. A role for mathematics in the physical sciences. *Noûs* 41: 235–275.

Resnik, Michael D., and David Kushner. 1987. Explanation, independence and realism in mathematics. *The British Journal for the Philosophy of Science* 38: 141–158.

Rota, Gian-Carlo. 1997. *Indiscrete thoughts*. Cham: Birkhäuser.

Schopenhauer, Arthur. 2010. *The world as will and representation*, vol. 1. Cambridge: Cambridge University Press.

von Wright, Georg Henrik. 1971. *Explanation and understanding*. London: Routledge & Kegan Paul.

Wang, Weijia. 2019. Artistic proofs: A Kantian approach to aesthetics in mathematics. *Estetika: The Central European Journal of Aesthetics* 56 (2): 223–243.

Zelcer, Mark. 2013. Against mathematical explanation. *Journal for General Philosophy of Science* 44: 173–192.

Zukav, Gary. 1980. *The dancing Wu Li masters: An overview of the new physics*. New York: Bantham.

Chapter 15
Beauty

Abstract According to heuristic philosophy of mathematics, one of the tasks of the philosophy of mathematics is to give an answer to the question: What is the nature of mathematical beauty? The chapter argues that a piece of mathematics is beautiful if it provides understanding, namely recognition of the fitness of the parts to each other. Beauty is relevant not only to finished mathematics but also, and primarily, to the making of mathematics, because it has a significant role both in finding solutions to mathematical problems and in choosing the mathematical fields and problems to pursue.

Keywords Views on mathematical beauty · Beauty as intrinsic property · Beauty as projected property · Aesthetic induction · Enlightenment · Understanding · Top-down beauty · Bottom-up beauty · Beauty and solutions to problems · Beauty and choice of fields and problems · Innate and acquired sense of beauty

15.1 The Relevance of Beauty to Mathematics

As we have seen in Chap. 3, according to heuristic philosophy of mathematics, one of the tasks of the philosophy of mathematics is to give an answer to the question: What is the nature of mathematical beauty?

An answer to this question is necessary, because beauty is relevant to mathematics. Therefore, as Wells says, beauty "must be incorporated into any adequate epistemology of mathematics. Philosophies of mathematics that ignore beauty will be inherently defective and incapable of effectively interpreting the activities of mathematicians" (Wells 1990, 41).

Beauty is relevant to mathematics in two respects. On the one hand, beauty is relevant to finished mathematics, because it has a significant role in the understanding of theorems and demonstrations. On the other hand, even more importantly, beauty is relevant to the making of mathematics, because it has a significant role both in finding solutions to mathematical problems and in choosing the mathematical fields and problems to pursue.

C. Cellucci, *The Making of Mathematics*, Synthese Library 448,
https://doi.org/10.1007/978-3-030-89731-4_15

That beauty is relevant to the making of mathematics is acknowledged by several mathematicians.

Thus, Thurston says: "My experience as a mathematician has convinced me that the aesthetic goals and the utilitarian goals for mathematics turn out, in the end, to be quite close. Our aesthetic instincts draw us to mathematics of a certain depth and connectivity" (Thurston 1990, 848).

Sinclair says that, "as soon as aesthetics is mentioned in connection to mathematics, most mathematicians are quick to assume that a reference is being made to beautiful theorems and elegant proofs" (Sinclair 2002, 46). But "the aesthetic also insinuates itself significantly during the process of inquiry (prior to final evaluations)" (ibid.).

Gowers says that, "if one is searching for a proof and is forced to make guesses about how it might look, it is a very good strategy to make guesses that are natural, elegant and interesting, if one possibly can" (Gowers 2007, 37). For, "it is remarkable how important a well-developed aesthetic sensibility can be, for purely pragmatic reasons, in mathematical research" (ibid.).

15.2 The Objection of Sensory Properties

Some objections, however, have been raised against the claim that beauty is relevant to mathematics.

A first objection is that aesthetic properties necessarily depend on sensory properties, while demonstrations do not necessarily have any sensory embodiment or manifestation.

Thus, Zangwill says that "aesthetic properties are properties which something has only if it has sensory properties" (Zangwill 1998, 66). But proofs "do not necessarily have any sensory embodiment or manifestation. One can consider a proof" in "one's head. And many say that we can appreciate" its "beauty in purely intellectual contemplation" (ibid., 78). Therefore, when we say that a proof is beautiful, this "is not genuine aesthetic appreciation. So aesthetic terms are metaphorically applied in these cases" (ibid., 79). Indeed, "if the valuable properties are aesthetic, then they depend on sensory properties, and if they are not aesthetic, then they do not" (ibid., 75).

This objection, however, is invalid because aesthetic properties need not depend on sensory properties. The objection would imply not only that there is no genuine mathematical beauty, but also that there is no genuine literary beauty, because literary beauty does not depend on sensory properties. Even if, to read a novel, I must be able to perceive the text in front of me, my aesthetic experience in reading the novel can hardly be said to consist in the perception of the words on the page in front of me. Nor do I literally see the facts narrated in the novel, I only imagine them, and imagination is the capacity to represent something even when it is not itself present before my eyes. So I can appreciate the beauty of the novel in purely intellectual contemplation.

In this regard, Zangwill says that "if a literary work has aesthetic properties, they derive from the particular choice of words, because of the way they sound," and "if a literary work has values which are not linked" to "the sonic properties of words, then they are not aesthetic values" (ibid., 75).

But this would imply that the aesthetic values of, say, Tolstoy's *Anna Karenina* are lost in an English translation, because the sonic properties of words in English are different from the sonic properties of the corresponding words in Russian. However, as any reader of an English translation of *Anna Karenina* can confirm, one can very well appreciate the aesthetic values of that novel even in translation.

Of course, mathematical beauty and literary beauty are different kinds of beauty. But this only means that there are different kinds of beauty even in the realm of beauty involving intellectual elaboration.

15.3 The Objection of Masked Epistemic Judgments

Another objection against the claim that beauty is relevant to mathematics is that, while aesthetic appreciation concerns taking disinterested pleasure in some object for its own sake, the supposed aesthetic judgments in mathematics are concerned with truth and utilitarian ends, so they are not genuinely aesthetic but in fact are masked epistemic judgments.

Thus, Todd says that "aesthetic appreciation concerns taking disinterested pleasure in some object for its own sake" (Todd 2018, 213). Therefore, "aesthetic judgements are unconcerned with truth and utilitarian ends: they must not be based on cognitive interest in attaining knowledge about something nor with assessing an object in relation to some purpose which it serves" (ibid.). But "the supposed aesthetic judgments appealed to in mathematics" in fact "refer to epistemic values in mathematics" (ibid., 216). So they are concerned with truth and utilitarian ends, therefore they "are really 'masked' epistemic rather than aesthetic assessments" (ibid., 211–212).

This objection, however, is invalid because it contrasts with the testimony of many mathematicians.

Thus, Poincaré says that "mathematics" has "an esthetic aim," its "adepts find therein delights analogous to those given by painting and music," so "the joy they thus feel" has "the esthetic character, even though the senses take no part therein" (Poincaré 2015, 280).

Weyl says: "My work always tried to unite the true with the beautiful; but when I had to choose one or the other, I usually chose the beautiful" (Dyson 1956, 458).

Von Neumann says that the mathematician's "criteria of selection, and also those of success, are mainly aesthetical" (von Neumann 1961, 8). And, as a mathematical discipline develops, "it becomes more and more purely aestheticizing, more and more purely 'l'art pour l'art'" (ibid., 9).

Borel says that "mathematics is to a great extent an art, an art whose development has been derived from, guided by, and judged according to aesthetic criteria" (Borel 1983, 11).

Some mathematicians even maintain that the aesthetic values are in conflict with the utilitarian ends.

Thus, Hardy says that "the 'real' mathematics of the 'real' mathematicians, the mathematics of Fermat and Euler and Gauss and Abel and Riemann, is almost wholly 'useless' (and this is as true of 'applied' as of 'pure' mathematics)" (Hardy 1992, 119). But, while almost useless, this is "the mathematics which has permanent aesthetic value," and "is eternal because the best of it" may "continue to cause intense emotional satisfaction to thousands of people after thousands of years" (ibid., 131). Conversely, the "parts of mathematics" that have "considerable practical utility" are "just the parts which have least aesthetic value," indeed "it is not possible to justify the life of any genuine professional mathematician on the ground of the 'utility' of his work" (ibid., 119–120).

15.4 Two Different Traditions about Mathematical Beauty

After examining some objections against the claim that beauty is relevant to mathematics, we consider the question of the nature of mathematical beauty. There are two different traditions about it.

(A) Mathematical beauty is an intrinsic property of mathematical things, so it is independent of subject, place, and time.
(B) Mathematical beauty is a property projected by the subject onto mathematical things, so it is dependent upon subject, place, and time. If some mathematical things have properties that are valued by the aesthetic criteria proper to a certain subject at a certain place and time, such properties will be called 'aesthetic properties', and the subject will describe those mathematical things as beautiful.

Tradition (A) goes back to antiquity, tradition (B) has been widespread in the modern and contemporary period.

15.5 Mathematical Beauty as an Intrinsic Property

An eminent representative of the tradition according to which mathematical beauty is an intrinsic property of mathematical things, is Plato.

Indeed, Plato says that "straight lines and circles, and the plane and solid figures which are formed out of them by means of compasses, rulers and squares" are "not, as other things are, beautiful in a relative way, but they are by their very nature forever beautiful by themselves" (Plato, *Philebus*, 51 c 3–d 1). They are "not beautiful at one time and not at another, or beautiful by one standard and ugly by another, or beautiful in one place and ugly in another," or "beautiful to some people but ugly to others" (Plato, *Symposium*, 211 a 3–5). They are of a kind of beauty that "always is and does not come into to be or perish, nor does it grow or wane" (ibid., 211 a 1–2).

On the other hand, however, according to Plato, beauty admits degrees. For example, he says that there are infinitely many right-angled scalene triangles, but "there is one that is the most beautiful, and surpasses all other scalene triangles, and that is the one a pair of which composes the equilateral triangle" (Plato, *Timaeus*, 54 a 5–7).

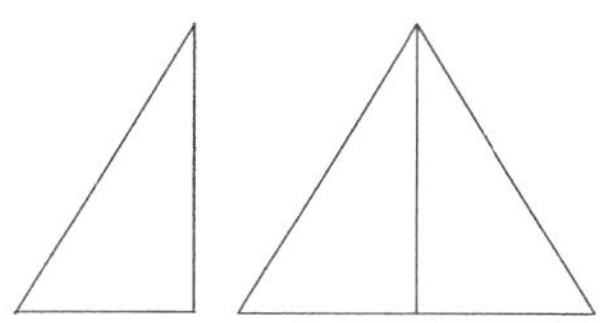

The tradition according to which mathematical beauty is an intrinsic property of mathematical things has had many followers, from antiquity until recently.

Thus, Dirac says that "it is quite clear that beauty does depend on one's culture and upbringing for certain kinds of beauty, pictures, literature, poetry, and so on," but "mathematical beauty is of a completely different kind and transcends these personal factors. It is the same in all countries and at all periods of time" (Dyson 1992, 305).

But the claim that mathematical beauty is independent of subject, place and time is invalid. Evidence for this is provided by the responses of readers of *The Mathematical Intelligencer* to a questionnaire that asked them to give each of 24 theorems "a score of 0 through 10 for beauty" (Wells 1988, 30). The responses show that "the idea that mathematicians largely agree in their aesthetic judgments is at best grossly oversimplified" (Wells 1990, 40). From them, it appears that "beauty, even in mathematics, depends upon historical and cultural contexts" (ibid., 39).

15.6 Mathematical Beauty as a Projected Property

An eminent representative of the tradition according to which mathematical beauty is a property projected by the subject onto mathematical things, is Kant.

Indeed, Kant says that "beauty is nothing by itself, without relation to the feeling of the subject" (Kant 2000, 103). In fact, beauty "is not a property of the object outside of me, but merely a kind of representation in me" (ibid., 237). The representation is "rational but related in a judgment solely to the subject (its feeling)," and is “to that extent always aesthetic” (ibid., 90). Its only “determining ground lies in a sensation that is immediately connected with the feeling of pleasure and displeasure” (ibid., 26). Thus, when we “call a demonstration” of certain “properties beautiful,” the “satisfaction, although its ground lies in concepts, is subjective” (ibid., 238–239).

Since, when we call a demonstration of certain properties beautiful, the satisfaction is subjective, there is the risk that subjectivity entails arbitrariness. To avoid this risk, judgments of mathematical beauty must be formed starting from the examples of “the ancient mathematicians” (ibid., 163). For, “among all the faculties and talents, taste is precisely the one which” is “most in need of the examples of what in the progress of culture has longest enjoyed approval” (ibid., 164).

This does not mean that the examples of the ancient mathematicians are “a posteriori sources of taste,” nor that appeals to them “contradict the autonomy of taste in every subject” (ibid., 163). It only means that they can “put others on the right path for seeking out the principles in themselves and thus for following their own, often better, course” (ibid., 164).

15.7 Mathematical Beauty and Aesthetic Induction

After Kant, the tradition according to which mathematical beauty is a property projected by the subject onto mathematical things, has had many followers. Of course, they do not necessarily share all of Kant’s views about mathematical beauty, but at least share the view that mathematical beauty is a projection of the subject. A follower of this tradition is McAllister.

Indeed, McAllister says that beauty is “a value that is projected into or attributed to objects by observers, not a property that intrinsically resides in objects” (McAllister 2005, 15). Therefore, it is incorrect to say that a mathematical object “has beauty,” one can only say that, if a mathematical object “exhibits properties that are valued by an observer’s aesthetic criteria, the observer will project beauty into the object and describe the object as beautiful” (ibid., 16).

Aesthetic criteria are not fixed, but “show historical evolution,” indeed “the community’s aesthetic tastes change with time” (ibid.). The historical evolution of aesthetic criteria is governed by “aesthetic induction,” the procedure by which mathematicians “attach aesthetic value to an aesthetic property roughly in proportion to the degree” of “success scored” by the mathematical objects “that exhibit the property” (ibid., 28).

Evidence for this “is provided by the gradual acceptance of new classes of numbers in mathematics, such as negative, irrational, and imaginary numbers” (ibid., 29). Initially, each of them “was regarded with aesthetic revulsion,” but, “as it demonstrated its empirical applicability in mathematical theorizing,” it came “to be attributed growing aesthetic merit” (ibid.).

In particular, "the evolution of aesthetic criteria applied to mathematical proofs" is "governed by the aesthetic induction" (ibid.). In the case of classical proofs, mathematicians have "regarded a proof as beautiful if" it lends itself to being grasped in a single act of mental apprehension" (ibid., 22). But, in the last few decades, new kinds of proofs have been developed which cannot be grasped in a single act of mental apprehension, such as computer-assisted proofs, "the most striking example" of which is the Appel and Haken "proof of the four-color conjecture" (ibid., 23). Therefore, some mathematicians "have argued that computer-assisted proofs do not deliver understanding" and hence "cannot be beautiful" (ibid., 24).

But "conceptions of understanding" are not fixed, they "evolve in accord with the aesthetic induction" (ibid., 29). Therefore, we "can forecast that the computer-assisted proof" such as the Appel and Haken proof, being successful, "will remold the concept of understanding and aesthetic criteria in mathematics," and eventually will gain "acceptance" (ibid., 28). For, by aesthetic induction, "mathematicians's aesthetic preferences evolve in response to" the success scored by "mathematical constructs" (ibid., 29).

Thus, according to McAllister, the beauty of mathematical objects, theorems, or proofs depends on their success, although with a time delay, because success requires time and tenacity.

McAllister's view, however, has shortcomings.

McAllister says that mathematicians's aesthetic preferences evolve in response to the success scored by mathematical constructs. But this conflicts, for example, with the fact that, although the calculus of infinitesimals of Leibniz and Newton was the backbone of the development of modern science and engineering in the industrial revolution, this did not increase the mathematicians's preference for infinitesimals. Cantor even said that infinitesimals were the "cholera bacillus of mathematics" (Cantor 1965, 505). They "belong in the waste-basket as paper numbers" (ibid., 507).

As another example, Hersh and John-Steiner say that, "at an elementary level" the "quadratic formula," namely the formula "which solves the general quadratic equation," is "one of the most memorized formulas in math," but it is "not beautiful!" (Hersh and John-Steiner 2011, 61).

As a further example, Hardy says that proof by cases is something "which a real mathematician tends to despise" (Hardy 1992, 114). For, it "is one of the duller forms of mathematical argument. A mathematical proof should resemble a simple and clear-cut constellation, not a scattered cluster in the Milky Way" (ibid., 113). Here Hardy refers to the fact that, in a proof by cases, often different reasons are given in the different cases, so the proof lacks unity, and in this sense is like a scattered cluster in the Milky Way.

In particular, McAllister says that the computer-assisted proof, such as the Appel and Haken proof, being successful, will eventually gain acceptance. But the Appel and Haken proof is a proof by cases which involves inspecting about 2000 cases, all of which are checked by computer. So there is the problem that the computer program may contain errors, and we cannot manually check the computer's results. Later on, the 2000 cases have been reduced to 633, and the risk of errors has been somewhat reduced by formalizing the proof in "the Coq formal system" (Gonthier 2008, 1382). But the problem remains that the proof lacks unity, and hence is like a scattered cluster in the Milky Way.

As Thurston says, the controversy evoked by the Appel and Haken proof "reflected a continuing desire for human understanding of a proof, in addition to knowledge that the theorem is true" (Thurston 1994, 162). Mathematicians "are not trying to meet some abstract production quota of definitions, theorems and proofs. The measure of" their "success is whether what" they "do enables people to understand and think more clearly and effectively about mathematics" (ibid., 163). Indeed, "what they really want is usually not some collection of 'answers' – what they want is understanding" (ibid., 162). But the Appel and Haken proof does not provide understanding, so mathematicians will not be content until they will find a better proof.

15.8 Mathematical Beauty and Enlightenment

Another follower of the tradition according to which mathematical beauty is a property projected by the subject onto mathematical things, is Rota.

Indeed, Rota says that "the beauty of a piece of mathematics is dependent upon schools and periods" (Rota 1997, 126). Admittedly, "given the historical period and the context, one finds substantial agreement among mathematicians as to which mathematics is to be regarded as beautiful," and, in this sense, "the beauty of a piece of mathematics does not consist merely of the subjective feeling experienced by an observer," it "is an objective property" (ibid.). But this claim of objectivity "needs to be tempered," because a piece of mathematics that today is considered to be beautiful, may no longer be thought to be beautiful "by future generations," so mathematical beauty is "inescapably context-dependent" (ibid.). Indeed, the appreciation of mathematical beauty requires "familiarity with a huge amount of background material" which is context-dependent, for example, "a proof is viewed as beautiful only after one is made aware of previous clumsier proofs" (ibid., 129–130). The familiarity with the background material "is arrived at the cost of time, effort, exercise and Sitzfleisch" (ibid., 128). So "the appreciation of mathematical beauty" is not "an instantaneous flash" (ibid., 130).

But what is "the sense of the term 'beauty' as it is used by mathematicians"? (ibid., 121). Mathematicians "are concerned with the truth" but they also "claim that beauty is the 'raison d'être' of mathematics" (ibid., 131). Indeed, "the lack of beauty in a piece of mathematics" is "a motivation for further research" (ibid., 128). This leads one "to suspect that mathematical truth and mathematical beauty may be related" (ibid., 131).

Now, truth is reached through proof, specifically axiomatic proof, namely deduction from axioms. But, when mathematicians "are puzzled by some mathematical assertion," it is "not because they are unable to follow the proof or the applications. Quite the contrary" (ibid.). They have been "able to verify its truth in the logical sense of the term, but something is still missing," namely "the sense of the statement that has been verified to be true. Verification alone does not explain" the "relevance of the statement" (ibid.). In short, "the logical truth of a statement does not enlighten" mathematicians "as to the sense of the statement," and "enlightenment not truth is what mathematicians seek" (ibid.). Indeed, "enlightenment is what keeps the mathematical enterprise alive" (ibid., 132).

Admittedly, "mathematicians seldom explicitly acknowledge the phenomenon of enlightenment" (ibid.). Nevertheless, they "say that a theorem is beautiful when they mean to say that the theorem is enlightening," and they "say that a proof is beautiful when" they mean to say that the proof is enlightening, because "it leads us to perceive the inevitability of the statement being proved" (ibid.). Therefore, when mathematicians speak of mathematical beauty, they mean to speak of enlightenment.

The reason why mathematicians speak of mathematical beauty rather than of enlightenment, is that "mathematical beauty and mathematical truth share one important property. Neither of them admits degrees" (ibid., 131). Conversely, "enlightenment admits degrees: some statements are more enlightening than others" (ibid., 132). Now, "mathematicians dislike concepts admitting degrees, and will go to any length to deny the logical role of any such concept" (ibid.). They "are annoyed by the graded truth which they observe in other sciences" (ibid., 131). Mathematical beauty is "the expression mathematicians have invented in order to obliquely" refer to "enlightenment" but "to avoid confronting enlightenment" (ibid., 132). Talk "of mathematical beauty saves" them "from having to deal with a concept" like enlightenment "which comes in degrees" (ibid.).

Rota's view, however, has shortcomings.

Rota says that, when mathematicians speak of mathematical beauty, they mean to speak of enlightenment. But he does not give any definition of enlightenment. So, he replaces an undefined concept with another undefined concept.

Also, Rota says that mathematicians talk of mathematical beauty instead of talking of enlightenment, because beauty does not admit degrees while enlightenment admits degrees, and mathematicians dislike concepts admitting degrees. But this contrasts with the fact that, as already Plato pointed out, mathematical beauty admits degrees. It is a common experience that we find something more beautiful than something else, and this also holds of mathematical beauty. Evidence for this is also provided by the above mentioned responses of readers of *The Mathematical Intelligencer* to a questionnaire that asked to give each of 24 theorems a score of 0 through 10 for beauty. Then, if mathematicians avoided talking of enlightenment because they dislike concepts admitting degrees, for the very same reason they should avoid talking of beauty.

Other shortcomings of Rota's view will be described below.

15.9 Mathematical Beauty and Understanding

Independently of the fact that Rota does not give any definition of enlightenment, 'enlightenment' is not an appropriate word to use in this context. For, one of the meanings of 'to enlighten' is 'to give insight', and insight suggests that the appreciation of mathematical beauty is an instantaneous flash. This contrasts with Rota's claim that the appreciation of mathematical beauty is not an instantaneous flash.

Instead of saying, like Rota, that when mathematicians speak of mathematical beauty they mean to speak of enlightenment, it seems more appropriate to say that, when mathematicians speak of mathematical beauty they mean to speak of understanding. Then, a piece of mathematics is beautiful if it provides understanding.

This may seem strange because understanding is not commonly associated with beauty. But if, as said in Chap. 14, understanding is the recognition of the fitness of the parts to each other, then understanding has an aesthetic dimension. For, one of the definitions of beauty in the modern and contemporary period is that beauty consists in the fitness of the parts to each other.

Thus, Alberti says that "beauty is a certain accordance and conspiration of the parts" (Alberti 1541, Book IX, Chap. 5, 136).

Berkeley says that beauty consists in the fact that "the parts" are "so related, and adjusted to one another, as that they may best conspire to the use and operation of the whole" (Berkeley 1993, 67).

Hogarth says that beauty consists in the "fitness of the parts to the design for which every individual thing is formed, either by art or nature" (Hogarth 1753, 3).

Heisenberg says that beauty is the "proper accordance of the parts with each other" (Heisenberg 1971, 98).

Now, insofar as beauty consists in the fitness of the parts to each other, and understanding can only come through the recognition of the fitness of the parts to each other, understanding has an aesthetic dimension.

15.10 Beauty of Theorems

It has been said above that a piece of mathematics is beautiful if it provides understanding. Then, in particular, a theorem is beautiful if it provides understanding.

In Chap. 14 it has been said that the understanding of a theorem is the recognition of the fitness of the terms employed in the theorem to each other. Then, a theorem is beautiful if it lets one recognize the fitness of terms employed in the theorem to each other.

There are degrees in beauty of theorems. The degree in beauty of a theorem is as much higher as the significance of the theorem is higher, so the degree of beauty of a theorem depends on the significance of the theorem.

Thus, the readers of *The Mathematical Intelligencer* gave the highest score for beauty to Euler's identity $e^{\pi i} = -1$. Also, in experimental studies aimed at finding the neural correlates of the experience of mathematical beauty, "the formula most consistently rated as beautiful" was "Euler's identity" (Zeki, Romaya, Benincasa, and Atiyah 2014, 4).

In fact, Euler's identity is highly significant because it involves three of the most important numbers in mathematics, e, π, and i, and establishes a basic relation between them in a transparent way.

15.11 Beauty of Demonstrations

Like a theorem, a demonstration is beautiful if it provides understanding.

In Chap. 14 it has been said that the understanding of a demonstration is the recognition of the fitness of the parts of the demonstration to each other. Then, a demonstration is beautiful if it lets one recognize the fitness of the parts of the demonstration to each other.

As there are degrees in beauty of theorems, there are also degrees in beauty of demonstrations. Indeed, Whitehead acnowledges that there is a "feeling, widespread among mathematicians, that some proofs are more beautiful than others" (Whitehead 1968, 60).

The degree of beauty of a demonstration depends on how vividly the demonstration lets one recognize the fitness of the parts of the demonstration to each other.

Beauty of theorems and beauty of demonstrations are not necessarily related. Specifically, beautiful theorems can have ugly demonstrations, while the converse seems more unlikely. As Atiyah says, to have a beautiful demonstration of an ugly theorem would be like to "have a beautiful road ending up at a dump. I don't think that is going to happen very often – you don't go down that road" (Atiyah 2009).

In Chap. 14 it has been also said that two kinds of understanding of a demonstration can be distinguished: top-down understanding, and bottom-up understanding. Accordingly, two kinds of beauty of a demonstration can be distinguished: top-down beauty, and bottom-up beauty.

A demonstration is top-down beautiful if it lets one recognize the fitness of the parts of the demonstration to each other, showing why a proposition is true. For example, Demonstration 2 of the Pythagorean theorem considered in Chap. 14 has this kind of beauty. Since the top-down beauty of a demonstration shows why a proposition is true, demonstrations that are top-down beautiful are relevant to finished mathematics.

A demonstration is bottom-up beautiful if it lets one recognize the fitness of the parts of the demonstration to each other, showing how to obtain hypotheses for solving a problem. For example, Plato's demonstration that, for any given square, the side of the square double in size is a diagonal of the given square (see Chap. 5) has this kind of beauty. Since the bottom-up beauty of a demonstration shows how to obtain hypotheses for solving a problem, bottom-up beauty is relevant to the making of mathematics.

15.12 The Disregard of Bottom-Up Beauty

Despite the fact that demonstrations which are bottom-up beautiful are relevant to the making of mathematics, current work on mathematical beauty disregards them, and only allows for demonstrations that are top-down beautiful.

For example, Lange says that "all explanatory proofs are beautiful (or, at least, not ugly). They derive their beauty from exactly what gives them their explanatory power" (Lange 2016, 45). But, by 'explanatory proofs', Lange exclusively means 'top-down explanatory proofs', hence axiomatic proofs. For, he says that he is concerned with "the distinction between explanatory and non-explanatory proofs of the same theorem from the same axioms" (ibid., 12). According to Lange, "two mathematical proofs may prove the same theorem from the same axioms, though only one of these proofs explains why that theorem is true," and the question is "to identify the ground of this distinction" by focusing "on the course that a given proof takes between its premises and its conclusion" (ibid., 9–10). Then, "the distinction between explanatory and non-explanatory proofs from the same premises must rest on differences in the way they extract the theorem from the axioms" (ibid., 10).

As another example, experimental studies aimed at finding the neural correlates of the experience of mathematical beauty only allow for demonstrations that are top-down beautiful. Thus, Zeki, Chén, and Romaya say that "at least part of the experienced beauty of a mathematical formulation lies in the fact that it adheres to the logical deductive system of the brain" (Zeki, Chén, and Romaya 2018, 7). The logical deductive system of the brain "is inherited and is therefore similar in mathematicians belonging otherwise to different races and cultures. It is in this sense that mathematical beauty has its roots in a biologically inherited logical-deductive system that is similar for all brains" (ibid.).

15.13 The Denial of Any Role to Beauty in Mathematical Research

A further example of the fact that current work on mathematical beauty disregards demonstrations which are bottom-up beautiful, and only allows for demonstrations which are top-down beautiful, is provided by Rota.

Indeed, Rota says that there can be beauty "in the presentation of mathematics" (Rota 1997, 128). Namely in finished mathematics. For, "a beautiful proof is more often than not the definitive proof" (ibid., 129). On the contrary, beauty has no role in the making of mathematics, because "mathematical research does not strive for beauty. Every mathematician knows that beauty cannot be directly sought" (ibid., 128). In fact, "mathematicians work to solve problems and to invent theories that will shed new light and not to produce beautiful theorems or pretty proofs" (ibid.).

This shows that Rota disregards demonstrations which are bottom-up beautiful, and only allows for demonstrations which are top-down beautiful.

But Rota's claim that, while there can be beauty in finished mathematics, beauty has no role in the making of mathematics, is unjustified. For, it is contradicted by many mathematicians, who state that mathematical research aims at beauty, hence beauty is inherent in mathematics as it is being developed.

Thus, Hardy says that "the mathematician's patterns" must "be beautiful," so "beauty is the first test: there is no permanent place in the world for ugly mathematics" (Hardy 1992, 85).

Kline says that one may "find the attitudes and efforts of mathematicians more intelligible by knowing that these men have sought beauty" (*Kline* 1964, *470).*

King says that "beauty" is "what motivates" mathematicians "to do mathematics in the first place" (King 1992, 29). So "the development of mathematics is primarily aesthetic, and not utilitarian" (ibid., 121).

Atiyah says that mathematicians "aim to produce and understand in an elegant and beautiful way," because "elegance and beauty are a sign of success, they are not just an add-on extra," therefore all mathematicians "search for beauty in mathematics" (Atiyah, Berry, Drury, Jaffe, and Goldsmith 2006, 50).

Evidence for the fact that mathematical research aims at beauty is also provided by the fact that new proofs are sought for theorems already correctly established. This has been pointed out by several mathematicians.

Thus, Gauss says that, "as soon as a new theorem is discovered," one "must not always consider the investigation closed and the search for other proofs as a superfluous luxury. For, in some cases one does not come at first to the most beautiful and simplest proofs" (Gauss 1863–1933, II, 159–160). And "it is just the insight into the wonderful concatenation of truths" that "often leads to the discovery of new truths. For these reasons, the discovery of new proofs for already known truths is often to be regarded at least as important as the discovery of the truths themselves" (ibid., 160).

Kline says that "much research for new proofs of theorems already correctly established is undertaken simply because the existing proofs have no aesthetic appeal" (Kline 1964, 470).

Krull says that mathematicians "in some cases prove a theorem not just once but many times; in fact there are famous mathematical theorems for which more than 10 or 20 proofs exist. I believe this state of affairs shows beyond doubt that aesthetic viewpoints play a large role in mathematics" (Krull 1987, 50).

15.14 Role of Beauty in Finding Solutions to Mathematical Problems

That current work on mathematical beauty disregards demonstrations which are bottom-up beautiful, and only allows for demonstrations which are top-down beautiful, is a serious limitation, because beauty can have a significant role in finding solutions to mathematical problems.

This is clear from the fact that, in the analytic method, the non-deductive rules by which hypotheses are obtained may yield so many hypotheses that it would be unfeasible to check all of them for plausibility. Beauty guides one to pick out some of them, so it can serve as a heuristic means to find a solution to a mathematical problem.

That beauty can have a significant role in finding solutions to mathematical problems is pointed out already by Poincaré. He says that finding a solution to a problem involves first "making new combinations" with ideas "already known," and then choosing "those which are useful and which are only a small minority" (Poincaré 2015, 386). To this aim, the self must first "form all the possible combinations" with ideas already known. For, "if it produces only a small part of these combinations, and if it makes them at random, there would be small chance that" the useful ones "would be found among them" (ibid., 392–393). Now, forming all the possible combinations with ideas already known cannot be the work of "the conscious self," since the latter "is narrowly limited" and hence cannot be "able in a short time to make more different combinations than the whole life of a conscious being could encompass" (ibid., 392). Instead, it is the work of "the unconscious," or "subliminal self" (ibid., 390). Then, among all the possible combinations, the unconscious self chooses the useful ones by "the feeling of mathematical beauty," which is "a true aesthetic feeling that all real mathematicians know" (ibid., 391). Thus beauty "plays the role of the delicate sieve" (ibid., 392). The combinations so chosen are then submitted to "verification" (ibid., 390).

Poincaré's account of the role of beauty in finding solutions to mathematical problems, however, has a serious shortcoming. He assumes that the useful combinations will be included among all the possible combinations of ideas already known, and hence among all the possible notions that can be defined in terms of ideas already known. But this conflicts with a result analogous to Tarski's undefinability theorem, by which, "for every deductive science in which arithmetic is contained it is possible to specify arithmetical notions which, so to speak, belong intuitively to this science, but which cannot be defined on the basis of this science" (Tarski 1983, 276). So, there is no guarantee that the useful combinations will be included among all the possible combinations of ideas already known.

The analytic method does not have this shortcoming, because the hypotheses for the solution to a problem are not combinations of ideas already known, they are obtained by non-deductive rules, which are ampliative.

Of course, that beauty guides one to pick out some of the hypotheses, does not guarantee that the hypotheses so chosen will yield a solution to a mathematical problem. Even when the hypotheses seem to yield a beautiful solution, they may turn out to be implausible. For example, Kempe 1879 gave a demonstration of the four-color conjecture which was beautiful but which was then shown to be incorrect. Dudeney 1926 gave a shorter and even more beautiful demonstration of the four-color conjecture, which was then also shown to be incorrect.

But this does not diminish the importance of the role of beauty in finding solutions to mathematical problems. By helping sift through the hypotheses generated by non-deductive rules and pick out some of them, beauty can significantly reduce the number of hypotheses to be considered, thus acting as a shortcut.

15.15 Role of Beauty in Choosing Mathematical Fields and Problems

That beauty can have a significant role in finding solutions to mathematical problems is not the only respect in which beauty is relevant to the making of mathematics. Beauty can also have a significant role in choosing the mathematical fields and problems to pursue.

This relates to what Sinclair calls the "motivational" role of the aesthetic, namely "the role of the aesthetic in attracting mathematicians to certain fields and, in turn, in stimulating them to work on certain problems" (Sinclair 2006, 89),

The motivational role of the aesthetic has been underlined by several people.

Thus, Hadamard says that the choice of subjects of research "is one of the most important things in research" (Hadamard 1954, 126). This "choice is directed by the sense of beauty" (ibid., 130). The "sense of beauty can inform us" that "such a direction of investigation is worth following," that "the question in itself deserves interest, that its solution will be of some value," whether "it permits further applications or not" (ibid., 127).

Penrose says: "How, in fact, does one decide which things in mathematics are important and which are not? Ultimately, the criteria have to be aesthetic ones," namely criteria "such as one has in music or painting or any other art form" (Penrose 1974, 266).

Tymoczko says that, in the choice of "research projects," the "aesthetic judgments – value judgments in the narrow sense – can serve as selection criteria for mathematics," indeed, "mathematicians can be seen as using aesthetic criteria in creating the 'mathematical canon'" which guides "research" (Tymoczko 1993, 68).

That beauty can have a significant role both in finding solutions to mathematical problems and in choosing the mathematical fields and problems to pursue, shows that beauty is relevant to the making of mathematics.

On the other hand, beauty has a significant role in the understanding of theorems and demonstrations. An aspect of this is the role of beauty in the learning of mathematics. As Sinclair says, "aesthetically-rich learning environments enable children to wonder, to notice, to imagine alternatives, to appreciate contingencies and to experience pleasure and pride" (Sinclair 2001, 26). This shows that beauty is relevant to finished mathematics.

Thus beauty is relevant both to the making of mathematics and to finished mathematics, and hence is relevant to mathematics generally.

15.16 Innate and Acquired Sense of Beauty

That beauty is relevant to mathematics generally, depends on the fact that human beings have a sense of beauty. More specifically, two senses of beauty can be distinguished: an innate sense of beauty and an acquired one. The innate sense of beauty is a result of biological evolution, the acquired sense of beauty is a result of cultural evolution.

That there is an innate sense of beauty is clear, for example, from the fact that newborn infants prefer attractive faces, the "preferences for attractive faces are present soon after birth, and are not the result of a long period of observing faces" (Slater and Kirby 1998, 92).

The innate sense of beauty is relevant to the ways in which we know the world and adapt to it, because it determines the kind of attention we pay to elements of the environment, the value we place on them, and our actions in response to them. Therefore, the innate sense of beauty has a significant role in our problem-solving activities, it helps us to adopt appropriate behaviour towards feeding, breeding, and finding suitable mates and habitats.

On the other hand, on the basis of the innate sense of beauty, an acquired sense of beauty develops, as a result of cultural evolution. It concerns primarily cultural objects, such as those of art, literature, mathematics, and science.

In Chap. 17, two kinds of mathematics will be distinguished: natural mathematics, which is a result of biological evolution, and mathematics as discipline, which is a result of cultural evolution. The innate sense of beauty is involved in natural mathematics, the acquired sense of beauty in mathematics as discipline.

References

Alberti, Leon Battista. 1541. *De re aedificatoria*. Strasbourg: Jakob Cammerlander.

Atiyah, Michael, Michael Berry, Luke Drury, Arthur Jaffe, and Brendan Goldsmith. 2006. The interface between mathematics and physics: A panel discussion. *Bulletin of the Irish Mathematical Society* 58: 33–54.

Berkeley, George. 1993. *Alciphron, or the minute philosopher*. London: Routledge.

Borel, Armand. 1983. Mathematics: Art and science. *The Mathematical Intelligencer* 5 (4): 9–17.

Cantor, Georg. 1965. Letter to Vivanti, 13 December 1893. In Herbert Meschkowski, Aus den Briefbüchern Georg Cantors, 504–508. *Archive for History of Exact Sciences* 2: 503–519.

Dudeney, Henry Ernest. 1926. *Modern puzzles and how to solve them*. London: C. Arthur Pearson.

Dyson, Freeman John. 1956. Obituary of Hermann Weyl. *Nature* 177: 457–458.

———. 1992. *From Eros to Gaia*. New York: Pantheon Books.

Gauss, Carl Friedrich. 1863–1933. *Werke*. Göttingen: Königlichen Gesellschaft der Wissenschaften.

Gonthier, Georges. 2008. Formal proof – The four-color theorem. *Notices of the American Mathematical Society* 55 (11): 1382–1393.

Gowers, Timothy. 2007. Mathematics, memory and mental arithmetic. In *Mathematical knowledge*, ed. Mary Leng, Alexander Paseau, and Michael Potter, 33–58. Oxford: Oxford University Press.

Hadamard, Jacques. 1954. *The psychology of invention in the mathematical field*. Mineola: Dover.

Hardy, Godfrey Harold. 1992. *A mathematician's apology*. Cambridge: Cambridge University Press.

Heisenberg, Werner. 1971. Die Bedeutung des Schönen in der exakten Naturwissenschaft. *Physikalische Blätter* 27 (3): 97–107.

Hersh, Reuben, and Vera John-Steiner. 2011. *Loving + hating mathematics: Challenging the myths of mathematical life*. Princeton: Princeton University Press.

Hogarth, William. 1753. *The analysis of beauty*. London: Reeves.

Kant, Immanuel. 2000. *Critique of the power of judgment*. Cambridge: Cambridge University Press.

Kempe, Alfred Bray. 1879. On the geographical problem of the four colours. *American Journal of Mathematics* 2 (3): 193–200.

King, Jerry P. 1992. *The art of mathematics*. New York: Plenum Press.

Kline, Morris. 1964. *Mathematics in Western culture*. Oxford: Oxford University Press.

Krull, Wolfgang. 1987. The aesthetic viewpoint in mathematics. *The Mathematical Intelligencer* 9 (1): 48–52.

Lange, Marc. 2016. Explanatory proofs and beautiful proofs. *Journal of Humanistic Mathematics* 6 (1): 8–51.

McAllister, James. 2005. Mathematical beauty and the evolution of the standards of mathematical proof. In *The visual mind II*, ed. Michele Emmer, 15–34. Cambridge: The MIT Press.

Penrose, Roger. 1974. The role of aesthetics in pure and applied mathematical research. *Bulletin of the Institute of Mathematics and Its Applications* 10 (2): 266–271.

Poincaré, Henri. 2015. *The foundations of science: Science and hypothesis, The value of science, Science and method*. Cambridge: Cambridge University Press.

Rota, Gian-Carlo. 1997. *Indiscrete thoughts*. Cham: Birkhäuser.

Sinclair, Nathalie. 2001. The aesthetic 'is' relevant. *For the Learning of Mathematics* 21: 25–32.

———. 2002. The kissing triangles: The aesthetics of mathematical discovery. *International Journal of Computers for Mathematical Learning* 7: 45–63.

———. 2006. The aesthetic sensibilities of mathematicians. In Nathalie Sinclair, David Pimm, and William Higginson (eds.), *Mathematics and the aesthetic: New approaches to an ancient affinity*, 87–104. New York: Springer.

Slater, Alan, and Rachel Kirby. 1998. Innate and learned perceptual abilities in the newborn infant. *Experimental Brain Research* 123: 90–94.

Tarski, Alfred. 1983. *Logic, semantics, metamathematics*. Indianapolis: Hackett.

Thurston, William P. 1990. Mathematical education. *Notices of the American Mathematical Society* 37: 844–850.

———. 1994. On proof and progress in mathematics. *Bulletin of the American Mathematical Society* 30: 161–177.

Todd, Cain S. 2018. Fitting feelings and elegant proofs: On the psychology of aesthetic evaluation in mathematics. *Philosophia Mathematica* 26: 211–233.

Tymoczko, Thomas. 1993. Value judgments in mathematics: Can we treat mathematics as an art? In *Essays in humanistic mathematics*, ed. Alvin M. White, 67–77. Washington: Mathematical Associaton of America.

von Neumann, John. 1961. The mathematician. In John von Neumann, *Collected works*, vol. 1, 1–9. Oxford: Pergamon Press.

Wells, David. 1988. Which is the most beautiful? *The Mathematical Intelligencer* 10 (4): 30–31.

———. 1990. Are these the most beautiful? *The Mathematical Intelligencer* 12 (3): 38–41.

Whitehead, Alfred North. 1968. *Modes of thought*. New York: Free Press.
Zangwill, Nick. 1998. Aesthetic/sensory dependence. *British Journal of Aesthetics* 38: 66–81.
Zeki, Semir, John Paul Romaya, Dionigi M.T. Benincasa, and Michael Atiyah. 2014. The experience of mathematical beauty and its neural correlates. *Frontiers in Human Neuroscience* 8 (Article 68): 1–12. https://doi.org/10.3389/fnhum.2014.00068.
Zeki, Semir, Oliver Y. Chén, and John Paul Romaya. 2018. The biological basis of mathematical beauty. *Frontiers in Human Neuroscience* 12 (Article 467): 1–8. https://doi.org/10.3389/fnhum.2018.00467.

Chapter 16
Applicability

Abstract According to heuristic philosophy of mathematics, one of the tasks of the philosophy of mathematics is to give an answer to the question: Why is mathematics applicable to the world? The question has received several answers. The chapter discusses them and argues that they are inadequate. Then it offers an alternative answer, but also underlines the limitations of the applicability of mathematics to the world.

Keywords Unreasonable effectiveness of mathematics · Single intelligence account · Pre-established harmony account · Mathematical universe account · Mapping account · Applicability and Galileo's move · Limitations of the applicability of mathematics · Reasonable ineffectiveness of mathematics

16.1 The Unreasonable Effectiveness of Mathematics

As we have seen in Chap. 3, according to heuristic philosophy of mathematics, one of the tasks of the philosophy of mathematics is to give an answer to the question: Why is mathematics applicable to the world? This question must not be confused with the question: Is mathematics applicable to the world? The latter is not problematic because, from the seventeenth century, there have been many successful applications of mathematics to the world. Conversely, the question 'Why is mathematics applicable to the world?' is problematic.

The question was put on the philosophical agenda by the Pythagoreans, who answered it by saying that mathematics is applicable to the world because "number is the essence of all things" (Aristotle, *Metaphysica*, A 5, 987 a 19).

The question remained on the philosophical agenda until the nineteenth century, but then disappeared because, with some rare exception such as Hilbert, mainstream philosophers of mathematics were not interested in it. For, their primary aim was to give a secure foundation to mathematics, and the question of the applicability of mathematics was unimportant to it.

C. Cellucci, *The Making of Mathematics*, Synthese Library 448,
https://doi.org/10.1007/978-3-030-89731-4_16

Mainstream philosophers of mathematics continued to show scarce interest in the question also in the second half of the twentieth century. Thus, the second edition of the standard anthology in the philosophy of mathematics Benacerraf and Putnam 1983, had no article on the question, not by choice of the editors but by "lack of material" (Steiner 1998, 14, footnote 8).

Nevertheless, in the second half of the twentieth century, Wigner put the question back on the philosophical agenda by his article entitled 'The unreasonable effectiveness of mathematics in the natural sciences'. The title is descriptive of the article's thesis: mathematics is effective in the natural sciences, but there is no rational explanation for its effectiveness.

Indeed, Wigner says that "the enormous usefulness of mathematics in the natural science is something bordering on the mysterious" and "there is no rational explanation for it" (Wigner 1960, 2). It is "difficult to avoid the impression that a miracle confronts us here" (ibid., 7). The "miracle of the appropriateness of the language of mathematics for the formulation of the laws of physics is a wonderful gift which we neither understand nor deserve. We should be grateful for it and hope that it will remain valid in future research" (ibid., 14).

Several mathematicians and physicists have expressed similar scepticism as to the possibility of giving an answer to the question of why mathematics is applicable to the world.

Thus, Hamming says: "Wigner gives a large number of examples of the effectiveness of mathematics in the physical sciences," and "because of these successes of mathematics, there is at present a strong trend toward making each of the sciences mathematical" (Hamming 1980, 82). But all attempted explanations of the effectiveness of mathematics "when added together simply are not enough to explain what" they are "set out to account for" (ibid., 90). When all such "explanations are over, the residue is still so large as to leave the question essentially unanswered" (ibid., 82).

Kline says: "The effectiveness of the mathematical representation and analysis of the physical world is as unexplainable as the very existence of the world itself and of man" (Kline 1981, 471–472).

Dyson says: "One of the many virtues of" this "book is that it leaves the central mystery, the miraculous effectiveness of mathematics as a tool for the understanding of nature, unexplained and unobscured" (Dyson 1983, 54).

Barrow says: "However trivial it may sound to reaffirm that mathematics describes the forms of the natural world since it describes all possible forms, the mystery remains why such a small number of relatively simple models work in a way that is so rich of consequences and so powerful for the description and understanding of the Universe" (Barrow 2020, 126).

In contrast with the scepticism of these mathematicians and physicists, the question of why mathematics is applicable to the world has received several answers. The chapter discusses them and argues that they are inadequate. Then it offers an alternative answer, but also underlines the limitations of the applicability of mathematics to the world.

16.2 The Single Intelligence Account

An answer to the question of why mathematics is applicable to the world is the single intelligence account, according to which mathematics is applicable to the world because the human subjective reason, that invented mathematics, and the objective reason of nature, that is structured mathematically, are identical since they both originated in a single intelligence, namely God. No one can prove this fact, but it really appears in our world.

Thus, Pope Benedict XVI says: "Mathematics, which we invented, really gives us access to the nature of the universe," so "an invention of the human mind and the structure of the universe coincide," hence "the subjective reason and the objective reason of nature are identical" (Pope Benedict XVI 2014, 92). This is because "it is 'one' reason that links them both. Our reason could not discover this other reason were there not an identical antecedent reason for both" (ibid.). Namely God. Of course, "no one can now prove – as is proven in an experiment, in technical laws – that they both really originated in a single intelligence" (ibid., 93). However, "this unity of intelligence, behind the two intelligences, really appears in our world" (ibid.).

But the single intelligence account is inadequate. For, on the one hand, the single intelligence account claims that mathematics is applicable to the world because our subjective reason, which invented mathematics, and the objective reason of nature, which is structured mathematically, are identical since they both originated in a single intelligence, namely God. On the other hand, the single intelligence account admits that no one can prove this fact. So, the single intelligence account depends on an unprovable assumption.

The single intelligence account tries to overcome this difficulty by claiming that the fact that both reasons originated in a single intelligence, namely God, needs no proof, because it really appears in our world. But this claim is unfounded, because the fact in question does not really appear in our world. Our experience of the universe is extremely limited. As Hume says, "a very small part of this great system, during a very short time, is very imperfectly discovered to us" (Hume 2007, 25). And "can a conclusion, with any propriety, be transferred from parts to the whole?" (ibid., 24). Can "we thence pronounce decisively concerning the origin of the whole?" (ibid., 25). To "ascertain this reasoning, it were requisite, that we had experience of the origin of worlds" (ibid., 26).

As Bayle says, "the Schoolmen have an axiom, that a philosopher ought not have recourse to God, 'Non est philosophi recurrere ad Deum': they call this recourse the asylum of ignorance. And, in fact, what could you say more absurd, in a work of physics, that this: stones are hard, fire is hot, cold freezes rivers, because God has ordered it so?" (Bayle 1820, 53). But the single intelligence account disregards this axiom of the Schoolmen.

16.3 The Pre-established Harmony Account

Another answer to the question of why mathematics is applicable to the world is the pre-established harmony account, according to which mathematics is applicable to the world because there exists a pre-established harmony between thought and world. Evidence for this is that mathematical concepts and results, originally developed for reasons internal to mathematics, turn out to be applicable to the world.

Thus, Hilbert says that there exists a "pre-established harmony" between "nature and thought" (Hilbert 1996b, 1149). Evidence for this is that mathematical concepts and results, originally developed for reasons internal to mathematics, turn out to be applicable to the world, and "the older examples for this are conic sections, which one studied long before one suspected that our planets or even electrons move in such a course" (Hilbert 1996c, 1160). But "the most magnificent and wonderful example of pre-established harmony is Einstein's famous theory of relativity," where the "differential equations for the gravitational potential are constructed mathematically and uniquely" (ibid.). This "construction would not have been possible without the profound and difficult mathematical investigations of Riemann, which existed long before" (ibid.). Only by admitting that there exists a pre-established harmony between nature and thought, we can account for "the numerous and surprising analogies and that apparently harmony which the mathematician so often perceives" (Hilbert 1996a, 1099).

But the pre-established harmony account is inadequate. For, if mathematical concepts and results, originally developed for reasons internal to mathematics, turn out to be applicable to the world, this is no evidence that there is a pre-established harmony between nature and thought. On the contrary, it is the very fact to be explained.

Besides, the pre-established harmony account conflicts with the fact that physical laws formulated mathematically do not describe physical phenomena precisely. Hilbert himself admits it, since he says that "the application of a theory to the world of appearances always requires a certain measure of good will and tactfulness; e.g., that we substitute the smallest possible bodies for points and the longest possible ones, e.g. light-rays, for lines. We also must not be too exact in testing the propositions, for these are only theoretical propositions" (Hilbert 1980, 41). But this conflicts with the fact that the pre-established harmony account would imply that physical laws formulated mathematically should describe physical phenomena precisely. Therefore, they should not require a certain measure of good will and tactfulness, nor that we must not be too exact in testing the propositions.

Also, the pre-established harmony account does not explain why only certain mathematical concepts and results turn out to be applicable to the world.

Moreover, the pre-established harmony account would imply that only correct mathematical concepts and results should be applicable to the world. But this conflicts, for example, with the fact that, although the calculus of infinitesimals of Leibniz or Newton was inconsistent, this did not prevent it from being the backbone of the development of modern science and of engineering in the industrial revolution.

Furthermore, the pre-established harmony account would imply that the mathematical concepts and results used to treat physical phenomena should be uniquely determined by the latter. But this conflicts with the fact that different mathematical concepts and results are used to deal with the same physical phenomena.

16.4 The Mathematical Universe Account

Another answer to the question of why mathematics is applicable to the world is the mathematical universe account, according to which mathematics is applicable to the world because the world is a mathematical structure.

Thus, Tegmark says that the world "is mathematics (more specifically, a mathematical structure)" (Tegmark 2008, 107). For, it consists of "abstract entities with relations between them" (ibid., 104). The ultimate aim of science "is to find a complete description of it" (ibid., 102). Such a description is provided by mathematics, which is "the study of formal systems," and uncovers mathematical structures that are "'out there', completely independently of the discoverer" (Hut et al. 2006, 768). The applicability of mathematics to the world is "a natural consequence of the fact that the latter 'is' a mathematical structure, and we are simply uncovering this bit by bit," so "our successful theories are not mathematics approximating physics, but mathematics approximating mathematics" (Tegmark 2008, 107).

But the mathematical universe account is inadequate. For, the claim that the world is a mathematical structure conflicts with the fact that only some, not all properties of the world, are mathematical in kind.

Besides, if the world were a mathematical structure, then, by knowing its mathematical properties, we would know the essence of natural substances. But this would mean going back to Aristotelian essentialist science.

Moreover, Tegmark himself admits that, saying that the ultimate aim of science is to find a complete description of the mathematical structure that is the world, conflicts with the fact that, by "results of Gödel, Church and Turing," there are "questions that can be posed but not answered" (Tegmark 2008, 135). For a mathematical structure, "this corresponds to relations that are unsatisfactorily defined in the sense that they cannot be implemented by computations that are guaranteed to halt" (ibid.). This would mean that "our universe would be somehow inconsistent or undefined" (ibid., 133).

Tegmark's answer to this difficulty is that "the mathematical structure that is" the physical world "is defined by computable functions" (ibid., 131). So, there are "no physical aspects of our universe that are uncomputable/undecidable, eliminating the above-mentioned concern that Gödel's work makes it somehow incomplete or inconsistent" (ibid., 136).

But Tegmark himself recognizes that this claim is flawed, because "virtually all historically successful theories of physics violate" it, and "it is far from obvious whether a viable computable alternative exists" (ibid., 138).

16.5 The Model Account

Another answer to the question of why mathematics is applicable to the world is the model account, according to which mathematics is applicable to the world because scientific theories are applied to models, where a model is a simplified version of the phenomenon under consideration in which exact calculations are possible.

Thus, Gowers says that mathematics is applicable to the world because "scientific theories" are not applied "directly to the world but rather to models," where a model is "an imaginary, simplified version of the part of the world being studied, one in which exact calculations are possible" (Gowers 2002, 4). Of course, "when choosing a model, one priority is to make its behaviour correspond closely to the actual, observed behaviour of the world" (ibid., 5). But, on the other hand, "when devising a model, one tries to ignore as much as possible about the phenomenon under consideration, abstracting from it only those features that are essential to understanding its behaviour" (ibid., 16). For example, when devising a model of the behaviour of stones being thrown, stones are "reduced to single points" (ibid.). The "resulting mathematical structures" are "abstract representations of the concrete situations being modelled" (ibid.).

But the model account is inadequate. For, a model is not the world, it is only a mathematical representation of the world. So the model account does not actually account for the applicability of mathematics to the world, but only for the applicability of mathematics to a mathematical representation of the world, hence to a mathematical substitute of it. Then the question arises of why this mathematical substitute is applicable to the world.

The latter question is even more problematic because, as Gowers admits, "there are very useful models with almost no resemblance to the world at all" (ibid., 5). Why, then, do they work? The model account has no answer.

16.6 The Mapping Account

Another answer to the question of why mathematics is applicable to the world is the mapping account, according to which mathematics is applicable to the world because there are structure-preserving mappings between physical systems and mathematical structures.

Thus, Pincock says that mathematics is applicable to the world because there are structure-preserving "mappings between the physical world and a mathematical domain" (Pincock 2004, 135). More precisely, there are structure-preserving mappings "between physical systems and mathematical structures" (Pincock 2007, 257).

But the mapping account is inadequate. For, to speak of structure-preserving mappings between physical systems and mathematical structures makes sense only if the physical systems are themselves mathematical structures. But a mathematical structure is a set of objects and a set of relations on those. This conflicts with the fact that, as Bueno and Colyvan say, "the world does not come equipped with a set of objects" and "sets of relations on those. These are either constructs of our theories of the world or identified by our theories of the world," so they are "delivered by our theories, not by the world itself" (Bueno and Colyvan 2011, 347). Then the mapping account "does require having" a mathematical structure "in order to get started. There is no avoiding such an assumption" (ibid.). But the world is not a mathematical structure.

16.7 The First Reason of the Applicability of Mathematics

Since the above answers to the question of why mathematics is applicable to the world are inadequate, an alternative answer is necessary. This can be obtained by indicating two reasons why mathematics is applicable to the world.

The first reason is Galileo's change of the object of science. The latter is no longer to know the essence of things, but only to know phenomenal properties of things, mathematical in kind (see Chap. 4). Then, mathematics is applicable to the world because science limits itself to considering only phenomenal properties of things, mathematical in kind.

With Galileo's change, as Kant says, "in any special doctrine of nature there can be only as much proper science as there is mathematics therein" (Kant 2002, 184). Namely, there can be only "as much proper science as there is mathematics capable of application there" (ibid., 185).

Thus, mathematics assumes a central role in science. Such central role was totally alien to Aristotle's science because, according to Aristotle, natural things could not be demonstrated by mathematical means (see Chap. 4).

A central role of mathematics was totally alien not only to Aristotle's science but to the Greek spirit in general. As Sambursky says, "just as the 'dissection of nature' by experiment (to use Bacon's happy definition) was foreign to the Greek, so the corresponding theoretical process of describing nature in mathematical terms was alien to his spirit" (Sambursky 1987, 238). As a result, "while originating the scientific approach," the Greeks were "prevented, for a period of a thousand years, from making the rapid progress that came about in a few decades of the seventeenth century" (ibid., ix).

This rapid progress came about with Galileo's change of the object of science. If the object of science is to know phenomenal properties of things, mathematical in kind, then the application of mathematics to the world becomes possible (see Chap. 4).

16.8 The Second Reason of the Applicability of Mathematics

The second reason why mathematics is applicable to the world is that some parts of mathematics evolve out of the world.

As Atiyah says, "almost all mathematics originally arose from external reality, even numbers and counting" (Atiyah 2004, 26). Indeed, as already mentioned in Chap. 9, geometry arose from the land-measuring in Egypt made necessary by the Nile overflowing. On the other hand, arithmetic arose because in Egypt there was no money, payments for both goods and labour was made in kind, bread and beer were the most common standards of value. This made arithmetic necessary, for example, to distribute these goods among a given number of workers. Thus, Problem 4 of the Rhind Mathematical Papyrus is: "Divide 8 loaves among 10 men" (Clagett 1999, 135). The resulting mathematics was applicable to the world because it evolved out of the world.

At some point, however, mathematics "turned to ask internal questions, e.g. the theory of prime numbers, which is not directly related to experience," but nevertheless "evolved out of it" (Atiyah 2004, 26). Moreover, some mathematics was expressly created to deal with physical phenomena. Thus, the calculus of infinitesimals was created to account for the motion of planets and other moving objects. The resulting mathematics was applicable to the world because it evolved out of it.

In contrast, with the birth of the formal axiomatic method, much mathematics has not evolved out of the world, it has been made for other reasons, including career reasons (see Chap. 6). Of course, there is no guarantee that such mathematics is applicable to the world. Nevertheless, it may happen that, within the mass of mathematics so produced, some piece of it shows some consonance with some physical phenomenon and can be of some use in dealing with it.

16.9 The Unreasonable Effectiveness of Mathematics Revisited

In this perspective, we can revisit Wigner's view that mathematics is effective in the natural sciences but there is no rational explanation for its effectiveness. This view is based on the unwarranted assumption that mathematics is effective in the natural sciences. To see why this is an unwarranted assumption, it is useful to consider Eddington's parable of the ichthyologist.

An ichthyologist "is exploring the life of the ocean. He casts a net into the water and brings up a fishy assortment. Surveying his catch" he arrives at the generalisation: "No sea-creature is less than two inches long" (Eddington 1939, 16). For, this is "true of his catch, and he assumes tentatively that" the generalisation "will remain true however often he repeats it" (ibid.).

An "onlooker may object that" the ichthyologist's "generalisation is wrong," because "there are plenty of sea-creatures under two inches long, only" the ichthyologist's "net is not adapted to catch them" (ibid.).

But "the ichthyologist dismisses this objection contemptuously" (ibid.). His answer to the objection is that "anything uncatchable" by his "net is ipso facto outside the scope of ichthyological knowledge, and is not part of the kingdom of fishes which has been defined as the theme of ichthyological knowledge. In short, what" his "net can't catch isn't fish" (Eddington 1939, 16).

This, however, is absurd. For, what fish is, is not defined by the ichthyologist's net. There are plenty of sea-creatures under two inches long, only the ichthyologist's net is not adapted to catch them.

Similarly, one may object that Wigner's assumption that mathematics is effective in the natural sciences is wrong, because there are plenty of phenomena that are open to scientific study, only mathematics is not adapted to catch them.

But Wigner would dismiss this objection contemptuously. His answer to the objection would be that anything uncatchable by mathematics is ipso facto outside the scope of scientific knowledge, and is not part of the kingdom of nature which has been defined as the theme of scientific knowledge. In short, what mathematics can't catch isn't nature.

This, however, is absurd. For, what nature is, is not defined by mathematics. There are plenty of phenomena that are open to scientific study, only mathematics is not adapted to catch them.

As Colyvan says, "just like Eddington's ichthyologist, with only a specific net available to him to investigate the contents of the ocean, Wigner has only one tool available to him in the investigation of the world – mathematics" (Colyvan 2012, 106). And, just as the ichthyologist misses interesting sea-creatures because of the limitations of his investigative tools, "Wigner misses interesting phenomena because of the limitations of his investigative tools" (ibid.).

16.10 Geometrical Curves and Mechanical Curves

In fact, the assumption that mathematics is effective in the natural sciences, has been disputed several times in the past. Since the seventeenth century, several doubts have been raised as to whether mathematics is effective in the natural sciences. We will examine some of them.

Descartes raises doubts as to whether mathematics is effective in the natural sciences in relation to the question of geometrical curves and mechanical curves.

Geometrical curves are those which are capable of precise and exact description, mechanical curves those which not so capable, because it is "geometrical that which is precise and exact," and "mechanical that which is not" (Descartes 1996, VI, 389). Indeed, geometrical curves can be described by algebraic equations, because for a geometrical curve "it is always possible to find an equation determining all its points" (ibid., VI, 395). Conversely, mechanical curves cannot be described by algebraic equations.

According to Descartes, only geometrical curves are accepted within geometry and pure mathematics. Conversely, mechanical curves, such as "the spiral, the quadratrix, and similar curves," are "not among those curves that I think should be included here" (ibid., VI, 390). Thus, mechanical curves are not accepted within geometry and pure mathematics.

On the other hand, mechanical curves are encountered in the physical world, so physics should be able to account for them.

But, according to Descartes, the only principles admitted or required in physics are those of geometry and pure mathematics. Indeed, Descartes says: "I do not admit, or require, other principles in physics than those of geometry and pure mathematics" (ibid., VIII-1, 78). Only such principles "can explain all natural phenomena, and can give quite certain demonstrations regarding them" (ibid., VIII-1, 78).

Now, if mechanical curves are not accepted within geometry and pure mathematics, and the only principles admitted or required in physics are those of geometry and pure mathematics, it follows that physics is unable to account for mechanical curves. This makes one doubt that mathematics is effective in the natural sciences.

16.11 Vibrating Strings

D'Alembert raises doubts as to whether mathematics is effective in the natural sciences in relation to the question of vibrating strings.

Suppose we have a perfectly flexible string that, when made to vibrate, moves in the xy-plane, every point of the string moves in a straight line perpendicular to the x-axis, and the displacement y of the string at each point is small compared to the length of the string. The problem of vibrating strings is to formulate the equation of the motion of the string in terms of the displacement y as a function of x and time t, and to solve this equation by giving an explicit expression for y as a function of x and t.

D'Alembert 1749 deals with this problem formulating the equation of the motion of the string as $\frac{\partial^2 y}{\partial x^2} = \frac{1}{c^2}\frac{\partial^2 y}{\partial t^2}$, and giving it the solution $y(x, t) = f(x + ct) + g(x - ct)$, where f and g are everywhere differentiable functions.

Euler 1749 deals with the problem in the same way, but in his solution to the equation he does not require that the functions f and g be everywhere differentiable, they can be any curve drawn by hand. This makes a difference in the case of the plucked string, a curve with a corner, which in the corner point is not differentiable.

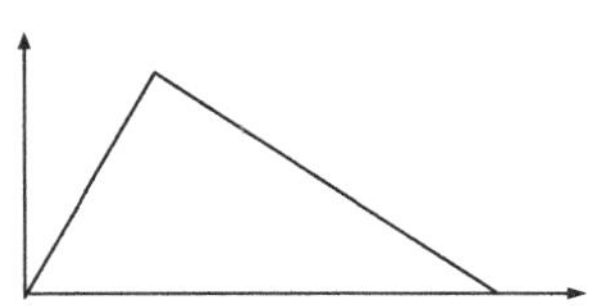

Thus, to deal with this case, it is essential not to restrict oneself to differentiable functions.

Against this, d'Alembert objects that, in the corner point, no meaning can be attributed to the function. For, the derivative of $y(x, t)$ with respect to x is different, depending on whether it is calculated for positive or negative increments of x. So, the second derivative $\frac{\partial^2 y}{\partial x^2}$ is not defined in the corner point, and hence it makes no sense to say that it is equal to $\frac{1}{c^2}\frac{\partial^2 y}{\partial t^2}$. Therefore, "the motion of the string cannot be subject to any analytic calculation, nor represented by any construction, when the curvature makes a jump at some point" (d'Alembert 1761–1780, I, 22).

But a plucked string is a physical configuration, so it must be possible to attribute a meaning to it. Since, on the contrary, no meaning can be attributed to the function in the corner point, this makes one doubt that mathematics is effective in the natural sciences.

16.12 The Notion of Function

These doubts of Descartes and d'Alembert are often treated with condescension, since they are supposed to originate from the use of an unduly restrictive notion of function.

Thus, Bourbaki says that, as the calculus has developed, it has become more and more clear that, "in spite of Descartes, algebraic curves and functions do not have, from the 'local' point of view which is that of the infinitesimal calculus, anything that distinguishes them from much more general functions" (Bourbaki 1994, 177). Indeed, "the functions and curves with a dynamic definition are functions and curves like the others, accessible to the same methods; and the variable 'time' is no more than a parameter, whose temporal aspect is purely a matter of language" (ibid.). Finally, in the nineteenth century, Dedekind introduced the modern notion of function.

Some presage of this notion can be found already in the seventeenth century, when "the notion of 'arbitrary curve' appears often" (ibid.). But "unfortunately this clear and fruitful" notion could not then "fight against the confusion created by Descartes" who, on the one hand, "banned from 'geometry' all curves not capable of a precise analytic definition," and, on the other hand, "restricted to algebraic operations alone the admissible procedures of formation in such a definition" (ibid.). Descartes's mechanical curves are functions in the sense of the modern notion of function, so they are susceptible of a mathematically precise and exact description. Therefore, the argument by which Descartes raises doubts about the absolute effectiveness of mathematics to treat all aspects of the physical world is invalid. The same can be said about the argument of d'Alembert.

This objection, however, is unconvincing, because Bourbaki assumes that all physical phenomena can be adequately described by functions in the sense of the modern notion of function. But this assumption is unjustified. As Zemanian says, "a physical variable is customarily thought of as a function, i.e., a rule which assigns a number to each numerical value of some independent variable. For example, if the independent variable is time *t* and the physical quantity is a force *f*, then one would say that the force is known if its value *f*(*t*) is specified at every instant of time *t*" (Zemanian 1987, 1). This means that, to know the force *f*, it would be necessary to specify its value *f*(*t*) at every instant of time *t*. But "it is impossible to observe the instantaneous values of *f*(*t*). Any measuring instrument would merely record the effect that *f* produces on it over some nonvanishing interval of time" (ibid.). So, the modern notion of function does not permit to adequately describe all physical phenomena.

The doubts of Descartes and d'Alembert may originate from the use of an unduly restrictive notion of function. But they pose a real problem, because the question of whether mathematics is effective in the natural sciences arises also with the modern notion of function.

16.13 Analytic Functions

Hadamard raises doubts as to whether mathematics is effective in the natural sciences in relation to the question of analytic functions.

By the Weierstrass approximation theorem, one "may always consider any functions as analytic, as, in the contrary case, they could be approximated with any required precision by analytic ones" (Hadamard 1923, 33). Therefore, at the beginning of the twentieth century, there was the widespread conviction that all physical phenomena can be modelled by some analytic function with an approximation as great as one chooses.

Thus, Poincaré says that "the numbers the physicist measures by experiment are never known except approximately; and besides any function always differs as little as you choose from a discontinuous function, and at the same time it differs as little as you choose from a continuous function" (Poincaré 2015, 288). Therefore, the physicist may "at will suppose that the function studied is continuous, or that it is discontinuous; that it has or has not a derivative; and may do so without fear of ever being contradicted, either by present experience or by any future experiment" (ibid.). In particular, the physicist may at will suppose that all functions he uses are analytic functions.

Against this, Hadamard objects that analytic functions do not permit to adequately model all physical phenomena. For, although all functions can be approximated by some analytic function with an approximation as great as one chooses, two approximations which differ very little from each other can yield completely different solutions to the same equation. Therefore, the question is not whether "an approximation would alter the data very little, but whether it would alter the solution very little" (Hadamard 1923, 33).

To see that two approximations which differ very little from each other can yield completely different solutions to the same equation, let us consider, for example, the equation of two-dimensional potentials

$$\frac{\partial^2 u}{\partial x^2} + \frac{\partial^2 u}{\partial y^2} = 0,$$

which concerns the distribution of temperature u on a surface of spatial dimensions x and y. Suppose we want to find a solution to this equation that, for $x = 0$, satisfies the following data

$$\begin{aligned} u(0, y) &= 0 \\ \frac{\partial u}{\partial x}(0, y) &= A_n \sin(ny), \end{aligned}$$

where n is a very large number, and A_n is a function of n assumed to be very small as n grows large (for example, $A_n = \frac{1}{n^p}$). These data differ from zero as little as one chooses. Nevertheless, the solution of the equation is,

$$u = \frac{A_n}{n} \sin(ny)\mathrm{Sh}(nx),$$

which, if $A_n = \frac{1}{n^p}$, is very large for any determinate value of x different from zero, on account of the mode of growth of the hyperbolic sine $\mathrm{Sh}(nx)$.

The presence of the factor $\sin(ny)$ "produces a 'fluting' of the surface" that, "however imperceptible in the immediate neighborhood of the y-axis, becomes enormous at any given distance of it however small, provided the fluting be taken sufficiently thin by taking n sufficiently great" (ibid., 34).

This example shows that data that differ very little from each other can yield enormous differences in solving the equation of a given phenomenon of nature. So, although all functions can be approximated by some analytic function with an approximation as great as one chooses, choosing one approximation rather than another can yield very different solutions. In particular, the equation of two-dimensional potentials for surfaces does not permit to make empirically useful predictions, because the data cannot be measured with an accuracy such as to keep the solution within given bounds.

Therefore, according to Hadamard, physics cannot be restricted to considering only analytic functions. Instead of assuming that all physical phenomena can be described by some analytic function with an approximation as great as one chooses, "making a rule not to assume analyticity of data agrees better with the true and inner nature of things" (ibid., 33). Since, on the contrary, there are equations of physical phenomena which only admit analytic functions as solutions, this makes one doubt that mathematics is effective in the natural sciences.

16.14 Renormalization

Doubts as to whether mathematics is effective in the natural sciences also arise from renormalization.

In quantum field theory, in calculating the magnetic moment of an electron, it is necessary to take into account the perturbation produced by the spontaneous creation of particles and antiparticles with an electric charge in the neighbourhood of the electron. The computation of the perturbation yields an infinite series of powers whose coefficients are divergent integrals.

This series is unsuitable for any numerical calculation. There is, however, a procedure, called renormalization, which, for each n, permits to replace the first n coefficients with finite integrals, that is converging. But the renormalization procedure "is not rigorous. In fact, though the substitution of series for magnitudes like charge or mass can be justified, the rest of the procedure, with its unsupported assumptions of cutoff and convergence, cannot" (Steiner 1992, 165). So, the renormalization procedure, "really, is pseudo-mathematics" (ibid., 166).

Despite this, however, in predicting the magnetic moment of an electron, renormalization has permitted to achieve an accuracy such that, if one "were to measure the distance from Los Angeles to New York to this accuracy, it would be exact to the thickness of a human hair" (Feynman 1985, 7).

How can we explain the great success of renormalized quantum field theory, "a conceptually unstable theory? A stubborn logician confronting this paradox would probably reject either renormalization theory or field theory (or both!), and some physicists such as Dirac, Landau, and Chew did so. But most field theorists reasoned differently" (Cao-Schweber 1993, 34). They did not let themselves be scared by the fact that the renormalization procedure is pseudo-mathematics, and continued to develop the theory.

Perhaps one day this pseudo-mathematics will be replaced by rigorous mathematics, but, at the moment, no such rigorous mathematics exists. This makes one doubt that mathematics is effective in the natural sciences.

16.15 Deterministic Chaos

Doubts as to whether mathematics is effective in the natural sciences also arise in relation to the question of deterministic chaos.

Let us consider a very simple physical system, whose evolution over time is completely described by a succession of discrete states $x_0, x_1, x_2, \ldots$, and which is deterministic in the sense that, for every n, $x_n = f(x_{n-1})$, where f is a given function. Then, for every n,

$$x_n = \underbrace{f(f(\ldots f}_{n}(x_0)\ldots)),$$

namely the evolution of the system is completely determined by the initial state x_0 and by the function *f*.

Now, let us take as *f* the tenfold function $x_n = 10 \times x_{n-1}$. Then, for each n, $x_n = 10^n \times x_0$. This means that a measurement error of the initial state of the system x_0 is tenfold at each subsequent state. So, if the initial state of the system x_0 is known with precision up to N decimals, then the next state will be known with precision up to $N-1$ decimals, . . . , the state x_n with precision up to $N-n$ decimals, . . . , the state x_N with precision up to $N-N=0$ decimals. So at the state x_N there is a total loss of precision, the behaviour of the system becomes absolutely unpredictable.

Since the initial state x_0 of the system can be known with a precision that, however large, is always limited to a finite number N of decimals, it follows that at the state x_N mathematics will no longer be able to predict the state of the system.

For example, if the physical system is the solar system, an error of less than 15 meters in the measurement of the Earth's position will make it impossible to predict the orbit of the Earth 100 million years into the future.

As Poincaré says, "it may happen that slight differences in the initial conditions produce very great differences in the final phenomena; a slight error in the former would make an enormous error in the latter. Prediction becomes impossible, and we have the fortuitous phenomenon" (Poincaré 2015, 397–398).

This makes one doubt that mathematics is effective in the natural sciences.

16.16 Mathematical Opportunism

The above examples, which raise doubts as to whether mathematics is effective in the natural sciences, suggest that there are limitations in the applicability of mathematics to the world. Mathematics is able to deal successfully only with the simplest of situations.

As Schwartz says, mathematics "is able to deal successfully only with the simplest of situations, more precisely, with a complex situation only to the extent that rare good fortune makes this complex situation hinge upon a few dominant simple factors" (Schwartz 1992, 21–22). Indeed, "beyond the well-traversed path, mathematics loses its bearings in a jungle of unnamed special functions and impenetrable combinatorial particularities" (ibid., 22). In fact, "the ability to keep many threads in hand, to draw for an argument from many disparate sources, is quite foreign to mathematics," typically, "science leaps ahead and mathematics plods behind" (ibid.).

Since mathematics is able to deal successfully only with the simplest of situations, with understandable opportunism scientists look for such situations and deal with them. But this does not cancel the fact that there are numerous complex situations that do not depend on a few dominant simple factors, which scientists do not know how to deal with by means of mathematics, and which therefore they opportunistically avoid.

When confronted with these situations, scientists leave mathematical models aside and take refuge in empirical modelling, such as computer modelling based on empirical observations, rather than on mathematically describable relationships of the system being modelled.

This is the case of biological and cognitive systems, which consist of a huge number of parts that do not interact linearly, and whose macroscopic, observable parts, behave in ways not derivable from simple microlevel principles. But difficulties arise also with certain physical phenomena, such as superconductivity, convection of complex fluids, or oscillations in reaction-diffusion systems (see, for example, Hooker 2011).

16.17 Limitations in the Application of Mathematics

The limitations in the application of mathematics to the world are of various kinds.

(1) Mathematics permits to formulate physical laws that do not describe physical phenomena precisely but only approximately. For example, according to the description of planetary motion given by Kepler's laws, the orbit of the planets has the shape of an ellipse and can be calculated precisely. But this is not literally true, because the orbits of the planets resemble only approximately an ellipse. In particular, the Earth's orbit would be an ellipse only if the Earth were the only planet in the solar system, if it were a sphere, and if it had no energy exchange with the Sun. Moreover, the orbits of the planets are not easily calculated, due to the confusing influence of Keplerian, Newtonian, Einsteinian, thermodynamic, and chaotic factors. Even when they can be calculated, the calculations permit accurate predictions only over a comparatively short period of time. This is sufficient for certain purposes, but does not permit accurate predictions over time scales of millions of years.

(2) Only a fraction of mathematics finds application to the world. Some mathematicians even take pride of this. Thus, Dieudonné says that, from the recent history, "it is perfectly clear that of all the striking progress" that mathematics has recently made, "not a single one, with the possible exception of distribution theory, had anything to do with physical applications" (Dieudonné 1964, 248). Even "in the theory of partial differential equations, the emphasis is now much more on 'internal' and structural problems than on questions having a direct physical significance" (ibid.).

(3) Often, what finds application to the world are not deep theorems. For example, what is used in quantum theory are not deep theorems about Hilbert space, but only the definition of a Hilbert space and little more. What is used in the so-called Grand Unified Theory is only some fairly elementary group theory. Some advanced mathematics is used in parts of physics such as string theory, but those parts of physics are still mostly empirically unconfirmed.

(4) The mathematics used in some parts of physics has been developed to a certain extent by physicists themselves. An eminent example of this is Newton, who developed the calculus of infinitesimals as a tool for his work in physics. Indeed, Newton says: "In writing" *Principia Mathematica* I "made much use of the method of fluxions direct and inverse," namely derivatives and integrals, "but did not set down the calculations in the book itself because the book was written by the method of composition" (Newton 1971, 296). Namely by the axiomatic method (for the reason explained in Chap. 3). In fact, as Truesdell says, Newton's *Principia Mathematica* is "a book dense with the theory and application of the infinitesimal calculus" (Truesdell 1968, 99, footnote 4).

(5) Sometimes, the mathematics used in some parts of physics and developed by physicists themselves is non-rigorous by current standards, so mathematicians look at it with suspicion. Thus, when Dirac introduced his delta function δ, he admitted that δ "is not a function" according to "the usual mathematical definition of a function, which requires a function to have a definite value for each point in its domain," but he claimed that the delta function δ "is something more general, which we may call an 'improper function'" (Dirac 1948, 58). But mathematicians were unconvinced. Thus, von Neumann said that the method of Dirac "in no way satisfies the requirements of mathematical rigor," because it "requires the introduction of 'improper' functions with self-contradictory properties," and they are merely "a mathematical 'fiction'" (von Neumann 1955, ix). Mathematicians remained unconvinced until Schwartz "generalized the notion of function, first by that of measure, then by that of distribution," and said that "δ will be a measure and not a function," and its derivative "δ' will be a distribution and not a measure" (Schwartz 1966, 4).

(6) More generally, in the last few decades physicists "have stumbled across a whole range of mathematical 'discoveries'" which "are beyond the reach, as yet, of mathematical rigour, but which have withstood the tests of time and alternative methods" (Atiyah, Dijkgraaf, and Hitchin 2010, 914–915). Their impact on mathematics "has been profound and widespread. Areas of mathematics such as topology and algebraic geometry," which "appear very distant from the physics frontier, have been dramatically affected" (ibid., 915). The meaning of this "is unclear" but "one may be tempted to invert Wigner's comment and marvel at 'the unreasonable effectiveness of physics in mathematics'" (ibid.).

16.18 The Reasonable Ineffectiveness of Mathematics

In view of what has been said above, it seems fair to say that it is unjustified to speak, like Wigner, of the unreasonable effectiveness of mathematics in the natural sciences.

As Steiner says, Wigner "ignores the failures, i.e., the instances in which scientists fail to find appropriate mathematical descriptions of natural phenomena," and such instances "outnumber the successes by far" (Steiner 1998, 9). Wigner "also ignores the mathematical concepts that never have found an application" (ibid.). What is more important, he ignores that "each success of applying a mathematical idea to physics is just that – an individual success of a mathematical concept. The success of the group concept, for example, might have nothing to do with a group being a mathematical concept" (ibid.).

Instead of speaking of the unreasonable effectiveness of mathematics, it seems more proper to speak of the reasonable ineffectiveness of mathematics in the natural sciences.

As Abbott says, "Wigner's thought that the effectiveness of mathematics is a 'miracle'," but the effectiveness of mathematics is "overstated" (Abbott 2013, 2149). We have the illusion that mathematics is effective because we focus on the successful mathematical models, but there are "millions of failed models in the minds of researchers, over the ages, which never made it on paper because they were wrong. We tend to publish the ones that have survived some level of experimental vindication," but they are the result of a "Darwinian selection process," the "successful models are merely selected out from many more failed ones" (ibid., 2150). In fact "there is no miracle," we must "remove the quandary of Wigner's 'miracle'" and, instead of speaking of the unreasonable effectiveness of mathematics, we must rather speak of "the reasonable ineffectiveness of mathematics" (Abbott 2013, 2147).

This must be complemented with the account of why the models that are selected are successful, which is provided by the two reasons why mathematics is applicable to the world considered above.

References

Abbott, Derek. 2013. The reasonable ineffectiveness of mathematics. *Proceedings of the IEEE* 101: 2147–2153.

Atiyah, Michael. 2004. Interview. In Martin Raussen, and Christian Skau, Interview with Michael Atiyah and Isadore Singer. *Newsletter of the European Mathematical Society* 53 (September 2004): 24–30.

Atiyah, Michael, Robbert Dijkgraaf, and Nigel Hitchin. 2010. Geometry and physics. *Philosophical Transactions of the Royal Society A* 368: 913–926.

Barrow, John D. 2020. *1+1 non fa (sempre) 2: Una lezione di matematica*. Bologna: il Mulino.

Bayle, Pierre. 1820. Anaxagoras. In Pierre Bayle, *Dictionnaire historique et critique*, vol. 2, 20–56. Paris: Desoer.

Benacerraf, Paul, and Hilary Putnam, eds. 1983. *Philosophy of mathematics: Selected readings*. Cambridge: Cambridge University Press.

Bourbaki, Nicolas. 1994. *Elements of the history of mathematics*. Berlin: Springer.

Bueno, Otávio, and Mark Colyvan. 2011. An inferential conception of the application of mathematics. *Noûs* 45: 345–374.

Cao, Tian Yu, and Silvan Sam Schweber. 1993. The conceptual foundations and the philosophical aspects of renormalization theory. *Synthese* 97: 33–108.

Clagett, Marshall. 1999. *Ancient Egyptian science: A source book*, vol. 3: *Ancient Egyptian mathematics*. Philadelphia: American Philosophical Society.

Colyvan, Mark. 2012. *An introduction to the philosophy of mathematics*. Cambridge: Cambridge University Press.

d'Alembert, Jean-Baptiste. 1749. Recherches sur la courbe que forme une corde tendue mise en vibration. *Histoire de l'Académie Royale des Sciences Et des Belles Lettres de Berlin Année 1747* (3): 214–219.

———. 1761–1780. *Opuscules mathématiques*. Paris: David.

Descartes, René. 1996. *Oeuvres*. Paris: Vrin.

Dieudonné, Jean. 1964. Recent developments in mathematics. *The American Mathematical Monthly* 71: 239–248.

Dirac, Paul Adrien Maurice. 1948. *The principles of quantum mechanics*. Oxford: Oxford University Press.

Dyson, Freeman John. 1983. Review of Yuri I. Manin. *Mathematics and physics. The Mathematical Intelligencer* 5 (2): 54–57.

Eddington, Arthur. 1939. *The philosophy of physical science*. Cambridge: Cambridge University Press.

Euler, Leonhard. 1749. De vibratione chordarum exercitatio. *Nova Acta Eruditorum, Series 2* 10: 512–527.

Feynman, Richard P. 1985. *QED: The strange theory of light and matter*. Princeton: Princeton University Press.

Gowers, Timothy. 2002. *Mathematics: A very short introduction*. Oxford: Oxford University Press.

Hadamard, Jacques. 1923. *Lectures on Cauchy's problem in linear partial differential equations*. New Haven: Yale University Press.

Hamming, Richard. 1980. The unreasonable effectiveness of mathematics. *The American Mathematical Monthly* 87: 81–90.

Hilbert, David. 1980. Letter to Frege, 29 December 1899. In Gottlob Frege, *Philosophical and mathematical correspondence*, 38–41. Oxford: Blackwell.

———. 1996a. From 'Mathematical Problems'. In *From Kant to Hilbert: A source book in the foundations of mathematics*, ed. William Ewald, vol. 2, 1096–1105. Oxford: Oxford University Press.

———. 1996b. The grounding of elementary number theory. In *From Kant to Hilbert: A source book in the foundations of mathematics*, ed. William Ewald, vol. 2, 1149–1157. Oxford: Oxford University Press.

———. 1996c. Logic and the knowledge of nature. In *From Kant to Hilbert: A source book in the foundations of mathematics*, ed. William Ewald, vol. 2, 1157–1165. Oxford: Oxford University Press.

Hooker, Cliff, ed. 2011. *Philosophy of complex systems*. Amsterdam: North Holland.

Hume, David. 2007. *Dialogues concerning natural religion and other writings*. Cambridge: Cambridge University Press.

Hut, Piet, Mark Alford, and Max Tegmark. 2006. On math, matter and mind. *Foundations of Physics* 36: 765–794.

Kant, Immanuel. 2002. *Theoretical philosophy after 1781*. Cambridge: Cambridge University Press.

Kline, Morris. 1981. *Mathematics and the physical world*. Mineola: Dover.

Newton, Isaac. 1971. MS Add. 3968, f. 402. In I. Bernard Cohen, *Introduction to Newton's 'Principia'*, 295–296. Cambridge: Cambridge University Press.

Pincock, Christopher. 2004. A new perspective on the problem of applying mathematics. *Philosophia Mathematica* 12: 135–161.

———. 2007. A role for mathematics in the physical sciences. *Noûs* 41: 235–275.

Poincaré, Henri. 2015. *The foundations of science: Science and hypothesis, The value of science, Science and method*. Cambridge: Cambridge University Press.

Pope Benedict XVI. 2014. *The garden of God: Toward a human ecology*. Washington: The Catholic University of America Press.

Sambursky, Samuel. 1987. *The physical world of the Greeks*. Princeton: Princeton University Press.

Schwartz, Laurent. 1966. *Théorie des distributions*. Paris: Hermann.

Schwartz, Jack. 1992. The pernicious influence of mathematics on science. In *Discrete thoughts*, ed. Mark Kac, Gian-Carlo Rota, and Jack Schwartz, 19–25. Boston: Birkhäuser.
Steiner, Mark. 1992. Mathematical rigor in physics. In *Proof and knowledge in mathematics*, ed. Michael Detlefsen, 158–170. London: Routledge.
———. 1998. *The applicability of mathematics as a philosophical problem*. Cambridge: Harvard University Press.
Tegmark, Max. 2008. The mathematical universe. *Foundations of Physics* 38: 101–150.
Truesdell, Clifford. 1968. *Essays in the history of mechanics*. New York: Springer.
von Neumann, John. 1955. *Mathematical foundations of quantum mechanics*. Princeton: Princeton University Press.
Wigner, Eugene Paul. 1960. The unreasonable effectiveness of mathematics in the natural sciences. *Communications on Pure and Applied Mathematics* 13: 1–14.
Zemanian, Armen H. 1987. *Distribution theory and transform analysis: An introduction to generalized functions, with applications*. Mineola: Dover.

Part V
Conclusion

Chapter 17
Knowledge, Mathematics, and Naturalism

Abstract According to heuristic philosophy of mathematics, one of the tasks of the philosophy of mathematics is to give an answer to the question: In what sense is mathematics knowledge? The chapter argues that mathematics is knowledge in the same sense as all other knowledge, namely it is a function of life that is essential to the survival of individual organisms and whole species, and, in the case of human beings, also meets other their needs, helping them fulfil their potentialities. For this reason, from antiquity, philosophers have felt the necessity to account for it.

Keywords Mathematics as knowledge · Knowledge as true justified belief · Knowledge as function of life · Biological role of knowledge · Cultural role of knowledge · Mathematical knowledge and naturalism · Natural mathematics · Mathematics as discipline · Mathematical knowledge and the a priori · Importance of mathematics to human life

17.1 Mathematics as Knowledge

As we have seen in Chap. 3, according to heuristic philosophy of mathematics, one of the tasks of the philosophy of mathematics is to give an answer to the question: In what sense is mathematics knowledge? This presupposes a positive answer to the question: Is mathematics knowledge? And generally presupposes an answer to the question: What is knowledge?

First, we consider the question: Is mathematics knowledge? Several people, not only philosophers but even mathematicians, have denied that mathematics is knowledge.

Thus, Croce says that, to establish its results, mathematics "requires certain fundamental principles" (Croce 1917, 365). But mathematical concepts, for example, "the point without extension, the line without superficies, and the superficies without solidity," are "unthinkable" (ibid., 367). And, as they "are unthinkable, so are the principles of mathematics unimaginable" (ibid., 368). Hence, mathematical concepts are "pseudoconcepts" altogether "void of representation," and the principles of mathematics are propositions "altogether void of truth" (ibid., 369).

C. Cellucci, *The Making of Mathematics*, Synthese Library 448,
https://doi.org/10.1007/978-3-030-89731-4_17

Therefore, "mathematics does not know," it "does not subserve knowing, but counting and calculating what is already known" (ibid., 365).

Wittgenstein says that "the propositions of mathematics" are "pseudo-propositions" (Wittgenstein 2002, 6.2). For, "a proposition of mathematics does not express a thought" (ibid., 6.21). In fact, "in real life" we "make use of mathematical propositions only in inferences from propositions that do not belong to mathematics to others that likewise do not belong to mathematics" (ibid., 6.211).

Mac Lane says that "it is "customary to ask of a piece of mathematics: 'Is it true?'," but "this issue of truth is a mistaken question" (Mac Lane 1986, 440). The "question 'Is mathematics true?' is out of place" (ibid., 442). Of course, "mathematics, through the decisive notion of formal proof, can appeal to an absolute standard of rigor," so "mathematics is 'correct'" (ibid.). But mathematics is "not 'true'" (ibid.). Indeed, mathematics is not knowledge, because it "makes no statement about the reality of some of its objects," it "is concerned not with reality" but only with "formal rules" (ibid.).

These claims, however, are unwarranted. They are based on the assumption that mathematical concepts and propositions, to be knowledge, must have reality. But this assumption is unjustified even in the case of the concepts and propositions of physics. Thus, Heisenberg says that, although "in the experiments about atomic events we have to do" with "phenomena that are just as real as any phenomena in daily life," on the contrary, "the atoms or the elementary particles themselves are not as real; they form a world of potentialities or possibilities rather than one of things or facts" (Heisenberg 1958, 186).

The concepts and propositions of mathematics or physics do not have reality, they are tools by which we make the world understandable to ourselves. Nevertheless, they give knowledge, because it is through them that we see and think about the world.

17.2 Knowledge as True Justified Belief

Now we consider the question: What is knowledge? In the past century, the standard answer to this question has been that knowledge is "a special kind of belief: belief that is both true and justified" (Williams 2001, 17). For, "knowledge excludes three things: ignorance, error, and (mere) opinion," and "the belief condition excludes ignorance, the truth condition excludes error, and the justification condition excludes mere opinion" (ibid., 16–17).

But the standard answer was shown to be invalid already by Plato. Indeed, Plato says: "Are you satisfied, and do you set it down, that knowledge is true belief accompanied by justification?" (Plato, *Theaetetus*, 202 c 7–8). No, "knowledge cannot" be "true belief accompanied by justification" (ibid., 210 b 1–2). For, justification is itself knowledge. Therefore, to say that knowledge is true belief accompanied by justification, would amount to saying that knowledge is true belief accompanied by knowledge. But "it would be utterly silly to say, when you are trying to find out what knowledge is," that knowledge "is true belief accompanied by knowledge" (ibid., 210 a 7–8). The circle would be too blatant.

The standard answer was again shown to be invalid at the beginning of the twentieth century, by several counterexamples of Meinong and Russell.

Thus, one of Russell's counterexamples is: "There is the man who looks at a clock which is not going, though he thinks it is, and who happens to look at it at the moment when it is right; this man acquires a true belief as to the time of day" (Russell 2009, 140). For, the time is exactly the time shown on the clock. The belief is also justified, because the man forms it by looking at a clock, and this is a standard way of knowing the time. But he "cannot be said to have knowledge" (ibid.). For, unbeknown to him, the clock is not going.

Even mathematical knowledge is not belief that is true and justified. For, by Gödel's second incompleteness theorem, mathematical knowledge cannot be said to be true and justified.

This, however, has not been enough to convince people to abandon the standard answer. Simply, several variants of it have been considered, according to which knowledge is belief that is true plus something else. But, like the answer that knowledge is belief that is true and justified, all these variants are shown to be invalid by counterexamples (see Cellucci 2008, Chap. 4).

Indeed, Zagzebski argues that counterexamples "are inescapable for virtually every analysis of knowledge which at least maintains that knowledge is true belief plus something else" (Zagzebski 1994, 65).

Therefore, it seems fair to say that we must abandon the answer that knowledge is belief that is true plus something else, and replace it with an answer of a different kind.

17.3 Human Knowledge as a Function of Life

The standard answer that knowledge is belief that is true and justified implies that knowledge is just a state of mind of a certain sort. The same can be said of all the variants according to which knowledge is true belief plus something else.

But to say that knowledge is just a state of mind of a certain sort is inadequate, because it involves that knowledge does not serve any function of life. We seek knowledge for its own sake, and its value consists quite simply in the fact that we enjoy it.

Thus, Schlick says that "knowledge, so far as it is science, does not serve any other of life's functions," it is "an independent function, whose exercise affords us immediate satisfaction" (Schlick 1974, 100). We "seek knowledge for its own sake, without thought to its application in life" (ibid., 96). The "value of knowledge consists quite simply in the fact that we enjoy it" (ibid., 101).

But it is not so. We do not seek knowledge merely for its own sake, nor its value consists quite simply in the fact that we enjoy it. We seek knowledge because it is essential to life. Indeed, knowledge is an important function of life.

The functions of life are usually described by saying that, "in order for any object to be considered alive, in addition to possession of DNA, it must perform various functions of life" (Hamilton 2006, 20). Namely, "it must be able to respire, or metabolize food into energy," to "regulate itself, or provide a stable internal environment," to "grow, to move, to take in or produce its own nutrition, to move materials around within the organism, and to manufacture materials that cannot be taken directly from its environment," and finally, "to reproduce with or without one of its own species" (ibid.).

But this description is essentially incomplete, because it omits an important item: knowledge. Knowledge is an important function of life because it is essential to life.

Indeed, knowledge is essential to human life. At the sunset of the Middle Age, Cecco d'Ascoli even deemed knowledge to be so essential to human life as to say: "Ma questa vita e l'altro mondo perde | Chi del savere ha sempre dispetto [This life and the other world are lost | To those who hold knowledge in contempt]" (Cecco d'Ascoli 1927, II, 7.28–30).

Generally, knowledge is essential to the life of all organisms. For, in order to be and stay alive, all organisms must to explore the ecological possibilities available to them, and to this end they must have knowledge of the environment.

17.4 Biological Role of Knowledge

Knowledge is essential to the life not only of individual organisms, but also of entire species. For, knowledge plays an essential part in natural selection.

Natural selection is usually described by saying that it "is a two-step process" (Mayr 2002, 131). At the first step, "consisting of all the processes leading to the production of a new zygote," a "new variation is produced" (ibid.). At the second step, "that of selection (elimination), the 'goodness' of the new individual is constantly tested, from the larval (or embryonic) stage until adulthood and its period of reproduction" (ibid., 132). Those "individuals who are most efficient in coping with the challenges of the environment and in competing with other members of their population and with those of other species will have the best chance to survive until the age of reproduction and to reproduce successfully" (ibid.). They "are the ones that are 'fittest to survive'" (ibid.). Such individuals are the result of "an adaptation," namely they have physiological traits that "were acquired," or whose "maintenance was favored," by "natural selection," and "the possession of which favors the individual in the struggle for existence" (ibid., 165).

But this description is essentially incomplete, because it omits an important item: knowledge.

Indeed, knowledge plays an important part in the natural selection of the human species. Its omission makes it inexplicable how our earliest human ancestors could be fit to survive, and hence how we, their descendants, can be here. For, compared to large mammals, our earliest human ancestors were weak and vulnerable creatures, so, in the midst of stronger competing or threatening species, they could not survive by virtue of physical strength. They could only survive by outsmarting those species, showing greater ingenuity in getting and using knowledge of the environment. To that aim, they made hypotheses about the environment and established their plausibility through a comparison with experience. So they acquired a superior kind of knowledge of the environment, through which they could carve out a niche for themselves in the scheme of things of biological evolution.

Generally, knowledge plays an essential part in the natural selection of all species. For, organisms that are unable to make appropriate hypotheses about the external world are more likely victims of the dangers of the environment.

17.5 Cultural Role of Knowledge

Since knowledge plays an essential part in the natural selection of all species, knowledge has a biological role. This, however, is not its only role, knowledge also has a cultural role. This is implicit in the very concept of culture. Culture is a shared body of knowledge that is transmitted from generation to generation non-genetically, namely not through DNA.

Since culture involves a transmission of a body of knowledge from generation to generation non-genetically, the cultural role of knowledge does not coincide with the biological role.

On the other hand, the cultural role and the biological role of knowledge are not opposed, because culture depends on the biological makeup of organisms.

Pinker says that "spiders spin spider webs because they have spider brains, which give them the urge to spin and the competence to succeed" (Pinker 1995, 18). Similarly, we can say that human beings think human thoughts because they have human brains, which give them the urge to think and the competence to succeed.

Moreover, even in its cultural role, knowledge can influence biological evolution. For, culture can permit individuals to modify the environment to a certain extent, and if several generations repeatedly modify the environment in the same way, this may lead to genetic changes.

In the case of human beings, a simple example of this is the culture of pastoralism, which dates back to some 10,000 years ago. It led to the activities associated with the production of milk, which are at the origin of lactase persistence, the ability of approximately one third of adults living in the world today, to digest the lactose of milk.

Lactase persistence varies widely in frequency across human populations, but is concentrated in geographical regions with a history of continued dairy farming. This is because the development of continued dairy farming for enough generations led to a genetic change in human populations in those regions. Indeed, lactase persistence "represents an adaptation to the domestication of dairying animals and the subsequent consumption of their milk" (Ségurel and Bon 2017, 297).

17.6 Biological Evolution and Cultural Evolution

In the case of human beings, an important character of culture is cultural evolution, namely the fact that, not only a shared body of knowledge is transmitted from generation to generation non-genetically, but it may undergo changes or expansions in the succession of generations.

It has been argued that there is evidence for cultural evolution also in non-human animals because, not only there is "cultural transmission of behavior through social learning in numerous vertebrate and invertebrate species," but there are also "cultural changes within a lineage" (Whiten 2019, 27). Here, however, we will confine ourselves to considering cultural evolution in human beings.

The origin of cultural evolution in human beings can be traced back to the agricultural revolution of some 10,000 years ago, which marked the transition from hunter-gatherer communities to settled agricultural communities, bringing about the domestication of animals and the cultivation of crops. This led to a dramatic change in lifestyle, a rapid increase in population, and a drastic acceleration in the rate of innovation, which is known as the "10,000 year explosion" (Cochran and Herpending 2009, 226).

With the agricultural revolution, survival no longer depended on the capacity to find and exploit wild resources, it became increasingly dependent on the ability to control the environment. Rather than the nomadic moving to new locations, the crucial factor for survival became the capacity to solve problems concerning the control of the environment. Such problems prompted the search for hypotheses to solve them, which gave rise to other problems and hypotheses, and so on. This led to a continued growth of knowledge, which enabled human beings to address more and more complex tasks in shorter and shorter time.

Once that stage was reached, biology became only part of the story. Cultural evolution assumed a primary role in creating means by which human beings could reach a higher level of development. Human beings were no longer only a result of biological evolution, but a result of the joint action of biological evolution and cultural evolution.

Indeed, cultural evolution does not reduce to biological evolution. The latter has predisposed human beings to face only situations similar to those that have already occurred in their evolutionary past. But the world presents ever new situations, requiring means more powerful than those deriving from nature, and they are provided by cultural evolution. Therefore, between biological evolution and cultural evolution there is continuity but not identity.

Then, it is unjustified to say, as Popper does, that "from the amoeba to Einstein is only one step" (Popper 1972, 246). From the amoeba to Einstein there are many important steps. Cultural evolution determines a substantial difference between human beings and amoebas. In their effort to solve the survival problem, both human beings and amoebas are constantly faced with the problem of having limited resources. But, while amoebas have little control over the environment, thanks to cultural evolution human beings are capable of exercising considerable control over it.

Admittedly, for a good part of their development, human beings had fairly limited control over the environment, and hence had to devote a very large part of their efforts to survival. Later on, however, their condition changed and today, thanks to cultural evolution, they exercise a fairly large control over the environment. Owing to this, they can devote only a relatively limited part of their efforts to survival, and can meet other their needs, helping them fulfil their potentialities.

This does not mean, however, that knowledge no longer serves survival. Knowledge still plays the biological role of serving first of all to solve the problem of survival, and cultural evolution also contributes to solving that problem, providing means to exercise control over the environment and to modify it.

17.7 Mathematical Knowledge and Naturalism

On the basis of what has been said above, knowledge is a function of life that is essential to the survival of individual organisms and whole species, and, in the case of human beings, also meets other their needs, helping them fulfil their potentialities.

This view of knowledge is a naturalistic one, meaning by 'naturalistic' a view according to which, as Dewey says, "knowledge is not something separate and self-sufficing, but is involved in the process by which life is sustained and evolved" (Dewey 2004, 50).

This also includes mathematical knowledge. Mathematical knowledge too is a function of life that is essential to the survival of individual organisms and whole species and, in the case of human beings, also meets other their needs, helping them fulfil their potentialities.

It may seem strange to say this, but it is not really so. Indeed, mathematical knowledge was already essential to the survival of our earliest human ancestors.

It has been said above that our earliest human ancestors could survive only because they were able to outsmart stronger competing or threatening species, showing greater ingenuity in getting and using knowledge about the environment.

Specifically, in order to survive, our earliest human ancestors, on the one hand, had to avoid becoming food for great predators, and, on the other hand, had to capture animals for food. To that aim, they had to solve several problems. Thus, they had to determine the location of animals in the environment, in order to escape from them or to pursuit them. They had to recognize the shape of animals in the environment, in order to distinguish preys from dangerous animals. They had to evaluate the number of animals in the environment, in order to decide whether to flee from them, or to pursuit them.

These problems were mathematical in kind. Indeed, determining the location of animals and recognizing the shape of animals in the environment were geometrical problems, evaluating the number of animals in the environment was an arithmetical problem. Thus, in order to survive, our earliest human ancestors had to solve mathematical problems.

Our earliest human ancestors were able to solve them, as it is shown by the fact that we are here. They solved them making hypotheses by non-deductive inferences, either from direct observation or from indirect observation, such as the observation of footprints, crushed or bent grass and vegetation, bent or broken branches or twigs, mud displaced from streams, even excrements. And they established the plausibility of their hypotheses through a comparison with experience. Therefore, they discovered properties of the environment by the analytic method, and accordingly produced successful behaviours.

17.8 Difference from Another Naturalistic View

This naturalistic view of mathematical knowledge is different from other naturalistic views, in particular, from that of Maddy.

Maddy shares Quine's view that the naturalist philosopher "begins within natural science – that is, with both the methods and the theories of science ready to hand – and she sets out to explain how human beings" come "to knowledge of the world, as it is characterized by science" (Maddy 1997, 183). Now, "at the outset of this investigation, it quickly becomes obvious that mathematics is central to our scientific study of the world and that the methods of mathematics differ markedly from those of natural science" (ibid.).

With regard to this, Quine would say that "the true justification" for the methods of mathematics "derives from the role of mathematics in science" (ibid.). But this would mean "subordinating mathematics to science," because it would amount to "identifying the proper methods of mathematics with the method of science" (ibid., 184). This runs "counter to the fundamental spirit that underlies all naturalism: the conviction that" mathematics "should be understood and evaluated on its own terms" (ibid.). On the contrary, "mathematics is not answerable to any extra-mathematical tribunal and not in need of any justification beyond proof and the axiomatic method" (ibid.). Mathematics must be understood and evaluated independently of any "extra-mathematical standpoint," including natural science, because natural "science is such an extra-mathematical standpoint" (ibid., 201).

But Maddy's naturalistic views of mathematical knowledge is unsatisfactory. For, the claim that mathematics must be understood and evaluated independently of any extra-mathematical standpoint, implies that the only evaluation criterion of mathematics is the consensus of the mathematical community. This contrasts with the fact that, in the mathematical community, there are often disagreements. There are even two alternative mathematics, infinitary mathematics and intuitionistic mathematics. The choice between them depends, in particular, on the fact that, as already mentioned in Chap. 2, intuitionistic mathematics makes it impossible to construct certain mathematical objects that are important for physics, so intuitionistic mathematics is inconvenient for physics. But this is an evaluation of mathematics from an extra-mathematical standpoint.

Also, Maddy says: "Suppose mathematicians decided to reject the old maxim against inconsistency – so that both '2 + 2 = 4' and '2 + 2 = 5' could be accepted – on the grounds that this would have a sociological benefit for the self-esteem of school children" (ibid., 198, footnote 9). If mathematicians maintained that "this goal overrides the various traditional goals," then, Maddy claims, "I find nothing in the mathematical naturalism presented here that provides grounds for protest" (ibid.). But Maddy's claim conflicts with the fact that the decision to accept both '2 + 2 = 4' and '2 + 2 = 5' would be made not from a standpoint internal to mathematics, but from a standpoint patently extra-mathematical, the alleged sociological benefit for the self-esteem of school children.

Moreover, Maddy says that mathematics is not answerable to any extra-mathematical tribunal and not in need of any justification beyond proof and the axiomatic method. But then, much in the same way, one could say that astrology is not answerable to any extra-astrological standard and not in need of any justification beyond astrological almanacs and the astrological method. Maddy answers this objection by saying that "mathematics is staggeringly useful, seemingly indispensable, to the practice of natural science, while astrology is not," so "there is a strong motivation for the scientific naturalist to give an acceptable account of mathematics with no parallel in the case of astrology" (ibid., 204–205). But Maddy's answer is based on an evaluation of mathematics from the standpoint of its usefulness to natural science, so from an extra-mathematical standpoint.

17.9 Space Sense

It has been said above that, in order to survive, our earliest human ancestors had to solve mathematical problems, and were able to do so. Now, their ability to solve them was a result of natural selection. This has been pointed out by Poincaré and Dantzig.

In particular, Poincaré points out that the ability of our earliest human ancestors to solve geometrical problems was a result of natural selection.

Indeed, Poincaré says that newborn babies have space sense, namely a system of spatial associations which forms a "primitive geometry" (Poincaré 2015, 428). These spatial associations "are not, for the most part, conquests of the individual, since their trace is seen in the new-born babe; they are conquests" of the species, which have been brought about by "natural selection" (ibid., 420).

Our system of spatial associations "has three dimensions" because "it has adapted itself to a world having certain properties," and "it is in order to be able to live in this world that" our system of spatial associations "has been established" (ibid., 427).

That our system of spatial associations has three dimensions is embodied in our physical makeup. This is apparent from the vestibular system of the ear. The three semicircular canals, filled with a viscous fluid, which make up the vestibular system, are placed on three planes, approximately perpendicular to each other, which form a Euclidean reference system.

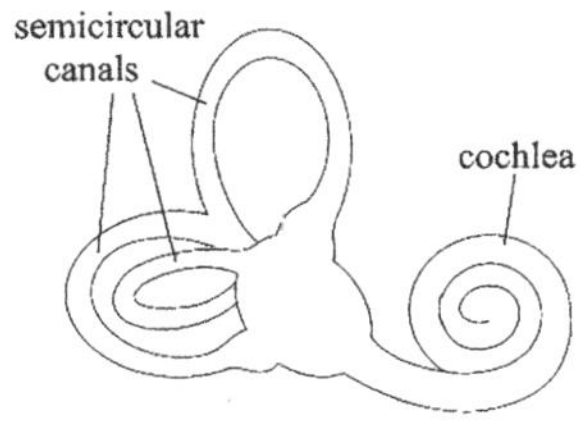

Any movement of the head results in shifts in the viscous fluid in the semicircular canals that are detected by hair cells. The resulting information is integrated with information from the other senses higher in the brain, and "we deduce, from it, by an unconscious integration, the final orientation of the head, referred to a certain initial orientation taken as origin" (Poincaré 1958, 74). The circular canals contribute to "inform us of the movements that we have executed, and that on the same ground as the muscular sensations" (ibid.).

Findings in the last few decades provide evidence for Poincaré's view that newborn babies have space sense. Indeed, newborn babies, as early as a few hours after birth, can "discriminate between simple forms" (Slater et al. 1991, 397). Also, newborn babies ranging from 32 to 61 hour old "are capable of detecting an invariant spatial relation between two elements and of using that information to discriminate a novel arrangement of the same elements" (Antell and Caron 1985, 21).

17.10 Number Sense

On the other hand, Dantzig points out that the ability of our earliest human ancestors to solve arithmetical problems was a result of natural selection.

Indeed, Dantzig says that "man, even in the lower stages of development, possesses a faculty which, for want of a better name, I shall call 'number sense'" (Dantzig 2005, 1). The latter "should not be confused with counting" and the number concept, "which is probably of a much later vintage, and involves" a "rather intricate mental process" (ibid.). Nevertheless, "a rudimentary number sense" was "the nucleus from which the number concept grew" (ibid., 5).

Dantzig's view that man, even in the lower stages of development, possesses number sense, means that this faculty is a result of natural selection, so infants have number sense. This contrasts with the view of the empiricist tradition that the infant mind is tabula rasa, or blank slate, and all knowledge comes from sensory experience. According to that view, infants do not have any number sense, and they do not form it until they are four or five years old.

Contrary to the view of the empiricist tradition, findings in the last few decades provide evidence for Dantzig's view that infants have number sense. Indeed, newborn babies ranging from 21 to 144 hour old can discriminate among "small number sets (2 to 3 and 3 to 2)" (Antell & Keating 1983, 695). Also, "5-month-old infants can calculate the results of simple arithmetical operations on small numbers," such as "1 + 1" and "2 − 1" (Wynn 1992, 749).

17.11 Space Sense in Non-human Animals

Space sense and number sense are not restricted to human beings, certain kinds of non-human animals have them, also as a result of natural selection.

Indeed, certain kinds of non-human animals have space sense.

For example, suppose you stand with your dog on a shoreline at *A*, and you throw a ball into the water to *B*. To retrieve the ball, the dog could swim directly from *A* to *B*, or could run along the shoreline to the point on shore closest to the ball, *C*, and then swim from *C* to *B*. But the dog does neither, he runs along the shoreline up to a certain intermediate point *D*, and then swims from *D* to *B*.

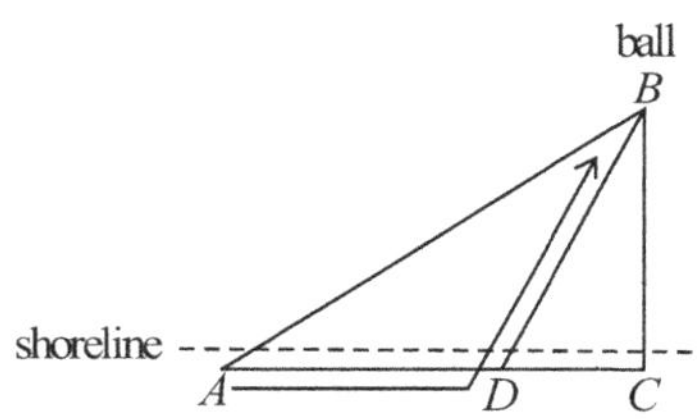

As Pennings argues, the dog does so because "he runs considerably faster than he swims," so he takes "the option of running a portion of the way, and then plunging into the lake at *D* and swimming diagonally to the ball" (Pennings 2003, 178). The point along the shoreline which minimizes the retrieval time can be found by calculus, and the point *D* chosen by the dog very nearly coincides with it. Of course, the dog "does not know calculus," but his "behavior is an example of the uncanny way in which nature (or Nature) often finds optimal solutions" (ibid., 182). Indeed, the ability of the dog to choose the point *D* is a result of natural selection.

As another example, suppose you observe a cat falling from a height of at least 30 cm. You will see that the cat will rapidly adjust his position in air and land on his four feet. Cats start developing this ability at "3 weeks of age," and fully possess the ability by the time they are "about 6 weeks of age," when they succeed "in righting quickly and landing on" their four "feet on every trial with no evidence of effort" (Sechzer et al. 1984, 502).

Cats have this ability owing to the vestibular system of their ear, which is similar to that of the human ear. It enables cats, when falling, in a fraction of a second, to figure out which way is up. On this basis, cats rotate their head towards the ground, they bend at the waist and extend their front limbs while tucking the rear ones.

So cats manage to land on their four feet. This ability of cats is a result of natural selection.

17.12 Number Sense in Non-human Animals

As certain kinds of non-human animals have space sense, certain kinds of non-human animals have number sense.

For example, in the absence of any specific training, "rhesus monkeys spontaneously compute addition operations over large numbers" (Flombaum et al. 2005, 315). They also "spontaneously compute," namely by a "single trial, no training," the "outcome of subtraction events" involving "three or less objects on each stage, even when the identity of the objects is different" (Sulkowski and Hauser 2001, 239). In particular, "when presented with two food quantities, rhesus monkeys select the larger quantity following subtractions of one piece of food from two or three; this preference is maintained when subjects must distinguish food from non-food subtractions, and when food is subtracted from either one or both initial quantities" (ibid.). This ability of rhesus monkeys to spontaneously compute is a result of natural selection.

As another example, "in the absence of any specific training, chicks spontaneously discriminated between two and three, in both cases preferring the larger stimulus set," their "behaviour seemed to indicate an ability to perform additions" by "combining two or more quantitative representations (addends) to form a new representation (i.e. the sum)" (Rugani et al. 2009, 2457). This ability of chicks to spontaneously discriminate is a result of natural selection.

17.13 Natural Mathematics and Mathematics as Discipline

Space sense and number sense, in human beings or in non-human animals, constitute natural mathematics. Thus, natural mathematics is the mathematics embodied in organisms as a result of natural selection.

Natural mathematics has been the first kind of mathematics to arise. It primarily serves to ensure the survival of certain kinds of organisms, including the human ones. Biological evolution has hardwired these organisms to perform certain mathematical operations that are essential for their survival. This explains why natural mathematics is not unique to the human species. Serving to ensure the survival, natural mathematics is important to the life of such organisms in a very basic sense.

Natural mathematics is not trivial. Devlin even says that "mother Nature turns out to be the greatest mathematician of all. Through evolution, Nature has endowed many of the animals and plants around us with built-in mathematical abilities that, from a human perspective, are truly remarkable" (Devlin 2005, 35).

However, natural mathematics has a limited scope, it does not have the full power of geometry and arithmetic as disciplines. The "geometrical and numerical concepts that we possess as adults may not be given to us as infants" (Spelke 2011, 313).

Cooper claims that mathematics as discipline "must itself be evolutionarily reducible," namely "reducible to evolutionary theory," because "mathematical knowledge can be seen as an extension of internalized evolutionary processes" (Cooper 2001, 135). But this claim is invalid, because mathematics as discipline has arisen too recently to be a direct product of biological evolution. It makes use of cultural artifacts such as mathematical diagrams and symbol systems, so it is a result of cultural evolution.

Being a result of biological and cultural evolution, respectively, natural mathematics and mathematics as discipline are different. They, however, are not opposed, because mathematics as discipline is a product of brains that are a result of natural selection and depends on them. Rather than opposition, between them there is continuity, not in Cooper's sense that mathematics as discipline is reducible to natural mathematics, but in the sense that both natural mathematics and mathematics as discipline are a product of brains that are a result of natural selection.

Moreover, as natural mathematics, serving to ensure survival, is important to the life of certain kinds of organisms, including the human ones, in a very basic sense, mathematics as discipline is important to human life in a more advanced sense. For, it responds to human needs that are functional to reach a higher level of development. Indeed, all sciences make use, to a greater or a lesser extent, of mathematics as discipline, all algorithms are a product of mathematics as discipline, all technology employs mathematics as discipline, all products are manufactured utilizing, to a greater or a lesser extent, mathematics as discipline, all business and stocks operations utilize mathematics as discipline.

Furthermore, both natural mathematics and mathematics as discipline are based on the same method, namely the analytic method. As already said above, our earliest human ancestors solved the mathematical problems with which they were confronted making hypotheses by non-deductive inferences and establishing their plausibility through a comparison with experience. Much in the same way, mathematicians solve mathematical problems making hypotheses by non-deductive inferences, and establishing their plausibility through a comparison with experience.

17.14 Mathematical Knowledge and A Priori Knowledge

Since both natural mathematics and mathematics as discipline are based on the analytic method, all of mathematical knowledge is a priori knowledge. It is a priori knowledge not in the sense that it is absolutely independent of all experience, but in the sense that it is based on hypotheses which are not derived from experience.

In the analytic method, the hypotheses are obtained by non-deductive rules, which are ampliative, so the hypotheses are not contained in the premisses. Hence, even when the premisses are derived from experience, the hypotheses obtained by non-deductive rules are not derived from experience. Nevertheless, the hypotheses are indispensable for the possibility of experience.

This suggests to call 'a priori knowledge' knowledge which is based on hypotheses that are not derived from experience, and nevertheless are indispensable for the possibility of experience. Mathematical knowledge is a priori knowledge in this sense. And so is all scientific knowledge because, like the method of mathematics, the method of natural sciences is the analytic method.

Since a priori knowledge is knowledge which is based on hypotheses that are not derived from experience, a priori knowledge is genetically a priori. On the other hand, however, a priori knowledge is not valid a priori. For, it is valid only if the hypotheses on which it is based are plausible, and they are plausible only if the arguments for them are stronger than the arguments against them on the basis of experience. So, the validity of a priori knowledge depends on experience. Therefore a priori knowledge is genetically a priori but not valid a priori.

17.15 Kant's Concept of A Priori Knowledge

The concept of a priori knowledge formulated here is different from other concepts of a priori knowledge, in particular those of Kant, Lorenz, and Popper.

According to Kant, a priori knowledge is knowledge that has the following four characters. It occurs "absolutely independently of all experience" (Kant 1998, B3). It has the character of "strict universality," namely, "no exception at all is allowed to be possible" (ibid., B4). It has "the character of inner necessity" (ibid., A2). And it is "certain" (ibid.)

Kant's concept of a priori knowledge, however, is inadequate. No knowledge occurs absolutely independently of all experience, because all knowledge, including mathematical knowledge, arises from interaction with the external world. No knowledge has the character of strict universality, because knowledge is at most plausible, so exceptions are always possible. No knowledge has the character of inner necessity, because knowledge is the product of our limited experience, so it is always contingent. No knowledge is certain, because knowledge is at most plausible, so it is potentially fallible.

The concept of a priori knowledge formulated here is not subject to these problems. According to it, a priori knowledge does not occur independently of all experience, because it is based on hypotheses which derive their plausibility from experience. It has no character of strict universality, because the hypotheses on which it is based are only plausible, so exceptions are always possible. It has no character of inner necessity, because the hypotheses on which it is based are contingent, being possibly incompatible with future data. It is not certain, because there is no guarantee that no counterexample to the hypotheses on which it is based will ever be found.

17.16 Lorenz's Concept of A Priori Knowledge

Another concept of a priori knowledge is Lorenz's concept.

According to Lorenz, "the origin of the a priori" is "a posteriori" (Lorenz 2009, 233). Kant's a priori forms of intuition do "not develop anew in every human being as a result of individual experience," but "from the standpoint of the individual," they exist "a priori" (Lorenz 1997, 18). And they must exist a priori if the individual is to survive, because they are that which "enables living things to survive and orient themselves in the outer world" (Lorenz 1977, 9). But Kant's a priori forms of intuition originally "arose a posteriori" (Lorenz 1997, 18). For, they "evolved phylogenetically through confrontation with and adaptation to that form or reality which we experience as phenomenal space" (Lorenz 1977, 9). Therefore, they "have to be understood just as any other organic adaptation" (Lorenz 2009, 239). So, Kant's a priori forms of intuition exist a priori for the individual, because they are "fixed prior to individual experience," but exist a posteriori for the species, because they "are adapted to the external world" (ibid., 233).

Lorenz's concept of a priori knowledge, however, is inadequate. Since, according to it, Kant's a priori forms of intuition exist a posteriori for the species, this means that they existed a posteriori for our earliest human ancestors. So, they did not exist a priori for them. But Kant's a priori forms of intuition should have existed a priori for them because, according to Lorenz, they must exist a priori if the individual is to survive. Therefore, Lorenz's sense of a priori knowledge makes it inexplicable how our earliest human ancestors were able to survive, and hence how we can be here.

Moreover, Lorenz's concept of a priori knowledge would imply that since, according to it, Kant's a priori forms of intuition existed a posteriori for our earliest human ancestors, the latter could derive these forms and their properties, for example, Euclid's fifth postulate, from experience. On the contrary, since these forms exist a priori for us, we are unable to derive Euclid's fifth postulate from experience. Poincaré observes: "It has often been said that if individual experience could not create geometry the same is not true of ancestral experience. But what does that mean? Is it meant that we could not experimentally demonstrate Euclid's postulate, but that our ancestors have been able to do it?" (Poincaré 2015, 91). This is absurd, because it would imply that our earliest human ancestors had intellectual powers that we do not have.

The concept of a priori knowledge formulated here is not subject to these problems. According to it, our earliest human ancestors made hypotheses about the environment by means of non-deductive inferences, so the hypotheses were not contained in the premisses of such inferences. Hence, even if the premisses were based on experience, the hypotheses were not derived from experience. Therefore, the hypotheses do not exist a priori only for us, but existed a priori also for our earliest human ancestors.

17.17 Popper's Concept of A Priori Knowledge

Another concept of a priori knowledge is Popper's concept.

According to Popper, "all knowledge is a priori, genetically a priori, in its content. For all knowledge is hypothetical or conjectural: it is our hypothesis. Only the elimination of hypotheses is a posteriori, the clash between hypotheses and reality. In this alone consists the empirical component of our knowledge" (Popper 2001, 47). Induction has no part in knowledge, because "there is no such thing as induction" (ibid., 54). Knowledge is obtained by "trial and error," namely by "trying out solutions to our problems and then discarding the false ones as erroneous" (ibid., 3). However, there are no rules of successful trials, "the demand for a theory of successful thinking cannot be satisfied" because "success depends on many things – for example on luck" (Popper 2002, 50). We can only say that the success of trials "depends very largely on the number and variety of the trials: the more we try, the more likely it is that one of our attempts will be successful" (Popper 1974, 312).

Popper's concept of a priori knowledge, however, is inadequate. As already pointed out in Chap. 3, to say that the success of trials depends very largely on the number of the trials, amounts to admitting that trial and error depends on induction. This contradicts Popper's claim that there is no such thing as induction. Moreover, because of our physical and time limitations, the number of trials we can make is very small with respect to all possible ones, so the probability that a single trial can be successful is very low. This makes it inexplicable why, as a matter of fact, mathematicians often arrive at plausible hypotheses, and hence why mathematics is successful.

The concept of a priori knowledge formulated here is not subject to these problems. According to it, a priori knowledge is the result of solving problems by the analytic method, and in the analytic method hypotheses are obtained by non-deductive rules, including induction. So, in the analytic method the search for hypotheses does not proceed blindly but is guided by non-deductive rules.

17.18 The Importance of Mathematics to Human Life

It has been said above that natural mathematics is important to the life of certain kinds of organisms, including human beings, in a very basic sense, and mathematics as discipline is important to human life in a more advanced sense. This means that, generally, mathematics is important to human life. In fact, this is the reason why mathematics has attracted philosophical reflection since at least Plato, and most of the great philosophers since have had at least something to say about it.

On the contrary, Hart claims that, if "mathematics has attracted philosophical reflection since Plato, and most of the great philosophers since have had at least something to say about it," it "is not because of the importance of mathematics in human life" (Hart 1996, 1). For, "agriculture is at least as important, but there is no philosophy of farming. It is instead distinctive and articulable philosophical problems that have attracted philosophical attention to mathematics," such as the problem of "mathematical truth" (ibid.). Or the problem of "mathematical objects" (ibid., 3).

But this claim is invalid. The importance of agriculture in human life is so clear that there is no need for philosophy to account for it. On the contrary, the importance of mathematics in human life is obscured by the widespread belief that mathematics is an abstract body of knowledge, remote from life and concrete reality.

Thus, Peirce says that "mathematics is the most abstract of all sciences" because "the whole science of mathematics is a science of hypotheses; so that nothing could be more completely abstracted from concrete reality" (Peirce 1931–1958, 3.428).

But mathematics is not an abstract body of knowledge, remote from life and concrete reality. As already said above, it responds to many human needs, from the basic one of survival to those which are functional to reach a higher level of development. Moreover, to be a science of hypotheses does not make mathematics an abstract science, because every science is a science of hypotheses.

Contrary to Hart's claim, if mathematics has attracted philosophical reflection since at least Plato, and most of the great philosophers since have had at least something to say about it, it is just because mathematics is important to human life. For this reason, from antiquity, philosophers have felt the need to account for it.

References

Antell, Sue E., and Albert J. Caron. 1985. Neonatal perception of spatial relationships. *Infant Behavior and Development* 8: 15–23.

Antell, Sue E., and Daniel P. Keating. 1983. Perception of numerical invariance in neonates. *Child Development* 54: 695–701.

Cecco d'Ascoli. 1927. *L'acerba*. Ascoli Piceno: Cesari.

Cellucci, Carlo. 2008. *Perché ancora la filosofia*. Rome: Laterza.

Cochran, Gregory, and Henry Herpending. 2009. *The 10,000 explosion: How civilization accelerated human evolution*. New York: Basic Books.

Cooper, William S. 2001. *The evolution of reason: Logic as a branch of biology*. Cambridge: Cambridge University Press.

Croce, Benedetto. 1917. *Logic as the science of the pure concept*. London: Macmillan.

Dantzig, Tobias. 2005. *Number: The language of science*. New York: Pi Press.

Devlin, Keith. 2005. *The math instinct*. New York: Thunder's Mouth Press.

Dewey, John. 2004. *Reconstruction in philosophy*. Mineola: Dover.

Flombaum, Jonathan I., Justin A. Junge, and Marc D. Hauser. 2005. Rhesus monkeys (Macaca mulatta) spontaneously compute addition operations over large numbers. *Cognition* 97: 315–325.

Hamilton, Gina. 2006. *Kingdoms of life: Animals*. Dayton: Milliken.

Hart, Wilbur Dyre. 1996. Introduction. In *The philosophy of mathematics*, ed. Wilbur Dyre Hart, 1–13. Oxford: Oxford University Press.

Heisenberg, Werner. 1958. *Physics and philosophy: The revolution in modern science*. New York: Harper.
Kant, Immanuel. 1998. *Critique of pure reason*. Cambridge: Cambridge University Press.
Lorenz, Konrad. 1977. *Behind the mirror*. London: Methuen.
———. 1997. *The natural science of the human species: An introduction to comparative behavioural research. The Russian manuscript (1944–1948)*. Boston: MIT Press.
———. 2009. Kant's doctrine of the a priori in the light of contemporary biology. In *Philosophy after Darwin: Classic and contemporary readings*, ed. Michael Ruse, 231–247. Princeton: Princeton University Press.
Mac Lane, Saunders. 1986. *Mathematics: Form and function*. Berlin: Springer.
Maddy, Penelope. 1997. *Naturalism in mathematics*. Oxford: Oxford University Press.
Mayr, Ernst. 2002. *What evolution is*. London: Phoenix.
Peirce, Charles Sanders. 1931–1958. *Collected papers*. Cambridge: Harvard University Press.
Pennings, Timothy J. 2003. Do dogs know calculus? *The College Mathematics Journal* 34 (3): 178–182.
Pinker, Steven. 1995. *The language instinct*. New York: Harper-Collins.
Poincaré, Henri. 2015. *The foundations of science: Science and hypothesis, The value of science, Science and method*. Cambridge: Cambridge University Press.
Popper, Karl Raimund. 1972. *Objective knowledge*. Oxford: Oxford University Press.
———. 1974. *Conjectures and refutations: The growth of scientific knowledge*. London: Routledge.
———. 2001. *All life is problem solving*. London: Routledge.
———. 2002. *Unended quest: An intellectual autobiography*. London: Routledge.
Rugani, Rosa, Laura Fontanari, Eleonora Simoni, Lucia Regolin, and Giorgio Vallortigara. 2009. Arithmetic in newborn chicks. *Proceedings of the Royal Society B* 276: 2451–2460.
Russell, Bertrand. 2009. *Human knowledge: Its scope and limits*. London: Routledge.
Schlick, Moritz. 1974. *General theory of knowledge*. Dordrecht: Springer.
Sechzer, Jeri A., Susan E. Folstein, Eric G. Geiger, Ronald F. Mervis, and Suzanne Meehan. 1984. Development and maturation of posture reflexes in normal kittens. *Experimental Neurology* 86: 493–505.
Ségurel, Laure, and Céline Bon. 2017. On the evolution of lactase persistence in humans. *Annual Revue of Genomics and Human Genetics* 18: 297–319.
Slater, Alan, Anne Mattock, and Elizabeth Brown. 1991. Form perception at birth: Cohen and Younger 1984 revisited. *Journal of Experimental Child Psychology* 51: 395–406.
Spelke, Elizabeth. 2011. Natural number and natural geometry. In *Space, time and number in the brain*, ed. Stanislas Dehaene and Elizabeth Merrit Brannon, 287–317. London: Elsevier.
Sulkowski, Gregory M., and Marc D. Hauser. 2001. Can rhesus monkeys spontaneously subtract? *Cognition* 79: 239–262.
Whiten, Andrew. 2019. Cultural evolution in animals. *Annual Review of Ecology, Evolution, and Systematics* 50: 27–48.
Williams, Michael. 2001. *Problems of knowledge: A critical introduction to epistemology*. Oxford: Oxford University Press.
Wittgenstein, Ludwig. 2002. *Tractatus Logico-Philosophicus*. London: Routledge.
Wynn, Karen. 1992. Addition and subtraction by human infants. *Nature* 358: 749–750.
Zagzebski, Linda. 1994. The inescapability of Gettier problems. *The Philosophical Quarterly* 44 (174): 65–73.

Chapter 18
Concluding Remarks

Abstract The chapter recapitulates and makes some concluding remarks about the aim of the book: to highlight the limitations of mainstream philosophy of mathematics, which restricts the philosophy of mathematics to finished mathematics, and to offer an alternative approach to the philosophy of mathematics, heuristic philosophy of mathematics, which deals with the making of mathematics, in particular discovery.

Keywords Mainstream philosophy of mathematics · Finished mathematics · Justification · Axiomatic method · Heuristic philosophy of mathematics · The making of mathematics · Discovery · Analytic method

18.1 Shortcomings of Mainstream Philosophy of Mathematics

The philosophy of mathematics, considered as the aim to account for mathematics as a part of human knowledge in general, goes back to the beginning of philosophy. Many major philosophers have made substantial contributions to it, and their work remains important even today.

But mainstream philosophy of mathematics, namely the philosophy of mathematics that has prevailed for the past century, has a different standpoint. It considers the philosophy of mathematics as a new independent subject introduced by Frege, which aims to account for mathematics not as a part of human knowledge in general, but as a matter unto itself.

According to it, the philosophy of mathematics cannot concern itself with the making of mathematics but only with finished mathematics, so it cannot contribute to the advancement of mathematics.

The task of the philosophy of mathematics is primarily to give an answer to the question: How do mathematical propositions come to be completely justified? The philosophy of mathematics performs this task by claiming that the method of

C. Cellucci, *The Making of Mathematics*, Synthese Library 448,
https://doi.org/10.1007/978-3-030-89731-4_18

mathematics is the axiomatic method, and mathematics is theorem proving by the axiomatic method. Mathematical propositions come to be completely justified because they are deduced from principles that are true, in some sense of 'true'.

Despite its large following, however, mainstream philosophy of mathematics is faced with serious difficulties.

The claim that the method of mathematics is the axiomatic method, and mathematics is theorem proving by the axiomatic method, is invalid because it is incompatible with Gödel's first incompleteness theorem.

The claim that mathematical propositions come to be completely justified because they are deduced from principles that are true, in some sense of 'true', is invalid because it is incompatible with Gödel's second incompleteness theorem.

Mainstream philosophy of mathematics has also several other shortcomings. For example, it cannot account for the fact that new demonstrations, even hundreds of them, have been sought for theorems already demonstrated, and for the fact that demonstrating a theorem of a given part of mathematics may require hypotheses from other parts of mathematics.

This, however, has not been enough to convince mainstream philosophers of mathematics to abandon their approach. As a result, mainstream philosophy of mathematics, not only does not contribute to the advancement of mathematics, but is unable to perform its intended primary task, to give an answer to the question of how mathematical propositions come to be completely justified.

18.2 Advantages of Heuristic Philosophy of Mathematics

Heuristic philosophy of mathematics is an alternative approach to mainstream philosophy of mathematics. Like the philosophical tradition, it considers the philosophy of mathematics as the aim to account for mathematics as a part of human knowledge in general.

According to it, the philosophy of mathematics can concern itself with the making of mathematics, in particular discovery, so it can possibly contribute to the advancement of mathematics, just as philosophy has done in the past.

The task of the philosophy of mathematics is primarily to give an answer to the question: How is mathematics made? The philosophy of mathematics performs this task by claiming that the method of mathematics is the analytic method, and mathematics is problem solving by the analytic method. Mathematics is made starting from problems, formulating hypotheses for their solution by non-deductive inferences, and establishing their plausibility through a comparison with experience.

Heuristic philosophy of mathematics does not have the shortcomings of mainstream philosophy of mathematics.

The claim that the method of mathematics is the analytic method, and mathematics is problem solving by the analytic method, is unaffected and even confirmed by Gödel's first incompleteness theorem.

The claim that mathematics is made starting from problems, and arriving at plausible hypotheses that are sufficient conditions for solving them through non-deductive inferences, is unaffected and even confirmed by Gödel's second incompleteness theorem.

Heuristic philosophy of mathematics does not either have the other shortcomings of mainstream philosophy of mathematics. For example, it can account for the fact that new demonstrations, even hundreds of them, are often sought for theorems already demonstrated, and for the fact that demonstrating a theorem of a given part of mathematics may require hypotheses from other parts of mathematics.

Therefore, it seems fair to say that heuristic philosophy of mathematics provides an adequate account of mathematics.

18.3 Looking Ahead

Instead of dealing with the making of mathematics, mainstream philosophy of mathematics engages in debates about the status of finished mathematics, in particular, the status of mathematical objects and mathematical truth within finished mathematics.

But these debates have little or nothing to say about real mathematics. This is clear from the fact that one can very well understand a good deal of them even with little knowledge of mathematics. This is not because little knowledge of mathematics is enough to say what mathematics is, but because such debates are not about real mathematics. They are only about artificial issues, which have no connection with, and hence shed no light on, real mathematics.

But a philosophy of mathematics disconnected from real mathematics cannot offer any real understanding of mathematics, let alone contribute to its advancement.

In the past, on several occasions, philosophy has offered a real understanding of mathematics, and has even contributed to its advancement. There is no reason why, with a suitable philosophical framework in place, philosophy should not be able to do so once again.

Index

C. Cellucci, *The Making of Mathematics*, Synthese Library 448,
https://doi.org/10.1007/978-3-030-89731-4

U

V

W